T0133070

Darwin and the Making
of Sexual Selection

Darwin and the Making of Sexual Selection

EVELLEEN RICHARDS

The University of Chicago Press
Chicago and London

The University of Chicago Press, Chicago 60637
The University of Chicago Press, Ltd., London
© 2017 by The University of Chicago
Published 2017.
Printed in the United States of America

26 25 24 23 22 21 20 19 18 17   1 2 3 4 5

ISBN-13: 978-0-226-43690-6 (cloth)
ISBN-13: 978-0-226-43706-4 (e-book)
DOI: 10.7208/chicago/9780226437064.001.0001

Published with support of the Susan E. Abrams Fund

Library of Congress Cataloging-in-Publication Data

Names: Richards, Evelleen, author.
Title: Darwin and the making of sexual selection / Evelleen Richards.
Description: Chicago ; London : The University of Chicago Press, 2017. |
    Includes bibliographical references and index.
Identifiers: LCCN 2016036765 | ISBN 9780226436906 (cloth : alk. paper) |
    ISBN 9780226437064 (e-book)
Subjects: LCSH: Darwin, Charles, 1809–1882. | Sexual selection—History.
Classification: LCC QH31.D2 R45 2017 | DDC 576.8/2—dc23 LC record
    available at https://lccn.loc.gov/2016036765

♾ This paper meets the requirements of ANSI/NISO Z39.48–1992
(Permanence of Paper).

In memory of my mother,
who saw to it that I got the education she was denied

CONTENTS

ILLUSTRATIONS

Collections are listed below, as well as abbreviations for works that are cited frequently in the notes. See the bibliography for publication details of the titles below and in the notes.

Autobiography	Barlow 1969, *The Autobiography of Charles Darwin, 1809–1882*
BL	British Library
Calendar	F. Burkhardt and Smith 1994, *A Calendar of the Correspondence of Charles Darwin*
CCD	F. Burkhardt et al. 1985–, *The Correspondence of Charles Darwin*
CUL	Cambridge University Library
DAR	Darwin Manuscripts Collection, Cambridge University Library
Descent 1871	C. Darwin 1871, *The Descent of Man* (first edition)
Descent 1879	C. Darwin (1879) 2004, *The Descent of Man* (reprint of the second revised edition)
Diary	R. D. Keynes 1988, *Charles Darwin's* Beagle *Diary*
DRC	Darwin Reprint Collection, Cambridge University Library
ED	Litchfield (1904) 1915, *Emma Darwin*
Expression	C. Darwin 1872, *The Expression of the Emotions in Man and Animals*
Foundations	C. Darwin 1909, *Foundations of the* Origin of Species
ICL	Imperial College Library
Marginalia	Di Gregorio and Gill 1990, *Charles Darwin's Marginalia*
NS	Stauffer 1975, *Charles Darwin's* Natural Selection
Notebooks	Barrett et al. 1987, *Charles Darwin's Notebooks*

Origin   C. Darwin 1859, *On the Origin of Species*

Variation   C. Darwin 1868, *Variation of Animals and Plants under Domestication*

Variorum   Peckham 1959, The Origin of Species *by Charles Darwin: A Variorum Text*

ACKNOWLEDGMENTS

This book has been almost as long in the making as sexual selection and has almost as checkered a history. Over the years I have accumulated many debts of gratitude, some too deep to be acknowledged adequately here.

My earliest debt must be to my parents, great readers both, who encouraged a love of books and kept a battered old copy of *The Descent of Man* (I have it still) in the family bookcase. I'd like to say it excited an early interest in sexual selection, but though my doctoral thesis was in the history of evolutionary theory, that specific impetus did not come until much later, when I became involved in setting up the first women's studies course at the University of Wollongong. I then published my first paper in the area, "Darwin and the Descent of Woman" (1983). I revisited the field periodically, but family and my major research commitment to the sociology of clinical trials preoccupied me. When I eventually returned to matters Darwinian, it was with the idea of a more detailed examination of Darwin and the "woman question." As I read and wrote, this focus came to seem too constricting, and the project expanded ambitiously into a history of the making of sexual selection. Along the long and tortuous road of research and writing, I have been encouraged and assisted by many companions and colleagues.

First and foremost, I must thank Adrian Desmond. It was he who initially inspired the book, gave unstintingly of his encyclopedic knowledge and experience, egged me on when I flagged, read and commented on early drafts of chapters, and generously encouraged the ways in which the work unfolded, even when it ran counter to his own interpretations. I can only hope, in emulation of my celebrated subject, that the finished product "will not quite kill me in [his] good estimation." I am also extremely grateful to James Secord for unfailing intellectual, professional, and personal support. Jim discussed critical issues, pointed me toward vital sources, and found time in his overcrowded schedule to read through the whole manuscript; his perceptive comments and advice helped shape its final structure. I owe special thanks to Jon Hodge, who in long-distance telephone calls and e-mails dissected various aspects of draft chapters, shared his vast knowledge of things Darwinian, sharpened my understanding, and saved me from several egregious blunders. Particularly helpful and insightful comments on the entire manuscript were also provided by John Schuster and by referees for the University of Chicago Press. Gowan Dawson, Paul White, and Gregory Radick looked critically over various chapters, which are much improved as a result. Others who offered suggestions and contributed advice and references include Harry Allen, Ruth Barton, Peter Bowler, Janet Browne, Roderick Buchanan, John Docker, James Endersby, Mario di Gregorio, Ian Hesketh, David Kohn, Bernard Lightman, James Moore, Jane Munro, James Paradis, Diane Paul, Anne Secord, and John van Wyhe. At a critical stage, Gillian Fuller came to the rescue with a stern subedit and some much-needed suggestions for pruning. The responsibility for what remains is mine.

I have benefited considerably from invitations to present my ideas at conferences and seminars. These include the Dyason Memorial Lecture at the Annual Conference of Australasian Association for History, Philosophy and Social Studies of Science, 2015; the Darwin 2009 Festival, University of Cambridge; and the "From Generation to Reproduction" Seminar Series, University of Cambridge, 2007. Early drafts of chapters 1 and 4 underwent lively criticism at two workshop sessions under the auspices of the "Past versus Present" Project of the Cambridge Victorian Studies Group in 2007.

A number of institutions gave crucial support. In its very earliest incarnations, the project was supported by two Australian Research Council grants that enabled international travel and leave for research. The Unit for History and Philosophy of Science at the University of Sydney has been a stimulating and supportive place to pursue research and to debate related issues with staff and students. I wish especially to thank Rachel Ankeny, Paul Griffiths, Ofer Gal, Hans Pols, and Stephen Gaukroger for helpful comments and searching questions. Over the years, the Department of History and Philosophy of Science at the University of Cambridge has been a generous foster home for research and collegial interaction and dialogue.

Many libraries and archival collections have provided material and assistance over the course of the project. The two collections I have relied on most and their always helpful staff require particular thanks: the University of Sydney Library, Rare Books and Special Collections; and, above all, Cambridge University Library with its unrivaled collection of Darwin manuscript material, correspondence, annotated books, journal articles and pamphlets.

I owe special thanks to the editors of the Darwin Correspondence Project, especially Shelley Innes, Alison Pearn, and Paul White, who on occasion shared morning tea, ungrudgingly lent their expertise to deciphering Darwin's more cryptic jottings, and helped access vital transcriptions and illustrative material. I owe thanks also to Nick Gill, who obligingly chased up and explained some obscure marginalia.

I would like also to thank staff of the British Library; the Wellcome Institute for the History of Medicine, London; the Mitchell Library and the State Library of New South Wales; the Australian National Library; and the University of New South Wales Library. I am very grateful to those libraries and archives that allowed access to unpublished manuscript material in their possession; notably the Bradlaugh Papers in the Bishopsgate Institute; the Huxley Papers in Archives, Imperial College London; the Richard Owen Papers and Alfred Russel Wallace Family Papers in the Natural History Museum, London; and the Ethnological Society of London minutes and related material, Archives, Royal Anthropological Institute, London.

For their supply of and permission to use illustrative images, I am indebted to the Fitzwilliam Museum, Cambridge, to the Syndics of Cambridge University Library, to the Tate Gallery and the National Portrait Gallery, London, and to the Wellcome Library, London. My warmest thanks go to my daughter Morgan Richards, who, with her partner Ili Barré, organized and Photoshopped the illustrations.

My copyeditor, Johanna Rosenbohm, disentangled nightmares of bibliographical inconsistencies and errors with unflagging grace and expertise. Marta Steele compiled a thorough, intelligent, and user-friendly index. The University of Chicago Press has exhibited great patience in waiting for this book, and I wish to thank all involved—notably Karen Merikangas Darling for editorial support, and Jenni Fry, Evan White, and Yvonne Zipter, who saw it through production. My special thanks go to the late Susan Abrams for her belief in the project and inspirational guidance through its initial phase. I am especially pleased to acknowledge support from the fund established in her name: the Susan Elizabeth Abrams Fund in History of Science.

My greatest debt is to Warwick and the family, who know far more about the history of sexual selection than they ever thought necessary.

Finally, I should like to thank Charles Robert Darwin. It has been an absorbing experience and an education to get to know him a little. I shall miss the man who, with all the contingent constraints of his time and place—the social attitudes and cultural values that I have been so committed to uncovering and interrogating—gave us the theory of sexual selection. Without him, this book could never have been written.

"An Awful Stretcher"

Charles Darwin's concept of natural selection has been endlessly written about; his "secondary" principle of sexual selection has received far less historical attention. Its sources and the conceptual pathways to its formulation remain largely unknown or misunderstood. The extraordinarily wide range of Darwin's investigations and the depth of his readings in this connection have never been explored.

Yet sexual selection (the struggle for mates) was of considerable strategic importance to Darwin's theory of evolution. It explained what natural selection (the struggle for existence) could not; and it offered a naturalistic, as opposed to divine, account of beauty and its perception. It was thus vital to the defense of evolution against the established theory of special creation.

In sexually reproducing species, there are often striking differences in what are termed *secondary sexual characteristics*. These species are called sexually "dimorphic," and their differences may be divided into two basic categories: (1) those that include structures (such as horns or spurs) with which males fight among themselves, and (2) those that include ornamental structures and behaviors (fanned tails, inflatable bladders, elaborate or colorful plumage, songs, dances and so on) exhibited (usually but not always by males) during courtship.

In the *Origin of Species* (1859), Darwin explained the evolution of such differences through the action of sexual selection. He recognized two aspects of the sexual struggle, corresponding to the above categories: male combat for possession of the females; and a "more peaceful" contest, female choice, whereby males compete with one another by means of "charms" (brilliance of plumage, song, etc.) in their wooing of the female during courtship. Sexual selection shored up natural selection in two ways: female choice accounted for the persistence of seemingly useless or disadvantageous ornamental characteristics (such as the bright plumage of many male birds that renders them more conspicuous to predators) that could not be explained in functional survival terms; and male combat not only explained the development of specialized "weapons" (spurs or horns), but also enhanced the action of natural selection by ensuring that the fittest males were reproduced.[1] However, in *The Descent of Man, and Selection in Relation to Sex* (1871), sexual selection took on a much wider role. There, Darwin's exhaustive elaboration of sexual selection throughout the animal kingdom (fully two-thirds of the work) was directed to substantiating his view that human racial and sexual differences—not just physical differences, but certain mental and moral differences—had evolved primarily through the action of sexual selection.

It was the culmination of a lifetime of intellectual effort and commitment, a "tremendous job," as Darwin described it, long and hard in the making.[2] Yet sexual selection never quite made it. Darwin constantly was forced to contest its validity with a great array of critics, and with his death in 1882, sexual selection went into abeyance, not to be revived until late in the twentieth century. Today it remains a controversial theory, subject to ongoing dispute and reinterpretation. However it may be read, though, sexual selection—for those who concern themselves with it—still means Darwin. And what Darwin meant still matters. He lives on in many contradictory (and some quite startling) guises in current narratives, justifications, or critiques of sexual selection.[3]

This book aims to resolve these contradictions. It examines the man, rather than the myth, and the social as well as the intellectual roots of his theory building. The history of sexual selection was never one of straightforward, unadulterated science, but from its beginnings was intertwined with cultural and social beliefs and shaped by professional and institutional power plays and the larger issues of the day. It is a history that reaches from the last decades of the eighteenth century to those of the nineteenth, from provincial England to the ends of the earth and back again, and through a period of extraordinary social and scientific transformation. It involved numerous narrative strands and characters, many of them little known in the history of evolutionary science, but its central actor was always Darwin. It was Darwin who worked to pull these strands together, to bind them into a tightly woven complex of theory, analogy, and practice, to make sexual selection *"altogether* [his] own subject."[4]

This phrase, with its recognition of Darwin's jealously guarded ownership of sexual selection, comes from Alfred Russel Wallace. It was Wallace, of course, who had in one week in 1858, as a struggling young naturalist and collector in the Malay Archipelago, put together the theory of evolution by natural selection and posted it off to a shattered Darwin, who had been obsessively honing the theory for some twenty years. In Wallace's absence, Darwin's leading allies, the geologist Sir Charles Lyell and botanist Joseph Dalton Hooker, famously settled the vital issue of priority by arranging the joint reading of extracts from Darwin's earlier writings on natural selection and Wallace's manuscript to the Linnean Society. While Darwin rushed into print and lasting fame with the *Origin*, Wallace modestly disclaimed ownership, deferring to "Mr. Darwin's celebrated theory of natural selection." But in the 1870s, to Darwin's great chagrin, the coauthor of natural selection emerged as the most insistent and damaging critic of sexual selection, determined to rid natural selection of this "abnormal excrescence."[5]

It all began in late May 1864, when Darwin sat down in the comfort of his study in Down House to pen a letter to Wallace, now returned to England from Borneo. It was Darwin's benign intention to assure the younger naturalist that natural selection was "just as much your theory as mine," and to praise his recently published "Man paper."[6] Darwin's letter, however, was not so much about their joint progeny, natural selection, as about Darwin's "own" theory of sexual selection. It marked the beginnings of their increasing theoretical estrangement. And it was Wallace's paper on "man" that first set the cat among the pigeons—more or less literally, as matters unfolded.

As early as 1857, Wallace had written to ask Darwin whether he would "discuss 'man'" in his forthcoming book on species. Darwin (who at one time had intended to do just that, but had stepped back from the abyss) replied that he would "avoid [the] whole subject, as so surrounded with prejudices." The dangerous subject was accordingly excised from the *Origin*, save for the one well-known sentence: "Light will be thrown on the origin of man and his history." Still, Darwin could not resist hinting in the *Origin* that he had the solution to another highly contentious and related problem, the "differences between the races of man, which are so strongly marked." These had evolved through "sexual selection of a particular kind, but," he cautioned, "without here entering on copious details my reasoning would appear frivolous."[7]

Now, five years down the track, Darwin, having kept even his closest confidants, Lyell and Hooker, in the dark, divulged his "frivolous" reasoning to the junior coauthor of natural selection. Wallace seemed fearlessly inclined to "go the whole orang" in taking on the issue of human evolution, risking reputation and the respectability so valued by the circumspect and gentlemanly Darwin.[8] As an added inducement, Darwin went so far as to offer Wallace his notes on the subject:

I suspect that a sort of sexual selection has been the most powerful means of changing the races of man. I can shew that the diff[erent] races have a widely diff[erent] standard of beauty. Among savages the most powerful men will have the pick of the women & they will generally leave the most descendants.

I have collected a few notes on man but I do not suppose I shall ever use them. Do you intend to follow out your views, & if so would you like at some future time to have my few references & notes? . . .

[PS] Our aristocracy is handsomer (more hideous according to a Chinese or Negro) than middle classes from pick of women.[9]

Whatever Darwin's hopes of Wallace were, they were dashed by return of post. Wallace had his own views on natural selection in human evolution, and it seemed on closer inspection that they were not entirely compatible with Darwin's. As far as sexual selection went in the formation of the different races, Wallace (who, unlike Darwin, had actually lived among indigenous peoples in South America and the East Indies) was skeptical of its effects: "In the very lowest tribes there is rarely much polygamy & women are more or less a matter of purchase—There is also little difference of social condition & I think it rarely happens that any healthy & un-deformed man remains without wife and children." As for Darwin's foray into the class effects of sexual selection, Wallace, true to his egalitarian values, disputed the "often repeated assertion" of the greater beauty of the aristocracy. "Mere physical beauty . . . is quite as frequent in one class of society as the other." In any case, the cash-strapped Wallace was engaged in writing up the "Narrative of my travels" for popular sale. He might "possibly some day go a little more into this subject (of 'Man') & if I do will accept the kind offer of your notes."[10]

Darwin was seriously affronted by this casual dismissal of his much-prized theory of sexual selection and the high honor of his "kind offer." He responded shortly that Wallace was "probably . . . right,"

except as I think about sexual selection which I will not give up. My belief in it, however, is contingent on my general belief in sexual selection. It is an awful stretcher to believe that a Peacock's tail was thus formed, but believing it, I believe in the same principle somewhat modified applied to man. I doubt whether my notes w[ould] be of any use to you, & as far as I remember they are chiefly on sexual selection.[11]

This exchange sets the scene for this book. It foregrounds the racial, class, and gender dimensions of Darwin's conception of sexual selection, as well as its inextricability from Darwin's views on human evolution. The difficulties Darwin faced in attributing an almost human aesthetic discrimination to a mere bird are acknowledged with his admission of its being an "awful stretcher." At the same time, he affirmed his determination to "not give up" his belief that the peacock's tail had indeed been formed in this way. We should note also that Darwin's "somewhat modified" application of this "same principle" to human sexual selection entailed the overturning of the principle of female choice, which he claimed predominated among animals. Among humans, Darwin assumed that male choice was the norm in differentiating both the sexes and races. This role reversal was to cause problems and contradictions for Darwin and sow discord and confusion among his readers. What is more, as Darwin here frames his conceptual shift from animals to humans, it is historically incorrect. It is incorrect because Darwin came to sexual selection not from his study of the sexual differences and mating behaviors of birds and other animals, as this exchange suggests, but the other way round: from his very Victorian interpretation of the human practices of wife choice, courtship, and marriage, which he then extended to animals. Finally, there is Wallace's positioning as chief critic and devil's advocate, which was crucial to the making (and unmaking) of Darwin's theory of sexual selection.

The two men could not have been more different. Darwin was the last and greatest of the gentlemen naturalists, independently wealthy, readily able to command resources and negotiate social and scientific networks for the promotion of his unorthodox views. His science may have been unorthodox, but Darwin himself was in many ways entirely conventional, a man of his time and place, with an ingrained respect for the social conventions and a strong and saving sense of scientific caution. Wallace, by contrast, was a self-made man, a scientific and social maverick, intrepid in controversy, who struggled throughout his life to finance his science and support his family.

To complicate matters, Wallace was one of those paradoxical Victorians who confound modern minds with their seemingly gullible endorsement of table rapping, spirit manifestations, and all the paraphernalia of the séances that became so fashionable in the late nineteenth century. Again, unusual for a Darwinian, Wallace was also a defender of women's rights and a socialist (the two did not necessarily go hand in hand). He made Darwin "groan" with his 1869 published declaration that natural selection alone could not explain crucial aspects of human evolution, asserting that "an Overruling

Intelligence" had guided the development of human intelligence and morality. Yet in old age he emerged as the great defender of Darwinism when the Darwinians themselves were in disarray, championing natural selection against the proliferation of rival evolutionary theories. He spent twenty-five years relentlessly critiquing Darwin's theory of sexual selection, only to turn round and expropriate it for his mature version of evolution, giving women the power to control social advance in a socialist society by their informed choice of only those males who possessed traits worthy of transmission to posterity.[12]

All this was to come. But the intimations of their future differences are there in their written exchange of 1864.

Reluctantly, the aging and chronically ill Darwin wrote the long-awaited work on "Man" and, as he had insisted to Wallace, the necessarily interconnected theory of sexual selection. The writing of *The Descent of Man* took place as the Darwinians, led by the master strategist Thomas Henry Huxley, fought for cultural authority and institutional dominance in the politically important field of anthropology, the new "science of man," where conflicting theories of "race" were at stake. The *Descent*'s larger context was a capitalist and increasingly secular society. It was a society fully embarked on imperialist expansion, while staring down homegrown trade unionism and the socialist threat and dealing with growing demands for reform, suffrage, and better educational opportunities for women and the working class. It was during this critical period that Darwin polished and refined his theory of sexual selection, largely through his intensive and disputatious—but mutually respectful—correspondence with Wallace. Wallace, with his apostasy, was a thorn in the Darwinian side. But Darwin needed and esteemed the opinion of the coauthor of natural selection, even when directed against him, telling Wallace: "I grieve to differ from you, & it actually terrifies me & makes me constantly distrust myself."[13]

Wallace was far from the only critic of sexual selection. Yet Darwin held to his theory to the very end, defiantly asserting its "truth." His last defense of sexual selection was read at a meeting of the Zoological Society just hours before he died. With Darwin dead and Wallace ascendant as the "greatest living champion" of Darwinism, and with Darwinism itself under siege, by the turn of the century sexual selection was in crisis, deemed "practically discredited" by evolutionary theorists.[14]

To understand to the full Darwin's commitment to sexual selection and its ensuing neglect, we must retrieve its history. This is essential if we are to know just how and when Darwin formulated his principle, how he distinguished it from natural selection, and more important, why Darwin was so convinced of its "truth" and of its necessity to his exposition of human evolution and racial divergence. Only then may we understand what exactly Darwin meant by "sexual selection," how it was made compatible with other critical components of his evolutionary theorizing, and how his understanding and presentation of the principle shaped its history.

In his preface to the second edition of the *Descent* (1874), Darwin exasperatedly refuted the allegation that he "invented sexual selection" when he found that "many details of structure in man could not be explained through natural selection." He pointed out that he had given a "tolerably clear sketch" of the principle in the first edition of the *Origin*, and that he had there stated that it was "applicable to man."[15] Indeed, the genesis of sexual selection goes back more than twenty years before the *Origin*, to Darwin's secretive evolutionary jottings in London in the late 1830s in what are known as his "transmutation notebooks" and his related notebooks on "Mind, Man and Materialism."[16] It had its origins in the repertoire of culturally freighted biological, aesthetic, and ethnological writings, as well as breeding literature and practices, that Darwin drew on in his early notebook constructions. The history of sexual selection stretches yet further back, to the erotic evolutionary writings of Darwin's grandfather, Erasmus Darwin, doctor, poet, radical, and womanizer—writings that had a much greater impact on the younger Darwin than is generally admitted.

Sexual selection thus has a very long history. It was not simply dependent on Darwin's observations of and readings into the literature on the sexual dimorphism and courtship behaviors of a huge range of sexually reproducing species, but was drawn from a great complexity of sources. And Darwin justified it with a mix of culturally laden constructions of sexuality, gender, and race, together with his interpretations of heredity and development (themselves culturally inflected). To date, those few historians who have examined sexual selection in any detail have largely ignored these dimensions of Darwin's conception.[17] The notable exceptions to this have been those literary critics who have centered their interpretations on Darwin's fictive encounters with "pretty women" in the courtship plots of the sentimental Victorian novels he enjoyed so much, and more recently, the historians Adrian Desmond and James Moore, who have represented sexual selection as the pro-abolitionist Darwin's evolutionary solution to the racially motivated claims of those contemporary apologists for slavery who argued that the human races were separate species with separate origins.[18] While these reinterpretations have brought valuable new insights—and Desmond and Moore have put Darwin's moral commitment to the abolition of slavery definitively on the historiographical agenda—the history of sexual selection is far more complex than these partial accounts suggest. Most significantly, a vital visual dimension is missing.

It cannot be overemphasized that Darwin's sexual selection predominantly relies on the visual perception of beauty. Bird and human alike must be able to perceive and discriminate among differences in external appearance—even down to the minute differences that Darwin insisted on—in making the essential choice of mate. For Darwin, it was absolutely fundamental that animals share an aesthetic sense with humans. This was primarily because of his need to demonstrate continuity from animals to humans and, therefore, the evolutionary development of this highest human faculty—the appreciation of beauty—against the rival theory of special creation, which held that the great diversity and beauty of nature had been designed by the Creator for his own or for human appreciation. Likewise, Darwin had a compelling need for a noncreationist explanation of nonfunctional, nonadaptive male ornamentation and conspicuous coloring, especially in birds, that could not be accounted for by natural selection or by his alternative theory of use-inheritance (contrary to popular belief, Darwin never abandoned some aspects of the inheritance of acquired characters).

Darwin's disclaimers of appreciation for or cultivation in the higher arts have been made much of, but they are misleading.[19] Like other well-to-do young gentlemen of his class and time, he was early schooled in the visual arts while at the University of Cambridge. His five-year stint as gentleman companion to Captain Robert FitzRoy (and unofficial ship's naturalist) on board the *Beagle* (1831–36) exposed the still youthful Darwin to an explosion of vivid visual experiences that resonated with him to the end of his days and was critical to the formation of his evolutionary views. Of all these novel experiences, the most remarkable—one that Darwin returned to again and again in his writings—was his first sighting, at the uttermost ends of the earth, of the "naked painted . . . hideous savage" of Tierra del Fuego.[20] The sensory shock of the "savage"—who seemed to him destitute of all the accustomed attributes of humanity, almost a member of another species—was an assault on his eyes that thoroughly challenged Darwin's family values of racial unity and the brotherhood of man, and all Christian notions of the unbridgeable gulf between human and beast.

On his return to London, Darwin read extensively in contemporary aesthetics (or theories of the beautiful) in the early stages of theory building. He underwent a further program of "visual re-education" through studying the selective practices of pigeon breeders, who examined each tincture of feather and minute difference of beak, to an understanding of the great "accuracy of eye and judgement" necessary to become a skillful breeder.[21] Many pigeon fanciers were silk weavers who bred pigeons on rooftops for pleasure, and for food and income through the downturns in their highly skilled trade. They brought their exacting, artistic "weaver's eye" from the handlooming of the delicate fabrics of their craft to the scrutiny and selective reshaping of the living bodies of their birds.[22] Their practices epitomize the connections that Darwin was to draw among fashion, aesthetics, and breeding.

Above all, Darwin lived in a culture that was profoundly visual. New developments in media and printing technologies, in the mass marketing of cheap prints, greeting cards, illustrated books and magazines, promoted Victorian visual acuity and acquisition. These were augmented by a staggering upsurge in the commercial display and advertisement of household goods and fashions, as decorative goods and ready-made clothing and accessories flooded the marketplace. Victorian fashions—feminine fashions, to be specific—were, to say the least, eye catching. The introduction of aniline dyes in the mid-1850s led to a startling intensity and novelty of color in fabrics. The innovation of the crinoline around the same time, a by-product of the new steel technology that brought high fashion within the reach of servants and factory workers, also made women (including women of the streets) much more noticeable and visible. Victorian visual culture was far from the drab and dreary we have come to expect, but colorful, flashy, often as trashy and titillating as it was obedient to the middle-class canons of taste and propriety.

Historians have only recently begun to look at the ways in which Darwin drew on the available visual and aesthetic resources of this culture.[23] A good deal, for instance, has been made of Darwin's response to the romantic and the sublime in literature and art, particularly through the influence of Alexander von Humboldt's perceptions of tropical nature.[24] Yet little attention, if any, has been paid to the role in Darwin's evolutionary thinking of the very British empiricist aesthetic tradition—exemplified by the views of his grandfather Erasmus—that was closely allied with a wider range of social, political and ethical debates that relate to the middle-class rise to political power.[25] Darwin's views on sexual selection owe more to this homegrown aesthetics than to Humboldt's romanticism. Similarly, the centrality of aesthetic rankings in early writings on race and the significance of this to Darwin's views on racial differences and origins have been consistently overlooked or underplayed. Historians have yet to acknowledge the extent to which Darwin's interpretations were colored by contemporary Eurocentric notions of the ugly and the beautiful and by the racial hierarchy they endorsed—a "visual ideology" that promoted a politics of race, colonialism, and extinction, that was in place well before the "scientific racism" of the later part of the nineteenth century, and that permeates Darwin's *Beagle*-based observations and ranking of indigenous peoples.[26]

Again, Victorian fashion and its representation in the popular press is the subject of a large literature that has yet to be linked with Darwin's theorizing.[27] Yet the "extremes of fashion" recurs as an explanatory motif in Darwin's expositions of the relation of breeding practices to sexual selection, and he explicitly compared women's decorative practices with those of "savages" and birds. New feminine colors, shapes, and decorations impinged on Darwin's everyday field of vision, as much as those of the varieties of pigeons he bred and studied (he was, after all, a husband and a father with daughters who took a keen interest in dress). Unlikely as it may seem, Darwin drew inspiration from the pronounced sexual dimorphism of nineteenth-century dress, in particular from the vagaries and extremes of the fashion choices of Victorian women, the visual expression of their identity, which was trivialized and satirized in the popular press.

Of still greater import for Darwin's theory building—as well as for his personal comfort and posterity—was the choice of a wife. This was a serious business for the men of the emergent middle classes who looked to a "good woman," the iconic "angel in the house," the personification of Victorian womanliness, modesty, self-denial and purity, who would safeguard their property, their propriety, and their posterity. Or as Darwin put it, as the well-fledged bachelor newly returned from the voyage of the *Beagle* and on the "sharp lookout" for a wife, one who was "better than an angel & had money."[28]

This leads me to my next point: sex. A history of sexual selection can hardly be written without confronting the history of sex.[29] Darwin's *Descent of Man*, as was then conventional in works directed to a wider reading audience of both sexes, is devoid of any overt discussion of human sexuality or erotic desire. Darwin had to negotiate with his publisher, John Murray, even to include what Murray termed the "objectionable adjective" in the subtitle of a work devoted to sexual selection.[30] Yet Darwin's evolutionary writings are saturated with sex: from his explication of the sex life of barnacles and the sexual contrivances of orchids, to his principles of natural and sexual selection where sex and superfecundity are the driving forces of organic change.[31]

Since the foundational work of Michel Foucault, we can no longer think of sex as ahistorical. Foucault showed that sexuality was a historically inflected activity that required study as much as any other social or cultural activity.[32] Understandings of and attitudes toward sexuality, as well as its practices, are mediated (among other things) by religion, class, politics, gender, and race. Gender and race are themselves similarly and even more obviously inflected, and well before Foucault, work had begun to retrieve their historical meanings and functions. Like sex, these terms are problematic; they change and shift around and are constantly reconstructed. They are not easy to pin down, and they underwent some significant changes during the course of Darwin's long life. Fortunately, we now have a wealth of studies of nineteenth-century sex, race, and gender that may be brought to bear on Darwin's theory building. For one thing, we know that Victorian sexuality was more diverse than the repressive stereotype we have inherited from its late-Victorian and Edwardian critics. At the same time, we must take account of those points where stereotype and its variations intersect in the making of sexual selection.

In all this, the historian must not neglect the close study of the observational and intellectual components of Darwin's theory building as he worked to pull the elements of sexual selection together, to differentiate it from natural selection, and to make it compatible with his observations and with his theories of inheritance, embryology, and divergence. Of particular significance was Darwin's interpretation of embryology (the study of animal development). Its centrality to his construction of sexual selection has been ignored by historians, who have focused on the relation of Darwin's embryology to natural selection and to contemporary evolutionary theory and become mired in ongoing controversy.[33] Its reconstruction here, as a critical element of his theorizing on sexual selection, clarifies much that has been misunderstood in Darwin's embryological argument for evolution. It also highlights the necessity of moving beyond the close textual analysis of the technical core of sexual selection in Darwin's published and unpublished writings. Complex as it is, this is not sufficient. Embryology had resonances beyond the museum or laboratory, well beyond Darwin's experimental researches on his prized pigeons, and it is these that largely make comprehensible its role in his evolutionary theorizing.

The traditional image of Darwin as the objective observer and theoretician of Down House—remote from the social and political concerns of his fellow Victorians—has been exploded. Today, few historians would dispute the connection of Darwin's theoretical constructions with Victorian politics and ideology. All else aside, some fifteen thousand letters have been retrieved and the publication of a projected thirty-two volumes of meticulously edited and annotated correspondence is well under way. His letters show overwhelmingly that Darwin was a consummate networker, building a far-flung web of informants from all walks of life—breeders, gardeners, keepers, doctors, missionaries, travelers, naturalists, and collectors—who did their best to answer questions, send specimens, and carry out experiments for him. It was a "network of power" with Darwin at its center, collating, managing, and deploying the information he received. Darwin also maintained close touch with a select coterie of scientific peers who carried out the necessary social functions of a scientist for him while he remained ensconced at Down, pursuing his researches. He persuaded them, "pumped" them for information and tested his views on them, urging them on to the defense of Darwinism in the professional and public arenas that he was too diffident (or famously ill) to enter. He was never a professional scientist, but he kept abreast of institutional and professional politics. He read the reviews, the lampoons, even collected cartoons of himself as a hairy ape. He also kept up to date with the greater political issues of his day, and it has become clear that they impinged more on his science and his published writings than was conventionally acknowledged.[34]

In addition to the correspondence, we now have a treasure trove of archived notes and manuscript material—even the marginalia in Darwin's library of books, journals, and pamphlets—much of it transcribed and published (much now available on the Internet), along with a great mass of "raw" untranscribed notes in Darwin's characteristic scrawl. He read omnivorously, not just works deemed scientific, but a slew of pamphlets, catalogs, and handbooks on breeding, horticulture, and beekeeping, as well as works of popular fiction, travelogues, history, and biography (which were mostly read aloud to him by Emma, his devoted wife). He used his books as working tools, ruthlessly tearing large volumes in half if they were too cumbersome to read while reclining on his study sofa, marking passages and making notes as he went, indexing pertinent pages and subjects. He kept "from 30 to 40 large portfolios" on different subjects into which he filed relevant notes, abstracts, and letter pages from his correspondents: "Before beginning on any subject I look to all the short indexes & make a general & classified index, & by taking the one or more proper portfolios I have all the information collected during my life ready for use."[35]

Lest this convey a picture of easy access and reassembly for the historian, it should be noted that Darwin's thrifty practice of cannibalizing notes and recycling bits of paper for other projects and uses has kept archivists and historians on their toes as they try to piece together what goes where (and there are still exciting discoveries to be made, as I shall tell). Darwin's paper trail is far from complete, has gaps just as the historian seems on the verge of significant discovery, and is at times indecipherable or so cryptic as to challenge comprehension. Yet, together with the support of the correspondence and Darwin's published works, these surviving notes and annotations do make possible, if difficult, the tracking of Darwin's path to sexual selection and the shifts it underwent as he worked to pull its elements together and present it to his Victorian readers, both lay and scientific.

This book, then, explores the intellectual and social roots of Charles Darwin's theory of sexual selection, analyzes its stages of theory building in his unpublished and published writings and its elaboration in *The Descent of Man*, and reviews its contemporary reception, reinterpretations and applications. I am not concerned with present-day theory, except insofar as it will become clear that there are significant differences between sexual selection as Darwin conceived it and as it is currently understood in evolutionary biology, and that these differences help to explain its long hiatus as well as to illuminate the social and cultural contingencies of the history of a significant, if controversial, biological concept.

I have chosen to explore the major sources or strands of sexual selection thematically in part 1 of this book. These having been established, my account then reverts to a more conventional chronological history of sexual selection in part 2. It reviews the major stages of theory building from the London notebooks on, but is primarily concerned with the period from the writing of Darwin's "big species book," which became the *Origin*, up to the publication of the *Descent* and readers' responses to it.

This is an ambitious project, and must cover a lot of ground from the late Enlightenment period of Erasmus Darwin's writings to the "eclipse" of sexual selection in the last decades of the nineteenth century, a period of great social, scientific, and institutional change. My thematic approach permits the exploration of Darwin's major ideas and their contextualizations in a way that does justice to these significant shifts and their interconnections with Darwin's formulation of sexual selection. Many, if not all, of these ideas (even where Darwin seems at his most original) were borrowed from other thinkers or practitioners, and were either adopted outright or adapted to his theorizing by Darwin—that is, they have earlier histories and practices that we must look for and analyze in order to understand them and their roles in the formation of sexual selection.[36] When this plurality of (sometimes contradictory) historical sources, themes, and practices is integrated with the fine-grained analysis of Darwin's notes, correspondence, and publications, it makes possible a multilayered, richly detailed history of the genesis of what was, for its time, a radical theoretical concept that is back in play in current evolutionary thinking.

I should add a note on terminology: Darwin and those on whose writings he drew, or those with whom he corresponded, at times used a vocabulary that offends us now. In a historical study of this kind, it is important to bear in mind that terms such as *savage, mankind, man, race, negro,* and so on, had a range of available meanings and that their general usage did not necessarily connote the meaning the enlightened now would attach to them. Their usage requires careful attention to context if their historical understanding is to be achieved. Also, although I have done my best to defer to modern sensibilities, it is not always possible to supply gender-free or nonracial alternatives.

The Outline of This Book

Part 1, "Beauty, Brotherhood, and Breeding: The Origins of Sexual Selection," consists of nine chapters, each of which examines a major source or theme in the making of sexual selection.

Chapter 1, "The Ugly Brother," analyzes the defining moment of Darwin's conception of sexual selection: his lived experience of savage encounter with the indigenes of Tierra del Fuego. It is argued that his early adoption of the view that the different human races had different inborn, heritable standards of beauty or aesthetic taste—a view that underpinned his notion of aesthetic choice in animals—was founded in Darwin's reconciliation of his liberal abolitionist background with a pervasive contemporary "visual ideology" that had some commonalities with later biological racism.

Chapter 2, "Good Wives," discusses the ways in which Darwin's family background and his experiences as Victorian husband and father entered into the making of sexual selection. Emma Darwin's role as nurturing wife and mother and Darwin's dependence on her support and his domestic comforts are viewed against the gender relations implicit in Victorian expectations of a "good wife," as well as in the code of "manliness" and the construction of "masculinity."

Chapter 3, "'Bliss Botanic' and 'Cocks Heroic'" examines the erotic evolutionary writings of Erasmus Darwin and how his grandson took the notion of sexual combat as a means of species benefit and as an explanation for the development of specialized male weapons from this source. The younger Darwin's early notebook theorizing on the sexual struggle is discussed in relation to his earliest formulation of natural selection and the ways in which he distinguished the struggle for mates from the struggle for existence. It is argued that his grandparent's naturalistic, sexualized aesthetics also influenced Charles Darwin's understanding of beauty and aesthetic choice.

Chapter 4, "Beauty Cuts the Knot," extends this analysis through an examination of Darwin's early readings in British empiricist aesthetics—notably the writings of Joshua Reynolds and Edmund Burke—and Darwin's adoption of a naturalistic notion of beauty as strongly gendered, as race and class specific, and weighted with moral and political meaning.

Chapter 5, "Reading the Face of Race," looks at Darwin's readings in early ethnological literature, which had tight associations with aesthetics and physiognomy. William Lawrence's "blasphemous" *Lectures*—which offered a uniquely materialist, monogenetic theory of racial origins, developed by analogy with domestication, and which located mental and moral differences in anatomy—was a particularly important source for Darwin's thesis of the role of aesthetic selection in the formation and stabilization of the human races. The vital role of the naturalist Edward Blyth in integrating the two forms of sexual selection and bringing Lawrence's views to Darwin's attention at two critical stages of theory formation is emphasized.

Chapter 6, "Good Breeding," deals with Darwin's immersion in the practices of animal breeders and the reconstitution of the notion of aesthetic preference by artificial selection.

Chapter 7, "Better Than a Dog Anyhow," deepens this analysis with a discussion of Darwin's close reading in early 1839 of a key text, Alexander Walker's *Intermarriage*. Walker placed explicitly before Darwin the reiterated comparison of wife choice with the selective procedures of animal improvers and made the masculine judgment of woman's beauty—a visually based, aesthetically informed process of "choice"—the agency of human progress or racial improvement.

Chapter 8, "Flirting with Fashion," examines the critical role of Victorian fashion in Darwin's thinking. It looks at how Darwin's exposure to the perceived vagaries and excesses of the dress choices of Victorian women and his consistent recourse to the metaphor of fashion enabled his transfer of the masculine, manipulative art of breeding to the sexual preferences of female animals. It analyzes his formulation of his fundamental threefold analogue of the selective practices of pigeon fanciers, the frivolous dress choices of fashion-conscious women, and the aesthetic choices of female birds, all of whom capriciously pushed their selections to nonfunctional extremes.

Chapter 9, "Development Matters," introduces the crucial embryological understandings underpinning Darwin's threefold analogue, sketches their political and social implications, and shows their functioning in Darwin's integration of the critical theoretical components of sexual selection: development, inheritance, and divergence. It indicates how this developmentally centered understanding confirmed the agency of female choice in animals, powered Darwin's dispute with Wallace, and gave him the confidence and self-belief to write *The Descent of Man*.

Part 2, "'For Beauty's Sake': The Making of Sexual Selection," consists of five chapters covering the major stages of theory building from the period of Darwin's "big species" book, through the writing of the *Origin*, up to the publication of the *Descent* and its reception.

Chapter 10, "Critical Years: From Pigeons to People," discusses Darwin's evolving views on sexual selection in a period of mounting racial and sexual tension in which a heightened sense of national identity reinforced the racial and gender superiority of the white middle-class male. It examines how Darwin drew on the writings of the new racial determinists—notably the race theorist Robert Knox and the polygenists Josiah Nott and George Gliddon—in reaffirming his earlier Lawrence-derived thesis of an aesthetic factor in the differentiation of gender, class, and race, and discusses the relation of Darwin's pigeon-keeping to this.

Chapter 11, "Putting Female Choice in (Proper) Place," takes the analysis from the "big species" book up to the publication of the *Origin*, focusing on female choice and the arguments and evidence Darwin adduced in its support. Two rediscovered pages of his "big species" book (thought lost) show that Darwin had put female choice in theoretical place by, at the latest, mid-1858. This chapter stresses the importance of embryology in Darwin's decision to publish his views on the determining role of sexual selection in human racial differentiation in the *Origin*. His naturalization of female choice in birds and normalization of male aesthetic choice in human racial divergence is discussed in relation to the "good woman" of Victorian domestic ideology and the beginnings of "first wave" feminism.

Chapter 12, "The Battle for Beauty," picks up the history of sexual selection in the late 1860s as Darwin set about researching and writing *The Descent of Man*. It analyzes Darwin's prolonged dispute with Wallace over the issues of female choice, protective coloration, and human evolution in the contexts of the Huxley-led drive for Darwinian cognitive and cultural authority; institutional conflict between the Darwinian "ethnologicals" and the racialist "anthropologicals"; the American Civil War; the Eyre affair; the social and professional implications of the "irrepressible" woman question; and John Stuart Mill's championing of women's rights.

Chapter 13, "Writing the *Descent*," deals with Darwin's final assembly of the components of sexual selection and human sexual and racial divergence into *The Descent of Man*. It is argued that the *Descent* was structured by Darwin's differences with Wallace, by his conflicts with Mill and the "new" evolutionary anthropologists, and the enduring resonances of savage encounter. This chapter also assesses the problems posed by Darwin's detailed reconstructions of plumage and attribution of high aesthetic appreciation to female birds, and the contradictions engendered by his key thesis of the primacy of beauty-based male choice as the determinant of human racial divergence.

Chapter 14, "The Post-*Descent* Years," explores the response to the *Descent* through Darwin's late Victorian readership—not simply his scientific readers, but the wider reading public who variously adopted, adapted, or rejected his theory of sexual selection. As sexual selection (along with natural selection) went into eclipse, female choice was seized on by an array of social purists, eugenicists, sexual reformers, birth controllers, feminists, and socialists, notably Wallace, who in a volte-face in 1890 advocated a post-socialist world in which free and informed female choice would guide the future direction of humanity. Huxley's reaction to all this is analyzed through the content and context of his famous Romanes Lecture.

The subsequent history of sexual selection is briefly reviewed in an epilogue, and Darwin is given the last word.

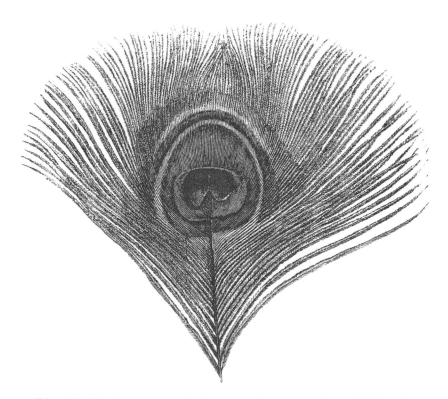

"The sight of a feather in a peacock's tail, whenever I gaze at it, makes me sick!" Darwin told Asa Gray in 1860. By 1864 he was convinced of the solution to the problems it posed. Peacock feather from *Descent* 1871, 2:137. University of Sydney Library, Rare Books and Special Collections.

PART ONE

Beauty, Brotherhood, and Breeding: The Origins of Sexual Selection

ONE

The Ugly Brother

I believe if the world was searched, no lower grade of man could be found. . . .

Their appearance was so strange, that it was scarcely like that of earthly inhabitants. . . .

I never saw such miserable creatures; stunted in their growth, their hideous faces bedaubed with white paint & quite naked.—One full aged woman absolutely so, the rain & spray were dripping from her body; their red skins filthy & greasy, their hair entangled, their voices discordant, their gesticulation violent & without any dignity. Viewing such men, one can hardly make oneself believe that they are fellow creatures placed in the same world. I can scarcely imagine that there is any spectacle more interesting & worthy of reflection, tha[n] one of these unbroken savages.

—Darwin on the Fuegians, 1832, 1834

So Darwin recorded his encounters with the "savages" of Tierra del Fuego in the diary he kept on board the *Beagle*. His immediate account registers both fascination and revulsion. At the same time, his shock at the profound difference of the Fuegians, the enormity of their distance from civilized man, was tempered by his assumption that these "unbroken" people may yet be "improved" by the taming hand of civilization.

At that point in time, he had reason to think so. Three of his fellow travelers on the *Beagle* were Fuegians, dubbed with the objectifying, Anglicized pseudonyms of Jemmy Button ("whose name" Darwin claimed, "expresses his purchase-money"[1]), York Minster, and Fuegia Basket. Their real names were, respectively, O'run-del'lico, El'leparu, and Yok'cushly. They had been taken from a tribal state in Tierra del Fuego to England by Captain FitzRoy on a previous voyage and given the rudiments of a Christian education.

They were now being returned to their native land along with a naive young missionary and assorted impedimenta of British civilization—described by FitzRoy as "serviceable articles," but including, according to Darwin, "wine glasses, butter-bolts, tea trays, soup turins, mahogany dressing case, fine white linen [and] beavor [sic] hats."[2] It was FitzRoy's personal project of improvement that they would form the nucleus of a Christian civilization on these inhospitable shores.

In the *Descent*, Darwin harked back to his "continual surprise" at "how closely" these domesticated *Beagle* Fuegians who could speak a little English "resembled us in disposition and in most of our mental faculties."[3] Yet their "barbarian" relatives, brutish, to his eyes devoid of all the expected attributes of humankind, pushed the boundaries of humanity to its very limits.

During his travels on the *Beagle*, Darwin met with a number of other races and saw many awe-inspiring sights. But at the conclusion to his diary, which became the basis of his best-selling *Journal of Researches* (rewritten in 1837, first published in 1839), Darwin returned to what had most struck him on his five-year voyage: the primal spectacle of "man in his lowest and most savage state":

> One's mind hurries back over past centuries, and then asks, could our progenitors have been men like these?—men, whose very signs and expressions are less intelligible to us than those of the domesticated animals; men, who do not possess the instinct of those animals, nor yet appear to boast of human reason, or at least of arts consequent on that reason. I do not believe it is possible to describe or paint the difference between savages and civilized man. It is the difference between a wild and tame animal.[4]

Words failed Darwin in attempting to describe the unbridgeable gulf between the Fuegian and his civilized observer; nor could this difference even be represented in paint, so far removed from iconography, from all aesthetic conventions, was the alien appearance of the Fuegian (whose own rudimentary aesthetic was limited to the appreciation of gaudy baubles such as blue beads and scarlet cloth). This unforgettable, unpaintable, indescribable difference (which Darwin did in fact describe over and over again in terms of the difference between wild and domesticated animals) was still with him at the end of his life.[5] But his best-known reprise of his Fuegian encounter was in the conclusion to the *Descent*. There, Darwin, in an attempt to counter the Victorian horror of bestial descent, asserted his personal preference for a brave monkey or a plucky baboon as ancestor, rather than a bestial Fuegian progenitor. It repays close reading:

Figure 1.1. "Fuegian" by Conrad Martin, symbol of the "ugly brother," displacing the earlier, idealized image of the "noble savage." Frontispiece to FitzRoy 1939. University of Sydney Library, Rare Books and Special Collections.

The main conclusion arrived at in this work, namely that man is descended from some lowly-organised form, will, I regret to think, be highly distasteful to many persons. But there can hardly be a doubt that we are descended from barbarians. The astonishment which I felt on first seeing a party of wild Fuegians on a wild and broken shore will never be forgotten by me, for the reflection at once rushed into my mind—such were our ancestors. These men were absolutely naked and bedaubed with paint, their long hair was tangled, their mouths frothed with excitement, and their expression was wild, startled, and distrustful. They possessed hardly any arts, and like wild animals lived on what they could catch; they had no government, and were merciless to every one not of their own small tribe. He who has seen a savage in his native land will not feel much shame, if forced to acknowledge that the blood of some more humble creature flows in his veins. For my own part I would as soon be descended from that heroic little monkey, who braved his dreaded enemy in order to save the life of his keeper; or from that old baboon, who, descending from the mountains, carried away in triumph his young comrade from a crowd of astonished dogs—as from a savage who delights to torture his enemies, offers up bloody sacrifices, practises infanticide without remorse, treats his wives like slaves, knows no decency, and is haunted by the grossest superstitions.[6]

Here, the captive monkey and the wild baboon are represented as more like Darwin and the reader, more moral—indeed, more civilized—than the alien, wild Fuegian who is nevertheless presented as closer in kin (only just) to the reader and to Darwin. In this rhetorical set piece, Darwin plays on the disgust felt for the indecent, filthy, cruel savage in order to lessen the disgust for the ape ancestor. Again he conveys his shaken sensibilities; once again he likens the Fuegians to wild animals; yet again he presents the striking, if distasteful realization: "Such were our ancestors."

It has been remarked of this passage that here Darwin is situating himself in relation to the central anthropological question of his time: Were the different human races of one common origin, as the old orthodox biblical view would have it (monogenism), or did they have a number of separate origins (polygenism)? Here, at the very end of his sustained theoretical defense of monogenism, of his carefully reasoned argument that, over the ages, the different races have originated from a common ancestor through the cumulative action of sexual selection, Darwin identifies himself as an emotional polygenist, as retaining his sense of absolute human difference, of antipathy to other races. To this end, he deploys his experiential authority by invoking a historically specific, intensely lived, still vivid encounter with wild, inhuman Fuegians.[7] At the same time, in the very same paragraph, Darwin affirms his evolutionary, monogenist position—"there can hardly be a doubt that we are descended from barbarians." "We" (Darwin and the Victorian reader) and the disgusting, uncivilized Fuegians, however repellent the notion, share a common origin.

That Darwin never doubted the common origin of the human races is beyond dispute. He came from a Unitarian and liberal Whig, Wedgwood-Darwin line of active opposition to slavery. It was the one social issue guaranteed to make his "blood boil."[8] His writings emphasize the essential humanity of the black men and women, enslaved and free, whom he encountered. He was righteously revolted by the cruelties and injustices to slaves that he heard of and witnessed during the voyage of the *Beagle*. Their screams and groans haunted him down the years. While back in Britain the abolitionist movement campaigned to outlaw slavery in the name of a universal human brotherhood, Darwin in Brazil—where Portugal was still transporting African slaves—praised the efforts of the antislavery agitators. He recorded his abhorrence of the system and those prejudiced "polished savages in Britain" who supported it, who ranked the slave as "hardly their brethren, even in Gods eyes."[9] He hoped for the day when the slaves with their "fine athletic figures," "underrated intellects," industrious habits, and sheer force of numbers would "assert their own rights" and "ultimately be the rulers" of the "ignorant, cowardly, & indolent" Brazilians.[10] On board the *Beagle*, Darwin jeopardized his comfortable relations with FitzRoy in an emotional confrontation over the aristocratic captain's paternalistic view of Brazilian slavery as a "tolerable evil."[11] In a later "explosion of feeling," provoked by Charles Lyell's published anti-abolitionist sentiments, Darwin went public with a passionate denunciation of the "sin" of slavery in the revised second edition of his *Journal of Researches* of 1845.[12] At the onset of the American Civil War, he declared himself prepared to accept a "million horrid deaths" for the greater good of an end to this "greatest curse on earth."[13]

It is possible, as Desmond and Moore claim, that it was Darwin's revulsion against slavery that underwrote his monogenism and, ultimately, his belief in human and biological evolution, a "concern that would lead to the emancipation of humanity from creationist bondage in the *Descent of Man*."[14] Yet it is not difficult to see that Darwin's unsettling, unforgettable Fuegian encounter, his shattering realization that such bestial-seeming beings might be our forebears, his readiness to exploit that savage encounter to emphasize the utter otherness of the savage as kin in order to make an animal kinship more acceptable to his Victorian readers, registers more than a mere abolitionist crusade, a ready-made, humanitarian assumption of racial unity that fueled his evolutionary theorizing.

Even when Darwin is at his most sensitive to the plight of the slave, he still plays on the uneasy conjuncture of man and beast. He gives an account of gesturing in attempting to communicate with a "negro, who was uncommonly stupid." To Darwin's consternation, the man assumes that Darwin is about to strike him:

> I shall never forget my feelings of surprise, disgust, and shame, at seeing a great and powerful man afraid even to ward off a blow, directed, as he thought, at his face. This man had been trained to a degradation lower than the slavery of the most helpless animal.[15]

Here the slave is a pitiable, servile, über-domesticated animal. At the same time, the Fuegians are untamed, untrustworthy, with the irrational instinctive ferocity "of a wild beast . . . each individual would endeavour to dash your brains out with a stone, as a tiger would be certain under similar circumstances to tear you."[16] Even the more tractable semi-civilized Indian gauchos of the Bahia Blanca "whilst gnawing bones of beef, looked, as they are, half-recalled wild beasts."[17]

There is a reiterated bestial imagery in Darwin's *Beagle* renderings of other races, a confusion of the classes of human and beast, that is more than mere metaphor, and that comes strikingly to the fore in his descriptions of the savage Fuegians.[18] It hints at the animality of Fuegian and hence our own ancestry, a proto-evolutionary transformation that marks the transition from the bestial to the human. It presages Darwin's argument in the *Descent* that there are no characteristics that absolutely distinguish humans from animals. More immediately, this crystallized for Darwin in 1838, when, back in England, he argued to himself the defensibility of his recently formulated transmutationist views:

> Nearly all will exclaim, your arguments are good but look at the immense difference. between man [and animals],—forget the use of language, & judge only by what you see. Compare, the Fuegian & Ourang outang, & dare to say difference so great.[19]

As an extension of this collapsing of Fuegian into orangutan, there is the compulsion of Darwin's comparison between the Fuegians on shore and the civilized Fuegians on board the *Beagle*, insistently typified as the difference between wild and domesticated animals. Again this is more than simplistic farmyard analogy. Darwin "saw what he thought was an authentic transformation of personality from aboriginal brutishness to the softer, tamer, more civilized nature of Western humanity, 'domestic' in all senses."[20]

By Darwin's account, the wild Fuegians "knew no government" or "domestic affection"; they lived in primitive communism; went about naked and shelterless in one of the worst climates in the world; were darkly suspect of cannibalism, infanticide, and worse; could barely communicate with one another in a primitive gargling language; and possessed the minimum of unchanging skills "like the instinct of animals."[21] Jemmy, Fuegia, and York, on the other hand, had been schooled in modesty and morality[22] and now knew the reforming powers of British dress, cleanliness, religious observance, and the ownership of property.

Optimistically, Darwin jotted, "3 years has been sufficient to change savages, into, as far as habits go, complete & voluntary Europeans." The older, taciturn York, Darwin was certain, would "in every respect live as far as his means go, like an Englishman." Yet he registered his reservations about the plump, vain, good-natured Jemmy, who, as a result of his civilizing experience, allegedly had become so Anglicized as to forget his own language and was "thoroughly ashamed of his own countrymen."[23] Even so, the privileged situation and easy manners of the *Beagle* Fuegians formed a salutary contrast with Darwin's disturbing experience of the cringing slave, where this improving domesticating process had been perverted by the abominable "training" of slavery to an unnatural, uncivilized, propertyless servility.

1.1 The return of the native

Then, FitzRoy's civilizing project absurdly unwound. The missionary, who, with the assistance of the *Beagle* Fuegians, was to educate their people in basic farming and building practices, distribute clothing, and inculcate some notions of modesty, cleanliness, and Christian principles, was intimidated and expropriated of his supplies by savages, who had no sense of property rights. The *Beagle* returned just in time to rescue him from tribesmen who were intent on the forcible removal of his facial and body hair with mussel-shell pincers (a lesson in Fuegian bodily aesthetics for the young Darwin that he put to use in *The Descent of Man*[24]). The ungrateful York Minster, sloughing off his English persona as readily as his starched shirt points, decamped with Fuegia after conspiring to separate the hapless Jemmy from his treasured British possessions.[25]

When Darwin next saw him, Jemmy was indistinguishable from his wild companions. The once "clean, well-dressed stout lad" of the *Beagle* had reverted to a "naked, thin, squalid savage" and "so ashamed of himself that he turned his back" to the ship. FitzRoy "could almost have cried," while Darwin lamented, "I never saw so complete & grievous a change." Jemmy's abject regression to savagery was partially reversed by hastily produced clothing. He dined with the Captain and "ate his dinner as tidily as formerly." Inexplicably, "poor Jemmy" refused FitzRoy's offer of repatriation to England. The belated discovery that he had "got a young & very nice looking squaw" (nice looking, that is, Darwin qualified, "for a Fuegian"), cleared up the mystery of Jemmy's reluctance to return to the niceties of civilization. After an exchange of presents, he cheerfully shook hands all round in best British fashion and went back to his native life and his native wife.[26]

It does not seem to have occurred to Darwin that Jemmy might resent what was, for all FitzRoy's justification of it, his abduction, his prolonged estrangement from family and tribe, his enforced adoption of a strange, only part-comprehended culture and language—the experiences and alien customs that distanced him from his own people on his return. Nor that what appeared degraded to Darwin was, for Jemmy, his people, his culture, and his preferred mode of living. We cannot know Jemmy's version of events, though we do have modern anthropological and other accounts that testify to the vitality and cultural richness of pre-missionary life for the four separate indigenous tribes of Tierra del Fuego. Darwin preferred the romantic explanation for Jemmy's rejection of the benefits of British civilization—the bathetic climax to FitzRoy's grand scheme: "Jemmy & his wife paddled away in their canoe loaded with presents & very happy." True love, even savage love, conquers all.[27]

There is nothing at all romantic about the portraits of Jemmy and his "very nice looking" wife published in FitzRoy's volume of the *Beagle* narrative. These are conventional ethnographic images, coded for the savage condition. Labeled "Jemmy" and "Jemmy's Wife," there is little to choose between them. They have almost identical matted, shoulder-length hair; flat crania; heavy brows; narrow, deep-set, dull eyes; spreading noses; wide cheeks; thick lips; and weak receding chins. The major difference between them lies in Jemmy's hollowed, famished cheeks, a sign of his recent deprivations. Both are expressionless, shambolic, symbolic of FitzRoy's shattered dream of a civilized Tierra del Fuego. The fascinating thing is that we are able to compare this brutish Jemmy of 1834 with his dandified version of a year earlier, also published by FitzRoy. Civilized Jemmy not only has smart European clothing and well-cut hair, but his very features were, FitzRoy maintained, "much improved by altered habits, and by education" during his stay in England.[28] The Jemmy of 1833 has a higher brow, bright and well-spaced eyes, narrow and well-defined nose and lips, and a firm chin. It is hard to recognize in him the same man as the savage of a mere year later. His head shape is entirely different, with a larger, more rounded cranium, and his expression is open and alert to the advantages of a Christianized, Europeanized life. He has yet to regress to dull, squalid savagery, with its accompanying physiognomic coarsening and mental decline.

Figure 1.2. FitzRoy's published sketches of "Fuegia Basket, 1833" (top left), "Jemmy's wife, 1834" (top right), "Jemmy Button, 1834" and "1833" (center), and "York Minster, 1833" (bottom, full face and profile). The two representations of Jemmy—the plump dandy of 1833 and the "thin haggard savage" of 1834—are transposed to give the impression of civilized improvement from the savage state, a transposition that glosses over the depressing reality of Jemmy's recidivism. FitzRoy 1939, 2. University of Sydney Library, Rare Books and Special Collections.

FitzRoy held no brief for the romantic convention of the noble savage. All his depictions of indigenes confirm his aversion to the savage physiognomy. Male Fuegians were "Satires upon Mankind"; "she-Fuegians" were only "by courtesy called women . . . They may be fit mates for such uncouth men; but to a civilized people their appearance is disgusting." Like Darwin, FitzRoy recounted the alleged horrors of cannibalism and infanticide. Yet he also described the Fuegians as a "brave, hardy race," as "ignorant, though rather intelligent" and, men and women alike, as "remarkably fond of little children." He admired their "sagacity and extensive local knowledge." With so much staked in his Fuegian experiment, he was an anxious observer of the reactions of his crew to what was, for most, their first sight of Fuegians in the raw. He does not mention Darwin's stunned response, but took heart from the more favorable reaction of Lieutenant Hamond to "such fine fellows": "It told me that a desire to benefit these ignorant, though by no means contemptible human beings, was a natural emotion, and not the effect of individual caprice or erroneous enthusiasm."[29] However "disagreeable, indeed painful," the sight of an uncivilized savage was, "unwilling as we may be to consider ourselves even remotely descended from human beings in such a state," FitzRoy affirmed his belief in Fuegian kinship and improvability. Julius Caesar, he reminded his Victorian readers, had found their own ancient Briton forebears painted and clad in pelts, like so many Fuegians.[30]

Revealingly, in his published account of the *Beagle* voyage, FitzRoy arranged the two portraits of Jemmy in reverse chronological order. So arranged, they convey the impression of progress and improvement from the savage state to that of enlightened Europeanized man, a before-and-after sequence illustrative of the beneficent effects of civilization, one that glosses the depressing reality of Jemmy's post-*Beagle* regression. Even in the face of the collapse of his cherished project and the recidivism of his Fuegian protégés, FitzRoy clung to his humanist faith in Fuegian improvement. He still hoped that "some benefit, however slight" might result from the exposure of Jemmy, Fuegia, and York to Christian beliefs and English values:

> Perhaps a shipwrecked seaman may hereafter receive help and kind treatment from Jemmy Button's children; prompted, as they can hardly fail to be, by the traditions they will have heard of men of other lands; and, by an idea, however faint, of their duty to God as well as their neighbour.[31]

He appended a detailed biblical-based defense of his "certainty, that all men are of one blood" to his volume of the *Beagle* narrative. FitzRoy could embrace the ugly Fuegian in the sure knowledge that he might be redeemed from savagery through Christian benevolence and philanthropy; even his physical ugliness, along with his morality and intelligence, could be ameliorated.[32]

For Darwin, on the other hand, the utter failure of FitzRoy's Fuegian project undercut his hitherto unquestioning belief in the savage "power of improvement." For all the romantic gloss he might put on Jemmy's return to his wife and his own people, the best that Darwin could hope for Jemmy was that he might be "as happy as if he had never left his country."[33] It was to be supposed that even the brutish Fuegians enjoyed a "sufficient share of happiness (whatever its kind might be) to render life worth having." Nature, reflected Darwin, "by making habit omnipotent" had "fitted the Fuegian to the climate & productions" of his miserable country.[34] While he hoped that FitzRoy's "noble" efforts should not have been in vain, Darwin could not share his optimism. Civilization, asserted the scion of the propertied, entrepreneurial Wedgwoods and Darwins, was dependent on the accumulation of property and the development of a hierarchical system of government. The "perfect equality" of the lowly Fuegians, evidenced by their intractable redistribution of all they were given or stole, made their political improvement "scarcely possible."[35]

The distance between savage and civilized man was extreme—so extreme that Darwin doubted that it could ever be bridged. Back in London, he privately confessed his doubts.

> Having proved mens & brutes bodies on one type: almost superfluous to consider minds . . . yet I will not shirk difficulty—I have felt some difficulty in conceiving how inhabitant of Tierra del Fuego is to be converted into civilized man.[36]

Much has been made of the significance of the famous Galapagos finches for Darwin's conversion to the transmutation of species, but these did not catch his attention until after he returned home.[37] Among the overabundance of diverse and scattered observations and specimens he brought back to England to order and make sense of, we must count the cumulative effects of Darwin's direct experience of racial diversity, of the instability of the "domesticated" or civilized human condition, of its all too ready reversion to barbarity; above all, of the shock of the "savage" to his conventional Christian views, of its inescapable assertion of the animality of humankind. The case has been made that it was his Fuegian encounter that first shook Darwin's creationist assumptions. It "dramatized for him the animal nature (and eventually animal *origins*) of man."[38]

There is more to it than this. Underlying and contradicting these new speculations that mesh with Darwin's orthodox-derived assumption of genealogical community ("Could our progenitors have been men like these?"), there is the jarring undertone of the new racism, of antipathy to physical strangeness, of revulsion from almost unimaginable moral and mental difference that found its mature expression in the conclusion to the *Descent*. Darwin's very antipathy may have borne evolutionary fruit. Was it the shudder of revulsion, his recoil from the radically different, that impelled Darwin to begin to feel his way toward a new rationalization of that difference? A letter from Darwin to Charles Kingsley in 1862 almost, but not quite, retrospectively says as much. It makes explicit—as no other statement on the Fuegians does—Darwin's instantaneous and sustained loathing for the notion of the "hideous savage" as progenitor:

> That is a grand & almost awful question on the genealogy of man . . . It is not so awful & difficult to me . . . partly from familiarity & partly, I think, from having seen a good many Barbarians. I declare the thought, when I first saw in T. del Fuego a naked painted, shivering hideous savage, that my ancestors might have been somewhat similar beings, was at that time as revolting to me, nay more revolting than my present belief that an incomparably remote ancestor was a hairy beast. Monkeys have downright good hearts, at least sometimes.[39]

What we do know from his early notebooks is that even as Darwin asserted the continuity between humans and animals on all points, declaring that he would "never allow" a "different origin" for humanity because of the perceived "hiatus" between humankind and animals, he nevertheless found the perceived differences and inequalities between the human races so great as to be at variance with the Christian imperative to brotherhood:

> Two savages, two species.—discussion [useless], unless it were fixed what a species means civilized Man, May exclaim with Christian we are all Brothers in spirit—all children of one father.—yet differences carried a long way.[40]

Unlike FitzRoy, Darwin had his doubts. He took them seriously enough to promote certain inquiries bearing on the specific distinctness of the human races on his return to England.[41] As a by-then-convinced transmutationist, he could not doubt the common origin of the human races; the issue was more one of how far back that convergence lay. To put it another way, how far had the races diverged from one another in the course of their developmental history—far enough to become separate species? "Although essentially the same creature," the Fuegians were utterly and profoundly alien—so alien that they looked to Darwin like beings from another planet, let alone a separate species of humanity.[42]

Darwin was hardly unique among nineteenth-century travelers and ethnographers in expressing "ethnocentric abhorrence" when confronted by institutions and behavior that did violence to their own value systems. It was a period in which the orthodox Christian assumption of universal human brotherhood was under increasing pressure from the new discourse of racial difference that accompanied the global expansion of British imperial power, a context in which the new sciences of human difference—anthropology and ethnology—were in contemporaneous formation. Just as the abolitionists succeeded in having slavery outlawed throughout the British Empire in 1833, an old heterodox polygenism gained new theological and biological currency. In such a milieu, the "sheer visual impact" of encounter with beings at such variance with customary human appearance and behavior led otherwise humane men to question human brotherhood. It beggared belief that the "hideous" Fuegians were "fellow-creatures placed in the same world" as the civilized naturalist on board HMS *Beagle*. The question as to whether Darwin and the Fuegians might indeed inhabit the same world was acquiring both scientific and political urgency.[43]

1.2 A visual ideology

Darwin's response to the Fuegians was structured by the prevailing "visual ideology" based on the body, aesthetic categories, and culture. By the end of the eighteenth century, the romantic convention of the noble savage had begun to give way before the image of the primitive or "ignoble" savage, an image of abject hideousness that was "less than equal, less than civilized and less than human."[44]

This reimaging was shaped by domestic as much as by imperial experiences. It was partly a reaction against those philanthropists who had vigorously deployed the "noble savage" convention in their campaign against the slave trade and slavery. The abolitionist stereotype of the African was more that of a natural Christian of Europeanized appearance who practiced the virtues of humility, patient suffering, and brotherly love, a sentimental stereotype that was to culminate in Harriet Beecher Stowe's pious Uncle Tom. In opposition to this, the defenders of slavery erected a thoroughly antipathetic, derogatory image of the African that degraded him to the level and appearance of an ape and played on the xenophobic aversion of Europeans to dark-complexioned strangers.

Another factor was the social dislocation and class dynamics that accompanied industrialization. This meant that the large number of Africans brought to England as slave-servants by wealthy planters, and those black sailors and soldiers who had been co-opted to the loyalist side in the American War of Independence and had migrated to England in exchange

for promised freedom, were thrown out of work or excluded from the paternalistic system that had once protected them. They constituted an impoverished and easily recognizable lower-class community, concentrated mainly
in the slum areas of London. Many were forced into beggary, prostitution,
and thievery. While they seem to have been reasonably well tolerated by the
other urban poor among whom they lived, there was little actual contact
between British-based destitute blacks and the wealthy, literate Englishmen
and -women who had taken up the antislavery cause.

Darwin was something of an exception in that, while he was in Edinburgh, he had come to know a freed slave, John Edmonstone, who taught
him how to stuff birds. The "very pleasant and intelligent" Edmonstone
(brought from South America by the eccentric naturalist and expert taxider-
mist Charles Waterton, who taught him his trade) supported himself on a
freelance basis, mounting specimens for the university museum and giving
cheap lessons to students. Darwin instanced him in the *Descent*, along with
FitzRoy's Fuegians, as demonstrating the "numerous points of mental similarity" among the most disparate human races. Yet Darwin never referred
to Edmonstone by name, but in ways that denoted his race and class: as a
"blackamoor" in a letter to his sisters, as a "negro" and "full-blooded negro"
in his published writings.[45]

Even ardent abolitionists like William Wilberforce, leader of the anti–
slave trade activists, shrank from actual social encounters with the black
objects of their sympathy and benevolence. For many, a black skin excited
aversion akin to that elicited by contact with the dirty, smut-encrusted lower
orders. During the long and complex history of the push for abolition of
the slave trade and then slavery itself, humanitarianism was often compromised by commercial considerations or political opportunism. The high-
flown sympathies of the abolitionist, directed at banning the overseas slave
trade, were often blind (as was Darwin) to glaring domestic inconsistencies
of grinding poverty, the conditions of factory children, and indentured labor. Nor did abolitionist sympathy with the slave necessarily extend to the
emancipation of slaves in British colonies, still less to the identification of
the free "Negro" as a brother and an equal. At a time when few thought all
white men equal, it made little sense to debate the equality of the black man.
The inequalities within English society fostered assumptions of black inferiority and adverse reaction to dark skins and other distinctive physical traits.[46]

Having little or no direct contact with people of other races, the educated
metropolitan public was largely dependent on accounts by travelers who, at
best, recorded their tolerant superiority to other customs and appearances,
but, like Darwin, often expressed ethnocentric abhorrence. The other source

of information—missionary literature inspired by evangelical Christianity—all too often played on the available symbolism of opposing white to black and good to evil. Dark skins were emblematic of the forces of darkness, the embodiment of the "other" to the enlightening English missionary. Savages were sinful and degraded, engaged in disgusting "pagan" practices, and required conversion so that they might realize their innate humanity and Christian salvation.[47]

Those few representatives of ethnic groups other than the African who reached London were displayed like so many exotic animals. A few, such as the Tahitian Islander Omai, persuaded to London by Captain Cook, became fashionable celebrities, though never the social equals of those who lionized them. More usually, they were exhibited for cash by enterprising showmen.[48]

The most notorious instance of ethnological show business was the display of the Khoikhoi (or Khoisan) woman Sarah Baartman, better known as the "Hottentot Venus," brought from the Cape of Good Hope to London in 1810. Baartman was "exhibited like a wild beast, being obliged to walk, stand, or sit as [her keeper] ordered her." The promotion of her "brutal figure,"—notably her protruding buttocks (or steatopygia), which voyeuristic spectators might poke or pinch for an extra fee, and her purported excessive genitalia (an elongation of the labia, the so-called "Hottentot apron")—veered between the unique or anatomically abnormal and her presentation as a biological type, the typical female representative of the "Hottentot" (or, sometimes, Bushman) race. Court proceedings brought by abolitionists who claimed she was "little other than a slave or chattel" were dismissed when Baartman's promoters produced a contract that guaranteed her a percentage of the profits for her exhibition.[49]

The court case had more to do with current debates about the right to liberty versus the right to property than with moral outrage over Baartman's enforced exposure to the vulgar. It points up some of the tensions within the antislavery movement. Reformers, eager to attack slavery, were yet reluctant to infringe on the property rights of slaveholders; they also balked at ideas of personhood that had the potential to destabilize the relationship between capital and "free" labor, bound up as it was with long-term indenture and even lifelong service. The abolitionists who brought the court case were engaged in promoting the commercial exploitation of the Cape Colony by the transformation of the indigenous "Hottentots" into free laborers in the developing frontier economy. They wanted to determine whether Baartman was made a public spectacle of her own free will—that is, whether she was commodifying herself (which was acceptable), or was compelled to it through her captivity and enslavement (which was not). Humanitarian and

progressive commercial interests coincided in the merging of abolitionist
rhetoric with that of imperial expansion in the question of the ownership
of labor power.[50]

Baartman was only one of hundreds of human curiosities exhibited in
freak shows or as circus spectacles in London. Nineteenth-century London
was the center of entertainment, spectacle, and display. The available cultural and interpretative resources readily confused or conflated physical
abnormality or "monstrosity" with racial or ethnic difference, both being
indiscriminate crowd pullers. A number of unfortunates with unusual or
abnormal physiognomies were put on show for a variously titillated or
credulous public and were attributed fabulous ethnic origins. Certain of
these, like the so-called "Aztecque children," engaged the attention of ethnologists who arranged special viewings and learned publications on their
racial origins.[51] The boundaries between ethnic and pathological difference
were hazy.

Like FitzRoy's Fuegians, few ethnic exhibits came willingly or wittingly to
England. While FitzRoy's intentions were philanthropic and not entrepreneurial, York, Fuegia, and Jemmy were also exhibited in London as worthy
objects of curiosity and benevolence: to FitzRoy's aristocratic friends and
relations, to members of English missionary societies, to Their Majesties
William and Adelaide, and to those philanthropic ladies who donated the
ludicrously inappropriate equipment for the ill-fated mission in Tierra del
Fuego.[52]

Not only live ethnic specimens went on display: while the living Fuegians were put through their paces for select audiences, anatomists at the
Royal College of Surgeons examined the internal anatomy of their fellow
who had not survived the first *Beagle* voyage and was brought back pickled
in a barrel.[53] This was the period when museums began their collections
of ethnic crania. Baartman's final exposure and definitive sexual and racial
stereotyping came after her death in Paris, with her dissection by the eminent anatomist Georges Cuvier. Cuvier's memoir focused on the "anomalous" steatopygia and genitalia of this female of the "lowest" human race,
morphological features that he saw as sharing affinities with those of the
mandrill and orangutan. Baartman's genitalia and skeleton were preserved
and displayed in the Musée de l'Homme [sic] in Paris, where they remained
until, after prolonged agitation, they were returned to Africa for interment
in 2002. Her poignant story is the subject of a large literature, which, somewhat arguably, depicts it as a defining moment in the history of the medical
construction of sexuality and an iconic instance of the Western exploitation
of the female black body.[54]

In England, such sporadic commodification and objectification of ethnic bodies and body parts undoubtedly contributed to the growing racism of the later Victorian period. Yet for the most part, these early cultural constructions tended more to the ethnocentric or xenophobic rather than the truly racist, in the sense that they did not systematically locate moral and intellectual traits, as well as physical ones, in biology, nor assume the innate cultural and biological inferiority of other races. Skin color, hair textures and so on, were not stabilized as markers of racial difference until later in the century. Even then, there was as much focus on the physical and cultural differences of the troublesome Irish, the so-called Celtic Calibans, which allegedly marked them out as a separate "race" from the Anglo-Saxon English and affirmed their inferiority to their longtime colonizers. Robert Knox, one of the earliest exponents of "race science," could claim that for years he had the "whole question of race to [him]self." As late as 1850, he found his audiences far more interested in the subgroups, or the "fair races," that he characterized among the British and Europeans, rather than in the "dark races of men."[55]

Early nineteenth-century constructions of racial difference primarily affirmed a sense of British middle-class identity, of cultural and moral preeminence, a national self-satisfaction at the civilizing and humanitarian mission of British antislavery and evangelical efforts. It was a sense of self-righteous, patronizing racial identity that could be counterposed to those benighted nationalities that still practiced the abomination of slavery. The savage, if no longer seen in Enlightenment terms as "natural man," the foil against whom the corruptions of civilization might be evaluated, was amenable to civilization. What separated civilized from savage man was not a difference in inherent mental makeup, but the effects of civilization.

The concept of civilization itself reflected a "convergence of issues" of economics, politics, morality, class, and gender, drawn together by the Reform Bill crisis of 1831–32 that reached its peak as Darwin left England on the *Beagle*. It was an expression of newly powerful British middle-class cultural values: liberal political institutions; utilitarian and evangelical impulses; middle-class standards of domestic comfort, respectability, and the "gospel of work"; all linked in a sexual context by the Malthusian-inspired doctrine of prudential restraint. It was the "ability to delay gratification, to exercise rational control over one's baser instincts" that was the foundation of individual liberty, political responsibility, and the preservation of the family, the cornerstone of society. Savages—like the impoverished, unwashed, overly populous, and potentially insurrectionary underbelly of the British class system—deprived of the discipline of labor and delayed sexual gratification, were "at the mercy of the forces of nature."[56] James Cowles Prichard, the doyen of British ethnology in this period, declared: "In a barbarous state of society the passions are under no restraint." The civilization of savages required the repression of their instinctive passions through the inculcation of habits of work, cleanliness, and religious observance; in other words, the "domestication" of the savage impulse. Exposure to British culture, customs, morality and religion guaranteed savage improvement.[57]

This comfortable complacency was increasingly challenged by the frontier colonial encounters of the early nineteenth century that focused the gaze on physical difference as an indicator of cultural worth and insusceptibility to the improving powers of civilization. These new discourses of primitivism and exclusion rationalized the expulsion of indigenes from native land, their economic exploitation, the destruction of indigenous ecosystems, and, ultimately, genocide. Norbert Finzsch has emphasized the extent to which aesthetic judgment was made a distinctive component of the construction of a racialized and gendered "abject Other," and was linked with the assessment of indigenous economic, societal, and linguistic achievements. This was not a single, transhistorical ideology; it was not without its inner contradictions, and it varied in ways peculiar to specific colonial contexts. But, Finzsch argues, there are sufficient commonalities to constitute the basis for identifying the various components of this "visual ideology."[58]

The assessment as "savage" was based primarily on observation—that is, the European gaze directed at the indigenous body. Appearance was all. From it might be read the inner intelligence, morality and ethics of the indigene. The complexion, the shape of the limbs and breasts, the kinkiness of hair, the size and shape of eyes, nose, and mouth, the expression—all began to be seen as indicators of where in the scale of civilization the indigene might be located. The lower the position on the scale, the more the exclusion and dispossession of the indigene might be justified. The ranking of appearance and cultural status was calibrated according to the European standard. Colonial observations and depictions of indigenes were stereotyped by the conventions of the physical and moral traits attributed to the civilized European self and his binary savage opposite.[59]

The savage was the negation of all that the European was assumed to stand for. Beauty of complexion and features was a European prerogative that could not be extended to savages. Moreover, beauty in the European tradition resonated with the conception of virtue. Beauty signified goodness; ugliness meant sin. The moral worth of a race was bound up with an aesthetic evaluation that meant that those savages deemed hideous to European eyes were associated with the most depraved moral practices, such as that bogey of the nineteenth-century European traveler, cannibalism. It was but a short step from Darwin's demonization of the Fuegians on the basis of their appearance alone—"they looked like so many demoniacs who had been fighting"—to his credulity as to the preferential consumption of their useless old women rather than their more valuable dogs in time of need. This shocking indication of the depth of Fuegian depravity and inhuman lack of family feeling was replayed in Darwin's published *Journal*, in all six editions of the *Origin*, and in his *Variation of Animals and Plants under Domestication*. It appears to have been without foundation, other than the readiness of the Fuegians to go along with the expectations of their European interrogators.[60]

Language was another indicator of where a race might be placed in the aesthetic hierarchy by the English-speaking observer. It was language that distinguished men from beasts, but savages were assumed to lack a proper language—a deficiency that made them less than human, almost on a level with the primates. Darwin, in demoting the Fuegian to the level of the orangutan, might tell himself to forget language and "judge only by what you see" (an affirmation of the significance of the "objectifying" European gaze), but his attribution of gargling, discordant, minimal speech to the Fuegians sat squarely in this tradition of eliding the difference between the savage and the primates on the basis of language. Elsewhere in his transmutation notebooks he was more explicit: "Not understanding language of Fuegian, puts on par with Monkey." Both listening and looking enabled the European to locate the indigenous body in a "matrix of progress and civilization, morality and ethics, growth or extinction."[61]

Extinction was a crucial component of the colonial construction of the ugly savage. The ownership of the land required the denial or cleansing of any prior claim. This might be expressed in the idea that the settled country was either virtually uninhabited or uncultivated and therefore lacking an owner; or alternatively, that the indigenous inhabitants were "dying races" and so could have no future claim to their land. White settlers were therefore entitled to indigenous lands because the original occupants did not use them, could not possibly use them because of their utter incapacity to do so, and/or were inexorably, by reason of their cultural and constitutional backwardness, moving toward extinction. In a period when racial variation was widely attributed to environmental factors, a form of cultural racism gained ground by which it was assumed that the European settler, endowed with a constitution enabling him to gain his livelihood by agriculture, commerce, and the manufacturing arts, must inevitably outstrip or displace the indigene, whose constitution had been formed by a "savage" environment that conduced to mere subsistence hunting and gathering. Such people, lacking even a basic civilizing sense of property accumulation, were condemned to a primitive, animal-like state of existence, as Darwin concluded the Fuegians to be.[62]

If his Fuegian encounter shook Darwin's belief in the infallible domesticating power of British civilization, his subsequent contacts with Australian Aborigines confirmed the colonial sense of their inevitable extinction. While Darwin considered them a little less ugly and a trifle more advanced than Fuegians in the scale of civilization, Aborigines were natural nomads who had no work ethic and so no defense against the encroachments of white settlement. Darwin predicted their inevitable displacement by the "White Man" who seemed "predestined" to inherit their country. This state of affairs had already come to pass in Tasmania, where a few years before Darwin's visit the natives had been rounded up and were kept, he reported, "in reality as prisoners" in a guarded promontory: "I believe it was not possible to avoid this cruel step; although without doubt the misconduct of the Whites first led to the Necessity."[63] Even such an enlightened humanitarian as Darwin was prepared to accept, albeit regretfully, the unavoidable extinction of native peoples in the wake of British colonial expansion. This was the price of progress, of civilization. The same Darwin who might inveigh against the Portuguese in Brazil for their injustices and cruelties to their slaves could still be proud to be an Englishman: "To hoist the British flag, seems to draw with it as a certain consequence, wealth, prosperity, and civilization."[64]

Colonial encounters increasingly depicted the indigene as not simply ugly, but abjectly so, a being whose very humanity was called into question, one whose passive displacement by civilized, colonizing Europeans was inevitable or, in certain cases, might actively be promoted by an implicit or complicit genocide. Charles Sturt echoed Darwin's (and FitzRoy's) contemporaneous observations of the Fuegians when he wrote of the Australian Aborigines he encountered: "It is scarcely possible to conceive that human beings could be so hideous and loathsome." If the men were abjectly hideous, the importuning women were even more so: "An old woman, a picture of whom would disgust my readers, made several attempts to embrace me. I . . . at length got rid of her by handing her over to Fraser, who was no wise particular as to the object of his attention."[65]

Observations of indigenes were persistently gendered. Gone was the alluring image of the South Sea siren, the natural beauty with the innocent sexuality, the idealized feminine projection of the noble savage. The nudity or seminudity of indigenous women, their lack of dress or any of the customary feminine refinements or conduct of civilized women, their overt display of a confronting, full-frontal, sexualized body, was at once an index of their lowly, bestial position and an affront to respectable masculinity and the aesthetic discernment of a European gentleman. Yet, at the same time, it signaled their tempting sexual availability, their low morals, and the rampant promiscuity of savage society that might readily be taken advantage of by the less fastidious: "Disgust bears the imprint of desire." The allegedly easy possession of indigenous women was symbolic of the colonial conquest of their country.[66]

There was also a masculine understanding that indigenous women, despite the obligatory repulsion they evoked in the European male, would naturally prefer that same obviously superior male to their own brutal kind. Major Thomas Mitchell reported that following a tribal battle, Aboriginal women would "not infrequently go over . . . to the victors"; and "thus it was . . . after we had made the lower tribes sensible of our superiority, that the three gins followed our party, beseeching us to take them with us."[67] Paul Strzelecki pushed these perceptions of the sexual superiority of the white male to their limits: such was the potency of white semen that their exposure to it rendered Aboriginal women unable to reproduce with men of their own race. It was the ill effects of interracial sexual unions that were driving the Aborigine to extinction.[68]

Darwin had his own views on extinction, though he made a note, "Sterility of one race of Mankind with another," in his copy of Strzelecki's description of his travels.[69] He put Mitchell's observation to use as the affirmation among primitives of his dawning view of the evolutionary role of male combat among animals: "The passion of the doe to the victorious stag. Who rubs the skin off to fight is analogous to the love of woman (as Mitchell remarks seen in savages) to brave men."[70] By the time Darwin made this early note on sexual selection, he also had the beginnings of the presumption of the significance of mating preferences for the formation of the different human races. This was largely dependent on the same visual ideology, the matrix of civilization, morality, and aesthetic judgment, that structured his Fuegian encounter and his assessments of other "barbarian" races he observed during his *Beagle* voyage.

1.3 "My excellent judgement in beauty"

As his *Beagle* diary shows, Darwin was a keenly appreciative student of beauty in both the aesthetic and sexual senses. He observed and recorded the sublime beauties of nature alongside the feminine beauties (or lack thereof) in the various nationalities and races with which he came in contact. After all, he was young, sexually alert, and curious, one of a shipload of sexually deprived young men.

A few months before his Fuegian encounter, Darwin had been mortified to learn that his childhood sweetheart, Fanny Owen, had become engaged to a dashing aristocrat a few days after the departure of the *Beagle*; she was in fact already married by the time he received the news in Rio de Janeiro. He assuaged his sexual rejection and wounded sensibilities by immersing himself in the spectacular tropical scenery around Rio. The erotic charge to Darwin's aesthetic response is overt: "As a Sultan in a Seraglio I am becoming quite hardened to beauty."[71] By the time the *Beagle* arrived in Buenos Aires, his broken heart on the mend, Darwin's chief amusement was to ride about and admire the colonial Spanish ladies with their graceful, gliding walk and form-revealing dress. These "angels" so aroused his enthusiasm that he wrote disparagingly to his sister Caroline of Englishwomen who "can neither walk nor dress . . . it would do the whole tribe of you a great deal of good to come to Buenos Ayres."[72] In Chile he found the women had "not forgotten how to blush," and he flirted with some "very pretty," if pious, Catholic senoritas. In Lima, he could hardly keep his eyes off the alluring *tapadas* with their "elastic gowns" and expressive "black & brilliant" eyes. Such was their "very powerful" effect on the young sea dog, that he fancied himself among "nice round mermaids."[73]

But in spite of all temptation, Darwin remained the quintessential Englishman. His susceptibility to Spanish angels was tinged with cultural chauvinism. He found distasteful their custom as hostesses of offering him choice morsels on the ends of their forks. At a remote estancia, he recorded his amused condescension when asked by a captain of the army (who was amazed to hear from the young English traveler that the earth is round) for his opinion as to "whether the ladies of Buenos Ayres were not the handsomest in the world":

> I replied, "Charmingly so." He added, I have one other question—"Do ladies in any other part of the world wear such large combs." I solemnly assured him they did not.—They were absolutely delighted.—The Captain exclaimed, "Look there, a man, who has seen half the world, says it is the case; we always thought so, but now we know it." My excellent judgement in beauty procured me a most hospitable reception; the Captain forced me to take his bed, & he would sleep on his Recado [saddle].[74]

To his uxorious cousin William Fox, luxuriating in domesticity with his own "dear little wife" and urging the "manifold advantages of Matrimony" on

Darwin, he perfunctorily dismissed all possibility of local candidates for the post: "As for the women in these countries they wear Caps & petticoats & a very few have pretty faces & then all is said." He was New World weary and had to restrain himself from bolting on the next ship home. He pined for England and "an English lady," though he had "almost forgotten what she is—something very angelic & good."[75]

Young colonial ladies might be attractive in spite of their Catholicism; their elders were morally suspect. Older Brazilian women, Darwin jotted, were priest ridden and corrupted by the institution of slavery; they were full of "cunning, sensuality & pride." They denied their feminine natures in most ingenious cruelties to their defenseless slaves: "They are born women, but die more like fiends."[76] Women of the lower orders of society scarcely rated a glance. As for those of mixed race or of Indian blood, Darwin registered only disgust as he recorded the sexual advances of a "creature of a woman," who dressed and rode like a gaucho, whom he recognized as a woman only when she so far forgot her inferiority as actually to flirt with Darwin.[77]

Sexual liaisons by ship's gentlemen with their racial inferiors were to be condemned in the strongest terms. Augustus Earle, *Beagle* artist for the early part of the voyage, had previously lived in New Zealand with a Maori woman and was scathingly critical of the impact of colonialism and prudish high-handed missionaries on traditional Maori culture. His sensational book airing these views drew indignant rebuttal from Darwin, who saw everything to praise in the missionaries' civilizing efforts among the savage Maori and their Christian forbearance in the face of Earle's "open licentiousness" in living with a native woman.[78]

Nevertheless, Darwin himself had an eye for the pretty women among the Pampas Indians, whom he recognized (contrary to some authors) as belonging to the same race as the Fuegians ("it is easy to see the same countenance, rendered hideous by the cold, want of food & less civilization in the Fuegian savage"). Some of the young women, he recorded with surprise, "deserved to be called even beautiful" ("They have a high colour & eyes which glisten with brilliancy.—Their legs, feet & arms are small & delicately formed"). Their beauty, however, could not save them from the hard life of the "wives of all Savages useful slaves." Nor could it protect them from the bloody war of extermination led by the brutal General Juan Manuel de Rosas. "Why what can be done," his soldiers rationalized their genocide to Darwin, "they breed so."[79]

While Darwin, in the same high-minded style of his antislavery sentiments, was initially sympathetic toward the Indians, lauding their courage and feats of horsemanship in their struggle against Rosas's murderous troops, his English property-owning concerns came to the fore when he saw the damage the Indians had inflicted on pampas estates. The enthusiasm of the settlers for Rosas's campaign of genocide against the "barbarians" was "natural," he wrote in his diary. Only when "all the Indians are butchered," beautiful women included, would the ranchers and their livestock be safe and a "grand extent of country" gained for the production of cattle and corn. Rosas "ultimately must be absolute Dictator, (they object to the term king) of this country."[80]

Complacent in his "excellent judgement in beauty," the emergent connoisseur ranked the indigenous peoples he encountered by their physical appearance, which he consistently related to their cultural and intellectual level and environmental conditions. The Tahitians (the men, at least—Darwin was "much disappointed" by the fabled Tahitian "beauties," reformed from "licentiousness" by zealous missionaries and unbecomingly dressed) were at one with their beautiful tropical surroundings, and their good looks and gentle hospitable natures reflected this. They were children of nature whose inherent innocence could only be refined by their exposure to Christian morality and English values. Their orderly, hierarchical society also augured well for their civilized future. They formed a salutary contrast with the warlike Maori of New Zealand. Here the people were grotesquely tattooed with disagreeable, rigid faces; fierce, cunning eyes; ungraceful, bulky figures; and "filthily dirty & offensive" persons and houses. Darwin was "much amused" by a ceremony of nose rubbing where men and women uttered "comfortable little grunts, very much in the same manner as two pigs do when rubbing against each other." Yet even with such unpromising material, Darwin found room for hope through enlightened English administration and mission work. "The march of improvement, consequent upon the introduction of Christianity throughout the South Sea, probably stands by itself in the records of history . . . These changes have now been effected by the philanthropic spirit of the British nation."[81]

In Australia, Darwin bribed good-natured Aborigines to perform for him: their faces were "very ugly," their bodies hairy and "most abominably filthy." Their hunting and tracking skills were impressive, but their culture was rudimentary. Their "Corrobery" was a "most rude barbarous scene, & to our ideas without any sort of meaning." Darwin ranked them just a "few degrees higher in civilization, or more correctly a few lower in barbarism, than the Fuegians."[82] Like the hideous Fuegians, the ugly Aborigines were excluded from the benevolent march of improvement that accompanied British expansion. Their fate was to disappear before it.

The theodicy of civilization framed all Darwin's aesthetic judgments of other races and nationalities, particularly those he deemed "savage" or "barbarian."[83] The terms might be aesthetic, but they displaced a politics of race, colonialism, and extinction. Even where certain "barbarians," such as the Pampas Indians, might be considered handsome, they were in the process of extermination in the cause of colonial productivity.[84] But genocide, like slavery, was an atrocity perpetrated by other nations in other places, historical stages that the more enlightened British had left behind. Darwin did not recognize its role in the rounding up and confining of the Tasmanians.[85] In New South Wales, as sheep displaced indigenes, he waxed almost delirious over its "extraordinary prosperity": "Wool, Wool—is repeated & must ever be the cry from one end of the country to the other." Its economic success was a tribute to the greatness of British imperialism: "Ancient Rome might have boasted of such a Colony." He lunched with the Macarthurs—leaders of the new squattocracy, riding high on the sheep's back—in a "beautiful, very large country house." The Englishness of it all quite went to his head: "There was such a bevy of pretty lady-like Australian girls, & so deliciously English-like the whole party looked, that one might have fancied oneself actually in England."[86]

Which was where Darwin longed to be: "There never was a Ship, so full of homesick heroes, as the Beagle." His *Beagle* travel may have broadened his mind, but that mind was irrevocably Anglocentric: "It is necessary to leave England, & see distant Colonies, of various nations, to know what wonderful people the English are."[87] In this sense, Darwin never really left home. England and the English were the measure of all things: nations, colonies, races, manners, morals, and of course, beauty. It followed that the quintessence of female beauty and virtue—Darwin's own "beau ideal"[88]—could only be an Englishwoman, the only proper wife for an Englishman.

Back in London, while anxiously engaged in the business of mate choice—he proposed to his cousin, Emma Wedgwood, just two weeks after the following notebook entry—Darwin pondered the question:

> Is our idea of beauty, that which we have been most generally accustomed to:—analogous case to my idea of conscience.—deduction from this would be that a mountaineer . . . born out of country yet would love mountains, & a negro, similarly treated would think negress beautiful,—(male glow worm doubtless admires female. Showing. No connection with male figure)—As forms change, so must idea of beauty.[89]

There it is: Darwin's first clear, extended statement, dated October 27, 1838, of the concept that the different human races have different inherent, not merely inherent, but *inherited* (Darwin is explicit about this, here and elsewhere in his notes) standards of beauty or aesthetic "taste," the foundation of his later theory of the role of aesthetic preference in the formation of the different human races. And what a distance Darwin had traveled to reach it. How packed it is with the personal and the political, with notions of place, race, gender, and sexuality. How crammed, too, with theorizing—a bare month earlier, Darwin had first noted down the idea of natural selection, which he still supplemented with the notion of inherited habit—and the intensive reading and note-taking that stand behind the deceptively simple statement "As forms change, so must idea of beauty."

We shall unpack this reading and theorizing over the course of the next few chapters. For the moment, let us note the attribution of an inherited aesthetic sense even to the lowly male glowworm, who "doubtless admires" the shapeless, wingless female that yet glows more brightly than he. Above all, let us not fail to see that these are masculine worlds, natural and human, dominated by masculine "judgement in beauty," with females—glowworm or "negress"—the objects of that judgment.

Darwin's notebook references were to that textbook signifier of physical difference from the European, the "Negro."[90] But Jemmy the Fuegian was surely the real-life exemplar of such inborn aesthetic preferences as Darwin now began to attribute to the different races. It was, after all, Jemmy the dandy, the most Anglicized of FitzRoy's Fuegians, who chose to forgo the civilized blandishments of his pampered London life and endure all the disadvantages of his remote, squalid existence, to remain with his Fuegian "squaw." She might be pretty enough by Fuegian standards, but she was scarcely to be compared with the English beauties among the fashionable philanthropic ladies who came to gaze at the Fuegians in London. We can take the Fuegian out of his country, but can we take his country out of the Fuegian? Even if Jemmy had never before set foot in Tierra del Fuego, he would still find his beau ideal among the "hideous" (to a European eye) Fuegians. Beauty lay in the eye, or, more precisely, the mind, of the beholder.[91] And that eye and mind had not only been trained by custom or habit to see beauty, as the various aestheticians he now began to consult claimed, but, as Darwin was by now convinced, had evolved to see and appreciate it: as forms change over time, so must minds and aesthetic perceptions. His own experiences predisposed him to the view—reinforced, as we shall see, by his readings in contemporary aesthetics and ethnography—that the shifting notion of the beautiful was connected to the plurality of physical forms and the moral and cultural codes that he had judged and ranked as he traveled the world on the *Beagle*.[92]

Taken together, all this indicates how Darwin, whose conventional, liberal views of human brotherhood had been made problematic by his Fuegian encounter, might resolve the paradox of such an utterly unacceptable brother. Darwin's polygenist shudder of horror could be reconciled with a monogenist account that emphasized and rationalized the essential ugliness of the savage other, while the kinship of this ugly brother might be pushed back to an earlier stage of evolution not far removed from the ape, which Darwin might yet represent as a more acceptable brother. Darwin might thereby "polemically extend a monogenist thesis across the whole of nature,"[93] rejoicing to find himself "netted together" with the animals ("our fellow brethren in pain, disease, death, suffering and famine, our slaves in the most laborious works, our companions in our amusements"), yet retain the feeling of revulsion from kinship with the repulsive, savage Fuegian. As Darwin's confidence in "my theory" grew, as he rejected the divine origins of "Man—wonderful Man," knowing well the social and theological consequences of granting the evolution of such high humanity from such bestial beginnings ("whole fabric totters & falls"), the physically and morally repugnant Fuegian of Darwin's depiction became the foil against which Darwin could contrast the humanized qualities of Jenny, the captive orangutan he visited in the London Zoological Gardens. By the time of the writing of the Descent, Darwin's failure of sympathy with his Fuegian "brethren" had become a rhetorical and theoretical strength that played on his own and his readers' prejudices.[94]

1.4 "'Improved' to extinction"

In the late 1830s, all this was still taking shape in Darwin's notebooks. Aesthetic preference was just one of the many conceptual strands Darwin was attempting to weave into his evolutionary theorizing. It took many years for the cloth to be loomed to his satisfaction. In 1860, while attempting to gather information confirming his views on human evolution, Darwin's thoughts turned to the unforgettable Fuegians. He contacted Thomas Bridges, then a young missionary based in the Falklands who had some contact with the Fuegians. Among various queries on expression, Darwin slipped in one on beauty:

> What idea of feminine beauty have the Fuegians? Do they admire women with strong American cast of countenance, or such as at all approach Europeans in appearance?[95]

Bridges eventually replied to say that it was "very certain" that the Fuegians looked on those countenances closest to the European as the most beautiful.[96] Darwin, however, stuck to his guns, and while citing Bridges's claim in the *Descent*, he disclaimed it:

> I cannot but think that this must be a mistake, unless indeed the statement refers to the few Fuegians who have lived for some time with Europeans, and who must consider us as superior beings.[97]

What of the recidivist Jemmy, who, supposedly bought for a button, had indeed lived with "superior" Europeans, had allegedly forgotten his own language, and was ashamed to be a Fuegian, but had yet reverted to squalid savagery and found his "beau ideal" among his own people? He lived to bear out Darwin's reiterated post-*Beagle* prophecy of "utter failure" for any mission-led civilizing project among the Fuegians and to be the despair of FitzRoy's remaining hopes. Jemmy was suspected of being implicated in the 1859 massacre of all Europeans (except the cook) in another abortive attempt to missionize the Fuegians.[98] As for York—that thoroughly compromised "complete & voluntary" Englishman of the young Darwin's diary—he disappeared from view. Fuegia fulfilled Darwin's more jaded expectations by reverting to licentiousness. Darwin's last published reference to her was, appropriately enough, in a footnote to the second edition of his *Journal of Researches*, where he dryly reported that Fuegia had "lived (I fear the term probably bears a double interpretation) some days on board" a sealer's vessel around 1842.[99]

For Darwin, the Fuegians remained the "very lowest of the human race." Their savagery—their moral and physical ugliness—was virtually intractable. They challenged his liberal view that all humans were capable of some improvement, of "progress." They were "improvable," but "not improving." He was ostensibly convinced otherwise in 1870 while working on the *Descent*. "I daresay many will decry [it] as extremely wicked," he warned his old shipmate, now Admiral Sir Bartholomew James Sulivan, who had sent him glowing reports of renewed missionary activity in Tierra del Fuego. Declaring himself a convert to Fuegian redemption, Darwin donated five pounds to the South American Missionary Society and was "proud" to be elected an honorary member.[100] Nevertheless, Darwin was undeterred from depicting the Fuegians as utterly beyond redemption in the conclusion to the *Descent*, condemning them forever to play ugly brother to a plucky monkey and a noble baboon.

As Darwin's fame and fortune continued to soar, the Fuegians were on their way to being "'improved' to extinction," as another old *Beagle* shipmate put it.[101] The Yámana (or Yahgan), Jemmy's tribe with which the *Beagle* crew had most contact, had a prehistory of some six thousand years in Tierra del Fuego. Of some three thousand members at the time of Darwin's *Beagle* contact, only around a quarter survived by the time of Darwin's death. Of these, by the 1920s, only fifty were left, confined within the missions. A series of devastating epidemics, the commercial depletion of the seals and whales on which they depended, the expropriation of their land by colonizing pastoralists and gold miners, and the deliberate destruction of their culture and traditional way of life by Anglican missionaries—all contributed to their demise. When there was scarcely anyone left to speak it, Thomas Bridges compiled a dictionary of the language of the Yámana people that indicates something of the subtlety and complexity of their culture and contradicts the crudities popularized by Darwin. The anthropologist Anne Chapman, who has worked with their few living descendants, contests almost all of Darwin's (mis)perceptions of the Yámana.[102] In the end, it might fairly be claimed that, while his savage encounter shook the young Darwin's world, their European encounter destroyed the Fuegians' world.

After all, as Darwin's civilized European world asserted itself in ever more remote areas of the globe, the extinction of the native races could hardly be avoided. In his *Journal of Researches*, the extermination that Darwin had witnessed at first hand was represented as an impersonal, apolitical process that displaced responsibility onto "some mysterious agency" or, alternatively, subsumed it under some natural "extirpating" principle:

> Besides these several evident causes of destruction, there appears to be some mysterious agency generally at work. Wherever the European has trod, death seems to pursue the aboriginal . . . We may look to the wide extent of the Americas, Polynesia, the Cape of Good Hope and Australia, and we find the same result. Nor is it the white man alone that thus acts the destroyer . . . The varieties of man seem to act on each other in the same way as different species of animals—the stronger always extirpating the weaker.[103]

In his determination to treat colonialism (its British variant, anyhow) as a beneficent, progressive process, Darwin had drifted into morally unsettling territory. The signs are that he was still unsure how to handle the "difficult problem" of the "gradual decrease and final extinction of the races of man" by the time of the *Descent*. In the *Descent*, as Darwin formally incorporated all of humanity under the authority of the naturalist, the impending extinction of the savage races, which he and his contemporaries took for granted, was naturalized through natural selection—through that process so aptly described in the subtitle to the *Origin of Species* as "the preservation of favoured races in the struggle for life." As Darwin applied this law of nature to the human races, the civilized, colonizing races were the favored ones in the struggle for existence. He could coolly predict, with the dispassion of the natural historian who moves effortlessly from humans (those higher, but hierarchically ranked, animals) to their closest animal relations and back again:

> At some future period, not very distant as measured by centuries, the civilized races of man will almost certainly exterminate, and replace, the savage races throughout the world. At the same time, the anthropomorphous apes . . . will no doubt be exterminated. The break between man and his nearest allies will then be wider, for it will intervene between man in a more civilized state, as we may hope, even than the Caucasian, and some ape as low as a baboon, instead of as now between the negro or Australian and the gorilla.[104]

There is no room here for empathy with those about to be exterminated in the progress of civilization; and it seems clear who is responsible for their extermination.

Yet at other points in his discussion, Darwin was reluctant to confront the implications of the agency of the colonizing "civilized" in the extermination of the less favored colonized. The pruning and improving hand of natural selection was certainly at play in the intertribal conflicts among the less civilized. Humans, wrote Darwin in the *Descent*, could survive all kinds of unfavorable climates and conditions—witness the Fuegians—but they could not survive each other in the competition for resources: "Extinction follows chiefly from the competition of tribe with tribe and race with race." In tribal societies, "the contest is soon settled by war, slaughter, cannibalism, slavery and absorption." However, in the case of contact between civilized nations and "barbarians," where "the struggle is short, except where a deadly climate gives its aid to the native race," much still remained "very obscure." The mysterious natural process that Darwin had invoked in the *Journal* reappears as Darwin casts around for alternative explanations and exculpations: "We can see that cultivation of the land will be fatal in many ways to savages, for they cannot, or will not, change their habits." There is no suggestion here of the forcible imposition of an incomparably more powerful set of economic, social, and political relations on indigenous peoples, or indeed of the expropriation of their land or their deliberate slaughter; rather, the fault seems to lie with the savages themselves. They could not or would not adapt to changing circumstances; they self-destructed through their own vices of drink, licentiousness, infanticide, or cannibalism; they were mowed down by disease "mysteriously" generated by contact between such unequal races; or, inexplicably, they simply lost the will to survive and reproduce—the very basis of evolutionary success.[105]

The only explanation Darwin could fall back on was the timeworn one of "civilization": "The grade of civilization seems a most important element in the success of nations which come in competition." It was a "more curious fact," mused Darwin, "that savages did not formerly waste away . . . before the classical nations, as they now do before modern civilized nations." There was no recorded "lament" by the "old moralists" over the "perishing barbarians."[106]

By the 1870s, as Darwin understood matters, the barbarians were no longer at the gates. Their degraded survivors were to be found only at the remote frontiers of empire, at the uttermost ends of the earth. Their abject condition and rapidly diminishing numbers represented no threat to empire and its continuing prosperity. The *Descent* bears no traces of his youthful reservations about the cruelties of colonial conquest. Nowhere in his writings, private or published, does Darwin attempt to make any case for the survival of indigenous peoples.[107] As a latter-day moralist, the Darwin of the *Descent* was not inclined to lament the passing of these morally deficient living fossils—these persistent remnants of an earlier phase of human evolution—nor even to extend to them the sympathy he expressed for maltreated animals. Indeed, sympathy toward the lower animals, "apparently unfelt by savages," was, along with capital accumulation, an index of the grade of civilization.[108] In the *Descent*, slavery remained "the great sin," an unnatural practice that, despite its existence among certain ants and its "almost universal" prevalence in a "rude state of civilization," Darwin never attempted to naturalize.[109] Unlike slavery, a moral wrong that could be righted, there was nothing to be done about savage extinction. It was a somewhat obscure, yet inevitable, natural process, a by-product of natural selection and the progress of civilization, for which the civilized themselves were not really responsible.

It is an irony of history that the high Tory FitzRoy, whom we regard (thanks largely to Darwin) as blind to the cruelties of slavery, was pilloried for the principled "pro-Maori" stand he took in the war between rebellious Maori and land-hungry colonists during his abortive post-*Beagle* incumbency as governor of New Zealand. The outraged colonists who petitioned for his recall alleged FitzRoy's "preposterous civility" toward the savage Maori and total disregard for "received theory" for the "good government of savages." But Darwin's sympathies were all for the African slave; he had no sympathy for the savage who would lay waste those flourishing English-like farmsteads that he had so admired in his account of New Zealand, that inspired such "high hopes . . . for the future progress of this fine island," otherwise synonymous with "cannibalism, murder, and all atrocious crimes."[110]

At bottom of the moral indifference of such a humane and otherwise sensitive man to the sufferings of dispossessed, oppressed, even slaughtered indigenes lay Darwin's enduring revulsion against the "barbarous inhabitants of Tierra del Fuego"[111] whom he so stigmatized in his writings. Darwin's Fuegian encounter lay at the heart of his evolutionism. In his mature theorizing, it came to link both natural and sexual selection. More immediately, it was the site for the collision of contending, contradictory theoretical currents in early nineteenth-century anthropology and in related fields of heredity, physiognomy, and aesthetics. It provoked a confusion of images, insights, and questions that he carried with him back to England. It set Darwin searching for answers in this same literature.

But of equal importance to the budding evolutionist and to the history of sexual selection was Darwin's choice of a wife. Marriage was as much on his mind in his bachelor quarters in London, as his newfound conviction of species change. Alongside his notebook jottings on evolution, Darwin toted up two columns for and against the married state, summarizing his expectations of that angelic species, a "good wife," preferably one with money. He determined on marriage: "Marry—Marry—Marry Q.E.D."[112]

TWO

Good Wives

Live long & choose good wives.

—Darwin's advice to the middle classes, 1868

I am not the least afraid of death—Remember what a good wife you have been to me.

—Darwin's last words to Emma, 1882

In 1868, William R. Greg, Darwin's old acquaintance from his Edinburgh days, bemoaned in *Fraser's Magazine* the problem of the survival of the rapidly reproducing "unfit" at the expense of the provident and improving middle classes, who tended to marry late and produce fewer offspring. Darwin, twenty-nine years married and hard at work on *The Descent of Man*, jotted his eugenic remedy for pending middle-class extinction in the margin of his copy. This was an affirmation of his personal marital and reproductive success through informed sexual selection: "Live long & choose good wives."[1]

Emma was indeed a good wife to Darwin. And he certainly put some shrewd calculation into her choice. On the pro side he listed: "Children— (if it please God)—constant companion . . . who will feel interested in one, object to be beloved and played with—better than a dog anyhow—Home, and someone to take care of house—Charms of music and female chit-chat." If he were not to marry, the consequences would be "no children (no second life), no one to care for one in old age," but he would have the compensations of more cash and time for his scientific pursuits and the intellectual companionship not to be expected of a wife: "Freedom to go where one liked—Choice of Society—*and little of it*. Conversation of clever men at clubs." Nevertheless, the picture of a "nice soft wife on a sofa with good fire, and books and music perhaps" was a far more alluring image than the "dingy reality of Grt Marlboro St. [his bachelor digs in London]." On balance, Darwin's bottom line came out forcefully against bachelordom and emphatically for the perceived benefits of a "good wife."[2]

Many historians seem to have found these semifacetious but calculating musings as unexceptionable as did the man who wrote them. Yet they are a key indicator of Darwin's self-regarding view of the subservience of women to men. In this, as in much else, Darwin was a man of his time and class.

In *The Descent of Man*, Darwin was to assert the evolutionary basis of the maternal traits of the human female and the male's innate aggressive and competitive characteristics. Woman's maternal instincts, he wrote, lead her to be generally more tender and altruistic than man, who is competitive, ambitious, and selfish. But above all, man is more intelligent than woman: "The chief distinction in the intellectual powers of the two sexes is shown by man's attaining to a higher eminence in whatever he takes up, than can woman—whether requiring deep thought, reason, or imagination, or merely the use of the senses or hands." He could explain how, through the long-continued actions of natural and sexual selection, aided by use-inheritance, "man has ultimately become superior to woman."[3]

Darwin did not, as is sometimes thought, turn to these issues only late in life. They were there from the first. Only a few months after he drew up his balance sheet on the question of marriage, Darwin found his "nice soft wife on a sofa" in his cousin Emma Wedgwood, although throughout their life together it was the semi-invalid Charles who occupied the sofa, not Emma. Emma hardly had the chance. As their daughter Henrietta recorded:

> My mother had ten children and suffered much from ill-health and discomforts during those years. Many of her children were delicate and difficult to rear, and three died. My father was often seriously ill and always suffering, so that her life was full of care, anxiety, and hard work. But she was supported by her perfect union with him, and by the sense that she made every minute of every weary hour more bearable to him.[4]

Darwin's family background and domestic relations are as salient as his *Beagle* experiences to his views on evolution and sexual selection.

2.1 Drawing aside the domestic curtain: A "feminine Darwiniana"

Charles Darwin was born in 1809, the privileged younger son of a well-to-do provincial family with strong social and familial ties to landed gen-

Emma Darwin, aged 31
From the portrait painted by George Richmond, R.A.

Figure 2.1. "A specimen of the genus": Emma Darwin, age thirty-one, at the time of her marriage in 1839. Frontispiece to Litchfield 1904. By kind permission of John van Wyhe, editor, *The Complete Work of Charles Darwin Online* (http://darwin-online.org.uk).

try and wealthy manufacturers. His father, Dr. Robert Darwin, a successful physician and shrewd financier, was the oldest surviving son of the famous Erasmus Darwin, whose philandering ways and erotic verse did not sit easily with his sedate descendants. On his mother Susanna's side, Darwin's grandfather was the potter industrialist Josiah Wedgwood, who had made his fortune from disassembling the processes of pottery making and turning erstwhile artisans into "such *machines* . . . as cannot err." Both grandparents were founding members of the Lunar Society of Birmingham, famed for its flirtation with revolutionary politics, its religious dissent, promotion of science, technological hubris, and educating enthusiasms.[5]

The most detailed contemporary account of Erasmus and his circle was written by his sometime confidante, the Lichfield poet Anna Seward.[6] Seward had, according to family legend, aspired to become the second Mrs. Erasmus Darwin when the doctor's first wife died. Erasmus, however, confined his relations with the rather precious Seward to the courtly exchange of verses, not vows. A man of large appetites, he installed a mistress-housekeeper "of a lower social class," a village girl of seventeen. They had two daughters whom Erasmus amiably acknowledged and brought up with his legitimate offspring, though not as their equals. The "Misses Parker" were destined for money-earning careers, not polite society. They were "educated for the purpose of educating others," and when they came of age Erasmus set them up with their own school for girls. It was for this enterprise that he wrote his short treatise on female education.[7]

Seward was a canon's daughter, but she had a pre-Victorian tolerance for such sexual arrangements with the lower orders. She continued to play hostess and muse to the gallant, increasingly corpulent Dr. Darwin. Her hopes were not finally dashed until Darwin transferred his affections to a married patient, the rich young Elizabeth Pole. Following the timely demise of her inconvenient spouse, Elizabeth married her ungainly, middle-aged suitor in 1781. They set up household in Elizabeth's Derby mansion with their joint brood of eight offspring, which included the Misses Parker and Elizabeth's late husband's illegitimate child. Seward nursed her chagrin as Erasmus sired another seven children and leaped to national fame with his best seller, the scientific but satisfyingly erotic poem *The Botanic Garden*, published in two parts in 1789 and 1791.[8] Seward got her revenge with her *Memoirs of the Life of Dr. Darwin*, rushing it into print after his death in 1802, in the teeth of intense opposition from Dr. Robert Darwin.

Robert Darwin had intended that the "life" of his famous parent should be written by Richard Lovell Edgeworth, himself a Lunar Society member, inventor, "philosopher," and educationist, who would present a properly masculine appreciation of the elder Darwin's scientific achievements.[9] Seward admitted she "ha[d] not science" or "philosophy," but she insisted on her literary credentials and her longer and more intimate friendship with Erasmus.[10] More than this, she claimed a proprietary interest in his best-known work.

Seward had instigated his writing of *The Botanic Garden* when she presented Erasmus with a poem on his real-life botanic garden in Lichfield. She rejected his suggestion that she write the verse and he write the scientific notes for a proposed poetic interpretation of the Linnaean system of plant classification, because the subject was "not strictly proper for a female pen." Erasmus not only authored the proposed poem, but, to her rankling resentment, he also appropriated ("plagiarised," she said) Seward's original verses. Her biography of him was less the act of a woman scorned than an assertion of Seward's thwarted property and professional "rights" to his life and work—work from which she was excluded by her gender. She fought off Edgeworth and Robert Darwin to produce what she termed, in acknowledgment of her woman's perspective on the great man, a "feminine Darwiniana," one that would "draw aside the domestic curtain" and impartially present that "great mass of genius and sarcasm" to view—sexual peccadilloes, public Jacobinism, family "despotism," irreligion, and all.[11]

Her noisy rattling of the Darwin domestic skeletons was partly muffled when her publisher, responding to the forceful protests of Robert Darwin, prevailed on her to "soften" what Seward termed "the simple unexaggerated statement of certain facts." The biography was heavily cut—"mutilated," according to Seward.[12] Much, however, remained, and not only on Erasmus.

Seward could be, and was, self-important and malicious. But she has the advantage of having known at first hand Darwin and his provincial circle, the wealthy, well-educated, and influential industrialists, manufacturers, and gentlemen members of the Lunar Society. She lived through the same revolutionary events and read many of the same works that excited their attention: those of Rousseau, Tom Paine, William Godwin, and Mary Wollstonecraft. Seward was, in many respects, a "perfect barometer of middle class opinion."[13] She swung from an initial fashionable sympathy with the ideals of the revolutionaries and their struggle against the "oppressive and barbarous monarchy of France" to an equally fashionable revulsion against the "levelling extremes" of Jacobinism, both abroad and at home, as Paineite doctrines spread among provincial and London artisans and a native revolutionary movement threatened English property owners.[14] Yet, for all her middle-class piety and conventionality, she still responded to Wollstonecraft's radical critique of established forms of femininity. "Have you read that wonderful book, The Rights of Woman?" Seward enthused in 1792:

> It has by turns, pleased and displeased, startled and half-convinced me that the author is oftener right than wrong. Though the ideas of absolute equality in the sexes are carried too far . . . yet do they expose a train of mischievous

mistake in the education of females. . . . [It] affords much better rules than can be found in the sophist Rousseau.[15]

After Wollstonecraft's death from puerperal fever and the publication of William Godwin's *Memoirs* that acknowledged his wife's illegitimate daughter by the American adventurer Gilbert Imlay, Seward refused to join the chorus of abuse orchestrated by the reactionary *Anti-Jacobin*. She spoke up for the "character, sentiments, conduct, and destiny of a very extraordinary woman."[16]

Again, while conceding the "voluptuousness" of certain passages, Seward publicly defended Darwin's *Botanic Garden* and its "herbal intrigues" against the charge of undermining morality, as it and Erasmus came under fire in the conservative backlash against the French Revolution and homegrown radicalism. In her correspondence she affirmed the "female right to [such] literature and science."[17] When attempts were later made to censor Darwin's evolutionary poem, the *Temple of Nature*, on the grounds of its alleged indecency and profanity, Seward dutifully scanned it for such instances, finding references to "sexual intercourse" and "solitary reproduction." She was scathingly dismissive of the ability of such terms to excite "Eveish curiosity" in young ladies:

> I had heard it was not fit for the female eye. It can only be unfit for the perusal of such females as still believe the legend of their nursery, that children are dug out of a parsley-bed.[18]

Seward, a lifelong spinster, had a more robust approach to sexuality than was generally possible for nineteenth-century women, spinster or married.[19] By the time of her death in 1809 (the year of Charles Darwin's birth), the reaction against the more permissive mores of the eighteenth century was in full swing, as an emergent middle class began its long consolidation with the affirmation of an imperative moral code that culminated in the Victorian subordination of women and their sexuality through the inculcation of the ideology of "respectability" and feminine propriety.[20] There is a revealing contrast between Seward's acceptance of sexual irregularity in Wollstonecraft and Erasmus and the attitudes of Charles Darwin and his daughter Henrietta, two and three generations later, respectively.

In 1879, Charles wrote a short "Life" of Erasmus as the introduction to an essay on Erasmus Darwin's evolutionary biology by a German botanist, Ernst Krause. He focused on retrieving the reputation of a "grossly calumniated," once justly celebrated man, who was, after all, as he possessively put it, "my grandfather." He was intent on rescuing Erasmus from "utterly groundless" assorted charges—mostly promulgated by Seward—bringing the testimony of family letters to assert his redeeming virtues of belief in a deity; humanity to the insane, animals, and schoolgirls; temperance; near vegetarianism; and abhorrence of slavery. From the standpoint of his own Victorian respectability, Charles did concede the "blemish" of extramarital offspring. He mused on the old reprobate's mixed ménage: "In our present state of society it may seem a strange fact that my grandfather's practice as a physician should not have suffered by his openly bringing up illegitimate children."[21]

The straitlaced Henrietta, eight years married and equal to her daughterly responsibilities as copyeditor and moral and religious watchdog, penciled the Misses Parker out of the text, relegating her father's comment to a footnote.[22] References to mistresses and illegitimacies—even a century after the event and coupled with due recognition of their impropriety—were permissible only in the subtext and in small print. Zealously, she pruned out other references to Erasmus's too-warm regard for women and to his unorthodox religious views, reducing the text substantially.[23]

Henrietta's red pencil marked out the boundaries of late-Victorian sexuality and gender. And therein lay another complication in Charles Darwin's complex relations with his grandparent. Erasmus Darwin's writings poeticized and naturalized sexuality and desire in a way that Charles Darwin's renderings of the same subjects could barely come at, let alone be represented to his readers. To a thoroughly conventional, deeply uxorious man like Charles, his grandfather's evolutionary poems were rather embarrassing, outmoded celebrations of life, love, and reproduction that matched Erasmus Darwin's real-life womanizing and erotic sensibilities. Charles could not account for the enthusiasm of their earlier admirers, which was "quite incomprehensible at the present day."[24]

Even so, Erasmus Darwin's naturalized biology was a "biology for gentlemen," one that did not challenge the conventions common to Erasmus's class and his intellectual and social circle, those endorsed by the Lunar Society gentlemen.[25] These are brought out most clearly in their views on female education and their expectations of the married state, views that formed part of Charles Darwin's social and intellectual heritage.

2.2 The proper education of an ideal mate

The Lunar men had high hopes of education for the improvement of society and were keen to put their understanding into practice, usually on their own children, daughters included. As members of the Dissenting bourgeoisie, they wanted more for the women of their class than the empty-headed frivolity and coquetry they associated with a decadent aristocracy. Their educational schemes aimed to give girls as well as boys the fundamentals of a good education that would improve their minds along with their morals. Science figured prominently because of the assumption that it was inherently progressive and liberating. Natural knowledge would lead automatically to a more moral and just society. Middle-class women, as companions of men and the mothers and instructors of their children, should be exposed to the beneficial effects of education in the sciences.

However, the "natural" accord among knowledge, power, and morality that worked so well for men had to be reworked to accommodate feminine "natures." Certain subjects were unsuited to women's refined sensibilities or might endanger their essential modesty and virtue. When ladies were present at "philosophical" meetings, Erasmus Darwin adapted his experiments to the "capacities of young women."[26] In 1796, his friend Thomas Beddoes, a Lunar associate and admirer of Wollstonecraft, presented a "highly gratifying" and "select course" of anatomical lectures "adapted for a female audience," attended by about forty ladies "of great respectability." The radical Beddoes drained his lectures of every topic "which might, by possibility, afford room for a perverted imagination to lay hold of." He measured his success by the failure of his lectures to "give alarm to the most timid female delicacy or excite disgust in the most refined mind."[27]

The most extreme Lunar opinion on the necessity of a different, more circumscribed education for girls was espoused by the most enthusiastic educator of all, Thomas Day, a wealthy young country gentleman and ardent disciple of Jean-Jacques Rousseau.[28] In *Émile; or, On Education* (1762), Rousseau had postulated an ideal system of education for his pupil Émile away from the corrupting influences of society, so that his natural intellectual and physical capacities might be cultivated for his participation in the prospective free and egalitarian society. Rousseau had provided a worthy mate for Émile in Sophie, but, whereas Émile's "natural" education emphasized freedom and independence, Sophie's stressed dependence and constraint in order to prepare her for her "natural" dependence on her future mate. Rousseau, in many respects so profoundly radical, held entirely orthodox, even conservative, views on women.[29] Under the fascinated gaze of the Lunar Society gents and with their endorsement, Day, having done his best to turn himself into an Émile, undertook a bizarre Rousseau-inspired experiment aimed at producing a perfect mate.

Faithful to Rousseau's strictures, Day was "suspicious" of conniving and sexually manipulative women, had a "strong detestation of female education" and a veritable "horror" of female authorship.[30] What Day wanted in a wife was the personification of Rousseau's fictional Sophie: "simplicity, perfect innocence, and attachment to himself." Despairing of finding a ready-made Sophie, Day decided to "breed up" (the words are Richard Edgeworth's) a wife who would fit his exacting standards. In 1769, at the age of twenty-one, Day selected two likely candidates, aged eleven and twelve years, from foundling homes. One of them, the auburn-haired "Sabrina Sidney" (high-handedly renamed as such by Day on acquisition), was of "remarkably promising appearance." Day took the chosen Sabrina to Lichfield, where he pursued his plan of education under Darwin's patronage. The unusual ménage was accepted as "quite natural and free from impropriety," and Sabrina was a frequent visitor in the Darwin and Seward homes.[31] Nor did anyone, including Day, see any contradiction in Day's scathing indictment of slavery and his own purchase, enforced training, and capricious domination of a young girl. It was left to Sabrina herself to make the connection. In later life she complained of her treatment, alleging that she had been Day's "slave."[32]

It is to Seward that we owe the unforgettable picture of Day's educational efforts to instill the unquestioning courage and loyalty of a "Spartan wife" in the hapless Sabrina by firing pistols at her petticoats and dropping hot sealing wax on her arms. After a few years, during which he blew hot and cold on the project, Day abandoned the experiment, allegedly because Sabrina disobeyed his "strict injunctions" as to her dress. Her feminine love of finery would out, despite Day's best efforts to suppress it. Sabrina was sent to a conventional school and emerged "feminine, elegant, amiable"—a middle-class miss whom Seward might approve—proving that "those modes of education, which have been sanctioned by long experience are seldom abandoned to advantage by ingenious system-mongers."[33]

Sabrina eventually married Day's friend Thomas Bicknell, while Day's well-wishers finally found the reluctant—but extremely choosy—bachelor an appropriate wife in a wealthy heiress of such "charming temper, benevolent mind, and cultivated understanding" that she was nicknamed Minerva ("But has she white and large arms?" asked Mr. Day), and who, in the event, agreed to remove herself from the "corrupting taint" of society, to "banish her harpsichord," and retire with Day to the country. "I never saw any woman so entirely intent upon accommodating herself to the sentiments, and wishes, and will of a husband. [. . .] the most complete matrimonial obedience," Edgeworth marveled.[34]

Day's views on women were extreme, and his educational experiment extraordinary. But Day was simply pushing harder on a set of assumptions common to so-called enlightened men of the late eighteenth century. They permeate Darwin's and Edgeworth's amused but sanctioning attitudes toward Day's "philosophic romance." The system of "Practical Education" that Edgeworth (who had in all four wives and nineteen surviving children on whom he vented his educating enthusiasms) later developed in conjunction with his long-suffering daughter Maria, recommended that the education of females should provide them with the accomplishments and qualities that would "please and attach men of superior sense and characters." The Edgeworths, father and daughter, dismissed Wollstonecraft's "speculative rights," agreeing more with Rousseau on the necessity of a "different education for the two sexes" that would "inure [girls] to restraint" and was thus more consonant with their practical "happiness" in their future situations in society.[35]

These same attitudes are manifest in Erasmus Darwin's short treatise on female education of 1797. On the face of it, Darwin's is a progressive, liberal scheme designed to give the young ladies at the Misses Parker's school a better education than most such contemporary establishments. Its intentions were, as his grandson Charles commented, "thoroughly benevolent." The elder Darwin advocated physical exercise, fresh air, and no tight corseting, together with the study of both the fine arts and the sciences for his daughters' pupils. But, while Erasmus defended the "cultivation of the minds of the female sex" against their "illiterate" detractors, he was decided in his view that daughters and wives should not have too much of a good thing: "great eminence" in any field was injurious to females who should possess "mild and retiring virtues" rather than "bold and dazzling ones."[36]

Erasmus Darwin's progressive school for the Miss Parkers turns out to be little different from the fictional one satirized by Jane Austen in *Emma* (1816) as a "real, honest, old-fashioned Boarding school" which emphasized exercise, fresh air and wholesome food, "where girls might be sent to be out of the way and scramble themselves into a little education, without any danger of coming back prodigies."[37]

While Robert Darwin and the Wedgwood sons went to Edinburgh University (which, unlike Cambridge and Oxford, took Dissenters), Darwin's mother Susanna Wedgwood was sent to London for cultural polish, then to the educational expert Edgeworth, who praised her "good solid understanding" and "obliging behavior and good disposition." Josiah Wedgwood seems to have had a warmer relationship with his daughter, who sometimes acted as his "French clerk," than Erasmus Darwin had with his numerous offspring. Unlike the freethinking, womanizing Erasmus, the steady, sober Josiah was a Unitarian who kept the faith. The radical Unitarians had a reputation for emphasizing women's redeeming effects on society and encouraging their participation in the public spheres of the abolitionist and other reform movements. Susanna's spinster sisters, Sarah and Kitty, were exemplars, being great pamphleteers, distributors of improving tracts, and donors to various causes, particularly "the Blacks." Sarah kept up the family passion for the antislavery campaign into old age. She was an independent-minded woman who managed her very large fortune "as ably as a man."[38]

Yet the major accomplishments of Susanna's "privileged education" were her proficiencies as pianist and housekeeper, which she brought to her marriage to Robert Darwin. She died when Charles was eight, leaving his upbringing to his indulgent older sisters, while his father aged into an irascible widower given to "sudden black moods." The sensitive young Charles may have admired his impressively large father "for his powers of observation and his sympathy," but he also found him intimidating. There is a revealing entry in his early notes on expression, describing "my F" in a rage. Nor was his future wife, the young Emma Wedgwood, comfortable in the doctor's domineering presence: he was "kind, but you couldn't go on with your own conversation when he was there."[39]

Charles found more congenial company at nearby Maer Hall, Emma's family home, where her father, Josiah the younger—turning his back on the trade that had founded the family fortune—aspired to country squire and gentleman, taking on the "trappings and most of the social opinions of the upper classes." Josiah was "silent and reserved so as to be a rather awful man," but uncle and nephew found mutual ground in sporting and shooting, with which the young Darwin was obsessed. Josiah had married Bessy Allen, the eldest of a wealthy family of nine daughters and two sons. The Allen girls had gone in "nervous dread" of their father, who was in the habit of thumping his fist on the table and ordering them to talk when he wished to be entertained after dinner. Unsurprisingly, Emma's mother, an intelligent, well-read woman, considered men to be "dangerous creatures who must be humored," and treated her stern and reserved husband with circumspection. Josiah also inspired nervous awe in other female relations, one of whom described him as "always right, always just, and always generous." Charles Darwin's sisters, who had their own household patriarch to placate, were astounded at the ease and familiarity with which Charles treated Uncle Jos, "as if he was a common mortal."[40]

Those second- and third-generation Wedgwoods and Darwins, who so often intermarried, may have inherited some unconventional theological and political notions, but they were much more orthodox in their understandings and expectations of women's domestic and social roles, familial understandings that are refracted in *The Descent of Man*. These staunch supporters of "negro" emancipation would have been confounded by the suggestion that their wives, daughters, and sisters needed emancipating. The elaborate division of labor that underlay the successful pottery enterprise that founded the Wedgwood fortunes extended to the domestic sphere, where the respective roles of men and women were thoroughly understood and defined. A Wedgwood (Emma's father Josiah) required his wife to be "sensible to his pains and his pleasures, participat[e] in his hopes, . . . [strengthen] his good dispositions and gently discourag[e] his harshness and petulance, and more than all . . . become flesh of his flesh and bone of his bone, by bearing him children."[41]

Men might indulge in "philosophy"; women were bound by religious piety to their roles of moral preceptors of family life. A husband should guard his religious opinions lest he distress his wife. A high premium was placed on feminine vivacity, sweetness, and "goodness"; little on feminine intellect, education, or independence. Robert Darwin grew apoplectic at the mere hint that Harriet Martineau, the literary Whig "Amazonian," might be viewed as a prospective daughter-in-law. This was a possibility gossiped about at one stage when his elder son, the indolent, enervated Erasmus, was to be seen in her company, "noon, morning, and night." Charles Darwin, newly returned from his *Beagle* voyage, reported back to the family from London on his besotted brother:

> Our only protection from so admirable a sister-in-law is her working him too hard. He begins to perceive . . . that he shall be not much better than her "nigger."—Imagine poor Erasmus a nigger to so philosophical & energetic a lady . . . She already takes him to task about his idleness—She is going some day to explain to him her notions about marriage—Perfect equality of rights is part of her doctrine. I much doubt whether it will be equality in practice. We must pray for our poor "nigger."

Charles was surprised to find the famous exponent of Malthusian economics and advocate of women's property rights, so "little ugly," though Martineau dominated the conversation and appeared "overwhelmed with her own projects, her own thoughts and own abilities." Her threatening, unwomanly assertiveness and intelligence were "palliated" by Erasmus, who explained to Charles, "One ought not to look at her as a woman."[42]

A wife with a profession of her own, or intent on puncturing male self-esteem and superiority, would never have done for Charles. In choosing his wife from his Wedgwood cousins, Darwin could be as comfortable in his expectations of her assumption of his male supremacy and importance as he was of her substantial dowry.[43]

2.3 Emma, "a specimen of the genus"

Not that Darwin was in any sense a typical Victorian patriarch. Even his most cynical biographers have conceded the genuineness of the love and respect borne him by his family and his tenderness toward them. One cannot but warm to a man who could write, during a short absence from home, with such doting unselfconsciousness to his wife of his concern for her welfare (Emma was, as usual, pregnant) and that of their infant "chicks," William and Annie (their firstborns), signing off with, "Goodbye. I long to kiss Annie's botty-wotty." Nor can one read of Darwin's desperate vigil by that same beloved Annie's bedside and his memorial on her death at ten years of age without feeling the depth of his grief and loss.[44]

Darwin's gentleness, modesty, and good nature are an historical byword. Yet, so is Emma Darwin's historical legacy as "perfect nurse" to Darwin's "perfect patient." Even against the Victorian stereotype of feminine servitude, Emma stands out in her total submergence of self in Darwin's well-being. Utterly devoted to his interests, she created and preserved the orderly, quiet, entirely domestic environment Darwin desperately craved for his work and health. Her days were planned to suit him and the elaborate routine he devised to achieve the maximum of work with minimum distress to his delicate constitution. Emma was ready to read aloud to him during his periods of rest on the indispensable sofa, write letters at his dictation, go for walks with him, soothe him with music in the evenings (she was an accomplished pianist), and be constantly at hand to alleviate his daily discomforts. She nursed him through endless nights of violent retching and hysterical weeping and days of giddiness and fainting, when he could not bear for her to leave his side.

She helped proof the *Origin*; transcribed many of Darwin's notes, manuscripts, and letters; and, along with their children and servants, dutifully watched over his experiments. But she had little interest in science, only in the scientist. She was deeply religious, and many of his opinions were painful to her; yet it was Emma whom Darwin entrusted to carry out the publication of the preliminary version of his "Species Theory" in the event of his death. It proved unnecessary (he lived for another thirty-eight years), but there is no doubt that Emma would have loyally carried out his wishes.[45]

With the possible exception of her religious beliefs, there is no evidence whatever that Darwin was not more than content with Emma's circumscribed role of perfect nurse and loyal helpmate. Before their marriage, he defined her proper sphere: Emma was to "humanize" him, to teach him that there was greater happiness in life than "building theories & accumulating facts in silence & solitude." He had not expected intellectual companionship in marriage and discouraged it. While she was still his fiancée, he dissuaded Emma from reading Lyell's *Elements of Geology*, which she had begun, thinking she should "get up a little knowledge" for him. In Darwin's experience, science was an exclusively male preserve, which women entered, if they entered at all, only as spectators or amanuenses—at the most, as fashionable dabblers, not to be taken seriously. He did not expect or want women to converse intelligently about science, but rather to be tolerant of masculine preoccupation with it, like "poor Mrs Lyell" who sat by, a "monument of patience," while Darwin and Lyell talked "unsophisticated geology" for half an hour.[46]

The only occasion when Darwin set aside these conventional views of his "nice soft wife" was when, disregarding his father's advice, he discussed his loss of religious faith with Emma shortly before they married. The result was not happy. Emma was seriously distressed, setting down her concerns in a carefully phrased letter that Darwin preserved. She suggested that the scientific habit of "believing nothing till it is proved" should not apply to matters of faith, and affirmed the value of prayer. Fearful for his immortal soul, she diffidently opposed her "feeling" to his "reasoning": "I should be most unhappy if I thought we did not belong to each other forever." Darwin appended a sentimental but guilty note to Emma's letter: "When I am dead, know that many times, I have kissed & cryed over this." Only a few months earlier, in his secret notebooks, he had gone so far as to speculate sacrilegiously that "love of the Deity" was merely an "effect of organization"; religious feeling was a heritable instinct, like the love of dogs for their masters: "Oh you materialist!"[47]

Husband and wife were mutually concerned not to let their religious differences mar their domestic relations, henceforth confining themselves to their respective spheres. Darwin continued with his science and his skepticism while Emma busied herself with his person and not with his disturbing ideas and work, which she nevertheless loyally supported. Darwin's increasing ill health and absorption in his work dictated their move to Down in 1842, and Emma's life narrowed to one of "watching and nursing . . . cut off from the world" (Henrietta's description). She had her reward in his gratitude expressed in the fulsome tributes of Darwin's "Recollections". She was his "greatest blessing," his "wise adviser & cheerful comforter throughout life," so infinitely his superior in "every single *moral* quality" [emphasis added].[48]

This stereotype of Victorian feminine servitude, domesticity, and piety is given a bit of a jolt by Henrietta's ascription of "remarkable independence" to her mother's character and way of thinking.[49] True, there are glimpses of another Emma behind the facade of the perfect nurse, as more family letters have come to light. She was, for her time, an unusually well educated and cultivated woman. She and her sisters were homeschooled by a series of governesses and tutors and given the run of the extensive library at Maer, bequeathed by old Josiah Wedgwood. As a young girl Emma went to London for the seasons, and she traveled extensively in Europe before her marriage, meeting some of the leading intellectuals of the day.

Her favorite aunt Jessie Allen was married to the distinguished historian Jean-Charles-Léonard de Sismondi, an associate of the exotic Madame de Staël and her circle of distinguished literati and nobility. Emma is described as unaffected and tolerant, as having a "healthy disrespect for the trappings of fame"; yet she was conscious of her place in the social hierarchy and secure in her assumption of English superiority. Emma did not care for the hand-kissing Swiss-born Sismondi, thinking him "not very manly." She disparaged a Geneva ball as a "very democratic occasion" where she was obliged to dance with a "very disagreeable and vulgar man," but later had the "good luck to dance with one or two Englishmen." As late as 1873, Emma marveled at a "working man's ball" attended by Henrietta, where Henrietta had actually danced with "a grocer and a shoemaker, who looked & behaved exactly like everybody else and were quite as well dressed."[50]

The young Emma much preferred her other uncle by marriage, Sir James Mackintosh, prominent Whig barrister and writer. Mackintosh was another admirer of de Staël and a moderate progressive who dabbled in politics. A friend of the political economist Thomas Robert Malthus, whose writings on population were to play such a critical role in the formulation of Charles Darwin's theory of natural selection, Mackintosh stood with Emma's father as Whig candidates for Parliament in the great reform elections of 1831 and 1832. Mackintosh and his wife were separated (his admiration for de Staël and Lady Holland, "those two Jezebels," having proved too much), and their daughter Fanny kept house for him in London. Fanny Mackintosh (who was to marry Emma's brother Hensleigh Wedgwood, with Malthus's daughter as bridesmaid) was a "furious politician" who took an active role in the antislavery campaign and later took up the causes of Italian republicanism, higher education for women, anti-vivisectionism, and female suffrage. With Fanny's encouragement, Emma helped launch a female antislavery society and, like other women family members, became caught up in the excitement as the Anti-slavery Bill and the Reform Bill were fought through Parliament in the midst of riots and turmoil. She supported her father's campaign (which probably means that Emma attended rallies or watched the speeches from the sidelines), but when Josiah was finally elected in 1832 as the first member for Stoke-on-Trent in the reformed parliament, Emma wrote perceptively, "All of us were very pleased at his coming in so grandly, especially as he is become too Tory for these radical times."[51]

The young Emma's interest in political speeches had its limits. She preferred cultural events—concerts, opera, the theater, and exhibitions—although she had little appreciation of the visual arts: looking at paintings was "horrid staring work."[52] She made no secret of her boredom with the company of some of the leading scientists of the day, describing Lyell and Robert Brown, the botanist, as "those two dead weights." Emma occasionally accompanied Darwin to scientific meetings, but her devotion to his interests stopped short of finding them anything other than "very wearisome," but "not more than all the rest" (an admission Darwin found highly amusing and repeated as a good anecdote).[53]

In spite of her professed indifference to Darwin's work, Emma seems to have understood it and its implications pretty well. Again, for all her piety, she could on occasion dissent from conventional religious opinion, as when she defended the ethics of "this new breed of agnostics." Her letters show her to have had an astringent humor and a wide general knowledge. If Darwin's taste dictated the choice of the popular sentimental novels she read aloud to him, her own choice was wider ranging. In August of 1848 they read Mary Wollstonecraft's *Vindication of the Rights of Woman* and Godwin's *Memoirs* of Wollstonecraft, though without recorded effect. Emma stood in little danger of following her sister-in-law Fanny's commitment to women's causes or radical politics. In 1851 a scandalized Aunt Jessie reported on Fanny's passionate support for Giuseppe Mazzini, the exiled republican who campaigned for Italian liberation from his London headquarters: "She is of his committee . . . ! How could Hensleigh permit it? It is so contrary to the modesty of her nature."[54]

Although Emma knew she should care about the higher education of women, she did not. She took enough interest in the cause of Elizabeth Garrett's candidacy for a professorship at Bedford Ladies' College, to pass on a letter of recommendation to her brother-in-law Erasmus Darwin, who was then chairman of the council. Unlike Charles and Emma, Erasmus (undoubtedly through his close associations with Martineau and Fanny Wedgwood), was a keen proponent of women's education. As a sign of the times (1865), Garrett's referee, Edward Cresy, assured Emma that Garrett's medical and scientific attainments had in no way impaired the "charm of her manner or her social converse[;] she is neither masculine nor pedantic," but a "well-bred English Lady."[55]

Emma's political interests revived as her children grew. Darwin, with his lifelong hatred of slavery and cruelty to animals, might be the more passionate humanitarian, but Emma was the more politically minded, avidly following the elections and parliamentary debates in her favorite newspaper, the *Times*. Where Darwin's hypersensitivity to suffering made him "skim" the reports of bloody conflict during the American Civil War and his initial support for the North wavered before its "boasting" and "abuse of England" and its failure to free all its slaves, Emma took the pragmatic line that the "slaves are gradually getting freed & that is what I chiefly care for." She even gave up the consistently anti-Union *Times*, substituting the liberal *Daily News*, and thought that all England should "get up its Uncle Tom again" (a reference to Harriet Beecher Stowe's best-selling denunciation of the institution of slavery, *Uncle Tom's Cabin*).[56]

Yet the issue that most aroused Emma's sympathy and support was not human but animal suffering. Her sole independent foray into the humanitarian social activity that characterized so many of the Wedgwood women was a campaign for the development of a humane animal trap that she began in 1863 at the height of the American Civil War. Emma collected subscriptions toward a prize offered by the Royal Society for the Prevention of Cruelty to Animals, wrote (often in Darwin's name) to various publications, enlisted influential supporters, and distributed a pamphlet that she and Darwin had "concocted." This detailed the cruelties inflicted by the steel traps used by landowners and gamekeepers, the "greatest cruelty of modern times."[57] It was this burning issue that almost (but not quite) converted Emma to the notion of women's suffrage (an issue actively promoted by her niece, Julia—"Snow"—Wedgwood, daughter of Fanny and Hensleigh), on the grounds that women generally had no sympathy with the game laws, so their vote would be "an additional handle against them."[58]

For the first eighteen years of their marriage, Emma, as Henrietta hinted, was almost continuously pregnant, and this was for Victorian women a major inhibitor of social engagement. The Darwins, it seems, did not practice contraception, although the use of the douche and "sponge" as a means of family limitation was discreetly adopted by some middle-class Victorians. Darwin would have had easy access to such information, but as it later emerged, he was totally opposed to contraception on the irreproachable evolutionary grounds that any decrease in population would inhibit competition and thereby social advancement and the future of the world (he had in mind the beneficial British colonizations of America, Australia, New

Zealand, and South Africa); but he also expressed the very patriarchal view that, were the knowledge of contraceptive practices to become widespread and morally acceptable among the married, they would lead to "extreme profligacy amongst unmarried women," endangering the family unit, the cornerstone of society.[59] Civilization itself was at stake, dependent on a high birthrate and the perpetuation of feminine virtue and ignorance.

Emma, in any case, bore the brunt of this ideological imposition, though Darwin sensitively suffered along with her during her many "times of trial." Childbirth was then a fear-filled event, highly risky for both mother and child, and its pains were to be stoically endured; a "horrid affair at the best," said Darwin. It was not until her eighth child, Leonard, was born in 1850 that Emma received the "blessed" relief of the newly introduced but controversial chloroform. Darwin fearfully administered the dose himself, knocked Emma out for a good hour and a half, and found chloroform as comforting to himself as to the patient.[60] Their ninth child, Horace, was born the following year in 1851. By October 1852, Darwin, always fearful that his "wretched stomach" might be passed on to his children, was yet complaining to his cousin William Fox, "Emma has been very neglectful of late & we have not had a child for more than one whole year." Emma miscarried twice before, in 1856 and "after weary months of discomfort," she gave birth to their last child, Charles Waring, a baby with Down syndrome, when she was forty-eight.[61]

The other delimiting factor was most obviously Darwin's abiding and debilitating ill health, which required Emma's constant attendance and all the feminine sympathy and support she might spare from their numerous, frequently ailing children. In 1865, Darwin detailed his symptoms as follows:

> Age 56–57.—For 25 years extreme spasmodic daily & nightly flatulence: occasional vomiting; on two occasions prolonged for months . . . Vomiting preceded by shivering, hysterical crying[,] dying sensations or half-faint, & copious very pallid urine. Now vomiting & every paroxys[m] of flatulence preceded by ringing of ears, rocking, treading on air & vision. Focus & black dots[.] All fatigues, specially reading, brings on these Head symptoms[,] nervousness when E[mma]. Leaves me.[62]

And so on, and on. There has been a good deal of controversy about the nature of Darwin's ill health and suggestions range from those of specific etiology to the frankly Freudian, focused primarily on Darwin's uneasy relations with his father and God the Father. But in the murkiness of this posthumous diagnostic debate, Darwin's relations with Emma seem more to the point. It takes little psychology to see in his utter reliance on Emma an almost childlike dependency, expressed in his nicknames for her, "Titty" and "Mammy." "Without you when sick," he wrote Emma in 1848, "I feel most desolate. . . . Oh Mammy I do long to be with you & under your protection for then I feel safe."[63]

This was a dependency that Emma actively encouraged. Illness was where she came into her own, and her worth was measured in the extent of Darwin's gratitude. Revealingly, in a letter Emma wrote to Henrietta after her marriage, she offered advice on diet and supposed that Henrietta had "rather enjoyed" her new husband's headache: "Nothing marries one as completely as sickness."[64]

This raises the leading question of how well the domestic relations of the Darwins fit the conventional Victorian gender roles that are endorsed in the *Descent*: If Emma's role corresponded closely with the conventional feminine attributes of sympathy, religiosity, and domesticity that found expression in philanthropic gestures toward animals and the enslaved, but above all in the excessive nurturing of her spouse, how does that spouse—with his notorious tenderness, emotionality, enfeebled physical condition, and debilitating sensitivities—fare in the spectrum of Victorian "manliness" and the construction of masculinity? We have seen the stress that Emma put on Sismondi's failure to meet her expectations of the code of manliness; but Darwin's manliness seems never to have been called into question by Emma or, indeed, by anyone else.[65]

2.4 Darwin, "manliness," and masculinity

"Manliness" was a subject endlessly proselytized by preachers, schoolmasters and novelists, shifting in the course of the nineteenth century from the "earnest, expressive manliness of the Evangelicals to the hearty, stiff-upper-lip variant in the era of Kitchener and Baden-Powell." It was a code that mingled the moral and the physical, culminating in a "muscular Christianity" that promoted a masculine ideal of manly endeavor, muscularity, and morality. Closely linked with this was the notion that male sexual energy had to be properly channeled into material and marital productivity. Evangelical discourse sought to maintain the sexual economy through hard work, healthy outdoor physical activity, temperate habits, and abstinence from solitary or illicit sex, dissociating "manliness" from its older currency of sexual conquests and the indiscriminate sowing of wild oats. The quintessentially British ethos of sport, of "playing the game," conveyed the moral worth as well as the hard bodies and stiff upper lips of those best fit to lead the nation and build the empire. Sport itself was erotically charged through its demanding physicality and the exhilaration expressive of man's sexual nature—especially hunting, with its focus on the chase, the consummation of the kill, and the display of horned trophies. Male sexuality was not denied but, rather, managed, until it might be legitimately and productively discharged in the marital bed. Never mind that this respectable and responsible ideal of domesticated muscular manhood was in increasing tension with the values endorsed by the sporting and clubbing culture, with the rituals and rites imposed by the segregation of the sexes in the public school education of middle-class boys, and by the enduring Victorian double standard.[66]

Darwin had no need to establish his reputation for adventurous hardiness—this was well documented and widely known through his highly popular publications on the voyage of the *Beagle*; and his established place among gentlemen who had no need to earn a living largely exempted him from the dominant code of manliness associated with the professional and business classes that involved self-control (especially with respect to sexuality), hard work, and independence.[67] Still, there was never any suggestion that the young Darwin was ever anything other than thoroughly sexually respectable and abstemious, even in the midst of the renowned temptations of the sirens of Tahiti who were the highly publicized undoing of an earlier famous ship's naturalist, Sir Joseph Banks (Darwin, as remarked, found Tahitian women disappointingly unattractive). Respectable Victorian husbands "looked primarily to their wives for sexual fulfillment," and Darwin seems to have been no exception to the rule. John Tosh discusses the prevalence of Victorian husbands turning wives into comforting mother figures and its sexual implications, which, while it may have made a wife a "highly equivocal object of desire," was "no bad thing," according to conventional notions of sexual difference and the Victorian ideology of pure and innocent womanhood.[68]

And for all his deep dependency on Emma, Darwin achieved a great deal of hard work in the face of incapacitating illness, as measured by his many publications (not to mention his extensive correspondence—for which of course he depended on the help of Emma and the family) and his scientific and social standing. As early as 1860, the Darwin legend was in public formation. A lengthy and favorable review of the *Origin* in *All the Year Round*, the popular weekly journal conducted by Charles Dickens, praised Darwin's perceived gentlemanly qualities and presented them along with his ill health as conducing to his scientific credentials and credibility:

> Charles Darwin . . . is gifted with clear and passionless judgment, and with an amiable and gentlemanly disposition; it is doubtful whether he have an

enemy in the world. . . . He is blessed with a sufficiency of worldly riches, and has not strong health—the very combination to make a student. He is sincerity itself, thoroughly believing all he states, and daring to state what he believes. . . . He has circumnavigated the globe, and beheld the manners of many men, savage and civilized; of many birds, beasts, reptiles, and fishes. . . . For more than twenty years he has been patiently accumulating and reflecting on all sorts of facts . . . regardless of expense and labour, he has long searched for the truth.[69]

In short, Darwin's well-publicized ill health that exempted him from the professional appearances and the social engagements expected of a prominent man of science, did not compromise the public face of his manliness that stood him in such good stead in the promotion of unorthodox theory. Of course, the labor and support of Emma were essential to this, as were the labor and support of female family members in general to the public success of professionals and businessmen.[70] But this female contribution to his scientific success went as absolutely unacknowledged by Darwin as it was in the code of manliness. Darwin credited his great success to discipline, hard work, and perseverance, the very same middle-class masculine virtues to which, in the *Descent*, he attributed the intellectual superiority of men over women.[71]

It is only when we dig deeper into the notion of Victorian masculinity, not as a set of cultural attributes (as manliness is generally represented), but as a manifestation of social power demonstrated in particular social contexts, that the gender relations obscured by the code of manliness are uncovered. John Stuart Mill put it succinctly in *The Subjection of Women*: "I believe that [women's] disabilities elsewhere are only clung to in order to maintain their subordination in domestic life; because the generality of the male sex cannot yet tolerate the idea of living with an equal."[72] Victorian critics of female suffrage, higher education, or reforms in the law on the status of women—such as the custody of children or the ownership of property— would fall back on the assertion of women's physical and intellectual inferiority or women's "mission," rather than voice the threat they felt to their masculine identities by the loss of authority in the home, at work, or in the all-male associations that typified the Victorian period.

Of these three interlinked arenas, the home was by far the most vulnerable one for the assertion of Darwin's masculine authority. He had served his time in the all-male strongholds of public school and university—first in Edinburgh, where Darwin briefly studied medicine, then at Cambridge, where he studied to become a clergyman but allegedly spent most of his time honing his sporting prowess; but, above all, on the *Beagle*. These were places where men bonded and experienced their shared conviviality and gender privilege, while women were excluded or sexually exploited—save for those rare opportunities for genteel social interaction with "ladies," who were treated with the exaggerated respect and courtesy to which their status entitled them. On the domestic front, the location of authority within the household was the "key determinant of masculine status."[73] The idea that the household was a microcosm of the political order (reflected in Darwin's assumption that the preservation of the family was essential to social advancement) reinforced the importance of the man being master in his own house.[74] This was doubly the case for Darwin: home might be the "woman's sphere" where, ideologically anyhow, Emma held sway, but it was also Darwin's place of work.

"Domesticated husbands and supportive wives" had by the mid-Victorian period "become central to the self-image of the Victorians." Husbands increasingly were expected to spend their leisure time in the home with their families, rather than seek out the exclusive company of other men. This led to the "distinctively masculine privilege of enjoying access to both the public *and* the private sphere." Those husbands who worked from home usually had a room set aside for this purpose, the "study," which conformed to the notion of separate spheres by removing their work from the domestic context. Darwin's study was not—as was usually the case—out of bounds to the rest of the family. Nevertheless, it demarcated Darwin's work from domestic routine. It was his acknowledged place of work, and when he was well, he worked, to the exclusion of all else.[75]

Darwin might call himself Emma's "slave" and Emma might address him as her "dear old Nigger" (an endearment to make the modern reader wince), but for all his seeming weaknesses and dependencies, physical and psychological, there can be no doubt but that Darwin was master of the household. His privileged status as invalid is the key here. His illness, however defined, seems to have been real enough—but so is the undeniable fact that Darwin deployed it in the mediation of social and domestic relations. The prosaic explanation that Darwin turned himself into an invalid primarily to get on with his work cannot be excluded. This would explain his acquiescence in the excessive care Emma bestowed on him; the advantage he consistently took of his semi-invalidism to avoid the strains of a social life, which would have interfered with his work; and the enormous amount of scientific work, both experimental and literary, that he accomplished in spite of his ill health. It also ensured that he was surrounded by feminine sympathy and support, which effectively buffered him against the intrusions of domestic

Figure 2.2. Charles Darwin's study at Down, site of masculine comfort, domestic authority, and serious scientific work. Courtesy of Wellcome Library, London.

and social obligations and the larger consequences of controversial scientific opinions. His excessive gratitude and dependency had the further effect of tightening his control over Emma and the children.

The atmosphere of Down House has been often evoked as affectionate and homely, yet there is no question that Darwin's invalid status and work routine were dominant and that his family patterned their lives around the demands of his twin occupations. There was a splinter of ice at Darwin's core, a cold determination and ambition to succeed, whatever the cost to self and family. The "tyranny of illness" ruled over them all.[76] Without departure from his consistent gentleness, modesty, and good nature, Darwin nevertheless achieved what he wanted. His most diffident wishes were as much deferred to as the despotic demands of any table-thumping, awe-inspiring patriarch. His love and gratitude endorsed the narrow, entirely domestic lives he tenderly imposed on wife and daughters.

The unacknowledged stresses of that environment are suggested by Henrietta's mysterious breakdown between the ages of thirteen and eighteen years, when she too assumed the role of invalid, a role she continued to exploit for much of her life. When Henrietta was eighty-six, she told her niece, Gwen Raverat, that she had never made a pot of tea in her life, had never been out in the dark alone, had never traveled without her maid; and that since the age of thirteen, she had had breakfast in bed. It was Raverat's opinion that it was unfortunate that Aunt Etty had had no "real work" into which she might have channeled her unbounded energy and managerial talents: "As it was, ill-health became her profession and absorbing interest."[77]

Whereas it might be argued that Darwin resorted to illness in order to get on with his work, Henrietta retreated to it because she had no work. The young Henrietta had aspired to know something of science. She had to be content with acting as her famous father's amanuensis and literary stylist, and with editing her mother's life and letters. While, in conventional fashion, her brothers were educated at school and university, Henrietta was offered a succession of governesses chosen by Emma, who was not overly concerned with their qualifications or abilities. Henrietta's sole criticism of her mother was to voice her regret over the poor quality of her education.[78]

While Darwin worried endlessly over suitable professions for his sons, Henrietta's one known gesture at a profession (the management of a women's college—which one is not clear) was actively discouraged by her concerned parents on the grounds that she was not the "strong woman" they deemed necessary to the position. According to Emma, the efforts by feminists, "all in the high Radical 'woman' line," to win Henrietta to the cause were doomed to failure because Henrietta thought the woman question a "shibboleth." Henrietta's own historical legacy was to dwindle into the caricature drawn by Raverat: the childless, aging eccentric who dressed extravagantly for every occasion and found her true womanly worth in festooning her own hapless husband with shawls, guarding him from dangerous drafts, and force-feeding him gruel.[79]

Darwin's other surviving daughter, Bessy, never married or overcame her childhood dependency on the family. Not that the sons emerged unscathed; all developed their share of the "family hypochondria" and, as Raverat saw it, lived their lives in Darwin's long shadow.[80]

For all Darwin's forebodings, his sons lived long and had successful professions of one kind or another, and, according to Raverat, counted among their assorted wives several "feminists." These included her mother Maud (wife of George, Charles and Emma's second son), who encouraged women's entry into the professions. Ellen Crofts (second wife of their third son, Francis) was a fellow and lecturer at Newnham.[81] These "new" women entered the family after Darwin's death in 1882. Apart from Fanny Wedgwood, the only family "feminist" that Darwin knew, her daughter Snow Wedgwood, was an acute and critical observer of Darwin domestic relations.

She did not know of any wife "quite so absorbed in a husband" as her aunt Emma: "Her time is quite taken up by ministration to him."[82]

Snow, like other female family members (Darwin's "angels"), made observations for him and collected information, but, like Henrietta, was not acknowledged in Darwin's publications, though the contributions of his sons were carefully cited. This may have been because it was considered inappropriate to publicize a woman's name. Snow had earlier published a thoughtful critique of the *Origin* in *Macmillan's Magazine*. Like many articles and books by women in Victorian England, it was published anonymously. Darwin had then told her, "I think that you understand my book perfectly, and that I find a very rare event with my critics." Snow went on to become an accredited author of fiction and biography. Yet, while working on the *Descent*, Darwin did not look to Snow for her opinions on feminine intellect or abilities but for her religious views, consulting her on "defining religious feeling."[83]

Apart from matters of syntax, it would seem that religion was the one acceptable area in which Darwin deemed females competent to make a judgment (and one not so much intellectual as emotional), while deferring to their moral authority. In the event, it was feminine religious conventionality that overrode his sons' wishes when Darwin's late-written "Recollections of the Development of My Mind and Character" (1876) was published posthumously as his *Autobiography* with the deletion of his religious opinions. Emma intervened to prevent the publication of certain passages she deemed "not worthy of his mind"; they would give pain to Darwin's religious friends and family. According to Leonard Darwin, his elder sister Henrietta went so far as to threaten legal proceedings to stop its publication altogether because she felt that on religious questions, it was "crude and but half thought-out."[84] Yet it was Henrietta who proofed the *Descent*—in fact, edited it, for Darwin thanked her profusely for her rephrasing of various sections. She even urged him to expand the sections on the evolution of mind, morals, and religious belief. She seems to have found nothing to cavil at in the section on woman's intellectual inferiority, which of course gave due recognition to the notion of feminine moral superiority. Similarly, Emma's concern with the *Descent* was that she would "dislike it very much as again putting God further off"; otherwise she found it "very interesting."[85]

Henrietta married shortly after the *Descent* was published, and Darwin proffered the following cloying formula for marital happiness:

> I have had my day and a happy life, notwithstanding my stomach; and this
> I owe almost entirely to our dear old mother, who, as you know well, is as

good as twice refined gold. Keep her as an example before your eyes, and then Litchfield will in future years worship and not only love you, as I worship our dear old mother.[86]

It never seems to have occurred to Darwin to question the excessive maternal solicitude and protectiveness he evoked from wife and children, who conspired to shield him from his oversensitive self. He was eternally grateful, he was Emma's slave, he worshipped her, he was a selfish brute, but he could console himself with the reflection that woman was naturally more tender and less selfish than man. Emma was simply exhibiting her innate qualities, as he was. He wrote revealingly in the *Descent*:

> Man is the rival of other men; he delights in competition, and this leads to ambition which passes too easily into selfishness. These latter qualities seem to be his natural and unfortunate birthright.[87]

It was unfortunate, but it was the natural order of things. The concept of the innate mental differences between the sexes was as psychologically indispensable as it was theoretically consistent. Emotional comfort might be distilled from theoretical necessity.

Not that I am suggesting that this was in any way a conscious process on Darwin's part. He is not to be numbered among those Victorian antifeminists who consciously set themselves to oppose women's suffrage, higher education, and entry into the professions. Nor did Darwin engage actively in sexual discrimination, as did Thomas Henry Huxley, when in 1868 this prominent proponent of higher education for women excluded these "amateurs" from ordinary meetings of the Ethnological Society of London on professional grounds. This, despite the fact that Darwin, the greatest scientist of the nineteenth century and famous in his own lifetime, remained an "amateur" all his life. True, with Darwin's handsome income from his solidly invested inherited capital, he could remain comfortably outside the struggle for scientific professionalization and keep his liberal principles intact. Just as he could encourage some few women to make scientific observations for him and even to publish, so he could applaud the "triumph of the Ladies at Cambridge" when women were finally accorded the right to present themselves for the "Little-Go" and Tripos Examinations in 1881.[88]

In any case, as Darwin told those few women bold enough to confront him directly about his claims as to women's inferior intelligence and the social implications of his evolutionary argument for their continuing inferiority, women had moral superiority on their side. If they wished to gain intellectual parity with men, women must not only be highly educated, they must also become "breadwinners"; but he warned that this inevitably would impact deleteriously on the education of children and the "happiness of our homes."[89] It was not only outside Darwin's experience, but beyond his comprehension that a home might be happy in which the woman of the house worked outside it and did not devote herself entirely to husband and children. For Darwin, the moral and intellectual differences between the sexes were as self-evident as the differences in plumage between the peahen and the peacock, and both sets of phenomena were reducible to the same natural causes. Before their marriage, Emma wrote presciently to him:

> I believe from your account of your own mind that you will only consider me as a specimen of the genus (I don't know what[,] simia I believe). You will be forming theories about me & if I am cross or out of temper you will only consider "What does that prove." Which will be a very grand & philosophical way of considering it.[90]

There was, after all, no inconsistency between Darwin's personal experience and his theoretical argument. The women he had known most intimately conformed entirely to Victorian conventions of femininity and domesticity. Of his own part in reinforcing those conventions, he remained sublimely unaware.

Darwin's family background gave him much more than a ready-made set of social values and cultural attitudes. In later life he found Erasmus Darwin's erotically charged writings an embarrassment, and did his best to distance himself from them. Yet, all the evidence goes to show that, as a young man, Darwin resonated to their siren song of the evolutionary ends of sex and reproduction. The next chapter examines how the echoes of that song resounded in the grandson's theory of sexual selection.

THREE

"Bliss Botanic" and "Cocks Heroic": Two Darwins in the "Temple of Nature"

The Loves of the Plants, so deliciously sung,

Must have softened his heart, when his bosom was young,

And the Temple of Nature has prompted his tongue—

Which nobody can deny.

—"The Descent of Man: A Continuation of an Old Song," *Blackwood's Edinburgh Magazine*, 1871 (preserved among Charles Darwin's papers)[1]

As soon as *The Descent of Man* was published, the more knowing saw connections with certain themes in the writings of Darwin's grandparent, Dr. Erasmus Darwin.

This was not an intellectual debt that Charles ever acknowledged. It was a sensitive issue for one who was jealous of his originality and whose critics had been only too ready to invoke the specter of his godless evolutionist grandsire. Erasmus Darwin is entirely absent from the *Origin of Species* and *The Descent of Man*. When, prompted by Lyell, Charles Darwin attached the note on his evolutionary predecessors to later editions of the *Origin*, Erasmus was dismissively relegated to a footnote as having "anticipated the erroneous grounds of opinion, and the views of Lamarck, in his 'Zoonomia.'" In his "Recollections," he claimed that his early readings of *Zoonomia* (Erasmus Darwin's major theoretical work) had no effect whatever on the formation of his own evolutionary ideas.[2]

In fact, Charles had been familiar with *Zoonomia* from his student days in Edinburgh when he "admired [it] greatly" and heard its evolutionary views praised by his onetime mentor, the radical anatomist and evolutionist Robert Edmond Grant. He had been sufficiently interested to follow up some of its themes in other texts and debates of the period and to read Anna

Seward's racy and irreverent biography of his famous forebear.[3] It was first and foremost to his grandfather's evolutionary views that Darwin turned, when on his return from the voyage of the *Beagle* he began his systematic search for an explanatory theory that would support his newfound conviction of ongoing species change, or transmutation. He entitled the first of his transmutation notebooks "Zoonomia" in emulation of his grandfather ("Notebook B," 1837).

More than this, and more to my point, Erasmus Darwin's views were critical to Charles Darwin's early formulations of the principle of sexual selection.[4] This applies not simply to the younger Darwin's adoption of Erasmus's notion of sexual combat as a means of species benefit and as an explanation for the development of specialized male "weapons," but also to the impact on Charles of his grandfather's views on beauty. Furthermore, his early notebook theorizing on the sexual struggle, sexual dimorphism, and beauty relates closely to Charles Darwin's largely Erasmus-derived views on behavior, or "habits"; and more fundamentally, to the romantic metaphor of gestation that structured the evolutionism of both Darwins. I would even go so far as to suggest that what Charles Darwin later called the "law of battle" may have primed him for his famous Malthusian insight.

For these reasons, it is necessary to examine the evolutionary views of Erasmus Darwin in some detail.

3.1 "Bliss botanic": The loves of the plants

Erasmus Darwin lived and wrote in the critical transition period between the end of the Enlightenment and the beginnings of the dominant nineteenth-century ideology of progressive industrial capitalism. His writings celebrated the latter. At the same time they drew on the classic pre-Enlightenment empiricist epistemology of Locke, Hartley, and Hume. His vision of a progressive, self-regulating, self-generating nature was a secularization of David Hartley's theory of salvation.

Hartley had regarded human life as a moral progression to a state of grace: God had imbued virtue with pleasing ideas, vice with odious ones. Through the association of ideas, each individual's spiritual capacities inevitably increased, and in this way, God ensured individual moral perfection. Erasmus Darwin transferred this interpretation to a naturalistic framework in which each organism acquired and gradually developed its capacities in interaction with its environment through the sensations of pleasure and pain. These acquired capacities were transmitted to the next generation, and so the species as a whole must constantly improve. It was a theory of change through learning, a trans-generational progress in which animals in part produce their own transformations over time as a consequence of their own exertions, and the accumulated benefits of their experience are inherited by their offspring.[5]

Erasmus Darwin's evolutionism thus was subversive on two major counts: it was redolent of materialism, and it glorified sex and fertility. His natural world did not require divine intervention but worked according to the laws of organic life; of these, the most powerful, productive, and happiest was the organic impetus to sexual reproduction or generation. There could be no life without reproduction and it was the purpose of life to reproduce. Sexuality and desire secured the future of a species, while asexual modes of reproduction, which provided no opportunity for "joy" or new organic forms, led to degeneration over time.[6]

These themes of sexuality and desire were central to Erasmus Darwin's early and most admired work, *The Loves of the Plants* (1789), the second but first-published part of *The Botanic Garden*. *The Loves of the Plants* was inspired by Linnaeus's recently developed doctrine of the sexuality of plants and his taxonomic system, which classified plants by their reproductive parts. It personified the sexual organs of flowers (the stamens and pistils) as men and women, and poetically described their diversity of courtships, marriages, and methods of reproduction. It was as much about Erasmus Darwin's eighteenth-century society and gender relations as it was about plant sexuality.

Its suggestive imagery was not lost on its enthusiastic readers. They were charmed and titillated by its mild eroticism and lush verse, but reassured by its ostensible botanical instructiveness and classical allusions and by the lengthy and erudite footnotes. It may have been a trifle "warm," but (to begin with, at any rate), in a period when sexuality, having no language of its own, was approached through the vocabulary of gender and morality, it was permissible reading matter for late eighteenth-century misses who were discovering the delights of botany and poetry while negotiating the conventions of polite sexual conduct. That well-known literary figure Horace Walpole, while turning a knowing masculine eye to its "Ovidian" implications, could safely commend this "most delicious poem on earth" to the Misses Berry.[7]

For all their eroticism and multifaceted sexual arrangements, the flower women of *The Botanic Garden* were variants of reassuring stereotypes—polarized between the blushing virgin and the seductive coquette—who figured in the romantic novels, pastoral poetry, and dramatic arts patronized by the landed gentry. And, while women might read such a poem, it was not, as Anna Seward knew, "strictly proper" for a woman to write a poem about sexual conduct. Even such an unconventional thinker as Mary Wollstonecraft accommodated her writings to the prevailing image of feminine delicacy of taste. Her unladylike disquisitions on politics and the rights of women were contentious enough; sexuality was forbidden territory. In these senses, *The Botanic Garden* was very much "botany for gentlemen."[8] It had, too, a more challenging subtext. Darwin was intent on inculcating not only the botanical lesson that plants had a sex life, but also the deeper and more radical claim that human love and sexuality were biologically based, not God-given and ordered. Just as his plant people obeyed "licentious Hymen" and "laugh[ed] at all but Nature's laws," so too human emotions and manners obeyed only "natural" laws of sexuality and desire.[9]

With *The Loves of the Plants*, Erasmus Darwin began the elaboration of his "laws of organic life," the focus of his major theoretical works, *Zoonomia* (1794) and *Phytologia* (1800), culminating in *The Temple of Nature* (1803), his final, posthumously published paean to biological and social progress. All of these works reflect his early preoccupation with sexuality and reproduction as the central and most important features of organic life and the keys to its progressive evolution.

3.2 From nature's womb

In certain key respects, Erasmus Darwin's conception of nature and inevitable human progress has affinities with the burgeoning romanticism that flowered in German *Naturphilosophie* and inspired a generation of British poets, naturalists, and anatomists.[10] Recent scholarship has emphasized the nexus between early British evolutionary theorizing, radicalism, and romanticism.[11] In the hands of certain leading exponents, they made for an explosive mixture, and Erasmus Darwin was one of the earliest to touch off the fuse.

His romanticism is evident in his uncompromising developmentalism and organicism, and in his insistence on the fundamental unity of all nature; but above all, in his invocation of the metaphor of the gestation of nature. This fertile metaphor, which pictured the history of the world as one long gestation analogous to a normal human pregnancy, dominated romantic evolutionary speculation in the late eighteenth and early nineteenth centuries. Romantic literature abounds with references to this universal gestation of nature—to the "impregnation of the terrestrial womb," the "pregnancy" of the world, the "generation," "gestation," "growth," or "development" of nature. For the majority of biologists, physicians, or philosophers who toyed with the notion, it remained an abstraction, a kind of ideal process that explained the changing forms assumed by the developing embryo and their supposed similarity to the ascending series of mature forms constituting the animal series or scale of being. It was not materially realized in nature. Less idealistic biologists, however, were free to give the metaphor a literal interpretation.[12]

From his earliest to his last published references to progressive organic development, Erasmus Darwin relied explicitly on the powerful appeal of the gestation metaphor. The most detailed account of his evolutionary hypothesis was offered in his *Zoonomia* under the heading "Of Generation."[13] By "generation" Darwin meant not only the conception and progressive development of individual organisms, but also—and more significantly—the progressive production of life on earth. Individual development, with all its intricacies of form change, was both evidence and model for the "generation" of species. Against the prevailing theologically inspired doctrine of preformation,[14] Darwin argued that the individual is gradually transformed in the womb and after birth, and that these transformations are analogous to the production of animal life "from a similar living filament. . . . And all this exactly as is daily seen in the transmutations of the tadpole, which acquires legs and lungs, when he wants them; and loses his tail, when it is no longer of service to him."[15]

Erasmus pushed metaphor as far as it would go. He attributed a common origin not just for warm-blooded animals, but for all living things: "The whole is one family of one parent." By the third edition of *Zoonomia* in 1801, undaunted by the increasingly vociferous charges of materialism, atheism, and worse, he suggested that life had first originated under suitable conditions of warmth and moisture by the "coalescence" of organic "molecules" and "fibrils" that possessed attractive powers, and that this spontaneous generation of microscopic organisms resembled the process of "actual" generation, or sexual reproduction, of more complex, or "perfect," animals. This sacrilegious notion was given a poetic spin in *The Temple of Nature*:

> From Nature's womb the plant or insect swims,
> And buds or breathes, with microscopic limbs.[16]

Life was born from nature's womb. Put in this way, Erasmus Darwin's evolutionary ideas appear to flow from a number of romantic assumptions about nature. But romanticism was only one of the strands with which he wove the web of his evolutionism and his glorification of generation. Apart from his specific debt to the association of ideas tradition identified with Hartley, Darwin drew on an assembly of utilitarian economic and political notions. There are distinct Hobbesian echoes in Darwin's theorizing on the role of struggle and competition in organic evolution, especially as he ap-

Figure 3.1. Frontispiece by Henry Fuseli from Erasmus Darwin's *Temple of Nature*. The goddess of poetry unveils the many-breasted Artemis of Ephesus, the fertile goddess of nature. Erasmus Darwin's evolutionary theorizing was intensely sexualized: life was born "from Nature's womb" and followed the sequence of embryonic stages, while the development of the aesthetic sense was centered on the infant's experience of handling and viewing the female breasts. E. Darwin 1803. University of Sydney Library, Rare Books and Special Collections.

plied them to his conception of sexual conflict—echoes that resonated in his grandson's version of the same.

3.3 In the Temple of Nature: Sex, scarcity, and struggle

Thomas Hobbes, architect of utilitarian political theory, had contended that the animate world was in a state of constant struggle, rather than in harmonious balance. In *Leviathan* (1651), Hobbes argued that all men pursue their personal desires for gain, security, and reputation, but since the resources men want are scarce, not all desires can be fulfilled. The result is the classic Hobbesian "war of every man against every man" among humans in a state of nature. Men use violence "to make themselves masters of other men's persons, wives, children and cattle" or to defend them. Struggle and competition are the "natural condition" of humanity. Hobbes instanced the struggle of male animals for access to females as evidence of the natural combativeness of the primitive human state. Men conquer and subordinate women through superior strength and thereby create paternal rights by contract: by conquest, in Hobbesian terms. Human passions make good intentions and voluntary compliance totally unreliable.[17]

At the time Erasmus Darwin was writing, the most theoretically elaborated view in opposition to Hobbes was that of Jean-Jacques Rousseau, the French philosopher and political theorist. In his *Discourse on Inequality* (1761), Rousseau argued against Hobbes that humans in a state of nature were naturally good, or, rather, innocent. This was because the desires of primitive men and women were limited: there could be no jealousy or rivalry, no marriage, no desire for a specific partner. There is brute desire or "physical love" between the sexes, but there is no "moral love" that fixes this desire on a particular object. This is because

> the moral Part of Love is a factitious Sentiment, engendered by Society, and cried up by the Women . . . in order to establish their Empire, and secure Command to that Sex which ought to obey. This Sentiment, being founded on certain Notions of Beauty and Merit which a Savage is not capable of having, and upon Comparisons which he is not capable of making, can scarcely exist in him . . . Every woman answers his purpose.[18]

Natural man yields to his impulses "without Choice and with more Pleasure than Fury." There is no violent competition among the males for females. Rousseau repudiated Hobbes's animal analogy by stressing the significant difference in the reproductive physiology of human females. Women, unlike female animals, do not have alternating periods of "Passion and Indifference" and thus are always sexually available: "We cannot therefore conclude from the Battles of certain Animals for the Possession of their Females, that the same would be the Case of Man in a State of Nature." Natural man is peaceful and egalitarian. He is spared the "terrible Passion . . . which in its Transports seems proper to destroy the human Species which it is destined to preserve." This is the lot of social man, who is subject to the individuation of desire (construed as the creation of women). "What must become of Men abandoned to this lawless and brutal Rage," lamented Rousseau, "without Modesty, without Shame, and every Day disputing the Objects of their Passion at the Expense of their Blood?"[19]

Rousseau's complex views on women, sexuality, and "savages" were bound up with his radical revision of Hobbes's theory of the social contract, which Rousseau used to argue the inalienable sovereignty of the people and the legitimacy of democracy. In essence, nature becomes the source whereby society, morals, and education are to be reformed and purified, displacing the accepted traditional source of "right": God via Church and King.[20]

Erasmus Darwin, along with most of the Lunar Society fraternity, was an admirer of Rousseau. But, while there are Rousseauean elements in his conception of nature and society, Darwin sided with Hobbes in extending the animal analogy to humanity and in viewing struggle and competition as the natural state of humanity. This was consistent with his insistence on the evolutionary continuity of animals and humans. Furthermore, contra Rousseau, Darwin gave male combat a positive role: it "improves" the species, biologically and socially.

In the first edition of *Zoonomia*, Erasmus described how the forms of many animals have been changed through their "exertions to gratify" the "three great objects of desire . . . lust, hunger, and security." Thus some species have acquired special structures in order to satisfy their need for particular food sources, such as hard beaks to crack nuts, while the coat colors or shapes of many animals seem adapted to their purposes to pursue prey or to escape or conceal themselves from predators. The males of some animals, however, have acquired structures (such as the spurs of cocks and the antlers of stags) that are of no use in their struggle for subsistence or defense against other species. These structures are specifically for the purpose of combat with males of their own species, and they have evolved through the sexual desire of their owners "for exclusive possession of the females; who are observed, like the ladies in the times of chivalry, to attend the car of the victor." Moreover, this sexual struggle among the inherently possessive males produces not only the structural differences between the sexes, but it also ensures the progress of the species:

The final cause of this contest amongst the males seems to be, that the strongest and most active animal should propagate the species, which should thence become improved.[21]

These Hobbesian themes of innate sexual struggle, conflict, and possessiveness were given extended poetic treatment in *The Temple of Nature*. Here Erasmus described how with the development of sexual love came the "Demon, Jealousy," who "lights the flames of war" in the wake of "unsuspecting Love":

> Here Cocks heroic burn with rival rage,
> And Quails with Quails in doubtful fight engage . . .
> With rustling pinions meet, and swelling chests,
> And seize with closing beaks their bleeding crests . . .
> While female bands attend in mute surprise,
> And view the victor with admiring eyes.[22]

The "bestial war" is dissolved by the advent of Eros, the God of "sentimental love,"[23] who arrives on the scene immediately after Erasmus's description of the inception of the human aesthetic sense through the infant's suckling and adoration of the "IDEAL BEAUTY" of its mother's breast (of which more below). Eros domesticates the warring birds and bulls and "binds Society in silken chains." But the primitive conflict lies always just beneath the surface of society. It is, moreover, necessary to human progress, to civilization. Man alone of the animals has no structures for combat, no spurs or horns, no fur or feathers for protection or adornment, but he shares their irresistible desire for sex, subsistence, and shelter, and he has the capacity to struggle and advance through his efforts to satisfy these wants. He learns and improves by association and imitation, and through inheritance of these newly acquired abilities. Thus language, which distinguishes man from brute, evolved from the gestures and vocalizations excited by sexual passion or jealous rage.[24]

From such lowly origins, human reason and intellect develop to the point of foresight and sympathy. It is sympathy that "opes the clenched hand of Avarice to the poor / Unbars the prison, liberates the slave . . . And gives Society to savage man." But, warned Erasmus, "the sacred charm" of sympathy is in constant tension with the unending struggle for subsistence:

> From Hunger's arm the shafts of Death are hurl'd,
> And one great Slaughter-house the warring world![25]

By the time Darwin composed *The Temple of Nature*, the Hobbesian war of all against all had been reinforced by Malthusian social arithmetic, and the population question was beginning to gnaw at the underbelly of the prophets of progress.

3.4 "Shout round the globe": The triumph of love and desire

For utilitarians, the two factors that determined the quantity of happiness in a country were the number of inhabitants and their individual happiness. By the late eighteenth century, theorists were beginning to question the inherent desirability of a large population. Scarcity and the high price of provisions brought about by a succession of crop failures, the great increase of population in and around the new industrial towns, the war with France, the manipulations of grain distributors, the mass food riots of 1795, and the great numbers thrown on poor relief all ensured that the issues of food scarcity and population growth were at the forefront of public consciousness and exercised the ingenuity of utilitarians.[26]

William Godwin's utopian solution was to put his confidence in human reason. He argued that before the earth became overpopulated and food sources were exhausted, humanity would cease to propagate. There would be no more birth or death; humans would be immortal. Reasonable men and women would transcend destructive sexual passions and social conflict. This must increase the sum of human happiness, and society could approach perfect harmony. The new society would be one of social and economic equality.[27]

The economist and cleric, the Reverend Thomas Robert Malthus, vehemently disagreed. He directed his *Essay on the Principle of Population* of 1798 against this unjustifiable Godwinian optimism. The growth of population must inevitably outstrip that of food supply, and Malthus mathematized the relation: population increases geometrically, food supply only arithmetically. This scientifically demonstrated imbalance between nature's supply of sustenance and man's need for both food and sex was an insurmountable impediment to notions of human perfectibility or, indeed, to any radical restructuring of society. Against Godwin's view that a more equitable distribution of the good things of life was the way to improve public felicity, Malthus notoriously naturalized social inequalities and injustices. Poverty, war, and pestilence were the ineradicable natural checks to excessive population growth, and all attempts to alleviate these scourges and alter social relations were not only misguided but also immoral.[28]

The Malthusian law of population was integral to the nineteenth-century movement of scientific naturalism, which included natural theology and entailed a significant shift in theodicy—the justification of God to man. The Malthusian God was a "reasonable" and lawful god, one who worked by natural law, not an "arbitrary" potentate who worked miracles in violation of nature's laws. The new model of the divine was suited to an age of commerce and manufacture, a "practical artisan or merchant who must plan ahead for all contingencies."[29] He was preeminently a contriver, and nature and society bore the marks of his contrivance. In the hands of Malthus and his apologists, the God of the population law validated the propertied classes; privileged self-interest and competition; and made the poverty, "sloth," and uncontrollable sexuality of the propertyless sinful. It was "Parson Malthus" who, to the lasting opprobrium of his radical detractors, argued for the retention of the protectionist Corn Laws (and thus for higher food prices) and for the abolition of poor relief. Those improvident enough to marry and reproduce without adequate subsistence "should be taught to know that the laws of nature, which are the laws of God, had doomed [them] and [their families] to suffer for disobeying their repeated admonitions." Parish assistance should be denied to illegitimate children: "The infant is, comparatively speaking, of little value to the society, as others will immediately supply its place."[30] It was such hardheaded Malthusian logic that rationalized the Poor Law Amendment Act of 1834 that introduced the infamous workhouses and imposed new draconian measures on the "surplus population."

In the Malthusian scheme of things, sex is inherently problematic. It is necessary for reproduction, but reproduction itself has become a problem. Thomas Laqueur has argued that Malthus rehabilitated the body and the absolute irreducibility of its sexual demands against Godwin's devaluation of sexual passion. The rehabilitation of sex, however, was far from Malthus's intention: sex equaled "population," and "population" consisted of propertyless masses who owned "nothing but dangerous sexual—and illusory political—wishes. Only the severest counter forces, famine and epidemic, can check [its] world-shattering power."[31] By the second edition of the *Essay* in 1803, Malthus, in response to criticism of his God-sanctioned scenario of unrelenting disaster and retribution for the poor and improvident, introduced the palliative of "moral restraint." Through moral restraint, the biological imperatives of sexuality and reproduction might be overcome. Sexual abstinence was the only acceptable means (Malthus condemned abortion and contraception as "vice") by which the insatiable generativeness and prospective revolt of the impoverished might be controlled; in this way they might accrue moral improvement and thus a modicum of social progress. Civilization depended on sexual repression. In this, Malthus concurred with the radical Godwin.[32]

For Erasmus Darwin, however, sex and fertility were not problematic or socially and politically dangerous, but the very means of biological and social improvement. In the Darwinian scheme of things, nature was not constructed or engineered, but self-generated. Life was born from nature's womb and, "by its own inherent activity," underwent a state of "perpetual improvement." Darwin was unique among the prophets of progress in integrating his biological perspective with his faith in progress.[33]

Erasmus provided two answers to the population question. In *Phytologia* (1800) he argued that new developments in science and technology would increase agricultural yields and so increase population numbers and the "happiness of the country." However, by the time he wrote *The Temple of Nature*, Darwin's perception of the world as "one great slaughterhouse" had deepened. By this stage, he accepted scarcity, along with war and disease, as the natural and necessary checks to excessive population growth. Nevertheless, Erasmus was not underwriting a Malthusian charter. The inevitable deaths and destruction, he argued, were counterbalanced by nature's generativeness:

> The births and deaths contend with equal strife,
> And every pore of Nature teems with Life.

From the death and dissolution of organisms arise new forms of life, new opportunities for desire and pleasure, and the sum total of earthly happiness is thereby increased. Even the mountains and continents, formed from the shells and remains of once living organisms, "ARE MIGHTY MONUMENTS OF PAST DELIGHT";

> Shout round the globe, how Reproduction strives
> With vanquish'd DEATH,—and Happiness survives;
> How Life increasing peoples every clime,
> And young renascent Nature conquers Time.[34]

Sex and reproduction vanquish death, happiness is augmented, and life improves through the endless competition of burgeoning life-forms, through continuous destruction and struggle. These same "firm immutable immortal laws" of nature have also ensured social progress, measured by the achievements of the great artists, scientists, and, above all, by the inventors and industrialists, who, with their new technologies—their potteries, steam manufactories and cotton mills—were changing the world around Darwin.

3.5 Radical revolution and progressive science and industry

Contemporaries regarded Erasmus as a political radical—a "Jacobin," said Seward—who, initially at least, made no secret of his support for the French Revolution. The far-reaching reverberations of the French revolution can hardly be overestimated. Post-revolutionary politics in Europe throughout the nineteenth century were "largely the struggle for and against the principles of 1789." Darwin belonged to that wide circle of reforming Whigs and Dissenters, exemplified by the Lunar Society, who exulted in the fall of the Bastille.[35]

The Lunar Society (so called because its members met on moonlit nights, when they could see to travel home) drew its membership from the predominantly Dissenting self-made manufacturers clustered in and around Birmingham, a town forged by the Industrial Revolution and excluded from the structure of parliamentary representation. Denied access to traditional forms of political power, the Dissenting industrialists manufactured their own through their entrepreneurial, mechanical, and organizational skills and their promotion of a new machine-age culture. Their systems of production and work discipline were models for other manufacturers to emulate. Erasmus was fascinated by their "mechanic arts," contributed to their inventions, and was actively involved in their canal-building enterprises and political lobbying. His notions of progress and improvement were inflected by the mercantile management of natural and human resources advocated by the Lunar Society industrialists. In *The Botanic Garden*, the goddess of botany tutors the attendant nymphs and gnomes as much on the processes of production as reproduction, invoking images of whirling wheels, pounding hammers, explosive steam, and hissing steel, alongside the more traditional sensuous tropes of pastoral dalliance.[36]

Erasmus Darwin's writings epitomize that form of late eighteenth-century enlightened ideology that subscribed to a secular, rationalist, and progressive individualism, whose object was to set the individual free from the superstitious and irrational oppression of church and social rank. It viewed science and technology as inherently progressive and liberating forces, and it took for granted that the free society would be a capitalist society. In theory, Darwin and the Lunar men were radicals who welcomed the "glorious revolution." In practice, they were middle-class reformers who campaigned against slavery and for the political and social rights of Dissenters, but who mostly did not—even at the height of revolutionary fervor—contemplate seriously the "spreading of the holy flame of freedom" to the lower orders.[37]

High-flown idealism was sometimes compromised by base commercial considerations or might provide entrepreneurial opportunities. While

Wedgwood and Erasmus Darwin were prominent in the early campaign for the abolition of the slave trade in the 1780s and '90s, other Lunar members were more equivocal, fearing abolition would impact adversely on British trade. The cameo that Wedgwood mass-produced in his trademark jasper-ware, with its kneeling black slave below the famous plea "Am I not a man and a brother?," was as successful a commercial venture as it was in arousing antislavery sentiment. Their technological hubris meant that Lunar employers like Wedgwood and Mathew Boulton and their versifying propagandist, Darwin, might conveniently overlook the less than beneficial impact of the new technological processes on a "refractory" labor force. Moreover, by the mid-nineties, the Jacobin Terror in France and the counterrevolutionary "Terror" unleashed in Britain by the conservative Pitt government had frightened most of the Lunar men from the cause of reform, and the society itself was in disarray.[38]

In 1791, the house, library, and laboratory of Joseph Priestley, a Unitarian scientist and the most vocal Lunar radical, were destroyed by a "Church and King mob" that looted the houses and shops of other wealthy Lunar Society Dissenters and sympathizers.[39] Priestley fled Birmingham, while the Lunar fraternity, confronted by the social consequences of their radical ideology, pulled back from the dangerous cause of political reform and sought refuge in an avowedly apolitical image. As revolutionary fervor and Paineite propaganda inspired a new plebeian radical consciousness and counterrevolutionary repression grew, landowners and manufacturers united in a common panic that deflected the middle class from their agitation for social and political change and, in effect, consolidated "Old Corruption." Increasingly, from 1791 on, the erstwhile radicals of the Lunar Society channeled their reforming energies into less contentious educational schemes and, characteristically, into the promotion of science.[40]

Erasmus Darwin retreated from the political world to shelter "under the banner of science." Through his later writings, in *Zoonomia* and *The Temple of Nature*, he "projected his aspirations for change on to the natural world," making progress a guaranteed feature of nature, rather than something to be politically struggled for.[41]

3.6 The revolutionary threat of the "unsex'd females"

His retreat to science did not protect him in a period when science itself—certainly the kind of science Erasmus Darwin portrayed—was considered subversive. He was a freethinker and a "materialist" who deduced social progress and morality from nature's laws, and this was more than enough to identify him with the radical camp. In 1795, as fear of French invasion mounted, Pitt introduced his notorious anti-sedition acts amid a renewed frenzy of book banning and prosecutions. Not even gentlemanly poets ensconced in remote country mansions seemed safe. Darwin was surrounded by "profess'd spies" and considered following Priestley to America, "the only place of safety" where "potatoes and milk . . . may be had . . . untax'd by Kings and Priests."[42]

However, the most that happened to Darwin was the erosion of his earlier fame by the sustained campaign of censorship and repression that linked him with avowed radicals like Godwin and Wollstonecraft. From around 1794 on, his writings and views were parodied and attacked by politically motivated critics. The best-known attack came from the government-sponsored weekly, the short-lived but brilliantly successful *Anti-Jacobin*, the brainchild of George Canning, undersecretary for foreign affairs. A "gentlemanly miscellany of politics and poetry," it promoted the values dear to the counterrevolution—patriotism, piety, and "the decencies of private life." Canning and company were the "avowed, determined, and irreconcilable enemies" of "JACOBINISM in all its shapes . . . whether as it openly threatens the subversion of States, or gradually saps the foundations of domestic happiness."[43]

Erasmus Darwin's problem was not just his alleged materialism, nor his evolutionary or pro-French views. His writings—particularly *The Botanic Garden*, with its frank eroticism and its naturalization of sexual relations—were targeted as a special source of the sexual corruption of young women through their study of the fashionable botany. It was the supposedly anti-domestic shape of Jacobinism that most roused the witty ire of the *Anti-Jacobin* reviewers. Marriage and domesticity, along with religion, were the guarantors of social stability. The *Anti-Jacobin* led the virulent campaign against William Godwin, who earlier had condemned marriage as an affair of property, but belatedly married the archenemy and "prostitute" Wollstonecraft. The twice-married and propertied Darwin, who certainly did not share Godwin's views, was guilty by association. They were often the joint targets of the *Anti-Jacobin*, ridiculed for believing "Whatever is, is WRONG," and in the *"eternal and absolute PERFECTIBILITY OF MAN."*[44]

"The Loves of the Triangles," the *Anti-Jacobin* parody of *The Loves of the Plants*, lampooned both the revolutionary notion of perfectibility and eroticized evolutionary biology. It was funny and clever, and hence all the more damaging. "It is diamond cut diamond here," exulted Anna Seward, although she rallied to the defense of Darwin's "fine poetry," to everything "but his irreligion, and encomiums on the terrible and tyrannic democracy of France."[45]

A sometime contributor to the *Anti-Jacobin*, the Reverend Richard Pol-whele, dealt the nastiest blow.[46] Polwhele's 1798 poem, *The Unsex'd Females*, conjured up a depraved, Francophile "female band" that, under Wollstone-craft's direction, has revolted against "NATURE'S law" of womanly sub-mission, domesticity, and modesty. These "unsex'd" females impudently "sport" their breasts and have taken to botany, "Philosophy," gymnastics, and other unnatural pursuits. They learn how to "point the prostitution of a plant" and are roused to lust and "democratic" politics by "bliss bo-tanic." In copious footnotes, Polwhele expressed his inability to see how the "study of the sexual system of plants can accord with female modesty": "Our botanizing girls . . . are in a fair way to becoming worthy disciples of Miss W[ollstonecraft]" and "will soon exchange the blush of modesty for the bronze of independence."[47] The only preventative against such unnatu-ral and ungovernable conduct was, naturally, religion: "A woman who has broken through all religious restraints, will be found ripe for every species of licentiousness . . . But, burst the ties of religion; and the bands of nature will snap asunder!" The Reverend Polwhele infamously detected the "Hand of Providence" in Wollstonecraft's fitting end "in consequence of childbirth."[48]

To understand the tenacity and forcefulness of the vilification of Woll-stonecraft, Darwin, and any other writers and works that seemed to threaten the established relations between the sexes in this period, it is necessary to understand the tensions inherent in a developing middle-class code of propriety. It was a code that simultaneously defined women almost entirely in terms of their sexuality while insisting that their every public action deny that sexuality.[49] These tensions were greatly exacerbated by the moral panic that swept Britain toward the close of the eighteenth century. Radicals, coun-terrevolutionaries, and evangelicals alike were variously involved in assert-ing and reconstructing the idealized feminine attributes, seeing in women a focal point for the regeneration of society. The strains and contradictions of this critical transition period significantly shaped nineteenth-century at-titudes toward sexuality and the attempts by naturalists, ethnologists, and anatomists to interpret human sexual relations.[50]

Erasmus Darwin's system of female education (discussed in sec. 2.2), however nonconfrontational it may seem, was as much the target of Pol-whele's tirade against botanizing, philosophizing, unlaced gymnastic girls as was Darwin's more famous, "too glowing" *Botanic Garden*. The earlier uncontroversial advocacy of improved education for women in the "con-duct books" that proliferated in the latter half of the eighteenth century was being brought up short against the backlash of counterrevolutionary rhetoric and the concurrent rise of a militant and conservative Evangeliral movement that promoted a morality centered on the home and family and women's place therein. Its leading spokeswoman, Hannah More (much praised by Polwhele), reasserted traditional gender roles against the "monstrous compositions" of "French infidels," "new German enlighteners," and Wollstonecraft's "direct vindication of adultery." "Propriety" was the "first, the second, and the third, requisite" for women, and their only proper roles that of obedient, pious, and chaste wives and mothers.[51]

3.7 Beauty, breasts, and faces

Erasmus Darwin's views on the relations between the sexes infiltrated his theorizing. His evolutionary scheme of biological and social progress through the exertions of individuals and the inheritance of acquired learning and skills guaranteed unlimited improvement to men but permitted only limited improvement to women. It was essentially a masculine view of what was proper and therefore natural for women.

Darwin did not deny sexual desire to women, although he regarded female sexuality as more muted and not necessary for conception.[52] Nevertheless, unlike its masculine counterpart, female sexual desire in Darwin's evolutionary scheme was hardly a progressive or improving force. Woman's sexuality required the exercise of male power for its expression. This is explicit in his view of the role of male combat for possession of the female and his attribution of subsequent species improvement to the active and superior male, who emerges triumphant from the contest and mates with the admiring, receptive female. What little suggestion there is of the exercise of female preference is subordinate to prior contest among males, which determines the mate to be preferred.

In the first edition of *Zoonomia*, Darwin represented generation in patriarchal Aristotelian terms, with the female's role reduced to the provision of a nest and nutrition for the essentially male conception.[53] By the third edition (1801), he revised this view to give both partners a role in generation: the "sexual glands" of the female and the male secrete "molecules" or "fibrils," which unite to form the embryo. However, even these minute particles are gendered: the masculine fibrils are active and "possess appetencies to embrace," while the maternal molecules are passive, possessing "propensities to be embraced."[54] Moreover, in both editions, Darwin, in a reversal of the traditional doctrine of maternal impressions (the belief that the female's ideas at the time of conception or during pregnancy could affect her offspring), promoted the argument that the male imagination acted on the form of the embryo to determine its sex and most other attributes.[55]

The quality and contents of the male mind were thus of supreme importance in sexual reproduction, which provided the means whereby the "ideas" of the mind could be conveyed to the next generation and progress assured. So powerful was the role of the male imagination, that if a man strongly desired a woman other than his wife, then his legitimate offspring would more closely resemble the desired and imagined sexual partner than the actual one. The "art of begetting beautiful children, and of procreating either males or females" might be taught to men through the appropriate exercise of their imaginations on the "fine extremities of their seminal glands."⁵⁶

Inevitably, aesthetics in Erasmus's schema, being closely linked with sexuality, desire, and the male imagination, was very much a masculine affair. His account of the development of the aesthetic sense was centered on the female breasts. Humans developed their sense of beauty from a love of pleasure afforded by their lower senses—those of warmth, smell, taste, hunger, thirst, and touch—which were associated with suckling at the maternal breast. What is beautiful to adults stemmed from this original association "with the form of the mother's breast; which the infant embraces with its hands, presses with its lips, and watches with its eyes." Sight, the foremost sense for perceiving the beautiful in a civilized adult, came only at the end of the sequence of the play of the lower senses, mirroring the development of the individual, which itself recapitulates the evolutionary development of the human species:

And hence at our maturer years, when an object of vision is presented to us, which by its waving spiral lines bears any similitude to the form of the female bosom, whether it be found in a landscape . . . or in other works of the pencil or the chissel, we feel a general glow of delight, which seems to influence all of our senses.

The curves of the human breasts were synonymous with and explained William Hogarth's famous "line of beauty," the curvilinear line, the essence of the beautiful in landscape and art.⁵⁷

While many objects are commonly called "beautiful" (such as a Greek temple or music or poetry), such things are merely "agreeable." We have no desire to embrace or kiss them. The "characteristic of beauty . . . is that it is the object of love." "Sentimental love," which is distinguished from sexual love, "with which it is frequently accompanied, consists in the desire or sensation of beholding, embracing, and saluting a beautiful object," just as the infant embraces and salutes its mother's bosom with its lips and fingers. This narrows this breast-centered interpretation of "beauty" down to the masculine "sense of female beauty" that "directs" the adult to the "object of his new passion":

> This animal attraction is love; which is a sensation, when the object is present; and a desire, when it is absent. Which constitutes the purest source of human felicity, the cordial drop in the otherwise vapid cup of life.[58]

A gendered aesthetic was hardly peculiar to Erasmus Darwin, but was characteristic of most aesthetic theory of this period.[59] Erasmus was familiar with this literature, and he took his concept of beauty from a rather surprising source, Edmund Burke's *Philosophical Inquiry into the Origin of Our Ideas of the Sublime and the Beautiful* (1757). It was the archconservative Burke, the great enemy of abstract "systems" and radical revolution, who thoroughly associated beauty with the female body (particularly the breasts), with sexual pleasure and with emotion (see sec. 4.5). For Burke, it was his sense of beauty, sexually engendered and emotionally constituted, that allowed the human male to "fix his choice" on a mate.[60]

As an evolutionist, Erasmus Darwin extended reason to animals ("Go, proud reasoner, and call the worm thy sister!");[61] but the indications are that Erasmus, like Burke, was not disposed to concede a sense of beauty to animals. This seems to be the gist of his distinction between animals, directed by their senses of smell and taste "in the gratification of their appetite of love," and the "human animal," for whom the sense of vision is preeminent, "who is directed to the object of his love by his sense of beauty."[62] All else aside, it is not so easy to relate animal breasts to the essential Hogarthian line of beauty to which the idealized, paired human breasts, with their perfect, symmetrical, soft, flowing curves, lend themselves so well.[63] Furthermore, it is hard to see how beauty—being so thoroughly a prerogative of the male imagination that, according to Erasmus, it might be manipulated to bypass the normal processes of reproductive transmission or inheritance—could play a consistent evolutionary role in any scheme of aesthetic selection that might have been available to him.

By the last decade of the eighteenth century, there *were* the beginnings in religious-oriented ethnology of just such a notion to explain the divergence of the human races from a single biblical origin (see ch. 5). But Erasmus was not interested in validating the Mosaic record; nor, in spite of his abolitionist stance (or perhaps because of it), was he much concerned with human racial variety.[64]

A more significant source for speculation on the improving role of a sense of beauty in animals was *On the Generation of Animals* (1651) by the physiologist William Harvey, a work that Erasmus the physician knew well. Together with Hobbes's *Leviathan*, Harvey's *Generation* may have inspired Erasmus Darwin's theory of male combat. Harvey had not only explained male combat in terms of species benefit, but he had also suggested that beauty—*male* beauty, that is—played some part in the process, in that the cock not only fought with other males for his mates, but also "charmed" or seduced potential mates with his comb and beautiful feathers, "the subject of dispute being no empty or vainglorious matter, but the perpetuation of the stock in this line or that."[65]

Joseph Addison, the influential early eighteenth-century aesthetician, also attributed an aesthetic sense to animals, but framed aesthetic choice in the traditional gendered aesthetics of males selecting females for their beauty. Like Erasmus, who certainly knew his *Spectator* essays, Addison emphasized the power of the imagination and the primacy of vision in the aesthetic experience. He linked visually based beauty with the sexual drive and reproduction: "Unless all animals were allured by the Beauty of their own Species, Generation would be at an end, and the Earth unpeopled." He illustrated this kind of beauty with the example of the male bird who may be "determined in his courtship by the single grain or tincture of a feather."[66]

Erasmus, however, did not follow up these leads. He was sufficiently interested in plumage to discuss its role in protective coloration or camouflage, but not as an attractant. So far as sexual selection went, the plumage of warring cocks and quails was there to bristle, to stiffen, to swell, to indicate the size, aggression, and sexual passion of their possessors. In the same way, "Savage-Man" paraded his strength, clenched his fist, scowled, and rolled his eyes as he prepared for combat: "Association's mystic power combines / Internal passion with external signs."[67]

Erasmus was a keen advocate of physiognomy, the practice of deducing the inner emotions, morality, intelligence, and character from the facial features and expression. Johann Caspar Lavater's sensationally successful *Essays on Physiognomy* (English translation 1789) enjoyed a great vogue in England. Physiognomy held a special appeal for English radicals with its leveling potential as a classless intermediary in social relations. The technique of reading the face was accessible to all, regardless of rank or wealth.[68] For Lavater and his many followers, appearance was less interestingly aesthetically than for what it might reveal about the inner workings of the human mind—and, as Lavater himself interpreted his religious-oriented physiognomy, ultimately the soul.[69] The freethinking Erasmus was less concerned with the standardized physiognomic signs of facial proportions and features than with the gestures and expressions that were the indicators of different emotions, such as anger, grief, or pleasure, which he might explain through the theory of associations.[70]

Zoonomia contains detailed descriptions of the physical symptoms, signs, and facial expressions of the various emotional states in animals and humans, which Erasmus traced to the habitual association of certain muscular motions with particular sensations of pain or pleasure in the course of individual development of "this natural language of the passions." In his treatise on female education, he urged the study of physiognomy for girls, who might cultivate pleasing expressions for their future roles as agreeable companions. Physiognomy also would teach girls to judge character in preparation for choosing a husband.[71]

This was as close as Erasmus came to linking an aestheticized emotional expression with the notion of female choice—which was not close at all. His scheme was set. The aesthetic sense was centered on the female breasts and on the emotional/aesthetic associations they engendered; aesthetic choice was exercised primarily by males. Faces were more valued for what they might reveal about the inner passions or character than for their beauty. "Beauty" itself was sexually and emotionally constituted and reproduced established power relations of male dominance and female submission and dependency.

Erasmus Darwin portrayed a universe governed by the sexual principle, by love and desire. He rooted human sexual relations, morality, and aesthetics in biology, not Christianity. But the social and sexual relations that imbued his evolutionary biology and his naturalized aesthetics were those common to his class and his intellectual and social circle, those endorsed by the Lunar Society gentlemen.

3.8 Two Darwins: Cultural continuities and theoretical commonalities

It is a serious mistake to take at face value Charles Darwin's discounting of his grandfather's impact on his own evolutionary views. Early in his career, the grandson had cogent reason to distance himself from radical, materialistic interpretations that put human evolutionary history at center stage.[72] Later, with his scientific and social authority established, when he might safely have acknowledged a familial debt, Darwin's touchiness about the relation of his views to his grandfather's was greatly intensified by a bitter, unresolved public conflict with the disaffected Darwinian Samuel Butler. This erupted over the publication of the previously mentioned essay on Erasmus Darwin's evolutionary biology by Ernst Krause. It was a long-drawn-out and complicated "boggle," which, as it happened, was unwittingly precipitated by Henrietta's red-penciled deletions from her father's introductory memoir (see ch. 2). It involved the savaging of Charles Darwin by a rabid Butler in the letter pages of the *Athenaeum*, where, among other charges, Butler alleged plagiarism of his grandfather's views. It left Darwin bitten where it hurt most, made more protective than ever of his priority and even more dismissive of his grandfather's "speculations."[73]

We now know, from the evidence of Charles Darwin's transmutation notebooks and his annotations in the family copies of Erasmus Darwin's works, just how well versed Darwin was in his grandfather's writings and how significant they were to his evolutionary theorizing.[74] From the opening of his first notebook on species transformation, like his grandfather before him, Charles Darwin structured his evolutionary speculations around the complex phenomena of generation or reproduction, especially sexual reproduction. The very first entry encapsulates his grandfather's views on generation:

> Two kinds of generation[:] the coeval [asexual] kind, all individuals absolutely similar. . . . The ordinary [sexual] kind which is a longer process, the new individual passing through several stages (?typical or shortened repetition of what the original molecule has done).[75]

Charles Darwin was a "lifelong generation theorist."[76] The unifying theme that ran through his early notebook theorizing and impressed its mark on everything thereafter was his tendency to try to understand species change on the model of the development or evolution of individual organisms. He brought all the evidence and information he could find to bear on this essentially romantic analogy. There can be little doubt that it was his grandfather's *Zoonomia* that inspired his lifelong adherence to an evolutionized version of the romantic gestation of nature, and his annotations bear this out.[77] The younger Darwin constructed and reconstructed his ideas of species change around the model of individual embryological evolution.

Not only was it central to his formulation of natural selection, but it was absolutely crucial to his views on sexual dimorphism and hence to his theory of sexual selection. This latter understanding developed slowly over many years of theory building and will be fully discussed in chapter 9. All that is necessary to understand at this point is that Charles, throughout his early notebooks, linked embryogenesis with sexual difference in both humans and lower organisms.[78] His understanding at this stage was that the separate sexes had evolved from ancestral hermaphrodites, and that the male was more developed or evolved than the female, who more closely resembled their young. But the significance of this for sexual selection, and for female choice in particular, was not to become clear to Darwin until he was at work on the *Origin of Species*.

His more immediate and obvious debt to his grandparent's speculations was that aspect of sexual selection that Charles Darwin was to call the "law of battle." He scored those sections of *Zoonomia* and *The Temple of Nature* where Erasmus discussed male combat and its role in species "improvement."[79] His earliest notebook reference to the evolutionary role of male combat was written around April 1838, where Charles wondered,

> Whether species may not be made by a little more vigour being given to the chance offspring who have a slight peculiarity of structure. [H]ence seals take victorious seals, hence deer victorious deer, hence males armed & pugnacious (all order: cocks all warlike).[80]

Around this same time, he read two other accounts of the selective or improving role of male combat by authorities he respected—Edward Blyth, the promising young zoologist, and Sir John Saunders Sebright, the eminent animal breeder (see chs. 5 and 6 respectively). The versions of sexual combat offered by Blyth and Sebright were at best sketchy and, almost certainly in the case of Blyth, were derived from *Zoonomia*. Along with the improving effects of the struggle for mates, both Blyth and Sebright also recognized those of the struggle for existence, though neither gave them an evolutionary interpretation. It seems likely that Charles Darwin, already primed by his grandfather's writings, reading these endorsements of the improving effects of the competition for mates and for subsistence, was led to this speculation on the role of the sexual struggle in the "making" of species. It is notable that, like Erasmus, he linked this with the possession of special structures that would give advantage in such combat—antlers or spurs. He also associated the sexual struggle with "a little more vigour": with the superior health, strength, and vitality of the successful combatant who would then mate and transmit this advantage to its progeny.

This antecedent formulation of sexual vigor or the possession of some advantageous structure or adaptation as a driving force in species change through male combat possibly alerted Charles to the adaptive role of competition, preparing him for his more generalized Malthus-inspired formulation of natural selection.[81] Darwin scored those lines in *The Temple of Nature* where Erasmus had depicted the "warring world" as "one great Slaughterhouse."[82] Others besides Erasmus had, of course, invoked a universal struggle for existence; not just Sebright and Blyth, but also—and perhaps most important for Darwin—his geological mentor Charles Lyell, who explained the extinction of species by such a process. But Erasmus alone gave it a positive, progressive outcome, optimistically versifying perpetual species improvement through the triumph of nature's superfecundity over the continuous death and destruction of the Malthusian struggle. Charles Darwin was to develop his theory of natural selection by a similar positive focus on what happened to the winners rather than the losers.[83] Up to this point, the only winners he had considered were those engaged in the struggle for mates who would transmit their winning advantage to their descendants.

What the notebooks show is that over the next few months, prior to his reading of Malthus, as he attempted to bring together what he understood of the complex laws of generation around the central organizing principle of the analogy between individual and species development, Charles followed up his initial formulation of the improving effects of the sexual struggle with a scatter of entries that sought to relate male combat to reproductive success. In this critical period, it was the struggle for mates rather than the struggle for existence that preoccupied the younger Darwin.

Just two weeks before he read Malthus, he gave Erasmus's poetic allusions to combative knights and admiring ladies a more robust, contemporary interpretation, substituting primitive "savages" for courtly jousters.[84] On the same day, in the midst of a lengthy note on the plumage and habits of domestic and wild birds, he noted that fanciers matched their male birds against one another to find the best songster and that one bird was known to have sung itself to death in such a contest.[85] A few days later Darwin mulled over the "superior strength of make" in males, particularly in those structures directly employed in fighting, such as the thigh of the fighting cock and the neck of the bull. He again noted that "in [other] birds singing of cocks settle point" and wondered (in a singular departure from the conventions of mating behavior) whether "females then fight for male" or "are merely most attracted" to the best singer. "Singing," he went on, was the "best sign of most vigorous males," while "other birds display beauty of plumage." It was "most strange" that cocks and hens were either alike, such as guinea fowl, or very different, like "Peacocks!!" There was, reaffirmed Darwin, a strong correlation with the stages of development or maturation. Where there was a pronounced sexual difference, the females most resembled the young, while the male, having passed through this juvenile developmental stage, takes on his characteristic adult plumage and "recedes from the species."[86]

A week later Darwin, in the course of a long entry (much indebted to *Zoonomia*) on animal and human expressions of sexual arousal, pleasure, shame, fear, and anger, wrote: "Jealousy probably originally entirely sexual; first try to attract female, (or object of attachment) & then failing to drive away rival."[87] This promising speculation—which seems tantalizingly to hint at the integration of the notion of male combat with that of female preference in explaining the origins of the "primitive emotion" of jealousy—was overtaken (certainly for historians in their readings of Darwin's notebooks) by the crucial reading of Malthus and Darwin's first notebook entry, on September 28, 1838, on what was to become natural selection. Darwin's reading of Malthus's providential reproductive law gave him the explanation he was seeking for the etiology of both individual and species generation, even to the minute changes Darwin insisted on. Malthusian superfecundity was the irresistible force that powered the adaptation of structure to changing conditions. It was a force "like a hundred thousand wedges" whose "final cause" was "to sort out proper structure & adapt it to change."[88]

The legend of a single decisive "Malthusian moment" in Darwin's construction of natural selection has been thoroughly discredited. Historians now agree that the theory of natural selection took many months to emerge and that Darwin's new Malthusian insights were only slowly integrated with his earlier views on adaptation, inheritance, and embryogenesis. A key stage in this process of theory building and transformation was Darwin's incorporation of the analogy between nature's selective breeding of species and the artificial breeding of domestic varieties into his very conception of natural selection.[89] Similarly, I argue, it took Darwin even longer to sort out and work through the differences between the struggle for mates and the struggle for existence, to understand and develop the implications of his new theory of natural selection for what was to become sexual selection. While this long process of theory construction, integration, and demarcation was under way, Darwin continued to make notes on mating behavior and birdsong, on sexual dimorphism (particularly in birds) and its correlation with development or maturation, and on beauty.[90] And just as natural selection was reconstituted by the breeding analogy, so too was sexual selection, especially that aspect of it that Darwin came to understand as female choice. These developments will be detailed in the chapters that follow. Enough here has been said to emphasize how closely tied Darwin's account of male combat was to that of Erasmus Darwin, how it emerged from those notebook observations and speculations that were themselves largely derived from *Zoonomia*, and how male combat suffused the pre-Malthusian form of what was to become natural selection.

Nevertheless, there was an important difference in Charles Darwin's early interpretation of the sexual struggle, and that was his emphasis on both "war" and "charms." From the beginning of his notes on sexual combat, the younger Darwin had stressed an element of sexual preference that was subordinate to the male sexual struggle: females, animal and human, have "passion" or "love" for males successful in combat. A similar element of subordinate preference was present in Erasmus Darwin's conception of male combat. But by late 1838, as told, Charles Darwin began tentatively to extend mating success through combat to song and to beauty. By 1842 this had become the definitive statement that males compete for females by means of combat or charms.[91] Nevertheless, there was still no overt recognition of any element of female preference in this early formulation of sexual selection. It is there only by default. It has to be read into Darwin's formulation. It is a significant omission, one that relates to Charles Darwin's views on beauty— views that also drew inspiration from the grandparental precedent and were absorbed into the grandson's theorizing.

First, there was Erasmus's conviction that the sense of beauty was a conditioned or habituated response, an acquired characteristic open to investigation like any other; that it was not some mysterious, God-given, innate ability, as contemporary theology (and much aesthetic theory) would have it. Erasmus's theory of habituated aesthetics bridged the gap between mind and those physical forms deemed beautiful, signifying an interaction between the two.[92] This psychological/physiological interpretation freed up his grandson to investigate and integrate the aesthetic sense into his own developing evolutionary scheme. It paved the way for Charles Darwin's maturing view that beauty lay not so much in the eye of the beholder as in the evolving mind's eye.

Initially, like Erasmus, Charles denied a sense of beauty to animals, but by late 1838 his thoroughgoing evolutionism was pushing him toward the view that animals must share a sense of the beautiful with humans. His sporadic attempts to explain "superabundant" plumage in birds and human racial differences were critical to this process. He arrived at an explanation of beauty that took him beyond the restrictions of his grandfather's breast-centered aesthetics, one that enabled him to explain both beauty of plumage and the different skin colors and facial characteristics of the human races. The grandson was (theoretically, anyway) always more interested in faces and feathers than in breasts. Nonetheless, his interpretation still owed a good deal to his grandfather's conception of a conditioned sense of the beautiful.

Second, there is Erasmus Darwin's "ontogenetic paradigm"[93] of the baby at its mother's breast who recapitulates the evolutionary development of the human aesthetic sense, from smell, through taste and touch, to sight, which is the highest and last developed in the phylogenetic or genealogical sequence. This sequential, developing sensory aesthetics informed Charles Darwin's understanding of the evolutionary development of the human aesthetic sense from rudimentary animalistic beginnings and the relation of that sequence to the central metaphor of gestation. It underlay the grandson's recognition of the roles of smell, taste, and touch in the aesthetic experience that humans, especially "savages," share with animals.

Erasmus Darwin's aesthetics is, nonetheless, a visually oriented aesthetics that gives preeminence to sight in the perception of the beautiful, a view also adopted by Charles. This not only meshed with Charles Darwin's views on the relation of sexual selection to embryology or maturation, but was fundamental to his explanation of aesthetic choice, which he extended to animals. In order for there to be sexual selection by aesthetic choice, animals as well as humans must be able to recognize prospective mates and perceive differences among them. This applied particularly to gradations in color and shading and to the subtle, minute differences in pattern or shape that the younger Darwin was to make so much of. The exercise of aesthetic choice implied a well-developed visual aesthetic sense and judgment even in birds, a sense that Charles might compare with that of the "lowest savages" and trace to its highest evolved form in a "cultivated man."[94]

Next, there is Erasmus Darwin's rejection of the conventional relation between beauty and utility. His aesthetics, like Burke's, did not include fitness or utility in the analysis of taste.[95] This was to prove of great significance in the development of Charles Darwin's understanding of beauty and in his discrimination of the struggle for existence from the struggle for mates, and is discussed in the next chapter (see sec. 4.6).

Finally, like Erasmus, Charles Darwin consistently linked the aesthetic sense with the sexual drive and reproduction, and with the emotions. This interpretation was reinforced by Charles's independent reading of Burke and other aestheticians. The concept of female choice was something Charles Darwin came to from other directions. Yet fundamental to this notion was the essential Erasmus-derived relation of beauty to sexuality and the mating urge. And, aesthetics for Charles was as strongly gendered as for his grandparent. When it came to humans, aesthetic choice was exercised primarily by men, and women were the object of that choice.

This gendered aesthetic presented a considerable stumbling block to any notion of female aesthetic preference as a formative force in evolution. In the end, Charles Darwin was to resolve the difficulty in a way that left human evolution and racial divergence subject to this masculinized aesthetic tradition. Even among animals, where female aesthetic preference held sway, it was males who actively sought, battled, displayed, and mated. Along with this, Charles Darwin promoted a view of generation that was as male centered as his grandparent's, with males being more highly developed than females and transmitting their constitutional masculine superiority—strength, intelligence, weapons, or beauty—primarily to their male offspring, or to their grandchildren, as Charles tellingly specified in the notebooks.[96] His hypothesis of pangenesis (first formulated around 1841 but only published in 1868), that all the diverse phenomena of generation could be explained through the union of "gemmules" (minute elements contributed by all the parts or organs of both parents and presuming the predominance of the male elements), relates back to the gendered reproductive particles theorized by Erasmus.[97] And, in its turn, as we shall see, pangenesis was brought to bear on Charles Darwin's mature concept of sexual selection in *The Descent of Man*.

Charles Darwin's was a "gentlemanly generation"[98] that merged with the social and sexual relations of Erasmus Darwin's evolutionary biology and naturalized aesthetics. These were as familiar and acceptable to his grandson as we might expect them to be for one who came to maturity in a family of high social standing and wealth, who belonged to the ranks of the landed gentry, and who drew on that same land-based wealth and privilege throughout his career as the last of the great gentlemen naturalists. There is much in Jonathan Hodge's insistence on the continuities between the agrarian-based gentlemanly and aristocratic capitalisms that emerged in the mid-eighteenth century and the commercial, countrified, propertied capitalism that backgrounded Charles Darwin's notebook science of the 1830s. Within these economic and social continuities that underlay the veneer of industrialization, his "radical" grandsire's memory and writings were enshrined and conventionalized in family history, a prominent component of Charles Darwin's intellectual and cultural heritage.[99] His biographical memoir of Erasmus was his descendant's belated attempt to restore to his ancestor the social and intellectual stature that he saw as due to them both: the stature that had been compromised by Seward's malice and by the conservative backlash against his grandparent's unorthodox views—views that, in so many ways, were in accordance with his own.

Yet, for all their theoretical commonalities, their shared themes and sociocultural attitudes, it is ultimately misleading to collapse Charles Darwin's world and ideas into those of Erasmus Darwin. There *were* important differences, some of which I have touched on, in their outlooks and in the worlds they inhabited—differences that were to be crucial to Charles Darwin's evolutionism. Nothing brings out these differences so starkly as the younger Darwin's firsthand, *Beagle*-based experiences of colonial expansionism and his encounters with "savage" races. It was such frontier encounters and conflicts that were beginning to erode the ideals inscribed in the trope of "brotherhood" on Wedgwood's famous medallion. It was the grandson's encounter with the "hideous" Fuegian other that challenged his hitherto unquestioned Wedgwood-Darwin ideological legacy, and was to have profound implications for his theory of sexual selection. It triggered his early-notebook conclusion that the different human races have different inborn, heritable standards of beauty, or aesthetic "taste"—the basis of his theory of the role of aesthetic preference in the formation of the different human races, a theory that he then extended to the whole animal kingdom. We must look beyond the writings of Erasmus, as did Charles Darwin, if we are to analyze this important aesthetic shift in his early-notebook constructions of sexual selection.

FOUR

Beauty Cuts the Knot

Beauty is instinctive feeling, & thus cuts the Knot:—Sir J. Reynolds explanation may perhaps account for our acquiring <<the instinct>> our notion of beauty & negroes another; but it does not explain the feeling in any one man.

—Darwin's notes, July 1838

On July 23, 1834, Darwin, having plumbed the depths of civilization in Tierra del Fuego, reached the bright lights of Valparaiso—"a sort of London or Paris, to any place we have been to." There he wrote to his old Cambridge crony, Charles Whitley. Whitley was an "out & out Tory" to Darwin's Whig, a serious student who had taken Darwin to task for his obsessive shooting and hunting and his scandalous neglect of his formal studies. He also attended to Darwin's aesthetic education.[1]

Our conventional view of Darwin—one he promoted—is of a man with few pretensions to the appreciation of the higher arts. But in his younger days, poetry, painting, and music gave him "great delight." It was Whitley who "inoculated" Darwin "with a taste for pictures and good engravings." On visits to the Fitzwilliam collections in Cambridge, Whitley taught Darwin "how to look," how to find meaning below the painted surface, to search for allusions, and to understand technique. This early introduction to the visual arts had a more significant and enduring effect on Darwin's evolutionary theorizing than has been appreciated.[2]

Darwin had last written to Whitley before leaving England on the *Beagle* at the end of 1831. He then told Whitley of his purchase of a rifle and a brace of pistols in expectation of "plenty of fighting with those d—— cannibals."[3] Almost three years later, the *Boy's Own* gloss Darwin had put on the voyage had completely worn away: he was exhausted by protracted travel with no end in view, thousands of miles from familiar places, family, and friends, he was dazzled and dazed by New World vistas and environments, struggling to comprehend their fragmentary histories, their ancient and still-living plants and animals, and their exotic human inhabitants; still reeling from his encounters with supposed real-life "cannibals" in Tierra del Fuego. Valparaiso gave him a semblance of accustomed civilization and culture, arousing a fresh sense of deprivation and loss. His overdue letter to his old friend is an extraordinary sequence of vignettes whose juxtapositions are as revealing as the images themselves.

Darwin begins with a playful reminder of the existence of a "certain hunter of beetles & pounder of rocks." This banter gives way to romantic yearning for a lost life and a lost world, for the snug domesticity he imagines Whitley to be enjoying and that Darwin once imagined for himself—a vision of "green cottages and white petticoats." These nostalgic indulgences are interrupted by visual jolts from the New World, images evidently at the forefront of Darwin's mind:

> We have seen much fine scenery, that of the Tropics in its glory & luxuriance, exceeds even the language of Humboldt to describe. A Persian writer could alone do justice to it, & if he succeeded he would in England, be called the "grandfather of all liars."—
>
> But, I have seen nothing, which more completely astonished me, than the first sight of a Savage; It was a naked Fuegian his long hair blowing about, his face besmeared with paint. There is in their countenances, an expression, which I believe to those who have not seen it, must be inconce[i]vably wild. Standing on a rock he uttered tones & made gesticulations than which, the crys [sic] of domestic animals are far more intelligible.
>
> When I return to England, you must take me in hand with respect to the fine arts. . . . How delightful it will be again to see in the FitzWilliam, Titian's Venus; how much more than delightful to go to some good concert or fine opera. These recollections will not do. I shall not be able tomorrow to pick out the entrails of some small animal, with half my usual gusto.[4]

His letter to Whitley, written in a state of heightened emotional intensity, triggered a rush of associations that give a very early indication of the significance of notions of the ugly and the beautiful to Darwin and of the connections he then, still on the *Beagle*, only partly perceived, but that were to inform his later writings on aesthetic choice and sexual selection.

4.1 "I formerly thought a good deal on the subject"

Soon after the publication of the *Origin*, his evolutionary interpretation of beauty under attack, Darwin told Charles Lyell that he had "formerly thought a good deal on the subject & was led quite to repudiate the doctrine of beauty being created *for beauty's sake*."[5] This for Darwin was the crux of the matter. His views on aesthetics had been formulated some thirty years earlier, and he resisted further education on the beautiful. We may track the formation of these views through the record of his early notes and readings. And we may note how little those views changed over the course of Darwin's long working life.

It cannot be overemphasized that Darwin came to the notion of aesthetic selection and was prompted to his study of aesthetics primarily from his consideration of the differences among the human races. His very earliest notebook references are to the sexual repugnance of unlike creatures for one another that would prevent their interbreeding. Considering the isolating mechanisms that might keep them separate and so prevent backcrossing, Darwin noted that "wild men do not cross readily" and that the "distinctness of tribes in T. del Fuego" was a case in point. A little later this had become: "There is in nature a real repulsion amounting to impossibility [of interbreeding between different forms] . . . we see it even in men; thus possibility of Caffers [Kaffirs] & Hottentots coexisting proves this." The "strong odour of negroes" was "a point of real repugnance." At this time, Darwin was even prepared to step outside religious and abolitionist prohibitions and theorize the black and white races as separate "species" ("for species they certainly are according to all common language") that would "keep to their type." Nevertheless, some interbreeding did occur among the human races. This was because, Darwin rationalized, man, being a creature of reason rather than instinct, could overcome this instinctive repugnance against interbreeding and choose a desirable sexual partner from another race— just as, perhaps, Darwin had himself fancied the beautiful young women among the doomed Pampas Indians. Man had "no limits to desire, in proportion instinct more, reason less, so will aversion be." Following this train of thought, he made the link between the perception of beauty and such interracial matings: "Animals have no notions of beauty, therefore instinctive feelings against other species [for sexual ends], whereas Man has such instincts very little."[6]

Within the year, Darwin was certain that animals too shared a sense of beauty with humans. At the same time, his notes began to turn from the emphasis on selective sexual repugnance to its converse. His "sharp lookout" for a "nice soft wife on a sofa" most evidently turned his thoughts to selective sexual attraction.[7] From the beginning of his notes on transmutation, Darwin had viewed sexual reproduction as the key to the process. He now added observations on the sexual behavior of apes and of himself[8] and began to pester breeders for information on inherited variations and mating preferences in domestic animals. He also read his way into theories of the beautiful, or aesthetics.

Darwin's interest in aesthetics was initially concerned with demonstrating the continuity (and therefore the evolutionary development) from animals to humans of this highest human faculty—by reducing it, like other mental attributes such as the moral sense, to the exercise of inherited habit. This was one of his early notebook explanations, derived from the *Zoonomia* of his grandfather and reinforced by his reading of Jean-Baptiste Lamarck, of the mechanism of adaptation or species change. The habits an animal might adopt to cope with the shifting environment would, over many generations, slowly become instincts, or innately determined patterns of behavior. These would in turn gradually modify animal structure, adapting it to its changing circumstances. By theorizing that habits first become instincts that are unconscious or automatic, Darwin could dissociate his theory from that of Lamarck, which he (mis)interpreted as relying on the conscious willing of an organism to adaptive change. Over time and many generations, habits would modify brain structure, and this would predispose the development of more complex social instincts. Through the combined interaction of an evolving brain, the habitual purposeful pursuit of mating, parental, and social activities, together with changing environment (or, in the case of humans, advancing civilization), such instincts would become reason and, finally, moral judgment or conscience. This remained Darwin's explanation of the evolution of mind and behavior for some time after his formulation of the principle of natural selection, and some aspects of this argument he never abandoned.[9] It satisfied his requirements that instincts were not providentially implanted in the mind, that intelligence and the moral sense were not God-given but had evolved from instinct, and that animal intelligence therefore differs from human in degree but not in kind.

In his readings Darwin was clearly looking for explanations of beauty compatible with these early constructions of mind and the moral sense. The subsidiary theme of these aesthetic researches was Darwin's hunt for evidence or arguments relating to the differing standards of racial beauty that he might apply to the problem of human racial origins (a concept with which he was already well acquainted; see ch. 5). For Darwin, these two aspects of the beautiful were necessarily interconnected.

Darwin's Fuegian experiences were at the forefront of his mind from the beginning of his researches. In the opening pages of his very first notebook on transmutation, Darwin wrote that we "know" that "in course of generations even mind & instinct become influenced—child of savage not civilized man." Jemmy, for all his European training and exposure, had reverted to type, to naked, squalid savagery, once he was back in his "natural" environment. Perhaps there was "unknown difficulty with *full grown* individual with fixed organization thus being modified,—therefore generation to adapt & alter the race to *changing* world."[10] Civilization was not something that could be impressed on the savage mind or constitution in the course of one lifetime. The adult organization resisted remolding. As with the domestication of animals, it was not a matter of training, but of breeding over many generations. The habits of civilization had to be *bred* into the race.

In his *Beagle* diary, Darwin had accounted for Jemmy's recidivism and the brutish appearance and lives of the Fuegians with the reflection that "omnipotent" habit had "fitted the Fuegian to the climate and productions" of his country. This was a common assumption among early naturalists and moral philosophers concerned with the problem of race formation from a common origin. Generally speaking, they relied on environmentalist explanations that attributed racial peculiarities to the influence of climate or to diet and mode of life and that assumed the transmissibility of acquired characters. The American Indians, with their relatively small racial variation from Canada to Tierra del Fuego, presented difficulties with respect to climatic influences that varied from the extremes of the sub-Arctic and the Andes to the tropics. Their smaller differences were more readily attributable to the effects of modes of life or habit, rather than to climate.[11]

An explanation along these lines was carried by Darwin on the *Beagle*: his much-thumbed copy of Alexander von Humboldt's *Personal Narrative of Travels to the Equinoctial Regions of the New Continent during the Years 1799–1804*. This was the work that had inspired him with the desire to travel to exotic lands. Darwin's *Beagle* diary resonates with a Humboldtian romantic view of nature, both as an aesthetic experience (a "chaos of delight") and a comprehensive, holistic conception of the interrelatedness of all phenomena: "[Humboldt] like another Sun illumines everything I behold." The diary's more emotive evocations of the South American landscapes are emulations of Humboldt's romantic style, discerned and discouraged by Darwin's sharp-eyed sister Caroline: Humboldt's "flowery [F]rench expressions" were all right for him, "his being a foreigner," but Darwin should stick with the "simple straightforward & far more agreeable [English] style."[12]

Style apart, in more specific ways the *Beagle* diary and Darwin's notebooks and marginalia show the influence of the foreigner. Darwin noted Humboldt's writings on the intellectual similarities between humans and apes and on savage ornamentation, and echoed his condemnations of slavery as well as the rhetoric that brought animals as "brethren" into the abolitionist schema of suffering, enslaved beings.[13] Above all, he absorbed Humboldt's more general perceptions of the relation between states of civilization and their links with the landscape and modes of existence. Humboldt's *Personal Narrative* was a model for viewing and describing the tropics that influenced naturalists and artists for decades. After Humboldt, "seeing the tropics meant looking at nature through the proper affective, scientific and colonial lenses."[14]

Darwin's letter to Whitley invokes Humboldtian aesthetic sensibilities in his appreciation of tropical verdure (which, however, really requires more erotic, oriental treatment—recalling Darwin's earlier libidinous diary image of himself as the "Sultan in a Seraglio . . . grown quite hardened to beauty"). Yet, with its very next compelling image, his letter signals a turn from his enthrallment with the Humboldtian tropical aesthetic to a new fascination, his "first sight of a Savage," which, paradoxically, he connects with a growing nostalgia for home, with English ways of seeing and hearing. His evocation of the figure of the naked, painted, "inconce[i]vably wild" solitary figure uttering his meaningless cries gives way to Darwin's looking to Whitley for further education in the fine arts, culminating in his desire to see again the sumptuous image of the goddess of love in Titian's painting in the Cambridge Fitzwilliam and to hear some good music.[15]

Tierra del Fuego marks the cusp where the *Beagle* turns to begin the Pacific leg of its voyaging, where Darwin will discover many "little embryo Englands" and congratulates himself on being "born an Englishman."[16] It is the site of the disjuncture with the Humboldtian aesthetic in Darwin's aesthetics.

Where the tropics inspired blissful beauty along with the temptations of surfeit, Tierra del Fuego excites in Darwin only the terrible, unendurable sublime. Its "death-like" impenetrable *Nothofagus* entanglements, its awesome glaziers and desolate peaks, are in awful contrast with the fertile excess of Humboldt's tropical aesthetic. For Darwin thereafter, beauty would not only signify complexity but also be held in tension with the ever-present danger and threat of the sublime.[17] As for the Fuegians themselves, they challenge all Darwin's assumptions of racial unity, all notions of Christian brotherhood or the holistic interrelatedness of his Humboldtian model. Darwin's "savage" encounter delivers a sensory shock, an assault on the eyes and ears that takes him outside the eloquent Humboldtian aesthetic and leaves him floundering for a means to express it.

The unaesthetic, besmeared, naked Fuegian exploded all aesthetic conventions for Darwin. The Fuegian's sheer otherness, his astounding, confronting difference, defies painted representation of the kind Darwin and his readers were accustomed to in the artistic rendering of other races. Here is no noble savage, but, as Darwin fumblingly begins to see it, a primal, morally and physically repugnant, bestial being, a putative ancestor who is the antithesis of everything that he, a cultured Englishman, holds best and most beautiful in the appreciation of the human body and mind. The aesthetic, racial, and sexual dimensions of this perception are made patent in the leap in his letter to Whitley from the image of uncouth, unbearable Fuegian barbarism to a longing for the familiar civilized culture of home and Cambridge, and especially for the sight of the beautiful naked *Venus*. The *Venus* is simultaneously the highest aesthetic and the most desirable and erotic expression of the naked female body in the canon of Western art that the young men have jointly gazed upon. It is a culturally sanctioned but sexually laden image, and its appearance and positioning in his letter to Whitley are weighted with meaning.

Put bluntly, Darwin's views on beauty owe far less to Humboldtian romanticism than to the very British empiricist aesthetic tradition that consolidated toward the close of the eighteenth century, a tradition exemplified by the views of his grandfather Erasmus and closely allied with the "wholly non-romantic theory of English political economy" and to the middle-class "muses" of science and technology.[18] Darwin's juxtaposition of the ugly Fuegian with Titian's beautiful *Venus and Cupid with a Lute-Player* is directly traceable to that exemplary English arbiter of taste, the premier portraitist and academician Sir Joshua Reynolds. Reynolds's *Discourses*—which Darwin, even in old age, recalled reading "with much interest" while still a student at Cambridge—were reread with great thoroughness during the notebook years and brought directly to bear on his Fuegian experiences and the related issue of racial perceptions of beauty. During this critical period, Darwin read his way through a representative collection of leading eighteenth-century British writers on aesthetics, including, besides Reynolds, David Hume, Edmund Burke, Dugald Stewart, and Archibald Alison (through the highly influential interpretation of Alison's work by Francis Jeffrey).[19]

By this time, Charles Whitley had gone from his life. Darwin, then in London, might look closer to home for aesthetic guidance. His "dear good old brother" Erasmus, the family aesthete, too indolent or enervated to work or to marry, was the obvious choice. Erasmus's coterie of heterodox intellectuals, writers, actors, and critics, with their largely Whig utilitarian concerns and attitudes, their participation in literary and artistic production,

Figure 4.1. The Titian *Venus—Venus and Cupid with a Lute-Player, 1560*—displayed in the Fitzwilliam Museum, Cambridge, so vividly recalled by Darwin during the voyage of the *Beagle*. © The Fitzwilliam Museum, Cambridge.

their preoccupation with matters of morality and religious criticism, their interest in scientific and social issues, represented aspects of an aesthetics that was not confined to discourses on art or "culture," but articulated with a wider range of social, political, and ethical discourses that relate back to the middle-class struggle for political power.

4.2 Portraits and politics

Aesthetics—or the "science of sensuous cognition," as it developed in the eighteenth century—had wider connotations than any modern understanding of "beauty" in art or nature, being assumed to embody fundamental issues of morality and human feelings, or "sensibility." Its central concern was the human body, both as perceiver and perceived. The perceived beauty or ugliness of a man or woman was closely linked with notions of desirability and reproduction, as well as with moral assessments that generally equated beauty with goodness, and ugliness, or "deformity," with depravity or evil. Aesthetics also had close historical connections with the development of liberalism, with the turn from autocratic and monarchical regimes toward liberal and republican ones. Aesthetic judgment, as its theoreticians interpreted it, was to produce men [*sic*] who enacted their freedom in a moral and responsible manner, who were unified by individual consent and judgment rather than by constraint and subordination, into a new political "community of taste." All social life was to be aestheticized. Those who spoke of art were speaking of these other matters too. Nor was its definition or discussion confined to aestheticians—just about every thinker of note engaged with aesthetics.[20]

Sir Joshua Reynolds, founder and first president of the Royal Academy of Arts, was among the most prominent of "civic art" theorists. He was also a professional artist, keen to exhibit art to the best commercial and cultural advantage in a newly wealthy (but not as yet powerful) urban, upper-middle-class society. His widely read and respected *Discourses on Art* promoted a "properly civic and republican" theory of art, a theory that was by intention egalitarian (within the limits of civic discourse) but was both "self-consciously political and unashamedly masculine." Taste, or the appreciation and judgment of beauty, was for Reynolds a manly endeavor. It required "long laborious comparison" and constant vigilance lest taste be polluted by the caprices of fashion, affectation, or the false refinements of an effeminate politeness or softness. The man of taste was a man of breeding. He cultivated a masculine aesthetic that ennobled and elevated, that promoted worthy action and civic virtue. The assumption was that women might embody the aesthetic, as in Reynolds's portraits of society beauties, but not define it.[21]

The *Discourses* were based on the Humean empiricist dictum that beauty is "no quality in things themselves," but "exists merely in the mind which contemplates them; and each mind perceives a different beauty."[22] Nevertheless, Reynolds (like most eighteenth-century aestheticians) was concerned with demonstrating the reality of a standard of taste that was not intuitive but was experientially based and might be rationally disputed. Taste was the "power of distinguishing right from wrong" in art, and relied on reason, learning, and common sense. It entailed a special way of looking: it depended on "our skill in selecting, and our care in digesting, methodizing, and comparing our observations." In effect, the connoisseur deployed the skills and methodology of the naturalist. The best judge of beauty was one who extended his judgment to all ages and nations. He would best appreciate those works of art that had stood the test of time and fashion. These timeless exemplars, or "authentic models"—such as the classical *Venus de' Medici* and the *Apollo Belvedere* or the productions of the Italian masters (such as, indeed, the Titian *Venus*)—were "proportioned" and "accommodated" to something constant and enduring in human nature. Universal exemplars, authenticated in this way, were empirical proof of a single, universal, true standard of taste. They were representations of "general nature" or "ideal beauty," or, in the phrase consistently used by Darwin, the "beau ideal."[23]

If beauty was understood as general or universal nature (which accorded with the artistic canon ratified by the polite "conversable world" inhabited by Hume and Reynolds), its opposite was the particular, the temporary, or the local, which Reynolds characterized as "deformity" or "blemish": "Those who have cultivated their taste can distinguish what is beautiful or deformed, or, in other words, what agrees with or deviates from the general idea of nature."[24] Nevertheless, through his attention to deviant or culturally determined perceptions of beauty, Reynolds went beyond most of his contemporaries toward a relativist, or "customary," notion of beauty.

In the seventh discourse, which Darwin singled out for special attention,[25] Reynolds discriminated between the objective "primary," or "real truths," of universal nature and the "secondary," or "apparent truths," of culturally conditioned beauty, or "fashion." As these secondary truths often had a long history and were widely disseminated, they approached "nearer to certainty, and to a sort of resemblance to real science," and therefore must not be trivialized. Reynolds thus allowed a degree of subjective response into aesthetic theory—a line pushed by associationists and taken up variously by Burke, Erasmus Darwin, and Archibald Alison. There is a resulting ambivalence in Reynolds's aesthetics, a tension between universal and relativist conceptions of beauty, between "natural" beauty (the "immutable verity") and "custom" ("local and temporary prejudices, fancies, fashions, or accidental connections of ideas"). In this discourse, Reynolds gave a famous instance of a powdered, bewigged European and a face-painted Cherokee Indian in illustration of the difficulty of determining "which of the different customs of different ages or countries we ought to give the preference, since they seem to be equally removed from nature": "Whoever of these two despises the other for this attention to the fashion of his country, whichever first feels himself provoked to laugh, is the barbarian."[26]

However, in an earlier essay, published in the *Idler* in 1759 (usually republished in conjunction with the final edition of the *Discourses*), Reynolds had gone further toward a relativist notion of beauty with an example that was even more compelling for Darwin. Stressing the need for prior experience in order to evolve adequate criteria for the discrimination of beauty from "deformity," Reynolds there argued that for those who "were more used to deformity than beauty, deformity would then lose the idea now annexed to it, and take that of beauty." "Habit and custom," he cautioned, might make "white black, and black white."

Figure 4.2. *The Voyage of the Sable Venus, from Angola to the West Indies*, 1793. The black Botticellian Venus rides her seashell chariot with Neptune as standard-bearer for the Union Jack. The image of an alluring black beauty who enslaved white men was a recurrent fantasy of those engaged in the slave trade and plantation culture of the West Indies. Edwards 1793, 2:27. University of Sydney Library, Rare Books and Special Collections.

It is custom alone determines our preference for the colour of the Europeans to the Ethiopians, and they, for the same reason, prefer their own colour to ours. I suppose nobody will doubt, if one of their Painters were to paint the Goddess of Beauty, but that he would represent her black, with thick lips, flat nose, and woolly hair; and it seems to me, he would act very unnaturally if he did not. for by what criterion will any one dispute the propriety of his idea! We, indeed, say, that the form and colour of the European is preferable to that of the Ethiopian; but I know of no reason we have for it, but that we are more accustomed to it.[27]

Here we have a closer fit to Darwin's leap in his letter to Whitley from the painted savage to the Titian *Venus*. There is an absolute fit to his notebook references to the different racial standards of beauty. In Darwin's words, after reading Reynolds: "we" (white Europeans) have "our notion of beauty & negroes another."[28]

4.3 Aesthetics and race

Reynolds was not alone among eighteenth-century aestheticians in relating judgments about beauty to custom or fashion and considering non-European notions of beauty—especially the black viewpoint—from this perspective. Those who inserted a black perspective included William Hogarth and Lord Kames (Henry Home). This was more than a mere academic exercise. A black notion of beauty challenged universalist aesthetic assumptions; but it also carried a political charge. Recognition of the validity of black beauty was bound up with recognition of the humanity of Africans and opposition to the slave trade. The moral or antislavery dimension of the aesthetic issue was made most explicit by those vehement defenders of slavery who alleged the deformity or sheer ugliness of the African, relating this to black inferiority in intelligence and taste.[29]

However, matters were not as clear-cut as this might suggest. Kames was a polygenist who favored the separate creation of the black and white races; while Hogarth, whose empathetic portrayals of black slaves and servants peopled his canvases and prints, ultimately rejected aesthetic relativism for his well-known universalist "line of beauty," a standpoint endorsed by the pro-abolitionist Erasmus Darwin.

In part, Reynolds's 1759 essay was a rebuke to Edmund Burke, whose *Philosophical Inquiry into the Origin of Our Ideas of the Sublime and Beautiful* (1757) was reread by Darwin around the same time as he was rereading Reynolds. Burke's *Inquiry* points up some of the contradictions in a simplistic association of aesthetic relativism with an abolitionist agenda. Burke famously argued that darkness, obscurity, and indistinctness were emblematic of the sublime and the most powerful of aesthetic emotions. In promoting this, he claimed that the fear of blackness was innate, not learned, instancing the well-known account of a boy born blind who, upon recovering his sight, was "struck with great horror" at his first sight of a "negro woman." In other words, for Burke, an outspoken abolitionist, the African generates the sublime through the innate horror engendered in the mind of the European by the very appearance of blackness.[30]

Reynolds sought to counter Burke's interpretation by invoking the link between experience and aesthetic perception, by structuring the perception of beauty as a learned response, and by stressing the relative nature of perception: blackness is not innately horrifying but can even appear beautiful to those accustomed to perceiving it. However, by the time Reynolds wrote his seventh discourse, Burke's interpretation had undergone a significant aesthetic shift. This is traceable to Gotthold Lessing's influential *Laocoon* (also read by Darwin on his brother's recommendation), where Lessing introduced the figure of the "filthy Hottentot" as representative of the interrelated ugly and the disgusting. In Lessing's "aesthetic of ugliness," the figure of the black African is no longer innately terrifying as Burke supposes, but simply disgusting; and this inherent disgustingness, when it is made the object of sexual passion and love, can only generate laughter in the racially and culturally superior European. So Reynolds, in his seventh discourse, upholding his earlier view that standards of beauty were relative to culture or habit, displaced Lessing's laughter with irony in his instance of the bewigged European who meets the painted Cherokee: whichever laughs first is the "barbarian."[31]

Darwin did, on occasion, laugh at the Fuegians—telling of one who responded to the interest of the *Beagle* crew by preening and showing himself to advantage ("for the rest of his days doubtless he will be the beau ideal of his tribe"), and of a "naked beauty" who bartered for bits of red rag, which she fastened around her black-painted face.[32] There is something in Darwin's first sighting of a "savage" of the Burkean symbolism of wild, dark, terrifying nature that he attempted to communicate to Whitley through his juxtaposition of the antithetical image of the most civilized conception of beauty he could call to mind. But there is more of the category of the disgusting (forcefully expressed in his other written references to indigenous Fuegians), a category absent from Burke's and Reynolds's aesthetics. Despite Darwin's later promotion of a relativistic notion of beauty as crucial to his concept of sexual selection and the evolution of racial difference, he radiated disgust for the "hideous barbarian" akin to Lessing's stereotyped image of the brutish Hottentot with his bestial customs. Darwin, at best, like Reynolds, adopted a position of modified relativism: the assumption of some type of privileged or universal aesthetic standard, based on European notions of the beautiful, remained central to his understanding of racial perceptions of beauty. In *The*

Figure 4.3. "Love and Beauty: Sartjee, the Hottentot Venus." One of many derisive depictions of white desire for the derriere of the "Hottentot Venus." Repository, Library of Congress Prints and Photographs Division, Washington, DC, digital ID cph.3c37332.

Descent of Man, he would claim: "It is certainly not true that there is in the mind of man any universal standard of beauty with respect to the human body."[33] Yet elsewhere in this same work, Darwin's perpetuation of conventional aesthetic categories and standards was made explicit:

> The taste for the beautiful, at least as far as female beauty is concerned, is not of a special nature in the human mind; for it differs widely in the different races of man. . . . Judging from the hideous ornaments and the equally hideous music admired by most savages, it might be urged that their aesthetic faculty was not so highly developed as in certain animals, for instance, in birds. Obviously no animal would be capable of admiring such scenes as the heavens at night, a beautiful landscape, or refined music; but such high tastes, depending as they do on culture and complex associations, are not enjoyed by barbarians or by uneducated persons.[34]

The pull toward relativist conceptions of beauty in Reynolds's writing did not, after all, lend itself to the notion that black aesthetics or beauty was the equal of European. This was made most explicit in Reynolds's thirteenth discourse, also closely read by Darwin. There, Reynolds reasserted the primacy of universal aesthetic values, making the point that great art was the province of the refined and civilized; the would-be egalitarian republic of taste that he envisaged excluded people of other races, along with all women, all children, and the vulgar, none of whom he deemed capable of the requisite intellectual labor of abstraction.[35]

In giving priority to culture or habit, Reynolds was following Hume's well-known essay "Of National Characters" in rejecting the more accepted attribution of climatic or geographic influences on concepts of "character" among the different nations or races.[36] Ultimately, for Reynolds as for Hume, civility and taste depended on the establishment of successful commerce. It was commerce that permitted the cultivated life in which the arts become "those arts by which Manufactures are embellished, and Science is refined." Aesthetic taste was thereby denied to common laborers and to those non-European nations or races that had not developed comparable political or commercial systems.[37] This was a judgment that Darwin, grandson and beneficiary of that leading manufacturer Josiah Wedgwood, shared. He was to state it unambiguously in the *Descent*: "Without the accumulation of capital the arts could not progress; and it is chiefly through their power that the civilized races have extended, and are now everywhere, extending their range, so as to take the place of the lower races."[38]

4.4 "Beauty is instinctive feeling, & thus cuts the Knot"

Reynolds's relation of habit to the differing racial perceptions of beauty could not have been more apposite to Darwin's purpose. Darwin thoroughly reread and abstracted the *Discourses* from around July to October 1838,[39] the period of his most intensive theory building that culminated with the reading of Thomas Malthus and the concept of natural selection.

Again, desolate Tierra del Fuego, where even the birds could not sing, was Darwin's jumping-off point, via Reynolds, to the very earliest notebook reference to the differing racial notions of beauty, written in July 1838:

> How strange so many birds singing in England, in Tierra del Fuego not one.—
> Now as we know birds learn from each other . . . Singing of birds, not being
> instinctive, is hereditary knowledge like that of man. . . . Beauty is instinc-
> tive feeling, & thus cuts the Knot:—Sir J. Reynolds explanation may perhaps
> account for our acquiring <<the *instinct*>> our notion of beauty & negroes
> another; but it does not explain the *feeling* in any one man.[40]

What Darwin at this stage wanted (against Reynolds, who held to the empiricist dictum that taste was learned and that we were born only with the capacity for cultivating it) was to make the sense of beauty—an instinct acquired through inherited habit—consistent with his early ideas on the evolution of reason and the moral sense. The birds of Tierra del Fuego could not sing because, unlike English birds, they had not learned to sing (there is an obvious parallel here with notions of savage deprivation of the benefits of education and civilization). If birdsong was not innate but could become hereditary through learning or habit (an idea earlier endorsed by Erasmus Darwin in *Zoonomia*), so might a particular sense of beauty, if habituated, become instinctive:[41] "Beauty is instinctive feeling, & thus cuts the knot." As Darwin now reinterpreted Reynolds, each race had its specific, habitual, inherited standard or ideal of beauty: white Europeans had their "notion of beauty & negroes another."

A few weeks later, Darwin followed up this insight with an entry on "national character" and its significance for his emerging views on the habituation of a distinctive racial aesthetic that was linked with intelligence and morality. Reflecting on his *Beagle* experiences, recent ethnological readings, and the Humean argument that climate could not explain racial or "national" differences, Darwin wrote:

> He who doubts about national character let him compare the American whether
> in the cold regions of the North,—the elevated table land of Peru the hot plains

of the Amazons & Brazil—with the Negroes of Africa . . . the American in Brazil is under the same conditions as Negro on the other side of the Atlantic. Why then is he so different—in organization. Same cause as colour & shape & ideosyncracy [*sic*].—Look at the Indian in slavery & look at the Negro—look at them both savage—look at them both semi-civilized.[42]

If the physiognomy and "character" of the different races could not be correlated with climate or environment, there had to be another "cause," perhaps related to stages of civilization and racial ideas of beauty, as at least one of the theorists he had been reading supposed (William Lawrence; see ch. 5).

In a sequence of notes on the *Discourses*, Darwin prodded Reynolds toward his own view ("my theory") that taste, or the judgment of beauty, like birdsong, was made instinctive through inherited habit.[43] He worried away at the connections among birdsong, the origin of language, the appreciation of music, and the development of the aesthetic sense. The "ignorant," on Reynolds's authority, had no aesthetic sense: "Hence pleasure in the beautiful. (distinct from sexual beauty) is acquired taste.—Whilst music extremely primitive.—almost like tastes of mouth & smell."[44]

Thirty years later, Darwin made a note, "Descent of Man," next to this old notebook entry. In the *Descent*, like his grandfather Erasmus before him, he argued that music (and poetry) had evolved from the sounds emitted by our semihuman progenitors during courtship and under the excitement of sexual passion. The tones and cries of the primitive savage—"to us hideous and unmeaning," as Darwin put it in the *Descent*—were finally transmuted into the "good concert or fine opera" that had thrilled the young Whitley and Darwin:

> The impassioned orator, bard, or musician, when with his varied tones and cadences he excites the strongest emotions in his hearers, little suspects that he uses the same means by which, at an extremely remote period, his half-human ancestors aroused each other's ardent passions, during their mutual courtship and rivalry.[45]

It was Darwin's memories of Tierra del Fuego with its songless birds and naked savages that still resonated when, a few weeks after reading Malthus, his naturalization of Reynolds's aesthetics reached its culmination in the extended notebook rumination of October 27, 1838. I have already touched on this significant passage (sec. 1.3). We are now in a better position to interpret this entry, and I give it in full:

Consult the VII discourse by Sir J. Reynolds.—Is our idea of beauty, that which we have been most generally accustomed to:—analogous case to my idea of conscience.—deduction from this would be that a mountaineer take[n] born out of country yet would love mountains, & a negro, similarly treated would think negress beautiful,—(male glow worm doubtless admires female. Showing. No connection with male figure)—As forms change, so must idea of beauty.—(Old Graecians living amongst naked figures, & observing powers common to savages???)—The existence of taste in human mind. Is to me clear evidence, of the general ideas of our ancestors being impressed on us.[46]

Here, Reynolds's views on the connection between habit and racial perceptions of beauty are given an explicit evolutionary interpretation and extended to the animal kingdom.

Darwin's language reflects his reading of Reynolds, but in his own recent experience, it was Jemmy the Fuegian—taken from his own country and reeducated, and who had, on his return, gone native with his native wife—who exemplified such habituated aesthetic preferences as Darwin now attributed to the different races. Note that Darwin initially went to write *"taken out of country"* [emphasis added], then crossed it out and substituted "born," thus clarifying his point that the aesthetic sense, once habituated to mountain scenery or to a different racial physiognomy, was inborn. Jemmy's adult organization was "fixed"; his mind and morality, his sense of beauty, correlated to his savage physiognomy. The various savage races Darwin had encountered across the Pacific, from Tierra del Fuego to Australia, reinforced the point. Physiognomy, mind, and aesthetic sense were interconnected. It was scarcely to be wondered at that Darwin, accustomed to a very different aesthetic, had instinctively recoiled from the savage.

Extrapolating from Reynolds's seventh discourse, Darwin wondered whether, just as the classic ideal of the naked form was reinforced by the exposure of "Old Graecians" to physically perfect near nakedness and its manifold marble representations, so savages were habituated to their particular sense of beauty through their unobstructed scrutiny of living, naked bodies. Reynolds had made the point that reformation of national "bad taste" was a work of time and could not be abruptly achieved by revolution.[47] It logically followed that only over many, many generations of reeducation and habituation to civilized standards, undertaken while the organization was still young and pliable, might the savage mind and aesthetic sense, and consequently mating preferences, be slowly altered.

This idea of the evolution of aesthetic judgment was, as Darwin claimed, analogous with his explanation of the evolution of moral judgment or conscience. He was by this stage beginning the long process of integrating his very recent Malthusian insight with his earlier views on moral behavior and inherited habit. He had not yet, however, moved beyond the claim that useful social habits of moral behavior—like those of parental nurture, social cooperation, and so on—would be practiced and sustained over the generations and inherited, a claim he could now extend by analogy, again via Reynolds, to the habit of perceiving and appreciating accustomed environments or physiognomies. The moral sense he might relate to utilitarian motives and ultimately to his new principle of natural selection.[48] He did not as yet have a coherent idea of aesthetic selection, nor of its relation to (or rather, distinction from) natural selection. This was to come only with his deepening understanding of domestication and artificial selection (see ch. 6). However, both habituated inherited practices—the moral and the aesthetic—could now be given a naturalistic, evolutionary explanation and dissociated from claims of their divine origin. The sense of beauty was not fixed or innate, as natural theology would have it, but had evolved over time through the inheritance of acquired notions of beauty that correlated with the sustained experience of viewing, or the habitual exposure to, particular environments or bodies and faces.

To a materialist, as Darwin was by now secretly calling himself, the idea that changes in bodily forms would bring about changes in aesthetics was compelling: "As forms change, so must idea of beauty." As the eye, brain, and mind evolved, so must aesthetic perception or judgment. Consistent with this, he extended the possibility to animals, even to the lowly male glowworm, who "doubtless admires" the wormlike, wingless, but glowing female—an instance of extreme sexual dimorphism—showing, Darwin intimated, that aesthetic sense was connected with the male mind but not with the male form. It was through the habitual masculine action of looking at and judging female forms that the aesthetic sense was developed and became instinctive. At this stage, Darwin's idea of aesthetic judgment or taste was as strongly gendered as Reynolds's (or, indeed, as his grandfather's). It was males, animal and human, cultured and savage, who exhibited aesthetic judgment in their admiration of females—females who were the objects of that admiration. Darwin's choice of the glowworm is indicative. It is one of the minority of instances in the animal kingdom in which the female outshines the male, where the male, being winged, seeks out the grounded, larviform female, guided to her by her luminosity.[49] More important for his developing views, this anomalous instance also exemplified Darwin's view that females were less developed than males: that among insects, where one was little developed, it was always the female that approached closer to the larval form, that the plumage of female and young birds was similar; that in humans, women were more childlike than men.[50]

It was another six months before Darwin began tentatively to extend aesthetic judgment to females, and then it was to that troubling instance of the peacock's tail, famously extolled as the exemplar of beauty created for human delight: "We must suppose [that peahens] admire [the] peacock's tail, as much as we do."[51] But that was as far as it went for many years to come. The notion of female judgment or choice did not come any more easily to Darwin than to those aestheticians he consulted.

4.5 Sex and sensibility

Reynolds's *Discourses* gave Darwin his basic grounding in aesthetics and helped to shape his early *Beagle* experiences of racial physiognomic and cultural difference. As he reinterpreted and adapted the *Discourses* to these same experiences and to the requirements of his early evolutionary theorizing, they explained a good deal. But there was still something wanting; the invocation of habit alone did not explain the "*feeling* in any one man." How to explain the specific desire or love for the singular woman among many women manifesting so many individual variations around the racial ideal? How indeed to explain the sentiment that might be attached to the physical features of, for instance, an Emma? Why this female and not that female? With his marriage to Emma just three weeks away, Darwin asked himself:

> January 6th. [1839] What passes in a man's mind. When he says he loves a person—do not the features pass before him marked, with the habitual expressemotions [sic], which make us love him, or her.—it is blind feeling, something like sexual feelings—love being an emotion does it regard[,] is it influenced by other emotions?[52]

While Reynolds's aesthetics allowed for a certain degree of subjective or emotional response, as an artist he was insistent on suppressing the physicality and emotionality of the body to make way for higher-level, allegorical meanings. What Reynolds's refined man of taste was supposed to look for in the Titian *Venus* was not her sexual attractiveness, but a disembodied idea of beauty and its associated allegory in art.

Titian painted various versions of the goddess of love, each showing the naked goddess in the company of a fully clothed male musician who has turned to gaze at her, directing the gaze of the viewer to the focal point of each painting: the lush body of the Venus. In the Fitzwilliam version, Venus holds a flute, an "instrument with palpable sexual associations," while her admirer fingers a lute, whose shape suggests the body of the woman on whose naked charms he fixes his gaze. They seem ready to engage in a duet, which might be interpreted (as it has been, for those wanting to subordinate the erotic symbolism of the painting to more abstruse aesthetic meaning) as the translation of desire into art, or of Titian's revision of the traditional aesthetic ranking of visual over the aural experience of beauty.[53] While Darwin was well aware of the distinction between "sexual beauty" and the higher aesthetic invoked by Reynolds, it is to be doubted that the young Darwin and Whitley managed this aesthetic disengagement from the inescapably erotic message of the painting.

The sexualized charms of the love goddess offered a "beau ideal" correlate to the domesticated, chaste "white petticoats" of the once-anticipated parsonage, or, as Darwin was actually experiencing it in early 1839, of the garishly decorated "Macaw Cottage" in London, where he and Emma were to set up house together. This was perfect beauty-as-desired versus imperfect beauty-as-loved.[54] As Darwin was by now representing it, beauty was not only complex and culturally relative, but sexual and emotive. Love was "something like sexual feelings," and it related to the "habitual expressemotions" of the loved one. Darwin's collapse of expression and emotions into the one hastily scribbled word strikingly reflects their interrelatedness in his thinking.

It was Edmund Burke, reinterpreted via his grandfather's *Zoonomia* and in light of Darwin's own evolutionary theorizing, who gave Darwin what he was looking for. Burke's conflation of the notion of the beautiful with sexuality and desire is notorious among aestheticians. His contemporary August Wilhelm von Schlegel scoffed that Burke's conception of the beautiful was a "tolerably pretty strumpet," while the sublime was a "grenadier with a big moustache."[55] Schlegel's sneer holds a certain truth. In the eighteenth century, beauty as applied to the human figure became associated with the female and lost ground to the masculine sublime. Sublimity had long been seen as an attribute of male beauty, exemplified in Greek statuary, as well as of a range of structures, man-made and natural, that excited pleasurable awe, wonder, and terror. Burke's contribution was to shift the emphasis from the objects that stimulated the sublime to the feelings or sensations they evoked and why, toward a psychological aesthetics. Burke's was also a profoundly political aesthetics.

In Burke's aesthetics, the terror of the sublime is necessary to instill awe and respect for power and authority, to restore order, while the beautiful elicits affection and loyalty and maintains social harmony and tranquility.

The reiterated message of his more political writings was that If the public display of terror was carried to excess, it would elicit only horror and negate the restorative effect of the sublime; the recourse to excessive state violence indicated that the underlying social bonds had fragmented, and its very exercise called into question the legitimacy of the system. This was the basis of Burke's famous crusade against the terror of the French Revolution and his conservative defense of the stabilizing effects of the British constitution and social hierarchy.[56]

Of greater import for Darwin's theorizing on sexual selection was Burke's relegation of the beautiful to the female domain: to the familiar, the domestic, the social, and the sexual. Beauty was, according to Burke, that quality or qualities by which bodies "cause love, or some passion similar to it," and it was "confined to merely sensible qualities." The sublime was characterized by pain and danger, the beautiful by "gratifications and *pleasures*."[57] This unambiguous association of beauty with sexually induced pleasure was centered on "that part of a beautiful woman where she is perhaps the most beautiful," her breasts, which are as enticing to the touch as they are visually beguiling: "the smoothness; the softness; the easy and insensible swell; the variety of the surface."[58] The eroticism of the beautiful is undeniable. But Burke made a clear separation of lust, a "brutish desire," from love, and in association with this, he denied Addison's attribution of an aesthetic sense to animals:

> The only distinction [brutes] observe with regard to their mates, is that of sex. It is true, that they stick severally to their own species in preference to all others. But this preference, I imagine, does not arise from any sense of beauty which they find in their species, as Mr. Addison supposes, but from a law of some other kind to which they are subject; and this we may fairly conclude, from their apparent want of choice amongst those objects to which the barriers of their species have confined them. But man . . . connects with the general passion, the idea of some *social* qualities, which direct and heighten the appetite which he has in common with all other animals; and as he is not designed like them to live at large, it is fit that he should have something to create a preference, and fix his choice; and this in general should be some sensible quality; as no other can so quickly, so powerfully, or so surely, produce its effect. The object therefore of this mixed passion, which we call love, is the *beauty* of the *sex*.[59]

Men might succumb to lust, but they were "attached to particulars by personal *beauty*." Because beauty was a social quality, it was associated with such things as "inspire us with sentiments of tenderness and affection." Burke's is an overtly feminized, sexualized, and Europeanized definition of beauty:

beautiful objects were small, smooth, curvaceous (women's breasts), delicate ("The beauty of women is considerably owing to their weakness or delicacy, and is even enhanced by their timidity"), and fair in color ("the colours
of beautiful bodies must not be dusky or muddy, but clean and fair").[60] But
for Burke, the overweening emphasis was on what was felt: on the sexually
engendered, emotionally constituted concept of beauty that guides a man
to the woman of his choice.

We begin to see more clearly those dimensions of Burke's aesthetics that
met with Darwin's slowly evolving views on beauty and aesthetic choice.
Darwin had owned his copy of Burke's *Inquiry* since his student days at Cambridge, and there are some half dozen specific references to it in the notebooks.[61] Initially, following Burke, Darwin had drawn a sharp distinction
between animal and human sexuality and interracial matings on the basis of
the animal's lack of an aesthetic sense and its instinctive repugnance against
interbreeding with members of another species. Inside the back cover of
his copy of the *Inquiry*, Darwin wrote: "[Burke] can see reason why instincts
(sexual) of animals stronger than in man—because not having any notions
of beauty to keep them in right line."[62]

However, by the time of his later-dated notebook references to Burke's
aesthetics, Darwin was becoming convinced that animals shared both reason and aesthetic sense with humans, and certain of these later entries were
directed to reinterpreting and extending Burke's aesthetics in this light and
giving them an evolutionary spin. As stressed, Darwin read Burke through
the familiar medium of his grandfather's evolutionary aesthetics, which also
owed much to Burke.[63] At first, Darwin, rereading Burke, queried his grandfather's theory of breast-centered beauty on the grounds that animals do not
have an aesthetic sense: "Stallion licking udders of mare strictly analogous
to men's affect[ion] for women's breasts. [Therefore] Dr Darwin's theory
probably wrong, other wise horses would have idea of beautiful forms." A
few weeks later, however, Charles changed his mind:

> I cannot help thinking horses admire a wide prospect.–The very superiority of
> man perhaps depends on the number of sources of pleasure & innate tastes,
> he partakes, taste for musical sound with birds. & ? howling monkeys—smell
> with many animals—see how a dog likes smell of Partridge–, man's taste for
> smell of flowers, owing to *parent* being fruit eater.—origin of colours?[64]

Smell and the associations it aroused were particularly evocative and
connected with the aesthetic experience. In near caricature of this, the whiff
of a painting in the National Gallery transported Darwin, after an absence of

seven years, straight back to the Fitzwilliam Museum. an "association with much pleasure immediately thrilled across me." In a more direct correlation of scent with sexuality and animality, from watching the orangutans in the Zoological Gardens, Darwin put his own sexuality into evolutionary/ aesthetic context: "We need not feel so much surprise at male animals smelling vaginae of females.—when it is recollected that smell of one's own pud[enda] not disagree[able]."[65]

Burke's psycho-physiological aesthetics led him to try to displace the primacy of vision from its central position in aesthetic theory. He imported hearing, smell, touch, even taste, into the aesthetic/sensual/sexual experience: the perception of female beauty was as much focused on touch, on the enticing softness and smoothness of the breasts, as on their visual perception—a lead followed and extended by Erasmus Darwin in his breast-centered theory of beauty. Charles Darwin, for his part, integrating the views of Burke and his grandparent into his own evolutionary theorizing, extended this to licking, salivating, and biting the desired object and, by analogy, to animal courtship. He was, by this stage, experiencing the heightened emotional and sexual intensity of his own courtship, exploring the pleasures and noting the physical symptoms of desire and sexual arousal:

> Sexual desire makes saliva to flow[,] yes *certainly*—curious association: I have seen Nina [the dog] licking her chops . . . one's tendency to kiss, & almost bite, that which one sexually loves . . .—Lascivious women. Are described as biting: so do stallions always. . . . The association of saliva is probably due to our distant ancestors having been like *dogs* to bitches.—How comes such an association in man.—it is bare fact, on my theory intelligible.[66]

For Darwin in the throes of love and passion, everything—sight, smell, hearing—was sexually inflected and traceable to hereditary habit. A cascade of entries around this time recorded his shudder of pleasure at music; the trembling that accompanied "any great mental affection"; the sexual and emotional connotations of blushing, its connection with anxiety about one's appearance in the presence of the opposite sex, the flushing of a woman's bosom being "like erection" in man; the convulsive nature of crying, the spurting tears of grief, joy, and "sublimity," and the shiver of sexual ardor. Like Burke before him, he examined his own face for the signs of sexual desire and, like Erasmus, correlated his "protrusion of chin" with that of bulls and horses.[67]

Burke's aesthetics—which challenged the neoclassical embargo placed by Reynolds on an excessive show of feeling even in the face of extreme suffering, brought forcefully before Darwin the physicality and emotionality of the body and facial features to a degree that the medically oriented interpretations of *Zoonomia* could not. He hooked the irrational, involuntary convulsions and grimaces of the insane from *Zoonomia* into his notebook entries on expression and hereditary habit. But it was Burke who licensed Charles to kiss and cry, to dwell on the features of the beloved, and to sentimentalize their affective associations rather than to acknowledge the calculating accounting that had directed him to the woman of his choice; the "heart is the seat of the emotions," it "feels."[68] Darwin responded to the "homespun psycho-physiology" of Burke's aesthetics as to the manner born, as indeed he had been. Following Erasmus, Charles traced the analysis of expression back to Burke's claim that the mind and body are so interconnected that one can enter into the interiority of another's mind by simulating their outward expression and mannerisms.[69]

The overly sensitive Darwin might resonate with Burke's "politics of pain," but at the same time, as an evolutionist and putative materialist, he put Burke's physiological grounding of emotion and mind to work. As Darwin reinterpreted Burke, aided by the insights taken from *Zoonomia*, the involuntary muscular starts, flushes, accelerations of the pulse, expulsions of tears, and facial expressions that accompanied emotional states were "hereditary habitual movement consequent on some action which the progenitor did, when excited or disturbed by the same cause, which now excites the expression."[70]

The observations he made of the expressions and behaviors of captive orangutans and other members of the monkey tribe were applied directly to tribal humans. In a flash of notebook humor, Darwin recorded his brother Erasmus's citation of Plato's argument that the "*necessary ideas*" of the beautiful and the good "arise from the pre-existence of the soul, are not derivable from experience." "Read monkeys for preexistence," mocked Charles and went straight out to the Zoological Gardens to analyze the emotional state of the pouting young orang and compare it with the human expression of disgust and defiance.[71] A little later he was noting that New World monkeys showed no sexual interest in human females, while their Old World relatives, structurally closer to humans, took every opportunity to look up women's skirts: "The monkeys understand the affinities of man, better than the boasted philosopher himself."[72] Not long after that, he was putting the Fuegians (who had never actually left the picture) back into center frame: "Judge only by what you see. Compare, the Fuegian & Ourang outang, & dare to say difference so great."[73]

The link Darwin might make between these observations and his Reynolds-inspired notion that the aesthetic sense was habituated to a particular physiognomy, was that the "great principle of liking," whether for looks (for particular expressions and features) or for scents, tastes, or sounds was "simply *hereditary habit*."[74]

By March 1839, Darwin was certain that all the "tastes of man [are the] same as in Allied Kingdoms." This included smell, music, colors (the peacock's tail being admired by both humans and peahens), and touch (the orangutans in the Zoological Gardens were not only given to fondling and sniffing each other's genitalia, but were also "very fond of soft, silk-handkerchief," while dogs and cats liked gentle tickling). Smell, hearing, touch, and taste were all involved in the aesthetic experience and might be traced to our animal affinities and animalistic "savage" ancestry.[75]

But in the end, appearance was all: "Judge only by what you see." It was vision rather than smell that was preeminent in the instinctive recognition of and pleasure in their kind by social animals, who cannot know their own smell or look. "No doubt," wrote the well-versed student of *Zoonomia*, "it may be attempted to be said that young animal learns parent smell & look so by association receives pleasure." Darwin, deciding in favor of Erasmus against Burke, gave priority to the visual sense: "Man, a socialist, does not know other men by smell, but by looks. Hence. Some obscure picture of other men. & hence idea of beauty."[76]

4.6 For beauty's sake

In Darwin's notebooks, the sense of beauty was acquiring a naturalistic genealogy. Through the exercise of inherited habit, it had emerged from animalistic sensual and instinctual perceptions to increasingly abstract, cerebral levels, to the higher aesthetic of the civilized, cultivated adult human being. His emphasis on the emotional construction of beauty meant that Darwin might connect the expression of the emotions, which was assuming increasing importance in his notebooks, with the notion of an evolving aesthetic. Ultimately, however, Darwin was not so interested in expression as a means of communication or as indicative of underlying emotional states, as Erasmus Darwin and Burke had been; nor did it figure in the notebooks as a means of sexual signaling, as might be expected of the architect of sexual selection. Rather, Darwin was most concerned with refuting the creationist claim by Sir Charles Bell (expert on expression and leading exponent of natural theology) that certain muscles had been bestowed on man for the sole purpose of communication "by that natural language, which is read in the changes of his countenance." For Bell, the communication of expression had a higher purpose. Human smiles, frowns, and sighs were to be referred to the divine designer, who had "laid the foundations of emotions that point to Him, affections by which we are drawn to Him, and which rest in Him as their object."[77]

For Darwin, intent on naturalistic, evolutionary explanation, there was no higher purpose. Human expression was instinctive and mostly attributable to inherited habit. It was comparable to, and had evolved from, animal responses, facial movements, and similar muscular contractions or relaxations under the stimulus of primary emotions of fear, anger, pain, grief, desire, pleasure, and so on. He set out to show that there were no "emotional" muscles or expressions unique to humans, as Bell argued: the corrugator muscles that made frowning possible were present in orangutans and chimpanzees; blushing (which for Bell showed most clearly the spiritual and moral side of human nature) was exhibited even in Fuegians (both Jemmy and Fuegia blushed when teased or made self-conscious about their appearance) and in darker races where it was not easily visible and served no purpose for the blusher or the beholder. Darwin could find no evidence for "believing that any muscle has been developed or even modified exclusively for the sake of expression."[78]

In short, Darwin, in dismissing Bell's attribution of a higher purpose to expression and emotional response, threw out the utilitarian baby with the creationist bathwater.[79] Expression had an evolutionary history, which Darwin put much effort into unraveling, finally producing a whole volume on the topic in 1872. But his primary concern was to undermine the creationist position and to defend his theory that humans had evolved from an animal ancestry. He also deployed the commonality of expression in the "most widely distinct races of man" to support his later claim that the human races had descended from a "single parent-stock, which must have been almost completely human in structure, and to a large extent in mind, before the period at which the races diverged from each other."[80] In Darwin's mature theorizing, then, emotional display was a more primitive response and must have preceded the development of the higher aesthetic sense that, through sexual selection, accounted for the divergence of the human races.

Significantly, his earlier notebook constructions, which tended to conflate the expression of the emotions with the aesthetic sense, had also dissociated beauty from utility.[81] Burke and Erasmus Darwin, through their emphasis on the association of ideas, had rejected the aesthetic convention that the experience of beauty was linked with utility, in the sense that the aesthetically appealing object pleased by virtue of its uses to the species as a whole.[82] Burke was most explicit, appealing to nature to argue against the "opinion" that "the idea of utility, or of a part's being adapted to answer its end, is the cause of beauty." He instanced the "wedgelike snout of a swine," the "great bag hanging to the bill of a pelican," and the monkey, who had the "hands of a man, joined to the springy limbs of a beast." All of these structures were useful to their possessors, yet singularly ugly to humans. As for that epitome of "extreme beauty," the peacock, it was not a good flier and lived like the swine in the farmyard. While it was true, Burke conceded, that the "infinitely wise and good Creator" had "frequently joined beauty to those things which he has made useful to us," this did not prove any necessary connection between use and beauty.[83]

In keeping with this, Burke also broke with the aesthetic tradition that opposed "deformity" to the beautiful, as exemplified in Reynolds's *Discourses*. The snout of the swine was not deformed but simply ugly. A thing might be "very ugly with any proportions, and with a perfect fitness to any uses." Beauty was a "*positive* and powerful quality" that was sexually attractive. Ugliness was the negation of all the sensations aroused by the beautiful; it was what repelled. Ugliness, however, should not be confused with the sublime; it was not sublime unless it was associated with terror.[84]

Burke's sharp discrimination of the beautiful from the useful and the ugly, and the cross-species examples he instanced, underscore Darwin's sporadic notebook references to the incipient problem of distinguishing between natural selection and what was to become sexual selection. As far as the human sense of beauty was concerned, Darwin's early adoption of an associationist, culturally relative aesthetic largely preempted utilitarian interpretation, although we may discern a certain concern with linking the perception of health, along with expression, to the finding of beauty in what might be otherwise considered physically ugly.[85] But this alliance of the perception of female beauty with that of health subsumes the traditionally contributory aspect of utility to beauty under the theory of association—both the perception of health and beauty lie in the mind of the beholder and depend on the association of ideas and not on any inherent beauty in health or any other utilitarian quality that might come to be associated with the perception of beauty. By the time of the *Descent*, Darwin was to deploy this nonutilitarian aesthetic in the rejection of any adaptive interpretation of human racial differences in physiognomy and color in relation to climate or environment, holding to the idea of the habituated perception of beauty as the explanation of racial differentiation through sexual selection.

Darwin's more immediate problems arose in extending the associationist or habituated aesthetic to animals, as his early evolutionary theorizing required. From an early stage of his notebook constructions, Darwin puzzled over how to bring the "abundant instances" of "superabundant" or non-

adaptive structures of birds—such as the tail of the peacock and the plumage of male birds of paradise—into his theorizing. They were certainly "not
connected with habits," for both males and females with such strikingly
different plumage, and juveniles whose plumage generally appeared most
like that of females, mostly shared the same habitats and habits.[86] Darwin's
fragmentary attempts to account for beauty of plumage through the sexual
struggle had not coalesced into anything definite (although by late 1838,
he had begun tentatively to extend mating success through combat to song
and to "charms"; see sec. 3.8). After his formulation of natural selection,
Darwin returned to the problem of such seemingly nonadaptive structures,
particularly to the difficulty of accounting for the persistence of potentially
disabling structures or plumage:

> All that we can say in such cases, is that the plumage has not been so injurious
> to bird as to allow any other kind of animal to usurp its place,—& therefore
> the degree of injuriousness must have been exceedingly small.—This is far
> more probable way of explaining, much structure, than attempting anything
> about habits.[87]

Darwin's adoption of a visually oriented aesthetic, which he extended to
animals and linked with the sexual drive and reproduction, brought with
it the implication that the visual perception of beauty played a part in the
choice of mate, that birds especially were not indiscriminate as to choice, as
Burke argued, but rather were "determined in [their] courtship by the single
grain or tincture of a feather," as Joseph Addison had claimed.[88] Addison,
as we saw, had assigned such aesthetic preference to male birds, generalizing from human aesthetic convention; Darwin had the further problem
of shifting such aesthetic discrimination and choice to females if he was to
explain the male predominance in color and superfluous plumage and other
structures, such as the exemplary beauty of the peacock's tail, although he
carefully noted such instances as he found where females were "more splendid" than males or sang as well as males.[89]

Darwin's answer to the problem of the peacock's tail and his claim that
such structures were purely ornamental and served no other functions ("the
most refined beauty may serve as a charm for the female, and for no other
purpose"[90]) devolved upon his immersion in the art of breeding or artificial
selection. Like the pigeon breeders he studied, Darwin was to argue that
female birds select for what pleases them aesthetically on the basis of externally visible, minute, subtly different characteristics and without regard to
use. Darwin's painstaking bird's-eye reconstructions of the evolution of the

ornamentation and color of the peacock and the exquisitely shaded "ball and socket" ocelli of the male argus pheasant were directed to turning the theological claim that these structures had been "created for beauty's sake" back on itself. As he wrote in the fourth edition of the *Origin* (1866):

> I willingly admit that a great number of male animals . . . have been rendered beautiful for beauty's sake; but this has been effected not for the delight of man, but through sexual selection, that is from the more beautiful males having been continually preferred by their less ornamented females.[91]

4.7 The taming of associationist aesthetics

Through his adaptations of Reynolds, Erasmus Darwin and Burke, Darwin had pieced together an evolutionary aesthetics that integrated associationist theory with cultural relativism. It was not until he read an article on taste in an old number of the *Edinburgh Review* that he found ready-made a theory of aesthetics that united these crucial aspects of the beautiful: "EXCELLENT," crowed Darwin, "very good article."[92]

Its author was Francis Jeffrey, editor of the *Edinburgh Review* from 1803 to 1829, and known to Darwin as a member of Harriet Martineau's literary circle.[93] The article that so excited Darwin (and surely he was directed to it by Martineau or brother Erasmus) was Jeffrey's review of the second edition of Archibald Alison's *Essays on the Nature and Principles of Taste*. The *Essays*, first published in 1790, had engaged little attention until Jeffrey's influential review of 1811 propelled Alison's empirical, associationist writings into the "new gospel" in British aesthetics. As aesthetics was integrated into the politics of liberalism, Jeffrey swung the "enormous critical, even juridical, power" of the Whiggish *Edinburgh Review* behind Alison. His review of Alison's *Essays* was expanded into the article "Beauty" in the eighth edition of the *Encyclopædia Britannica* and republished as an introduction to the numerous subsequent editions of the *Essays*. Alison, as repackaged by Jeffrey, promoted a "laissez-faire" aesthetic theory of the relativity of taste in the "free market for culture."[94]

According to this aesthetic, the perception of beauty was "not universal, but entirely dependent upon the opportunities which each individual has had to associate ideas of emotion with the object to which it is ascribed." There was no such thing as absolute or intrinsic beauty. One had only to consider the "different and inconsistent standards" of the idea of female beauty in the different regions of the world—Africa, Asia, Europe, Tartary, Lapland, Patagonia [!]:

If there was anything absolutely or intrinsically beautiful, in any of the forms thus distinguished, it is inconceivable that men should differ so outrageously in their conceptions of it: If beauty were a real and independent quality, it seems impossible that it should be distinctly and clearly felt by one set of persons, where another set, altogether as sensitive, could see nothing but its opposite.[95]

The "countenance of a young and beautiful woman," the "most beautiful object in nature," was not dependent on a mere "combination of forms and colours," but on a "collection of signs and tokens," shaped by one's habitual "associations," by one's moral sentiments or affections. The sensation of female beauty was invariably linked to youth and health, and was excited by the "higher and purer feelings" rather than the baser passions. The same considerations were to be extended to the appreciation of different landscapes. Jeffrey instanced the "frank confession" of "two Cockney tourists," who, transported out of their habitual cityscape, are quite unable to find any beauty in what others found so picturesque: the "naked mountains and dreary lakes" of the Scottish Highlands. Having yawned their way along the banks of Loch Lomond, they post thankfully (and risibly) back to the "beauty and grandeur—of Finsbury Square!" These same relativities, claimed Jeffrey, applied to music, to literature, art, architecture, and dress, "which always appears beautiful to the natives, and somewhat monstrous and absurd to foreigners."[96]

This was a theory of aesthetics that Darwin might embrace. It combined associationism with the theory of moral sentiments, giving proper weight, as Darwin required, to feeling or emotion, to the associations of sympathy or affection, the affectively internalized moral code that shaped the aesthetic judgment. It effectively integrated the more relativistic aspects of Reynolds's views on beauty with the emotional and sentimental associations evoked by Burke and Erasmus Darwin.[97] It reinforced Darwin's belief that beauty was no more fixed or static than species. The standard of taste or beauty was a labile category influenced by habit through the association of ideas and by moral criteria; it varied from person to person and race to race, and eluded fixed definition.

Still, he worried away at its implications for his theorizing. A major difficulty he discerned was that if there were no standards of beauty at all, no "beau ideal," what did this mean for aesthetic choice as applied to the human races: "Mem[orandum]. Negro, beau." And what of animals? It was in this connection that Darwin made his earliest explicit reference to the possibility of female choice: "How does Hen determine which most beautiful cock, which best singer"?[98]

It is not clear exactly when Darwin read Jeffrey's review.[99] There is a hint in Jeffrey's discussion of habituated landscape as well as racial physiognomies in Darwin's decisive notebook reference to Reynolds's seventh discourse of late October 1838, which supports an earlier reading. In any case, as we know from this and other notebook entries, Darwin was determined on a view of beauty that devolved upon the concept of the "beau ideal," or standard of beauty, as an instinctive appreciation of their own kind—tribes, races or species—that social animals acquired through their habituation to the "looks" of their particular social group. It is through this instinctively applied aesthetic standard that the hen determines which is the most beautiful cock (though Darwin was still a long way from generalizing this understanding as female choice).

In point of fact, despite their advocacy of a relativity of perceptions of beauty that gave due recognition to the "accidents" of cultural and social diversity, neither Jeffrey nor Alison (both middle-of-the-road Whigs) promoted a democratic aesthetics in which all tastes were "equally good or desirable." The quaint preference of Cockneys for the urban ugliness of Finsbury Square was not, after all, as aesthetically valid as that of their social superiors for the beauties of the Scottish Highlands. As Jeffrey endorsed Alison's position, the "best taste must be that which belongs to the best affections, the most active fancy, and the most attentive habits of observation." In other words, the best taste was practiced by the class that had the advantage of superior education and superior cultivation of the moral sense, that had access to political power and was able to exert its social and cultural authority in determining taste and its associated moral code. "Whole classes of men," other nations or races, might exhibit "outrageous" conceptions of female beauty because of the "accidental or arbitrary" associations habitual to them; true female beauty consisted in the "visible signs and expressions of youth and health, and of gentleness, vivacity and kindness"—those feminine qualities valued by the upper-middle-class reading public to whom Jeffrey directed his review.[100] Jeffrey himself was notoriously unresponsive to any suggestion that women might play a greater role in the progressive commercial society he advocated, writing that the "proper and natural business" of women was the "practical regulation of private life, in all its bearings, affections and concerns."[101]

There were, then, strong racial, class, and gender dimensions to the new aesthetic. It connected with contemporary debates about the nature and functioning of commercial society, with an emergent political economy and liberal political thought, with the philosophy of moral sentiments, and with more general concerns about manners, morals, and polite conduct, particularly with respect to women. This was the culmination of the period in which the beautiful and the feminine came to be regarded as identical, when the very definition of beauty was connected with female manners and feminine morality.[102] It is no accident that the new aesthetic coincided with a renewed emphasis on religion, with the concurrent reassertion of traditional feminine virtues and the advancement of a morality centered on the home and family. The overtly sexualized and "atheistic" aesthetics of the radical Erasmus Darwin had met with hostility and erasure. Only Burke's fame as the foremost conservative critic of the French Revolution protected his equally sexualized, contentious views on beauty from a similar expurgation.[103] Alison and Jeffrey domesticated associationism, muting its more sensuous and sexual aspects; and the great popularity of their interpretation was largely due to its compatibility with a religiously oriented moral aesthetic that meshed with contemporary social and political needs.

Alison's theory of aesthetics shared a common grounding with the related school of Scottish Common Sense philosophy, most often associated with the writings of Dugald Stewart,[104] Thomas Reid, and Lord Kames, all of whom located a similar amalgam of perceptual discrimination and moral judgment within a robust religious framework. Common Sense theory had close ties with natural theology, which deployed a similar religiously inflected moral aesthetic in the argument from design. It remained a significant intellectual force in Britain until the 1870s, influencing the aesthetics of, among others, John Ruskin and Frances Power Cobbe, both prominent critics of an emergent secular aesthetic that they sheeted home to a materialistic, utilitarian Darwinism.[105] Largely through Alison's writings on taste, Common Sense philosophy was also influential in shaping American aesthetic thought, particularly that variant of it that found popular expression in nineteenth-century sentimental discourse. The value that Alison placed on emotionalism or sensibility in terms of moral development, on human feeling or sympathetic imagination, rather than reason as the guide to the good and the true, reached its apotheosis in Stowe's *Uncle Tom's Cabin*, which expressed its antislavery rhetoric through a pietistic, emotional evangelism.[106]

Darwin may have had no problems with the abolitionist deployment of an associationist sentimental aesthetic, but, as an evolutionist, he had real differences with the conventional linking of such an aesthetic with the argument from design. The whole intention of his notebook constructions of beauty was to dissociate it from the conventional theological attribution of a higher purpose. Nevertheless, the emphasis that Alison and the Common Sense school placed on the moral value of the cultivation of the aesthetic sense and its relation to religious sentiment is evident in Darwin's mature evolutionary aesthetics.

In *The Descent of Man*, Darwin compared the fearful reaction of his little dog to a wind-blown parasol to the "savage" fear of invisible spirits who might cause storms or other such retributive natural phenomena. The primitive belief in spiritual agencies, Darwin thought, would, as the intellectual and moral faculties evolved, progress into belief in the existence of one or more gods and thence into religious devotion.[107] The Fuegians—the semi-civilized ones, anyway—were in an "intermediate condition." York Minster had connected certain culpable actions with bad weather, but Jemmy had "with justifiable pride, stoutly maintained that there was no devil in his land." Nor had Darwin been able to discover any native Fuegian belief in a god or any religious practices.[108] Later in the *Descent*, in the midst of Darwin's discussion of ideas of beauty in "semi-civilized and savage nations," he reverted to the subject of gods:

> We thus see how widely the different races of men differ in their taste for the beautiful. In every nation sufficiently advanced to have made effigies of their gods or of their deified rulers, the sculptors no doubt have endeavoured to express their highest ideals of beauty and grandeur. Under this point of view it is well to compare in our mind the Jupiter or Apollo of the Greeks with the Egyptian or Assyrian statues; and these with the hideous bas-reliefs on the ruined buildings of Central America.[109]

We are back with the ugly Fuegian, who knows no god and is the antithesis of the Titian *Venus*, the goddess of love and beauty. If gods represent a people's ideal of beauty, and if belief in a god or gods is dependent on moral and intellectual progress, then, in Darwin's aesthetics, beauty of body and morality are interlinked. It is the evolving, habituated notion of the beautiful that, through sexual selection, accounts for the differentiation of people into races or cultures with distinct physical differences and moral codes.[110] In this sense, the traditional aesthetic equation of beauty with goodness and ugliness with depravity or evil persists in Darwin's aesthetic hierarchy that ascends from the godless, immoral Fuegians—whose taste for valueless glitter and gaudy rags is comparable to that of animals and less than that of birds—to the higher levels of artistic representation of the beautiful in Greek mythology and statuary, with the further implication that the highest, so-far-attained association of the beautiful and the good is to be found in the aesthetic judgment of the cultured, civilized European.

Just how little Darwin traveled beyond his early notebook views on the habituation of the aesthetic sense and its relation to mind, morality, and the emotions—from the period of his intensive readings of Reynolds, Erasmus Darwin, Burke, and Jeffrey—when he had "thought a good deal on the subject" of beauty, is borne out by the testimony of his published *Autobiography*. Here, in a well-known passage, the aged Darwin lamented the loss of his once-keen aesthetic sense. He could not bear poetry, was "nauseated" by Shakespeare, and had entirely lost his taste for music and pictures, while fine scenery no longer inspired in him "exquisite delight." All that remained was a banal taste for novels with happy endings and "some person whom one can thoroughly love, and if it be a pretty woman all the better." He attributed his anesthesia to the "atrophy of that part of the brain . . . on which the higher tastes depend" through his cultivation of those parts concerned with "grinding general laws out of large collections of facts." If he were to live his life over again, he would have "kept [the aesthetic part of his brain] active through use," by regularly reading poetry and listening to music: "The loss of these tastes is the loss of happiness, and may possibly be injurious to the intellect, and more probably to the moral character, by enfeebling the emotional part of our nature."[111]

Darwin's notions of beauty were structured by the readings he adapted to the demands of his evolutionary theorizing, by the strongly gendered conventions of his society, and by his lived experiences of "savage" encounter. Like most of his contemporaries, he linked beauty to morality and culture and ranked it by a hierarchical, Eurocentric scale. He rejected the conventional relation between religion and aesthetics, substituting a naturalistic explanation for the beautiful. "Beauty," as Darwin constituted and applied it, was not simply aesthetic. Not only was it race and class specific, it was also strongly gendered and weighted with political and economic meaning. It was this notion of beauty that he brought to the making of sexual selection. And it was this notion of beauty that Darwin pursued in his concurrent readings into early theories on race and human variety—readings crucial to the inception of his early views on the role of aesthetic selection in racial differentiation.

FIVE

Reading the Face of Race

Connexions in marriage will generally be formed on the idea of human beauty in any country; an influence, this, which will gradually approximate the countenance towards one common standard.

—William Lawrence, *Lectures on Physiology, Zoology and the Natural History of Man*, 1822

Darwin's annotated copy of William Lawrence's "blasphemous" *Lectures* is bound between faded and stained cardboard covers. Its tattered pages indicate hard wear and tear. He had owned it prior to the *Beagle* voyage, but perhaps prudently, he did not take it with him on the *Beagle*. Darwin's copy was a pirated one, published in 1822 by the militant cobbler turned bookseller William Benbow, who financed his inflammatory politics with the sale of pornographic prints.[1] This cheaply bound copy with its radical imprimatur was critical to Darwin's early formulation of the concept of aesthetic selection as a formative factor in race.

Lawrence's *Lectures*, first published in 1819, were originally delivered before the Royal College of Surgeons, bastion of conservative anatomical doctrine and political privilege. They caused a furor. Lawrence recklessly argued against the ruling orthodoxy of vitalism, inherited from the great surgeon John Hunter himself, and urged the necessity of separating physiology and anatomy from the claims of natural theology. The proof of the existence of the soul, he declared, should not be sought in the "blood and filth of the dissecting room." Lawrence was forced to resign his post as lecturer at the college and recant his views after his materialist explanations of man and mind were attacked in the Tory press. With his positions at the hospitals of Bridewell and Bethlem at stake, Lawrence withdrew his *Lectures* from sale. This had the countereffect of ensuring the work's notoriety. It was seized on by the pauper press. Lawrence's efforts to suppress the pirating of his work resulted in a ruling by the Court of Chancery that it was blasphemous, thereby greatly enhancing its attractiveness for radicals and freethinkers. The book was kept in circulation for decades by some six pirated editions that were "hawked at the corners of the streets in sixpenny numbers." A surprising number of subsequent evolutionists besides Darwin—including Alfred Russel Wallace, Robert Chambers (author of the sensational *Vestiges*) and the young Asa Gray—read and drew inspiration from these cheap, readily available reprints.[2]

Yet Lawrence's *Lectures* remains a much underrated resource for the formulation of Darwin's early views, and not only on race and aesthetics.[3] The whole issue has been confused by problems of dating and by the fact that Darwin independently read many of the same works that Lawrence drew on in his *Lectures*. Certain of these (notably those of Samuel Stanhope Smith and James Cowles Prichard) also offered explanations of racial diversity in terms of differing racial standards of beauty. All three—Smith, Prichard, and Lawrence—drew promiscuously from an array of Enlightenment theorists on "man," from a wide variety of literature on travel, stock breeding, anatomy, and physiology; from historical, political, and philological treatises; and from the general cultural pervasiveness and practices of physiognomy and/or phrenology. Consequently, a high level of intertextuality characterizes their writings, although there are important differences in their interpretations and emphases.

Both Smith and Prichard have been singled out as the "crystallizing" source of Darwin's views on the significance of aesthetic choice in race formation.[4] But it was Lawrence's *Lectures* to which Darwin was primarily indebted for this crucial insight. Further to this, I shall argue the importance of the naturalist Edward Blyth (another underrated source for Darwin's views on sexual selection) in integrating and bringing Lawrence's views on domestication, variation, and inheritance and their bearing on the origin of the human races, forcefully before Darwin at two widely separated but critical stages of theory formation.

5.1 Faces, noses, and crania: The good, the bad, and the ugly

Historians are only beginning to appreciate the significant part played by aesthetics in the emergence of early ideas on race in those European theories of human variety that came to prominence in the late eighteenth and early nineteenth centuries.[5] These racial theories are generally labeled climatic or subsistence, depending on whether they made climate and environment the primary determinants of the physiognomy and nature of the people concerned, or whether they attributed such characteristics to modes of subsistence that determined the social stage of development of a people. Theorists commonly brought the perspectives of the moralist and the natural historian to their categorizations of the peoples, customs, and places that the great voyages of "discovery" were uncovering. Ideas on beauty routinely entered into these early racial constructions. There was a commonality of opinion that the "Caucausian" peoples inhabiting the temperate regions of Europe and Middle Asia were the most beautiful—this consensus being made compatible with both climate theory and judgments of morality and level of "civilization"—while the "ugliest" were to be found in the far northern reaches of Europe (as the Lapps), in the frozen steppes of Eastern Asia (Tartars or "Calmucks"), or in remote antipodean outposts, the product of "savagery" and extremes of climate ("Hottentots," Tasmanians, or Fuegians). The ability to make aesthetic judgments was linked with such determinations and was itself a means of sorting the "civilized" from the "savage."[6]

The privileging of climate in relation to facial and bodily beauty owed much to Johann Joachim Winckelmann's 1755 thesis that the supremacy of ancient Greek art and culture was determined by the ideal temperate climate of Athens. This conduced to the physical perfection of its people and to the social conditions that permitted the translation of this perfection into statuary of the highest aesthetic attainment, like the *Apollo Belvedere* and the *Laocoon*. Artists of other climates and nations reproduced in their art the physiognomies familiar to them, which Winckelmann denigrated by comparison with the Greek ideal—particularly with the ideal profile exemplified in representations of gods and goddesses where the "forehead and nose make a straight line."[7] Winckelmann's ideal profile was to have a long life in theories of human variety.

The other major inspiration for notions of the beautiful in early racial categorizations was Johann Caspar Lavater's extraordinarily popular *Physiognomische Fragmente*, issued in four large volumes from 1775 to 1778. Lavater's work was reissued in countless translations, revisions, abridgments, imitations, and piratings. Physiognomy entered into the common vocabulary and culture of Europe and America. Its intrinsic appeal lay in its promise of deciphering a person's true character on the evidence of the external appearance, above all by the face. By studying Lavater's volumes, readers might familiarize themselves with his anthology of facial types and apply his physiognomic method efficaciously to living faces. So highly regarded were Lavater's works that they were thought as essential to every family "as even the Bible itself."[8]

Lavater's method was essentially semiotic: the inner nature was revealed by signs visible on the face, comprising a natural language of the body. These could be decoded by the trained observer, who could detect or assess the authentic inner moral character, or, as Lavater represented it, the "soul." Lavater backed his physiognomy with claims of objectivity and scientific rigor, but his scientistic hubris invited attack from those committed to the defense of Enlightenment science against the incursions of "unreason." Lavater held to his project of a mathematized physiognomy, but sought to deflect criticism by retreating to the claim that physiognomy was an "artistic" discipline that might be subsumed under those arts termed the "sciences of beauty," or aesthetics.[9]

This was not mere wordplay. His physiognomic method relied on the traditional aesthetic equation between the good and the beautiful: a beautiful face was indicative of a beautiful soul, an ugly one being synonymous with vice. The physiognomist's skill in discerning character was referred to that of the connoisseur of art: the artistic representation of the human face—rather than the actual face, with its distracting play of expression—was the ideal object for physiognomic study. Lavater's "empirical physiognomics" distinguished between the "firm" or "permanent" facial features (the nose and skull shape), which were certain and reliable, and the "soft" or malleable (muscles, expressions, gestures), which were unreliable physiognomic signs. The skull was particularly significant, being shaped over time by the brain, and was therefore an absolutely reliable sign of inner character. It was this insistence on the empirical significance of the skull that linked Lavater's physiognomy directly with the derivative phrenology, which made similar claims to objectivity and scientific legitimacy.[10]

In practice, Lavater's physiognomic judgments were heavily reliant on "soft" features, like the mouth. But his emphasis on the firm and unalterable features of the human body (its "primordial form") as the "natural" signs of the individual's "pure predestination" identifies Lavater as a quasi-biological determinist. Individual character, intelligence, and ability were fixed and determined by anatomy, the infallible signs of the inner soul. In this sense, anatomy was destiny. No amount of education, training, or self-fashioning could equalize the primordial inequalities stamped on the face and cranium.[11]

Inevitably, physiognomics had a racial dimension. Skin color, nose shape, and skull shape were the somatic features that signified the supposed aesthetic, moral, and intellectual superiority of the European. Those who doubted the reality of "national character" were invited to compare the physiognomy of a "Negro and an Englishman, a native of Lapland and an Italian, a Frenchman and an inhabitant of Tierra del Fuego." Fuegian physiognomy signified "stupidity" and *"incapacity* to produce a culture"; while the apelike "low forehead and deep sunk eye" of the "Calmuck" denoted "cowardice and rapine."[12] The physiognomic traits of modern Europeans were the standard—the "rule of nature"—against which other nations and races were assessed and found wanting: only "Negroes admire a flat nose."[13]

Darwin was well acquainted with the pervasive physiognomic reading of faces and heads. He was famously almost refused passage on the *Beagle* because FitzRoy read the shape of Darwin's nose as signaling insufficient energy and determination for the voyage. In England, the huge popularity of Lavater's work and its adoption by radicals who saw in it the potential for leveling class relations (given the technique, the servant might read the face as readily as the master) meant that it was suspected of dangerous implications.[14] This suspicion was borne out by its subsequent adaptation to the materialist interpretations of race and gender of the radical anatomist Alexander Walker and, after him, Robert Knox, both closely read by Darwin, both important to his theorizing on sexual selection (see chs. 7 and 10). Darwin's leading authority on expression and its artistic renderings, Sir Charles Bell, explicitly dissociated his natural theological interpretations of expression from physiognomy and its offshoot phrenology (which had even more blatant materialist connotations and was popular among both the liberal intelligentsia and the working classes). Nevertheless, Bell's physiological explanations of mental processes and their connection with the expression of the emotions were underwritten by physiognomic practice, as were Darwin's.[15]

Darwin read Lavater's works with some circumspection ("I must be very cautious") but was clearly attracted by their evolutionary implications.[16] Physiognomics—with its incipient biological determinism, its nexus with traditional aesthetics, its emphasis on the necessary connection of the facial features with notions of moral worth and "national character"—is inflected in Darwin's early and late views on sexual selection. Furthermore, the practices of the physiognomist who deduced his judgments of inner character solely from external characteristics bear comparison with those of the specialist animal breeders who selected their prize breeding stock on the same basis—a comparison brought compellingly before Darwin by Walker, who applied these same physiognomic principles to wife choice. The interplay of these convergent practices in Darwin's construction of sexual selection is the subject of the following two chapters. We are here concerned with the anthropological trajectory of physiognomy.

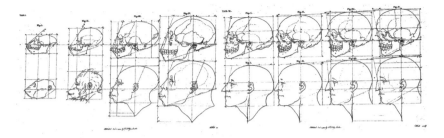

Figure 5.1. "From Ape to Apollo," Camper's famous facial angle. Camper 1794. Courtesy of Wellcome Library, London.

5.2 A measure of beauty: The facial angle

Where Lavater had only aspired to mathematize physiognomy, it was Petrus Camper—who joined his expertise in the anatomy of the orangutan and the various human "nations" with his passion for painting and sculpting—who produced the facial angle that became, in the hands of nineteenth-century race scientists, an infamous measure of racial beauty and intelligence. In his *Dissertation on the Natural Varieties Which Characterize the Human Physiognomy* (published in Dutch in 1791, quickly issued in French, German, and English translations), Camper provided a series of drawings representing different kinds of profiled skulls and heads of apes, humans of different races, and Greco-Latin statuary, each placed on a grid with superimposed facial angle.

In a sequence of increasing facial angles, Camper's iconic series culminated in the head of Apollo with its ideal (Winckleman-endorsed) facial angle of one hundred degrees. Among humans, Camper noted, the facial angle varied between seventy and eighty degrees: anything above belonged to the realm of art, anything below to the animal kingdom.[17]

Camper was not promoting a human hierarchy from ape to Apollo, with a godlike European at the top and a simian-like African at the bottom. He held staunchly to the "unity of man" thesis and was an opponent of rigid racial categories, and especially of the blurring of the human/animal distinction. He was vehemently opposed to slavery and an early critic of color discrimination. His facial angle was designed to demonstrate that the characteristic "national" physiognomies of humanity were natural, obeying morphological rules, and had not been produced by deliberate manipulations of skulls, noses, and lips, as many believed.[18] His audiences were tremendously impressed by the graphic metamorphosis of an African into a European and vice versa that Camper would sketch out in front of them. As the jaw protruded (or became more prognathous), the nose would appear to diminish, while the nostrils would spread and the lips become everted. Changes in one part of the body produced changes in others, so that the appearance was materially altered. It was all a matter of morphology, of correlation of parts, and had nothing to do with any alleged deformation practices or moral or intellectual inequalities.[19]

The problem was that, whatever Camper's "benign intention," his endlessly replicated series of profiles and crania from ape to Apollo made its own statement: "It 'states' that there is a hierarchy of humanity that privileges the European as the closest to the universal [aesthetic] ideal." Camper's facial angle was popularized not only through his many lectures to artists and anatomists in France, Germany, and England, but also by that indefatigable scavenger Lavater.[20] Lavater offered his own version of progressive metamorphoses, from the elongated snout of the frog, via the projecting jaws and sloping forehead of the ape, to a human profile, culminating in the laurel-crowned brow of Apollo. His intention was that the aesthetic template of the facial angle would affirm visually the separation of man from animal. But on the face of it, Lavater's graphic series is interpretable, like Camper's, as an aesthetic hierarchy from the animal to the human, a proto-evolutionary genealogy for man.[21]

The dangers implicit in such schemes provoked Johann Friedrich Blumenbach's critique of Camper's facial series. Blumenbach, a leading European anatomist and anthropologist, boasted the largest extant collection of skulls of "different nations," built up through his extensive contacts with naturalists and expeditionists. In his influential *On the Natural Variety of Mankind* (originally published in 1775), he classified humanity into five "varieties," using the methods of the natural historian and basing his classification on the whole body, including stature, color, skull shape, and physiognomy—which he called the "national face" peculiar to each variety. In the English translation (owned and annotated by Darwin), edited by Thomas Bendyshe on behalf of the defiantly racialist Anthropological Society of London, this inevitably became "racial face."[22]

It was Blumenbach who gave the name "Caucasian" to the "most beautiful race of men," the Georgians who inhabited the slopes of Mount Caucasus (identified by the French naturalist Georges-Louis Leclerc de Buffon as the cradle of humanity), on account of the pleasing symmetry of their skulls, the general whiteness of their skins, and their oval faces: "That kind of face, which according to our opinion of symmetry, we think becoming and beautiful."[23] Blumenbach was a persistent critic of contemporary pluralists or polygenists, minimizing the differences among the human varieties and

affirming their common origin. These concerns underlay his empirical criticisms of Camper's facial angle: it was not a precise measure of "national"
difference, which should be based on wider criteria than skull profiles.
Nonetheless, Blumenbach's own promotion of skull shape as an indicator
of "variety" did much to consolidate its primacy as the only reliable, scientific measure of racial difference. And, with his elevation of the Caucasians
to the most beautiful people—assigning to them his own ideal, prototypical
European face—Blumenbach reinforced the notion of European superiority
in an aesthetic hierarchy of races.[24]

The aesthetic dimension of Blumenbach's anthropology was made most
explicit by his selection of a Georgian female skull—rather than a male skull,
in line with usual practice—to represent the Caucasian race. Blumenbach
thereby made feminine beauty a major criterion in his endorsement of Caucasian aesthetic supremacy, a criterion that reflects both the contemporary
aesthetic identification of beauty with the female and the value (sexual,
aesthetic, and economic) historically assigned to Georgian women in the
slave markets of the Levant. Such women, the story went, were so valued by
the Turks that they were kept exclusively for Turkish harems. The legendary
beauty and desirability of Georgian slaves, coupled with the erotic fantasy of
the Oriental harem, was to have a long life in Western theories of aesthetic
choice, including Darwin's.[25]

Meanwhile, Blumenbach's criticisms notwithstanding, the template of
the facial angle was promoted by various anatomists and naturalists who
saw in it the promise of a definitive, quantitative method of race classification. It was adopted by the eminent French anatomist Georges Cuvier, who
correlated the degree of prognathism with brain size and intelligence; by the
influential Belgian statistician Adolphe Quetelet, whose statistical theories
became the basis of American Civil War anthropometric investigations; by
Samuel George Morton, whose craniological measurements underpinned
the American school of polygenists; and in Britain by the above-mentioned
Robert Knox, a major popularizer of "race science" (see ch. 10).[26]

But it was the visual impact of Camper's beautifully presented series of
faces and skulls, from ape to Apollo, that most inspired those who saw in
it confirmation of European aesthetic, intellectual, and moral superiority.
In England, Camper's series was the basis for the extended illustrative chart
from the "perpendicular face of the human European" to the "horizontal
one of the woodcock" in Charles White's *Account of the Regular Gradation in
Man* (1799). White, a Manchester physician, located the African furthest in
humanity from the European and inferred black inferiority from the proximity of the African to the ape in his visual reinterpretation of Camper's

series: the African was "nearer to the brute creation than any other of the human species."[27]

While his simianized African suggests an intermediate link between the animal and the human in his scale of being, White was most intent on driving a gap between the African and European. He categorized four separate "species" of humanity, rejecting the conventional belief that all the human varieties had descended from a single pair. Besides skull conformation and facial angle, White located these racial or "species" differences in skin color, hair color and texture, the presence or absence of a manly beard, and length of limb (the long arms of Africans demonstrated their apelike affinities). He also demarcated the African from the European on aesthetic grounds: the European is the "most removed from the brute creation . . . the most beautiful of the human race." White's paean to the beauty of the European "species" smacks as much of Lavaterian physiognomy as of Camper's facial angle: "that nobly arched head, containing such a quantity of brain," "the perpendicular face, the prominent nose, and round projecting chin." But when it came down to it, White's aesthetic ideal was not located in a Greek god, but in an English female who might have been sketched by Jeffrey or Alison—except for her recognizably Burkean attributes:

> In what other quarter of the globe shall we find the blush that overspreads the soft features of the beautiful women of Europe, that emblem of modesty, of delicate feelings, and of sense? Where that nice expression of the amiable and softer passions in the countenance; and that general elegance of features and complexion? Where, except on the bosom of the European woman, two such plump and snowy white hemispheres, tipt with vermillion?[28]

For White, the aesthetic shift from the identification of beauty with the masculine to the feminine was caricatured in his overtly sexualized fixation on the breasts of European women as emblematic of European superiority.[29]

Increasingly, the anthropological aesthetic ideal moved from the masculine to the feminine and from Greek statuary to contemporary European faces and heads. William Lawrence elevated Blumenbach's Georgian skull, as the embodiment of physical beauty and moral virtue, over the Medicean *Venus*—the conventional icon of female beauty—complaining that the head of the goddess of love was so small that she was likely an "idiot." In effect, the Greek ideal physiognomy became merged with a "civilized" European or "Caucasian" norm, one susceptible to measurement, which was to be contrasted with the inferior physiognomy, morality, and intelligence of other races.[30]

5.3 Back to the *Beagle*

By the time Darwin set sail on the *Beagle*, the visual ideology of the "ugly savage," explored in section 1.2, had acquired a certain anatomical and ethnographical legitimacy.

The second volume of Charles Lyell's *Principles of Geology* (1832), which Darwin absorbed while on board the *Beagle*, included Lyell's opinion that a low facial angle in certain races of men was "frequently accompanied with inferiority of mental powers." However, Lyell condemned as "mere visionary speculation" any "attempt to trace a graduated scale of intelligence through the different species of animals." The notion that an orangutan might aspire to the "attributes and dignity of man" was unthinkable. Lyell commended the "convincing proofs" offered by Blumenbach, Prichard, and Lawrence that the "varieties of form, colour, and organization" of the different human varieties were "perfectly consistent" with received opinion on their derivation from an originally created single pair.[31]

Lyell was much concerned with the recent upsurge of evolutionary speculation among English radicals and medical reformers on the humanlike morphology of the ape brain, which had been given a fillip by the observations of the French materialist Bory de Saint-Vincent. Bory was a transmutationist, a disciple of Lamarck, but also a pluralist who advocated the separate developments of the human races (which he designated as separate species—some fifteen of them) from a plurality of ape ancestors. His entry on "Man" in his coedited *Dictionnaire* of natural history, the assorted volumes of which had been hauled on board by Darwin, outraged FitzRoy. Darwin was no pluralist advocating separate racial origins and he rejected Bory's claim that the Patagonians and Fuegians were separate species. However shocking the savage Fuegians might be, they were the "same" people as the Pampas Indians, though much degenerated. Yet he quietly collected and pondered the implications of the different kinds of lice infesting the different races (including the English), an argument exploited by pluralists like Bory and Charles White. Many years later, in the *Descent*, Darwin agreed that this evidence favored the classing of the human races as distinct species, while maintaining the common origin of both humans and their parasites.[32]

Lyell and Bory aside, the poop cabin on the *Beagle* held a fair sampling of natural histories and travel narratives that touched on ethnological issues, the most extensive discussion being offered by Johann Reinhold Forster's *Observations Made during a Voyage Round the World* (1778).[33] A German pastor who came to England to teach at the Warrington Academy, center of radical and Dissenting thought, Forster traveled as naturalist on Cook's second expedition to the South Seas (1772–75). Forster's *Observations* set a new benchmark for ethnography, becoming the standard source for descriptions and categorizations of the peoples of the Pacific and Tierra del Fuego. He distanced his experientially based ethnology from "systems formed in the closet" and from the earlier influential Arcadian vision of the South Seas offered by the French explorer Bougainville (also in the *Beagle* library). Bougainville famously romanticized Tahiti as the "Nouvelle Cythère," an idyllic pastoral setting inhabited by a race of hospitable, godlike men and "celestial" nymphs who offered themselves freely to the ship's crew.[34]

Nonetheless, the vaunted "philosophical eye" that Forster turned on his ethnological subjects was primarily aesthetic—and somewhat erotic. His interpretations were not uninfluenced by his more than strictly ethnological barter with the "accomplished jilts" of Tahiti. These he described as having,

> an open, chearful countenance, a full, bright and sparkling eye . . . the features arranged with uncommon symmetry, and heightened and improved by a smile, which beggars all description . . . [They] are in general finely, nay delicately shaped. The arms, hands, and fingers of some are so exquisitely delicate and beautiful, that they would do honour to a Venus of Medicis.

Even so, Forster made it plain that these Pacific Venuses could not bear comparison with the "fairest beauty of Europe," whom, being more like an angel, he classified among the virtuous "heavenly sisterhood of Eve's fair daughters," as far removed from the "ugly, loathsome ouran-outang" as she was from the "deformed negro."[35]

Forster's ethnology constituted an aesthetic hierarchy of races in the unity of the species. He ranked the Oceanic islanders from the Tahitians down on the basis of a correlation between their relative beauty and degree of civilization (measured by their manners, "happiness," population, property, and political structure—including the situation of women). The Fuegians were located at the extreme end of this aesthetic continuum, as a type unrelated to the Oceanic islanders and the key instance of the "most wretched savage, removed but in the first degree from absolute animality."[36]

The correspondence between Darwin's and Forster's aesthetic and moral rankings is beyond coincidence. We need look no further than Forster's several lengthy descriptions of Fuegian physiognomy and culture for the stereotype of the ugly Fuegian that Lavaterian physiognomy was to incorporate and Darwin's own publications were to consolidate.[37]

Back in London Darwin did not hesitate to jump the physiognomic and scriptural gap between Fuegian and orangutan that Lyell insisted on and that

Figure 5.2. "Chimpanzee disappointed and sulky," from Darwin's *Expression of the Emotions in Man and Animals* (1872). The expressions and behavior of captive apes showed such affinity with human emotions and intelligence that Darwin viewed them as almost more human than the uncivilized Fuegians. *Expression*, 141. University of Sydney Library, Rare Books and Special Collections.

Forster, having invoked the dangerous comparison, had retreated back to.[38] He began to read his way into the available literature on "race."

This literature was located at the intersection of a complexity of positions on the origin and unity of the human "varieties," racial equality, slavery, the colonial system, the credibility of revealed religion, and materialism. It constituted a body of writings where no precise meaning attached to "race," where pluralists were not necessarily anti-abolitionists (even White expressed abolitionist sentiments[39]), where both polygenists and monogenists might draw on physiognomy or Camper's facial angle to support their opposing views. Lavater's insistence on the fixed and essential nature of individual identity was attractive to those like White who opposed environmental accounts of human variety (which tended to be associated with abolitionist defenses of human unity and brotherhood). On the other hand,

Samuel Stanhope Smith used Lavaterian physiognomy to defend the biblical unity of the human races against the heterodox pluralism of the Scottish polygenist Lord Kames (Henry Home); and it was Smith who first coherently enunciated the principle of aesthetic choice in forming the characteristic racial physiognomies.

5.4 The American connection: "Smith's Varieties of Human Race [well read]"

Smith (1751–1819) was an American professor of moral philosophy, president of the College of New Jersey (later Princeton) from 1795 to 1812. His *Essay on the Causes of the Variety of Complexion and Figure in the Human Species* was first published in 1786 and reissued in a much-expanded edition in 1810. Smith's writings were contingent on the significant social issues of American slavery and racial intermixture, but historians agree in locating his *Essay* primarily in a Scottish intellectual context. As the son-in-law of John Witherspoon (signatory to the Declaration of Independence), Smith had close connections with one of the most important Scottish moralists and theologians working in America. Physiognomy aside, Smith drew extensively on the writings of the Scottish Common Sense philosophers in forging his theological/ethnological agenda against Kames.[40]

Common Sense philosophy, as promulgated by Thomas Reid and Dugald Stewart, had a strong innatist component and, while tending toward cultural relativism, assumed a fundamental sameness among humankind. The related tradition of Scottish conjectural history, associated with Adam Ferguson, Adam Smith, and Kames, emphasized the progress of civilization and the conditions of social and economic life: subsistence modes, the division of labor, and the institution of property. Both schools questioned the environmental determinism they associated with continental philosophers and naturalists such as Buffon and Rousseau. In general they maintained that "moral" causes (which included forms of government or the economy) were the real determinants of human cultures. Most adhered to orthodox monogenetic assumptions of human unity, but Kames questioned the biblical account of the single origin and unity of the human races and provoked controversy among his Scottish contemporaries.[41]

Kames was an Edinburgh judge, a leading aesthetician and distinguished patron of Scottish literati. He intended his *Sketches of the History of Man* (1774) as a "natural history of man" that would chart the "History of the Species, in its progress from the savage state to its highest civilization and improvement." Kames consistently compared humans with the rest of the animal kingdom, rambling voluminously across the spectrum of social institutions: property, government, manners, the useful and fine arts, the moral sciences, and, interestingly, the status of women. Like Erasmus Darwin, Kames was an enlightened advocate of female education, arguing its necessity for "making women fit companions for men of sense." He articulated what was to become a familiar ethnological trope, accentuated by Forster and reiterated by Charles Darwin on the *Beagle*: that the status of women was a significant measure of social progress, their "low condition" among "savages and barbarians" leading to polygamy and other "brutish" practices. In a civilized society, women, while intellectually inferior to men, had, with the benefit of education, "sufficient understanding to make a decent figure under [the] good government [of men]." Nevertheless, Kames's naturalization of the sexes obeyed the conventions: in matrimony "it is the privilege of the male, as superior and protector, to make a choice: the female preferred has no privilege but barely to consent or to refuse. Nature fits them for these different parts; the male is bold, the female bashful."[42]

Kames's views on women went unchallenged; what aroused fierce debate was his capitulation to pluralism in disputing the environmentalist claim that varying climatic or environmental conditions could produce the observable physical and mental differences of the human races. Was Buffon, Kames asked rhetorically, "seriously of opinion" that "climate, or . . . other accidental cause, can account for the copper colour and smooth chin universal among the Americans, the prominence of the *pudenda* universal among Hottentot women, or the black nipple no less universal among female Samoides?"—a point incorporated by Darwin in a note on climate in his transmutation notebooks.[43] While Kames conceded some influence to social conditions in shaping moral and intellectual differences, he veered toward the contentious view that the "uniformity and permanency" of racial physical differences were such that they could only be the "offspring of design," and that God had originally created "different races of men."[44]

Smith's *Essay* featured the themes rehearsed by Kames's Scottish critics, but he added certain arguments of his own, seemingly confirmed by the "ethnological laboratory" that was America. His expanded edition of 1810 was intended to combat the more recent pluralist "infidelity" of Charles White and his American followers, who found White's facial angle–based demonstration of black inferiority specially appealing. Smith, in endorsing the unity of the species and a common human nature, was also concerned with underwriting the political stability of the new republic, with giving a naturalistic foundation to the common principles of public virtue embodied in its Declaration.[45] To demonstrate unity and refute the infidel pluralists, Smith had to explain how humanity had come to vary by natural causes. Assuming a single original stock, his intention was to show how the "state of society" or "manner of living," united with the effects of climate, could afford "sufficient principles on which to account for all the varieties that exist among mankind."[46]

He took his cue from physiognomic practice, positing the human countenance as the preeminent site for locating and interpreting the modifications wrought by climate and social custom: God's benevolence is reflected in the plasticity of human faces and natures so that they may adapt to different climates and states of society. Lavater, that otherwise admirable authority, had not "accurately considered" the extent to which the features, expression (which reflects the emotions and so modifies and fixes the features), and even skull shape are improved by agreeable scenes, clothing, cultivated pursuits, a refined diet, commercial enterprise, liberal government, good manners, religion, and moral habits.[47]

Against the insistence of White and Kames on the immutability of color and physiognomy, Smith asserted the "undoubted fact" that descendants of African slaves in America were "gradually losing those peculiarities so offensive to our eye." Slaves were slowly changing toward a European form— especially house servants, who, out of the sun and exposed to civilized society, were assuming a more agreeable complexion and regular features. Put simply: the American climate and civilization were transforming black slaves into white men and women. The social and aesthetic problem would go away as Africans gradually disappeared. It was a reassuring message that denied inherent inferiority while conceding present inferiority, and so rationalized existing prejudices and practices in the new republic.[48]

Although Smith invoked the authority of Blumenbach against White's defining index of the facial angle, he did not deny its anatomical and aesthetic legitimacy. His solution was rather to insist on the God-given plasticity of human physiognomy and on the efficacy of climate and culture, particularly culture, in shaping it toward the realization of the civilized European ideal— that state for which "man" was designed by his Creator, in which he attains the "greatest beauty of the human form, as well as the highest perfection of his whole nature." This perfection, naturally, correlated with the "perfect symmetry" and "beautiful proportions" of Greek and Georgian physiognomy, which "most nearly correspond with the original idea of the Creator."[49] It was a defense of human unity made at the cost of reinforcing the very same racial and aesthetic hierarchy that the template of the facial angle endorsed.

Culture, in Smith's ethnology, therefore assumed a critical role in racial modification. It was under this category that he introduced another and most important modifying cause: a "general and national standard of beauty," which was "various in different countries" and had "its effect in forming the features and fashioning the person": "A Laplander prefers the flat, round faces of his dark skinned country women to the fairest beauties of England."[50] In every country, people practiced various arts to achieve some favorite idea of the human form—plucking body hair, stretching ears, flattening heads, or binding feet. Such artifices "insensibly, in a long course of time, produce great and striking consequences" through the inheritance of acquired characters. The inferior animals afford many instances of this natural law: "By competent skill, and the application of proper pains, or, on the other hand, by neglect, or ignorance, the races of all our domestic animals may be almost infinitely varied."[51]

In similar fashion, human marriages were unconsciously shaped by the particular ideal of beauty peculiar to different races and nations; this would, through inheritance, tend to further differentiate and perpetuate racial and national complexions and facial and bodily features: "If men, in the union of the sexes, were as much under control as some of the inferior animals, their persons might be moulded, in the course of a few generations, to almost any standard."[52] However, these eugenic possibilities were too often compromised by fleeting passion, "tasteless caprice," or gross financial considerations. Another factor was the power of the "superior ranks" in appropriating to themselves the "most beautiful of the sex." Hence the Persian nobility had greatly improved their original "harsh" and "disproportioned" Tartar features by this exercise of choice; likewise the Turks.[53]

Beauty, as physiognomy taught, was also associated with moral values and mental capacity. Just as one could not find beauty among savages, so neither intelligence nor morality was to be found: "And the Hottentots, the Laplanders, and the people of Tierra del Fuego are the most stupid of mankind for this."[54] The brutish, inexpressive features of the savage reflected the brutishness and vacancy of his mind.

In essence, Smith offered a coherent, naturalistic explanation of racial diversity that depended largely on the human (but God-directed) agency of aesthetic selection. According to his records, Darwin read Smith's *Essay* in April 1840 and again in 1843.[55] He did not own a copy; he possibly read it in the extensive library at Maer (his Uncle Josh Wedgewood's residence, where we know he read Kames's essay). We do not have his notes or marginalia. Smith's association of beauty with moral values and mental capacity, his low ranking of the Fuegians in these, his relativistic yet hierarchical aesthetics, his views on expression and on the effects of inherited habit, the analogy with domestication—all are familiar themes in Darwin's London notebooks. The problem is that these notebook themes predate Darwin's recorded reading of Smith's *Essay*, and more tellingly, none of them explicitly can be related to the *Essay*; nor, unlike Lawrence and Prichard, is Smith cited in either the *Descent* or *The Expression of the Emotions*.

Smith's *Essay* was well known in British ethnological circles, especially in Edinburgh, where, a year after its initial Philadelphia publication, it was published with additional notes by a "Gentleman of the University of Edinburgh" (actually an American medical student, Benjamin Smith Barton, who became a leading American naturalist). A London edition was published in 1789. The *Essay* was well reviewed, particularly in radical circles, where its "oddly optimistic" melding of democratized physiognomic principles with climatological and cultural evolution appealed to those like Mary Wollstonecraft, who drew on aspects of it in her own writings on national difference.[56] The most important ethnological application of Smith's *Essay* was made by Prichard in the first edition of his influential *Researches into the Physical History of Man* (1813).

Prichard's is regarded as the single most important voice in the founding of British ethnology. In Britain, ethnology had a strong religious inflection and an early humanitarian focus, being linked with Quaker and evangelical philanthropy and the abolitionist movement. With the legal abolition of the slave trade in 1807, the political and philanthropic focus of the study of non-Europeans shifted to a concern with civilizing, Christianizing, and (to an extent) monitoring the colonial abuses and incipient extinction of indigenes. The Aborigines Protection Society was formed in 1837. But the tensions between humanitarian and scientific interests led to the foundation of the Ethnological Society in 1843 under the presidency of Prichard, whose successively reworked and expanded editions of his *Researches* had dominated British ethnological research for the previous four decades.[57]

5.5 Ethnology and analogy:
"Prichard. Physical Researches. [read]"

Prichard (1786–1848) came from a Quaker family in the Bristol iron trade. He studied medicine in Bristol and St. Thomas's Hospital in London, before going, like other Dissenting physicians, to Edinburgh in 1805 to take out his degree. Edinburgh was then at the height of its reputation, offering what was considered to be the best medical training in Britain, if not the world. There Prichard was directly exposed to Common Sense philosophical orthodoxy and its well-entrenched interest in racial variation. Edinburgh medical edu-

cation also kept abreast of continental developments in anatomy and anthropology.[58]

Prichard's 1808 Edinburgh dissertation, *De generis humani varietate* [On
the Origin of Human Variety], which became his *Researches into the Physical
History of Man* of 1813, reflects his exposure both to the comparative anatomy of Blumenbach and to the epistemology of Dugald Stewart's popular
lectures in moral philosophy. But, above all, Prichard's *Researches* demonstrates his debt to Smith's *Essay*, certain significant aspects of which he
adapted to his own argument for human unity.[59]

In Edinburgh, Prichard became involved with the rising Evangelical
movement, and, as with Smith, this strong commitment to biblical authority structured his scientific defense of the unity of the human species. Prichard's philanthropy was manifested in his strong opposition to the slave trade
and his concern with public health reform, while he shared the conventional Evangelical revulsion against the atheistic barbarities of the French
Revolution and repudiated political radicalism and materialism. He was a
staunch anti-transmutationist, strongly opposed to the evolutionary theories of Lamarck and Erasmus Darwin. These prejudices pushed Prichard into
a greater reliance on linguistic and historical sources and arguments—in
the tradition of Christian apologetic writing—that came to characterize the
ethnological method.[60]

Yet, like Smith, Prichard necessarily was heavily reliant on biological solutions in explaining human variation in terms consistent with the biblical
account. His "analogical method" of reasoning compared physical variability within the different species of animals (especially domesticated species,
which Prichard, like Blumenbach and Smith, found particularly suggestive)
to that evidenced by the different human races, which he argued on this basis
to be one species. The variations of skin and hair color, facial features, head
shape, and stature among humans were no greater than those to be found in
dogs; as all dogs were one species, it followed that all humans must be one
species. Furthermore, "in obedience to the most firmly established laws of
philosophical reasoning," it was requisite to refer these "strictly analogous"
variations to "similar causes." There must therefore be some "principle of
natural deviation" that explained both human and animal variability.[61]

When he came to consider this principle in the 1813 edition of his *Researches*, Prichard's Common Sense–inspired criticisms of what he saw as
the excesses of the prevailing environmentalism led to his decisive rejection
of climate as a mechanism in the differentiation of races. However, Prichard
followed Smith closely in correlating culture and color, featuring Smith's
claim that the appearance of domestic slaves had been "improved" over that

of field slaves. He radically extended this claim to conclude that "the process of Nature in the human species is the transmutation of the characters of the Negro into that of the European, or the evolution of white varieties in black races of men." In effect, "the primitive stock of men were Negroes."[62] As man had mentally progressed from savagery to civilization, he had physically "evolved" from the type of black African to white European.

It was possibly because readers found "repugnant" the notion that the original humans "created in the image of their Maker" were black, that Prichard dropped his unsettling hypothesis of primitive blackness from subsequent editions of his *Researches*. Other reasons lay in the increasingly negative evaluation of the cultural worth of non-European peoples that accompanied the economic expansion of Britain. As the ideology of the "ugly savage" took hold, cultural achievement came to be seen as the product of racial capacity. Civilization could not form or transform race; rather, race determined the capacity of a people for civilization. In the second edition of the Researches of 1826, Prichard subordinated civilization to climate and explained racial development in terms of the conventional correlation of climate and physical type.[63] This shift also meant the abandonment of Prichard's earlier emphasis on the significant role of aesthetic selection in the formation of race.

In this first edition of 1813—like Smith, and true to his analogical method—Prichard pointed to the "very considerable" changes wrought by skillful breeders of cattle, horses, and other domesticated animals. Male aesthetic choice had acted in the same manner to produce the different races, Prichard stated, citing Smith's "ingenious remarks on this subject":

> The perception of beauty is the chief principle in every country which directs men in their marriages. It does not appear that the inferior tribes of animals have anything analogous to this feeling, but in the human kind it is universally implanted. It is very obvious that this peculiarity in the constitution of man, must have considerable effects on the physical character of the race, and that it must act as a constant principle of improvement, supplying the place in our own kind of the beneficial controul which we exercise over the brute creation.[64]

In this same edition, consistent with his rejection of environmentalism, Prichard refuted the notion of the inheritance of acquired characters: the law of inheritance "has ordained that offspring shall always be constructed according to the natural and primitive constitution of the parents, and therefore shall inherit only their connate [i.e., congenital] peculiarities and not any of their acquired qualities."[65] Prichard documented several instances of the transmission of congenital peculiarities, including the Royal Society reports on the famous "porcupine" man and his warty offspring and the case of the six-fingered and -toed Ruhe family in Germany. If such curious variations had occurred in circumstances that favored their propagation, we would find "races of men much more different from ourselves than any which now exist, and therefore affording stronger argument of diversity of kind."[66]

Such peculiarities were not, however, generally selected and perpetuated. Too great a difference excited aversion or disgust. Man's innate sense of beauty would lead to the rejection of naturally occurring deviations that were too extreme or did not conform to this instinctive human aesthetics. It was the pious Prichard's Common Sense conviction that this aesthetics was God-given and directed to beneficial ends. Providence had implanted a single (European) standard of beauty to serve as a "constant principle of improvement" in the human species. Because the higher ranks of society could exercise this selective aesthetic more readily than the lower orders, the result was, even among barbarous Pacific Islanders, that "improvement of person . . . which is much more conspicuous in the higher than in the inferior classes."[67]

Yet, in order to explain racial diversity, Prichard had to allow for the existence of different aesthetics in different races. "It is probable," he wrote, "that the natural idea of the beautiful in the human person has been more or less distorted in almost every nation. Peculiar characters . . . accidentally enter into the ideal standard." So the "Negroes of Africa . . . are said to consider a flat nose and thick lips as principal ingredients of beauty." The effects of such different standards of beauty must play an important role in the initial production and subsequent widening of human physical differences among races.[68]

When, with the 1826 edition of the *Researches*, Prichard abandoned aesthetic selection in favor of climate, he modified his theory of heredity to accommodate this significant shift.[69] By the third edition of the *Researches*— the first volume of which was issued in 1836—many passages now suggested the heritability of acquired characteristics in the formation of race. By the time the fifth and last volume was issued in 1847, Prichard's argument was virtually indistinguishable from a traditional environmentalist account of racial differentiation and was so read by reviewers.[70]

Prichard's views on race were important to Darwin. How could they not be? Prichard was the most prestigious and influential British authority on race in the first half of the nineteenth century. In August 1839, Darwin heard Prichard speak "On the Extinction of Human Races" at the Birmingham meeting of the British Association for the Advancement of Science and joined the committee formed to collect information from travelers on endangered indigenous races. His first recorded reading of Prichard's *Researches* occurred in September 1839.[71]

But Darwin's reading of Prichard was not the key to the formulation of his theory of racial aesthetic selection, as has been claimed.[72] The difficulty with this interpretation is that all of Darwin's extant references are to the third and (posthumous) fourth editions of the *Researches*, assorted volumes of which he owned, annotated, and abstracted.[73] The most he could glean from these, as far as aesthetic selection went, was a scattering of references to notions of beauty among the different races. The vital formulation of aesthetic selection had been eliminated in favor of an environmentalist interpretation of racial differences. Darwin never read the crucial first edition.[74] He read the later editions, as he read so many other anthropological texts in the years before the *Descent*, already primed to look for and collect instances of different racial notions of the beautiful in support of the theory of aesthetic selection in the formation of race.

Prichard's writings on racial origins, with their major shifts and revisions, were never subordinated to a single unifying theme that allowed his readers easy access to the essentials of his argument for racial unity. Ironically, through its attempted suppression, Lawrence's *Lectures* remained unchanged and in print for almost fifty years (the last pirated printing was in 1866).[75] It was this work that kept in circulation the major themes of aesthetic selection, individual variation, and heredity that Prichard had assembled for the first edition of his *Researches*, but which were here repackaged by the materialist Lawrence.

5.6 "Oh you materialist!": "Lawrence read"

William Lawrence (1783–1867) was a professional medical man and comparative anatomist, an armchair ethnologist who based his *Lectures on Man* primarily on published material. His leading authority was Blumenbach (to whom Lawrence dedicated his *Lectures*), although he was well versed in other major ethnological writings, including those of Smith and, of course, Prichard.[76]

Lawrence's *Lectures* drew on the same matrix of aesthetics, anatomy, biblically inflected ethnological theory, and the politics of abolition and colonialism. However, his *Lectures* firmly relocated the discussion of human racial difference in a materialist context that emphasized the necessary connection of mind with physical organization. Lawrence was an early adherent of the new secularist phrenology that was rapidly overtaking physiognomy in popular thought and, particularly, in reform-oriented medical circles, though his *Lectures* also demonstrates the lingering influence of physiognomy.[77] Like Blumenbach, Smith, and Prichard, Lawrence also relied heavily on the analogue of domestication in animals. Man, he argued, was "in the true . . . sense of the word, more of a domesticated animal than any other." It followed that "the differences of physical organisation and of moral and intellectual qualities, which characterise the several races of our species, are analogous in kind and degree to those which distinguish the breeds of domestic animals; and must, therefore, be accounted for on the same principles."[78]

Like the early Prichard, Lawrence completely excluded the inheritance of acquired characters. The state of domestication conduced to the production of "new properties" or "congenital peculiarities of form, like those of colour," and only such congenital variations "are transmitted by generation."[79] He too pointed to the known cases of inherited albinism, polydactyly, and the famous "porcupine" family, and speculated:

> Let us suppose that the porcupine family had been exiled from human society, and been obliged to take up their abode in some solitary spot or desert island. By matching with each other, a race would be produced, more widely different from us in external appearance than the Negro. If they had been discovered in some remote period, our philosophers would have explained to us how the soil, sun, or climate, had produced so strange an organisation; or would have demonstrated that they must have sprung from an originally different race; for who would acknowledge such bristly beings for brothers?[80]

Yet brothers they would be, warts and all. Lawrence, for all his secular rhetoric, was a committed monogenist. He was convinced that all the human races were varieties of the one species whose differences had been produced over time through the accumulation of inherited congenital variations.[81]

Lawrence, however, went further than Prichard by totally rejecting both climate and culture as the causes of racial differentiation. Climate could not account for the fact that Native Americans were spread over every climate, likewise the Africans and the Mongolians.[82] As for culture, no degree of civilization, way of life, form of government, religion, or education could turn a black man into a white one, as Smith and Prichard assumed. Whatever their climate, country, or circumstances, the physical and moral characteristics defining the white and dark races remained distinct.

We must look deeper for their causes, and seek them in some circumstances inseparably interwoven in the original constitution of man . . . We shall find in the comparison of the crania of the white and dark races, a sufficient explanation of the superiority constantly evinced by the former, and of the inferior lot to which the latter have been irrevocably doomed.[83]

This did not mean that Lawrence endorsed the misery and inhumanity of the slave ship and plantation; nor did he deny the many instances of "strong intellect" and the "natural goodness of heart" of the "Negro."[84] But it did mean that the expectations of well-intentioned philanthropists for the "improvement" of the African and the American to European standards were "quite as unreasonable" as it would be "to hope that the bull-dog may equal the greyhound in speed; that the latter may be taught to hunt by scent like the hound; or that the mastiff may rival in talents and acquirements the sagacious and docile poodle."[85]

The differences in moral and intellectual character among the different races were grounded in their different physical organizations, and were to be explained by the same two principles as in domesticates: "namely, the occasional production of an offspring with different characters from those of the parents, as a native or congenital variety; and the propagation of such variety by generation."[86]

The perpetuation of such varieties depended on the prevention of "intermarriage," or backcrossing, into the population at large. This was easily accomplished in domestic animals whose matings were controlled to preserve the desired attributes. In humans, matters were more complex than Lawrence's hypothetical case of geographic isolation of the porcupine family. Here culture did come into play, and Lawrence followed Smith in arguing that in larger populations "differences of manners, religion, and language, and mutual animosities" would prevent intermarriage with surrounding people, so that over the course of time, individual peculiarities would be lost and a characteristic countenance and form would become established. These would be reinforced by a corresponding ideal of human beauty, and Lawrence quoted almost verbatim from Smith:

Connexions in marriage will generally be formed on the idea of human beauty in any country; an influence, this, which will gradually approximate the countenance towards one common standard. If men, in the affair of marriage, were as much under management as some animals are in the exercise of their generative functions, an absolute ruler might accomplish, in his dominions, almost any idea of the human form.[87]

Darwin marked many of Lawrence's observations on different racial notions of beauty. He double-scored the passage where Lawrence, again quoting Smith, discussed the power of the aristocracy in selecting the most beautiful women in marriage and the consequent "elegant proportions" and "beautiful features" that distinguished them from the inferior orders. He also marked the passage where Lawrence ventured into eugenic territory:

> A superior breed of human persons could only be produced by selections and exclusions similar to those so successfully employed in rearing our more valuable animals. Yet, in the human species, where the object is of such consequence, the principle is almost entirely overlooked. Hence all the native deformities of mind and body . . . are handed down to posterity, and tend by their multiplication and extension to degrade the race.[88]

Lawrence drew some sharp class distinctions. By his account, London-bred "Cocknies" came off the worse by comparison of facial expression and bodily strength and symmetry even with the savage Bushmen and Hottentots, who "have become almost proverbial for ugliness" but whose imperfect or malformed offspring either did not survive the rigors of their upbringing or failed to be selected as mates. The consequences for the ruling class, "those to whom the destinies of nations are intrusted," were even more pernicious, warned the republican Lawrence, if they were to continue their current practices of inbreeding or selecting wives for their wealth or looks without regard to their mental qualities. The "strongest illustration of these principles" was to be found in the parlous state of many royal houses in Europe, with their successive generations of "ideots" and other degenerates.[89]

Lawrence's materialist anatomy, views on racial improvement, and sideswipes at a degenerate and corrupt ruling class were consistent with the democratic politics and doctrines of self-development of the London medical reformers, who included Thomas Wakley (editor of the radical *Lancet*, with whom Lawrence was closely connected in the early 1820s) and the transmutationist Robert Edmond Grant. As Adrian Desmond's extensive researches have shown, the radical anatomists of the early nineteenth century, intent on institutional reform, deployed a naturalistic—essentially evolutionary—worldview against the entrenched Tory-Anglican privileges of the professional elite of the chartered medical corporations. Their political anatomy, charged with a subversive evolutionism and materialism, was picked up and promoted in the pauper press by assorted freethinkers, "infidels," radical artisans, Chartists, Owenite socialists, republicans, Benthamites, and others locked out of the corridors of power.[90]

The popular appeal of Lawrence's pirated *Lectures* to such an audience is obvious. Nor is it difficult to understand the appeal to Darwin of his well annotated copy of Lawrence's *Lectures*. Unlike Darwin's early mentor, Grant, Lawrence was no overt transmutationist. His evolutionism extended only to his speculations on the origin of new races, not species. But he applied them across the board—to humans, domesticated animals, and animals in a state of nature. For Darwin, busily blurring the boundaries between races and species, and between humans and animals, Lawrence's speculations could hardly fail to resonate. They had the further attraction of taking the debate on racial origins out of its established context of Christian apologetics. Lawrence's insistence on locating the moral and intellectual characteristics of the different races in their anatomy or physical organization was a harbinger of the new physical anthropology that was eventually to displace Prichard's theologically subservient ethnological method from the science of "man." It was a direction in which Darwin's own notes were taking him: "As forms change, so must idea of beauty." Lawrence's thoroughgoing materialism, his insistence that animal and human variation must be accounted for on the same natural principles, his rejection of climate and cultural correlations with race, his explanation of racial differentiation in terms of the inheritance of accumulated congenital variations, his arresting illustration of the case of the porcupine family, his discussions of geographical and cultural isolation, the highly suggestive analogy of domestication and the connection of aesthetic selection with racial differences—these were all issues and themes with which Darwin juggled in the early notebooks. It is possible to go further than this and make a direct connection between Lawrence's *Lectures* and the content of the notebooks.

It is highly unlikely that Darwin, having purchased his pre-*Beagle* copy of the *Lectures*, refrained from looking into such an attractively notorious work. When he began to keep such records, he listed it "to be read" some time in 1838 and subsequently noted "read."[91] In confirmation of this, he twice referred specifically to "Lawrence's example of the man with the scabrous skin" (i.e., the porcupine man) on both sides of a letter received on December 17, 1838.[92] But there is good reason to date his reading (or rereading) of Lawrence's *Lectures* from, at the latest, early June 1838. In a notebook entry made around then, Darwin summarized the major differences in the "races of men" as Lawrence had listed them, although he does not cite Lawrence: color, form of head ("hence intellect?"), hair, and forms of legs. He went on to speculate on this basis on the structure of the "father of mankind" or common ancestor, an "animal halfway between man & monkey." This led him to add, "Negro [or father of Negro] probably was first black at base of nails & over white of eyes."[93] This is an explicit reference to Lawrence's *Lectures*, where Lawrence stated that Negro children are born fair and gradually darken, becoming dark first around the skin of the nails, and where he also described the changes in eye color. Darwin marked this passage in his copy and made a notation inside the cover, "Eyes of Negros [*sic*] at Birth."[94]

It was not long after this notebook entry, that Darwin made his crucial notes on the innately different notions of beauty in the different races.[95] In an additional significant entry, paraphrasing Lawrence's extensive critique of the same, Darwin questioned the standard ethnological correlation of climate and civilization with race. There had to be another correlate or "cause" that would explain "colour & shape & idiosyncrasy."[96] Coincident with this, Darwin began reading intensively in aesthetics and flirted with materialism.[97] In one sequence of reasoning around late May or early June 1838, Darwin, pondering the inheritance of instinct in horses and dogs, reminded himself "Talent &c in man not hereditary, because crossed with women with pretty faces," but concluded a few entries later, "Thought (or desires more properly) being hereditary . . . it is difficult to imagine it anything but structure of brain." This led him to the irreverent thought that "love of the deity" was simply "effect of organization. [O]h you Materialist!," and then on to the rationalization "Why is thought, being a secretion of brain, more wonderful than gravity a property of matter?"[98] These almost lighthearted, closet heterodoxies, resonant of Lawrence's alleged heresy that "the soul is only the brain," took a more sober turn a few months later when Darwin (with the fate of Lawrence's *Lectures* in mind?) cautioned himself to "avoid stating how far, I believe, in Materialism."[99]

All this is persuasive enough. But just how significant his copy of Lawrence's *Lectures* was to Darwin's formulation of aesthetic selection may be adduced from the exclamation "found" that he wrote in the margin next to his double-scoring of Lawrence's discussion of the power of the aristocracy in selecting the most beautiful women for mates.[100] This can mean only that Darwin must have remembered this passage from an earlier reading; that he went back to Lawrence's *Lectures* and registered his triumph at having his recollection confirmed. Darwin had a long and retentive memory: "My memory is extensive, yet hazy," he wrote in his "Recollections."

> It suffices to make me cautious by vaguely telling me that I have observed or read something opposed to the conclusion which I am drawing, or on the other hand in favour of it; and after a time I can generally recollect where to search for my authority.[101]

The class basis of sexual selection was one of Darwin's pet themes. When he first introduced his idea that a "sort of sexual selection has been the most powerful means of changing the races of man" to Alfred Russel Wallace in 1864, he concluded with the jocular postscript: "Our aristocracy is handsomer? (more hideous according to a Chinese or Negro) than middle classes from pick of women."[102] Lawrence's views on the "beauty of the English aristocracy" (and on the developing color of Negro infants) were duly attributed in the *Descent* together with other of his observations bearing on sexual selection, such as love of ornamentation and beardlessness in certain races.[103]

Lawrence's *Lectures* is the earliest extant annotated source to which we are able explicitly to relate Darwin's early views on race and aesthetics.[104] In the one accessible package, Lawrence served up the essential arguments for racial divergence through sexual selection—including the all-important analogy with artificial selection—that Darwin was to rework and extend across the animal kingdom. Through Lawrence, Darwin could draw indirectly on the writings of Smith and the early Prichard. He subsequently seems to have read and absorbed Smith's *Essay*, and he garnered instances of sexual selection from the later Prichard. But, in a world where the traditional theological underpinnings of monogenism were coming under increasing strain, Lawrence's *Lectures* uniquely pointed the way to how a monogenetic account of the origin of the human races might be reconciled with a secular, materialist outlook.

In 1836, the sixth pirated edition of the *Lectures* had hit the streets with a "sneer" at the clergy and other assorted Tory targets.[105] In the meantime, Lawrence himself had traded his earlier radicalism for respectability, while the passage of time was making his *Lectures* less subversive than radicals may have wished. In the same year, Prichard, in the preface to the third edition of his *Researches*, lamented the explosion of pluralist views ("even Cuvier . . . Humboldt, Spix and Martius"). The only exception to this deplorable trend was "Mr. Lawrence's well known *Lectures*, in which the able author has maintained, with great extent of research, the unity of species in all human races."[106]

It is a curious historical twist: Prichard, having modified his earlier views under social pressure, in turn, under further social pressure, ratified Lawrence's *Lectures*, whose argument, as we have seen, owed a good deal to those same earlier views of Prichard's. This twist in the story is given yet another turn when we consider the role of Edward Blyth in bringing the ideas of Prichard and Lawrence—particularly those of Lawrence—to Darwin's attention.

5.7 Edward Blyth: "A very clever, odd, wild fellow"

Blyth was a rather tragic character. A self-made zoologist with an omnivorous knowledge of British and Indian fauna, especially birds and domesticated animals, Blyth "scraped along" as a writer and editor of natural history until financial exigency forced him to accept a poorly paid position as curator of the museum of the Asiatic Society of Bengal in 1841. He lived in virtual exile in Calcutta for the next twenty-one years in a "chronic state of discontent," bedeviled by frustrated career opportunities and want of books and companionship.[107]

His combative personality, along with his attempts to supplement his inadequate income with animal trading (a disreputable commercial enterprise with which no real gentleman would soil his hands), brought him into conflict with his employers. Blyth's difficulties were exacerbated by the death of his wife Elizabeth in 1857, after just three years of married happiness. The shock of her loss precipitated physical and mental ill health and, seemingly, excessive drinking. In 1862, Blyth was repatriated on a grudgingly granted pension. Back in England, he sporadically pursued his zoological writings and unprofitable dealings in animals, sinking into alcoholism punctuated by periods of mental instability, which led to at least one incarceration in an asylum. He died in 1873, shunned by the respectable scientists who had profited from his collections and observations. "Unfortunately he drinks," Joseph Hooker curtly informed Darwin in 1864, disclaiming all interest in Blyth's current whereabouts or intentions.[108]

To do him justice, although the gulf between their respective social and scientific positions loomed large, Darwin did attempt to help "poor Blyth" on several occasions, even, in spite of Hooker's warning, inviting him to Down House. And not without good reason: Blyth was of inestimable assistance in Darwin's investigations into breeding and sexual selection.

Darwin first met Blyth in London at a meeting of the Zoological Society in 1838 while Blyth's star was still rising and his published papers on natural history had established his early reputation as an acute field observer and expert on the varieties and breeds of those birds and other fauna he chose to study. "I liked all I saw of him," Darwin told Hooker in 1848, urging him to contact Blyth in Calcutta on Hooker's expedition to the Himalayas. Even then, Darwin had a shrewd estimate of Blyth's character: "He is a very clever, odd, wild fellow, who will never do, what he could do, from not sticking to any one subject."[109]

In the 1850s, Blyth became Darwin's chief consultant on the fauna, domestic and wild, of the Indian subcontinent. In a series of lengthy missives written in the evenings after his society work was done, Blyth assuaged his intellectual isolation by pouring out his extensive knowledge and theoretical views to an appreciative but demanding Darwin. On the publication of the *Origin*, Blyth publicly and unequivocally declared himself a Darwinian.[110] Darwin preserved some forty of Blyth's letters and "Notes for Mr. Darwin," dating from April 21, 1855, abstracting and annotating their contents. Blyth's importance is evident in Darwin's numerous citations to this "excellent authority" in the *Origin*, in *Variation of Animals and Plants under Domestication*, and the *Descent*.

Long before this, however, several of Blyth's early papers were a major source of information, if not inspiration, for Darwin's notes on transmutation. In one paper in particular, "An Attempt to Classify the Varieties of Animals" (published in the *Magazine of Natural History* in 1835), Blyth, in characteristic mode, drew on his extensive reading and observations on domesticated and wild animals—including the human animal—to group and display assorted stimulating hints for a perceptive transmutationist. They included the inheritance and perpetuation of acquired and congenital variations (both in a state of nature and under domestication); the significance of food supply and isolation (both artificial and natural) in the formation and maintenance of breeds and "true varieties"; the struggle for existence; and both variants of what would become sexual selection.[111]

Along with various sources on breeding and natural history and the more substantial works of Lyell and Prichard,[112] Blyth's views were most heavily indebted to Lawrence's *Lectures*. He cited the *Lectures* three times in the course of this paper, and closely followed Lawrence on the origin of the human races. These races Blyth classed as "true varieties" originating as a spontaneous variation or a "kind of deformities, or monstrous births," which "as Mr. Lawrence observes" are very rarely perpetuated in a state of nature, but which "by man's agency, often become the origin of a new race."[113] Like Lawrence, and against Prichard, Blyth argued that "temperature exerts no *permanent* gradual influence whatever" on color in the human races: "The coloring principle in black races is inherent in them, and is quite independent of external agency." He proceeded to speculate, qua Lawrence:

> Wherever a black individual was produced . . . the natural aversion it would certainly inspire would soon cause it to become isolated, and, before long, would, most probably, compel the race to seek refuge in emigration. . . . It is highly probable that analogous-born varieties may have given rise to the Mongolian, Malay, and certain others of the more diverse races of mankind.[114]

In speculating on how these new human varieties were perpetuated as races, Blyth distinguished between the struggle among males for mates, which occurs in the "brute creation," and aesthetic selection, which prevails among humans:

> As in the brute creation, by a wise provision, the typical characters of a species are, in a state of nature, preserved by those individuals chiefly propagating, whose organization is the most perfect, and which, consequently, by their superior energy and physical powers, are enabled to vanquish and drive away the weak and sickly, so in the human race degeneration is, in great measure prevented by the innate and natural preference which is always given to the most comely; and this is the principal and main reason why the varieties which are produced in savage tribes, must generally either become extinct in the first generation, or, if propagated, would most likely be left to themselves and so become the origin of a new race. . . . The inferior animals appear not to have the slightest predilection for superior personal appearance . . . the most powerful alone becomes the favourite.[115]

Earlier in this same article, Blyth had conflated the struggle for mates with the struggle for existence, arguing from domesticated to wild animals and back to the practices of artificial selection: in domesticated herds, the "strongest bull" drives away weaker rivals to remain "sole master" of the herd, so that all the young "must have had their origin from one which possessed the maximum of power and physical strength; and which, consequently, in the struggle for existence, was the best able to maintain his ground, and defend himself from every enemy." In nature, the strongest, the most agile, the "best organized" must always survive to leave the most progeny. This was a law "intended by Providence to keep up the typical qualities of a species," and could be "easily converted by man into a means of raising different varieties." However, unless these breeds were artificially maintained by "regulating the sexual intercourse," they would "all naturally soon revert to the original type." This same principle accounted for the "degenerating" effects of inbreeding; in a state of nature, the tendency to deviate to such disastrous effect is counteracted by unregulated crossings and by male combat that "causes each race to be chiefly propagated by the most typical and perfect individuals."[116]

While Blyth recognized the creative roles of artificial selection, the struggle for existence, and the sexual struggle in the emergence of varieties, he saw them, ultimately, as conserving forces: they do not lead to species change but to species preservation. The controlled breeding of domesti-

rates can be taken only so far before the accumulated deleterious results
of the practice of inbreeding render the animal sterile or unable to survive,
while in nature only the best or "most typical" survive to breed. Too great a
variation will disqualify the animal from its mode of existence in the wild
and from the sexual struggle, and so will not be perpetuated. Both varieties
in nature and domestic breeds oscillate around a mean; in other words, a
species. In humans, modifications that were too extreme—for instance (in
Blyth's view), a black skin—would cause aversion and could be propagated
only in isolation (like Lawrence's example of the porcupine family), remote
from the rest of humanity where aesthetic selection (the "innate and natural
preference" for the "most comely") acted to prevent "degeneration" and to
preserve the superior or "typical characters" of the species. At this stage (the
1830s), Blyth was an overt creationist who saw evidences of design in nature
and looked to an "omnipotent and all-foreseeing Providence" as the final
cause of the balance of nature and the localization and stability of species.[117]

Darwin annotated his copy of Blyth's 1835 article and made a number
of notebook references to it. The earliest of these dates around April 1838,
though Darwin reread this particular paper a number of times over the following years.[118] He was clearly impressed by Blyth's work and drew on it for
his theorizing and for information to support that theorizing.[119] On one of
his readings, Darwin double-scored the passage in which Blyth spelled out
the essentials of male combat and its role in reproductive success for the
stronger over the weaker. In the back of the volume containing Blyth's paper,
Darwin made two separate references to this passage, explicitly referring to
it as "sexual selection" (though these references probably date from a much
later period when he was at work on his "big book" on species in the 1850s;
see sec. 10.5).[120] On Darwin's 1838 reading, he evidently saw the import of
Blyth's distinction between the two forms of what became sexual selection,
including Blyth's denial of aesthetic selection to animals—a view Darwin
was to reverse within the next few months.[121] Darwin also double-scored
Blyth's reference to Lawrence's views on the inheritance of albinism and his
citation of Lawrence's *Lectures*; his earliest explicit notebook reference to the
Lectures occurs only a matter of weeks after his first recorded reference to
Blyth's paper.[122]

Although Blyth's understanding of sexual selection at this stage was very
different from that Darwin came to hold, Blyth's recognition of the two
forms of sexual selection and their respective deployments in animal and
human variation had implications for a convinced transmutationist who
might extend them beyond the limitations Blyth imposed (particularly if
he then went on to read Lawrence). In Blyth's hands, aesthetic selection

in humans was not a force for promoting racial difference, as it was for Lawrence (and Smith and the early Prichard), but rather acted negatively to maintain the "type" and prevent "degeneration." Crucially, Blyth's conception of aesthetic selection did not allow for different racial perceptions of beauty, as did Lawrence's, where a black skin might be deemed beautiful by an African even if repellent to a European—a position Darwin was to adopt within the next few months. Darwin, the emergent evolutionist, had already moved beyond Blyth's demarcation between humans and animals, arguing explicitly against Blyth's distinction between animal instinct and human reason.[123] It was only a matter of time before Darwin asserted the existence of an aesthetic sense and aesthetic choice in animals.

To a certain extent, then, Darwin developed his ideas in opposition to Blyth's. Blyth's contribution was to juxtapose a glittering collection of evidence, generalizations, and speculations (rather like some industrious bowerbird) displaying them in such a way that they could not fail to catch and hold Darwin's attention, stimulating his alternative interpretation. Down the long years, much to Darwin's benefit, Blyth continued to gather, rearrange, and display a staggering array of eclectic information and ideas on variation, descent, geographic distribution, and sexual dimorphism in domestic and wild animals and the human races. In one of his copious "Notes for Mr. Darwin" (read by Darwin in December 1855), Blyth offered a reprise of his 1835 views that had so interested Darwin and that owed so much to Lawrence.[124] Darwin returned to Blyth's earlier papers and to Lawrence's *Lectures*, and, following Blyth's suggestion, sought out Robert Knox's "curious" work and read it too. All were to play their part in crystallizing Darwin's views on sexual selection in the late 1850s, as I shall discuss in chapters 10 and 11. But by then—through his reading of Blyth and Lawrence around mid-1838, and his concurrent readings of Reynolds and other aestheticians on the different racial notions of beauty—Darwin had been conversant with the concept of aesthetic preference and its class, racial, and gender dimensions for almost twenty years.

Darwin's double citation of Lawrence on the hereditary nature of the scaly skin of the porcupine family was scribbled, together with other instances of inherited variation in animals, on a letter from William Yarrell, who was Darwin's preferred authority on animal breeding or artificial selection. The "beautiful" analogy[125] that Darwin drew between nature's and man's selective breeding, or between natural and artificial selection, is well known. Less explored is Darwin's reliance on the potent analogy between artificial and aesthetic selection that, as we have seen, Smith, Prichard, Lawrence, and Blyth had all successively deployed in explicating the concept of aesthetic selection in humans. Each of them structured humans as essentially domesticated animals and, even more explicitly, compared mating in humans to the art of breeding.

Good breeding, in both animals and humans, was something that Darwin's country-gentleman background had well prepared him to appreciate. It is to Darwin's extensive researches among the breeders—gentlemanly and not so gentlemanly—that I shall now turn.

SIX

Good Breeding: The Art of Mating

Man [is] an eminently *domesticated* animal.

—Darwin to Wallace, March 1867

In November 1844, having confessed to Joseph Hooker his almost criminal belief in the mutability of species ("it is like confessing a murder"), Darwin told his closest confidant that he knew of no systematic books on the subject, although "there are plenty, as Lyell, Pritchard [*sic*] &c, on the view of the immutability." Among those on mutability, Jean-Baptiste Lamarck's was "veritable rubbish," while the "Germanic" notion that "climate, food, &c sh[oul]d make a Pediculus [louse] formed to climb Hair" was no better. "All these absurd views," he told Hooker, "arise, from no one having, as far as I know, approached the subject on the side of variation under domestication, & having studied all that is known about domestication."[1]

Yet others before Darwin had seen in domestication a model for the transformation of natural forms. They included certain European naturalists (notably Lamarck),[2] and Samuel Stanhope Smith, James Cowles Prichard, William Lawrence, and Edward Blyth had all argued by analogy from domestic variability to the variability of the human races. Prichard had insisted on a strict "analogical method" that referred variability in both humans and animals to "similar causes"; that is, to some common "principle of natural deviation." For Prichard, the domestication analogy stopped short at races and breeds: no new species had been bred over thousands of years of domestication, just as no new species had evolved in nature. Only the materialist Lawrence, freed from the theological imperative, pushed the domestication analogy further. Claiming that humans were "more of a domesticated animal than any other"—a sentiment echoed by Darwin—Lawrence had drawn the most explicit parallel between the human races and the breeds of domestic animals, even to moral and intellectual differences that were "analogous in kind and degree." Domestication stimulated the production of congenital variations, and such variations were propagated by inheritance. Pointing to the enormous changes wrought by breeders, Lawrence argued that, by a process of similar "selections and exclusions," almost "any idea of the human form" might be achieved. Following Smith and Prichard, Lawrence had specifically linked aesthetic preference with artificial selection, as had Blyth after him.

It was this linkage that Darwin was to exploit to the full. The analogue of artificial selection was indispensable to Darwin's exposition of natural selection—but even more so to his explication of sexual selection. Natural selection was analogous to artificial selection; sexual selection, especially aesthetic selection, or so-called female choice, was constituted by artificial selection.

6.1 The domestication of beauty

Early in 1839, Darwin circulated a printed questionnaire on crossings, variation, and inheritance in animals. Its intended recipients were those with most practical experience in such matters: those involved in the creation of new domestic varieties by selective breeding. By this stage, Darwin's species work was consolidating around what he began to call "my Malthusian views," and he intensified his investigation of the crucial process of reproduction, or "generation," with its associated issues of variation and inheritance.

Among his queries about animal breeding, Darwin interpolated one on sexual selection: "Amongst animals (especially if in a free, or nearly free condition,) do the males show any preference to the young, healthy, or handsome females? or is their desire quite blind?"[3] The framing of this question, with its specific focus on male choice, is interesting, especially as, in an earlier draft, Darwin had posed his question as: "Idea of beauty in animals: do females prefer certain males? or vice versa.—when in a flock."[4]

Was Darwin's shift—from the possibility of female choice to that of unequivocal male choice—significant for his theorizing at this stage? Or was he, rather, merely reframing the formal printed version in terms he thought to be more familiar or acceptable to breeders? Either way, the idea of sexual preference and its evolutionary implications was sufficiently at the forefront of Darwin's thinking in 1838–39 for him to seek to test it among those best placed to observe it.

Darwin's systematic enrollment of commercial and hobbyist breeders strikingly inflected his species work. The selective breeding of new domestic varieties became the observable analogue for the hidden process of species formation in nature, and was fundamental to Darwin's exposition of natural selection. Domestication was the "best and safest clue" to the "coadaptations of organic beings to each other and to their physical conditions of life."[5] The "one long argument" of the *Origin* devolves on the sustained comparison between the agency of humans and the agency of nature, as Darwin leads the reader from the mundane processes of artificial selection and the production of new breeds and varieties to the wilder shores of natural selection and the origin of species.

By educating the reader in the minutiae of pigeon breeding, Darwin set out to show that nature too might be viewed with the practiced eye of a breeder. It was a "program of visual reeducation," one that Darwin himself had undertaken. It required an appreciation of the "great effect produced by the accumulation . . . during successive generations, of differences absolutely inappreciable by an uneducated eye." It took years of training, skill, and knowledge:

> Not one man in a thousand has accuracy of eye and judgement sufficient to become an eminent breeder. If gifted with these qualities, and he studies his subject for years, and devotes his lifetime to it with indomitable perseverance, he will succeed, and may make great improvements.[6]

Pigeon cognoscenti did not seek out large changes or sudden variations in shape or feather, but by "methodical selection" of minute differences, they aimed to realize an ideal type through the application of a set of exacting standards. Breeders in general, Darwin stressed, regarded an animal's organization as "something quite plastic, which they can model almost as they please." Selection was "the magician's wand, by means of which he may summon into life whatever form and mould he pleases."[7]

Besides such methodical or intentional selection, Darwin distinguished another and "more important" process of "unconscious" selection. This resulted from breeding from the "best individual animals" without any wish or expectation of permanently altering the breed. In this way animals and plants most valued by humans have been slowly improved and diversified. Even "savages" who had no notion of inheritance or breeding would unconsciously select by preserving their choice animals from accident and famine: "We see the value set on animals even by the barbarians of Tierra del Fuego, by their killing and devouring their old women, in times of dearth, as of less value than their dogs."[8]

The analogy between what humans have done, intentionally or unconsciously, in a relatively short time and what nature has done, only infinitely more slowly over eons, was of great explanatory value to Darwin in convincing his readers of the plausibility and efficacy of natural selection. But, as Darwin explained to Lyell, there was an "important distinction" between them: "Man can scarcely select except external & visible characters, & secondly he selects for his own good; whereas under nature, characters of all kinds are selected exclusively for each creature's own good."[9] By contrast with the self-serving, unnatural art of breeding new varieties of animals and plants, nature's selective breeding was represented as vastly more prolonged, more discriminating, and more comprehensive.

Sexual selection, on the other hand, had much more in common with the human art of selecting, particularly for those qualities that give humans pleasure; so much so that sexual selection "can almost be treated as a variant" of artificial selection. Darwin made the nexus overt in the *Descent*:

> In the same manner that man can improve the breed of his game-cocks by the selection of those birds which are victorious in the cockpit, so it appears that the strongest and most vigorous males, or those provided with the best weapons, have prevailed under nature, and have led to the improvement of the natural breed or species. . . . In the same manner as man can give beauty, according to his standard of taste, to his male poultry—can give to the Sebright bantam a new and elegant plumage, an erect and peculiar carriage—so it appears that in a state of nature female birds, by having long selected the more attractive males, have added to their beauty.[10]

The "same manner" of artificial selection smooths the transition from male combat to female choice, both aspects of the struggle for mates being reduced to aspects of the breeder's art of selectively breeding for strength or for beauty. There is, however, a tighter fit between breeding for beauty and female choice. The one morphs into the other. In order for there to be sexual selection by female choice, the animals must be able to perceive differences in their prospective mates. Like the pigeon connoisseurs so esteemed by Darwin, female birds select for what pleases them aesthetically on the basis of externally visible characteristics, even to the minutest, most subtle of differences. In the second edition of the *Descent*, Darwin qualified his attribution of the aesthetic sense and discriminatory powers of the human bird breeder to female birds, explaining that it must not be supposed that lower animals have a sense of beauty comparable to that of a "cultivated man, with his multiform and complex associated ideas." The ubiquitous "savage" did duty again: "A more just comparison would be between the taste for the beautiful in animals, and that in the lowest savages, who admire and deck themselves with any brilliant, glittering, or curious object."[11]

Yet, in Darwin's descriptions of human evolution, any distinction between the animal breeder and the sexual selector virtually disappeared, and he never saw reason to qualify this elision. The human male was made synonymous with the animal breeder, who exercises his caprice in varying the appearance of the breed:

> Each breeder has impressed . . . the character of his own mind—his own taste and judgment—on his animals. What reason, then, can be assigned why similar results should not follow from the long-continued selection of the most admired women by those men of each tribe, who were able to rear to maturity the greater number of children?[12]

This is not so much analogy as identification.[13] Although he went on to argue that this would be "unconscious" rather than intentional or "methodical" selection, Darwin's human application of sexual selection carries with it all the connotations of the powerful breeder who modifies his domestic productions to his will. As the breeder selects and shapes his domesticates, so man has molded woman to his fancy. It was this inseparability of sexual selection from the masculine, manipulative, competitive world of the Victorian breeder that gave Darwin's sexual selection its distinctive role in human evolution; and it was this same inseparability that caused Darwin considerable difficulty in coming to terms with female choice in animals.

One further introductory point remains to be made. From the beginning of Darwin's investigations into breeding, some intense personal concerns impinged on his theory building. Darwin and Emma Wedgwood married in January 1839. His heightened awareness of his own sexuality and Wedgwood-Darwin heritage, his anxieties about his increasing ill health and the risks of inbreeding, his views on the relations between husband and wife—all mingled with Darwin's early inquiries and thoughts about courtship, generation, and inheritance.[14]

6.2 The politics of breeding

Animal breeding was an economically and socially significant activity that engaged a broad sector of the British population in the first half of the nineteenth century. Breeders ranged from wealthy, often aristocratic, landowners who bred exemplary livestock, to more humble husbandmen and tenant farmers who raised less valuable but more essential sheep, cattle, and pigs for consumption. Breeders were not confined to rural areas: city dwellers bred a variety of pets, from pigeons and rabbits to dogs.[15]

Darwin was familiar from childhood with the practices of animal husbandry, pet keeping, and gardening. A number of his Wedgwood-Darwin relatives and neighboring Shropshire landowners were engaged in animal breeding and agricultural improvement. His move from London to Down House in 1842 ruralized him again and afforded a new circle of informants. Down House became the site for Darwin's own extensive experiments on the breeding of poultry, pigeons, and rabbits and the hybridization of orchids. From Down, Darwin could travel to the large London horticultural and poultry shows; he even, in the mid-1850s, joined two of the London pigeon fanciers' clubs, and rubbed shoulders with "Spital-field weavers" and pigeon enthusiasts from all walks of life.[16]

In addition, Darwin built a far-flung network of personal contacts and correspondence with notable breeders. When his lengthy printed questionnaire of 1839 failed to elicit the desired responses (only two are known[17]), Darwin began to tailor his questions to the expertise of individuals. In particular, he cultivated those connected with the breeding of poultry and pigeons, the subjects he studied intensively in the years immediately preceding the *Origin of Species*. They included the ornithologist William Yarrell (fellow of the Zoological Society and author of standard works on the birds and fish of Great Britain) and William Tegetmeier (journalist and expert on domesticated birds). Both of them obligingly sought information for Darwin from other breeders and introduced him to prizewinning fanciers such as Bernard P. Brent, who became one of Darwin's more important breeding informants. Darwin's breeding network reached across the British colonial empire, including, as we saw, Edward Blyth in the Indian subcontinent, and assorted missionaries, consular officials, medical men, and merchants.[18]

Darwin also read extensively and systematically in the literature of domestication. His reading ranged across the spectrum: from the handbooks on pig, sheep, and cattle breeding found in country houses and farms throughout England, through more specialist periodicals and books, to professional publications in physiology, ethnology, and zoology. Following his usual practice, he industriously abstracted, made further reading lists, and followed chains of references across the literature. These notes, together with his collated correspondence with a wide assortment of breeders, represented a huge database on breeding from which he drew selectively for his published works; notably the *Origin*, the *Variation of Animals and Plants under Domestication*, and the *Descent*.[19]

A major difficulty Darwin faced in dealing with such an accumulation—derived largely from outside the accredited scientific community—was "*to know what to trust.*" He dealt with the problem primarily by cultivating one or two select sources in each field (so Tegetmeier became his authority on bird breeding), with whom he might cross-check information gleaned from less trustworthy sources. Another criterion of reliability was the proximity of the source to the scientific community. Yarrell, until his death in 1855, was Darwin's chosen authority on general breeding issues relating to inheritance and variability. Sir John Saunders Sebright, gentleman agriculturalist and Whig politician, breeder of the famous Sebright bantam, was another who ranked high in Darwin's estimation.[20]

But, to a large extent, Darwin took on the breeders' own evaluative standards. Inevitably, commercial considerations, the monetary worth or value attached to a prime exemplar of the breeders' art, imbued Darwin's deployment of the analogues of artificial and natural selection: "Hard cash paid down, over and over again is an excellent test of inherited superiority." Consequently, Darwin's extensive analogy between artificial and natural selection depended crucially on "the existence of a group of men engaged in competitive struggle for prizes and individual success." The ethos of competitive struggle linked—and provided mutual corroboration for—selection by man and selection by nature.[21]

There was more to it than this. Darwin's borrowings from the breeding world brought with them much other social and cultural baggage. At the most fundamental level, "breeding" is resonant of notions of human rank and ancestry, of blood, lineage, and relationships, as much as it is of animal pedigree or success in artificial selection. For the aristocracy, "breeding" involved traditional economic and social considerations, the inheritance of titles, wealth, the entailment of property, primogeniture, connections in marriage and endowments, relations of power, and political influence. That more recent creation, the upwardly mobile "man of breeding" promoted by eighteenth-century aestheticians, was a man of cultivated taste, moral probity, polished manners, and conversation, a gentleman who might lay claim to a place in the "democracy of taste." Applied to the lower orders, "breeding" connoted a Malthusian superfluity of population, the twin specters of irresponsible, unbridled carnality—the Charybdis of costly public welfare or the Scylla of potential revolution—and their joint panaceas of religion and moral restraint; or, in secular reinterpretations, the radical and shocking remedy of birth control.

Animals shared and represented the class relations of their human owners and breeders. "Well bred" animals, such as the massive prize cattle exhibited by elite "improvers" at agricultural meetings, were as much a testimony to the rank and "good breeding" of their aristocratic owners as to their success in manipulating the size and elegant proportions of their stock. In a period when John Bull and roast beef were national symbols, these huge and costly cattle—bred for show, not use—were emblematic of the prestige and patriotism of their owners. The primary qualification for breeding an animal of such proportions and pristine bloodlines was the listing of the genealogically distinguished breeder in Debrett's peerage.[22]

The overweening emphasis on pedigree and pure bloodlines spilled over into breeding practices and notions of heredity. Sebright, in his *Art of Improving the Breeds of Domestic Animals* of 1809 (in a passage that Darwin quadruple-scored), cautioned against selecting for breeding on the basis of "apparent qualities." He defined a *"well bred"* animal as one "descended from a race of ancestors, who have, through several generations, possessed, in a high degree, the properties which it is our object to obtain."[23] In his role as Whig reformer, Sebright readily transferred the values embodied in his pamphlet on breeding to his campaign for the Reform Bill of 1832. He opposed the extension of the right to vote in county elections to short-term tenants: inevitably, the great landed proprietors would control such voters and "free, popular elections . . . would be entirely at an end." Sebright's prejudice against the ability of short-tenure landholders to participate in reform provoked sarcastic rejoinder: "And why not, good Sir John? Is a man's *capacity* and *independence* to be measured by the *length of his lease?*"[24]

Political considerations also attached to the common belief of breeders in the inheritance of acquired characters. For all his emphasis on the importance of ancestry in breeding, Sebright was equally insistent that defining breed characteristics—like the different instincts evinced by the different breeds of dog—might be acquired, albeit slowly and gradually, through the effects of inherited habit. Sebright's 1836 pamphlet "On the Instinct of Animals" was important in shaping Darwin's early views on the role of inherited habit in species change or adaptation, and especially in explaining the development of mental attributes like the moral and aesthetic senses in humans. Adding to Darwin's elation at finding his own early theorizing endorsed by so eminent a breeder ("admirable essay," "excellent authority"), Sebright, as Darwin particularly noted, extended his explanation of animal instinct to the formation of "national character" in the "human race."[25]

The congruity of the widespread belief in the inheritance of acquired characters with ideologies of self-help and social improvement and the benefits of education has been well documented. At one end of the spectrum, they might find expression in the adaptation of Lamarckian evolutionism to radical/democratic ends in the rhetoric of disaffected marginal medical men, artisans, and the emergent working class. The social transformation of the figure of the "gentleman"—once the prerogative of the inbred hereditary aristocracy—into the phenomenon of the morally upright, self-made man of the Victorian era drew on the notion of acquired character and self-fashioning, dissociated from traditional rank and privilege.[26] Earlier Whig reformers like Sebright might call on models of both inborn and acquired character as necessary, integrating them through the medium of the breeder's art, to explain animal instinct as well as historically and politically significant national and class differences in character and morality through the conserving effects of inherited habit.

Darwin—as an evolutionist concerned with the generation of individuals and species and the inheritance, transmission, and accumulation of the variations that determined evolutionary development—was much occupied with unraveling issues of inheritance in the early processes of theory formation. His transmutation notebooks also indicate a certain awareness of the political implications of selective breeding and the theories of hereditary transmission it inspired. The value attached to ancestry and the notion of "improving" the breed by preserving pure bloodlines and excluding upstarts was embodied in what Darwin called "Yarrell's law," after his leading authority: "What has long been in blood will remain in blood.—converse, what has not been, will not remain." And "an animal is able to transmit only those peculiarities, to its offspring, which have been *gained slowly*."[27] Darwin went to some lengths to confirm this generalization that guaranteed the stability of hereditary change and conformed with his early conceptions of species and geological change and generation—and with his defining *Beagle* experience of Jemmy Button's ready reversion to the savage state. Like Sebright, Darwin could assimilate "hereditary habits" to Yarrell's law. For Darwin, this meant that any major adaptive change could be acquired only gradually, through the slow accumulation of many small steps by mature organisms over many successive generations. The agents effecting change had to be "long in the blood." His insistence on explaining major structural change through the gradual accumulation of minute changes was an explicit rejection of radical or revolutionary change in the political as well as the biological senses, and this carried over into his Malthusian reinterpretation of species change. When he came to integrate his conservative Yarrellian generational gradualism with the geological gradualism he imbibed from Lyell, Darwin noted that his extension of the rule of the constantly acting Malthusian population law from the human to all species "baffles the idea of revolution" in nature as in "government" and "institutions."[28]

The liberal reformist platform of the 1830s, promoted by Sebright and endorsed by the Darwin of the notebooks, was still dominated by landed aristocratic and gentlemanly capitalist interests—by those concerned with agriculture, banking, colonies, trading, commerce, and property—rather than by the newer, urban, middle-class industrialists and machine manufacturers. For all the changes wrought by the Industrial Revolution, England remained largely an agricultural nation, with most wealth and power still secured in land. Land enclosures, involving the creation of large estates, the expulsion of unwanted laborers, and the destruction of their cottages, persisted into the 1870s, when it is estimated that some seven thousand proprietors owned more than four-fifths of the United Kingdom. Darwin participated in this process, buying up farms and land for the "sake of my sons," becoming, over the years, a substantial landowner.[29]

The political economy of land and labor meant that pigeon-keeping, once a mainstay of rural agriculture and a legal monopoly of British nobility and gentry, became largely the preserve of town-based laborers and artisans, particularly weavers. The highly specialized craft of handlooming practiced by the Spitalfields silk workers referred to by Darwin—which persisted in the face of the mechanization of the textile trades, the removal of trade protection, and fluctuating unemployment—was subject to the vagaries of fashion and commerce to the extent that pigeon breeding was an important source of supplemental income and food. Brent, one of its leading exponents (himself described by Darwin as a "very queer little fish"), lamented that pigeon breeding had unfortunate associations with "Costermongers, Pugilists, Rat-catchers, and Dog-stealers," all because the majority of fanciers were "artisans—men who lived in the courts, alleys, and other by-places of the metropolis." Particular breeds were associated with particular trades and locations: Spitalfields weavers were famous for their pouters, while racing pigeons were the favorites of miners in the North of England. Pigeon clubs were divided along class lines, ranging from the humbler meetings in beer shops and gin palaces to the socially exclusive Philoperisteron Society, formed by gentlemen fanciers of the West End of London.[30]

The politics of breeding were pervaded not only by class, but also by sexual and racial ideologies. First and foremost, breeding was predominantly a masculine occupation. Although women were involved in small-animal breeding and beekeeping, large-animal breeding—horses, cattle, and sheep—was carried out almost entirely by men. The clubbing, pub culture, and sporting dimensions of the breeding of pigeons, fighting cocks, and racehorses ensured their masculine dominance. It was men who wrote the breeding texts, formalized the practices, and kept the lore. Nearly all of Dar-

Figure 6.1. "I w[oul]d accept Wallace's view about Birds did I not remember pigeons." Diverse races of fancy pigeons, many bred by Darwin. The carrier (left front) and pouter (central) are among the few breeds of pigeon to exhibit sexual dimorphism, male carriers having more developed wattles over the eyes and beaks and male pouters having a greater "pout" or inflatable crop than the females. Darwin interpreted this as confirmation of his theory of male sex-limited inheritance from the first, which was critical to his theory of sexual selection and to his dispute with Wallace. *Illustrated London News,* January 18, 1851, 48. University of Sydney Library Collection.

win's breeding correspondents were men, a notable exception being Mary Anne Whitby, a pioneering English silkworm raiser. The system of recording and registering bloodlines that developed in the stud books kept by leading breeders of thoroughbred horses and cattle followed established traditions of property and title inheritance in fixing breed type through the male line. Even though breeders acknowledged the female contribution to pedigree, much greater significance was attached to obtaining a "choice" male animal for breeding (the terminology is Darwin's). The traditional belief that the male was more important in stamping his properties on the offspring predominated among breeders, and Darwin concurred with this.[31]

The general prejudice of breeders against crossing (the mating of different breeds), which Darwin also echoed, was consonant with cultural attitudes toward racial miscegenation with its allegedly degenerate, untrustworthy "half-breeds," "half-castes," "cross-breeds," "mixed-breeds" or "mongrels." It was this same prejudice that fed the polygenist view, in the face of all evidence to the contrary, that the different human races were not interfertile, or that mixed-race offspring, or "racial hybrids," were sterile or at least unviable—a claim that Darwin took seriously enough to investigate thoroughly. As a corollary of Yarrell's law, Darwin assumed, a blending of two established or very old breeds or races was not possible, and because of their different physical and mental constitutions, they would have an instinctive aversion to interbreeding. This was an assumption that he extended to the human races, even where there were no overt constitutional incompatibilities. The "smell of negro," Darwin wrote across the cover of Sebright's 1809 pamphlet on breeding, was a signifier of such racial incompatibility and aversion, "like differences between Australian [dingo] and common dog & some goats—Cashmere goat . . . said not to smell." In his notebooks, this was spelled out as the "strong odour of negroes, a point of real repugnance," and then amplified into "a real repulsion amounting to impossibility" of interbreeding between the different races of "men."[32]

The undesirability of miscegenation was counterbalanced by that other social horror, incest. The art of breeding was a balancing act (Sebright compared it to rope dancing) that avoided either extreme by preserving bloodlines by selecting from related—but not too closely related—stock that possessed the desired qualities. In practice, breeders of prize cattle valued not merely the length of the pedigree, but the length of time there had been a succession of the "best blood" without the admixture of any "inferior blood." The practice of breeding in-and-in (that is, the mating of closely related animals without the healthy injection of an occasional cross) persisted in the face of the physical deterioration of leading breeds, whose pedigrees were valued more than their beef. Cultural notions of what counted as incest might be redefined according to the exigencies of the breeders' art. Sebright did not consider breeding from father and daughter or mother and son to be breeding in-and-in, for the offspring are "only half of the same blood" of either parent; breeding from brother and sister was "a little close." While acknowledging the bad effects of inbreeding, Sebright deplored the "ridiculous prejudice which formerly prevailed" against breeding from animals having "any degree of relationship." There would be no new varieties, no improvements in the breed, if such considerations—which would prevent breeders from availing themselves of an animal of "superior merit"—had been put into practice.[33]

"This shows the whole principle of making varieties," Darwin penciled next to this passage in his copy of Sebright's "most important" pamphlet of 1809. Extrapolating from the artificial practices of animal breeding, a degree of inbreeding was the only way in which those minute variations that would bring about species change could be accumulated and perpetuated in the breeding population in a state of nature. As Darwin elaborated the point: "In plants. man prevents mixture, varies conditions & destroys, the unfavourable kind—could he do the last *effectively* & keep on same exact conditions for many generations he would make species, which would be infertile with other species."[34]

Darwin first read Sebright's pamphlet on the art of breeding around mid-1838, and maybe he drew some reassurance from it for the perpetuation of the superior merit of the Wedgwood-Darwin lineage, inbred though it would be if he were to marry Emma, although he noted "bad effects of incestuous intercourse," and, perhaps more anxiously, "does not take into account loss of desire."[35]

A notebook entry dated only a few days before Darwin's first formulation of his Malthusian insight offers an instance of the complexity of personal concerns, developing theoretical understandings, breeding lore, and contemporary attitudes to sexuality that permeated Darwin's domestication analogy:

> Sept. 25th [1838] Young man at . . . Hair Dresser, assures me he has known many cases of bitch going to mongrel, & all subsequent litters having throw of this mongrel.—I did not ask the question.—His bitch will not take & if she did take, probably would not be fertile without she know & LIKES HIM & then is actually obliged to be held.—like she-wolf of Hunter . . . there is great difference between hybrids & inter se offspring in latter being unhealthy.— *males* bred in & in *never lose passion.*[36]

Darwin's exultation at having this unknown "young man's" unprompted confirmation of Darwin's own belief in telegony (the widespread belief that an earlier mating could "taint" the offspring of a subsequent mating with a different sire—see below) and the sexual choosiness of his bitch (capitalized), together with the same informant's claims that inbreeding affects the health of offspring but (italicized) *not* male sexual desire, requires only minimal unpacking to uncover some of the underlying sexual tensions. The anecdotal evidence of female choice comes packaged with reassuring assumptions of masculine dominance and potent polygamy matched by traditional feminine monogamous coyness; this is qualified by an uneasy recognition of female willfulness and sexual preference for the low-bred, racially mixed mate who, even when physically excluded, could continue to influence subsequent offspring.

6.3 Do females have a choice?

The very idea of female choice required as much careful handling as the females who displayed it. If we turn to the *Descent*, thirty-three years down the track, we find the muted echo of the young man at the barbershop (and, of course, the young Darwin):

> The female . . . with the rarest exceptions, is less eager than the male. As the illustrious Hunter long ago observed, she generally "requires to be courted;" she is coy, and may often be seen endeavouring for a long time to escape from the male. Everyone who has attended to the habits of animals will be able to call to mind instances of this kind. Judging from various facts, hereafter to be given, and from the results which may fairly be attributed to sexual selection, the female, though comparatively passive, generally exerts some choice and accepts one male in preference to others. Or she may accept, as appearances would lead us to believe, not the male which is the most attractive to her, but the one which is the least distasteful. The exertion of some choice on the part of the female seems almost as general a law as the eagerness of the male.[37]

The young man has been transmuted into the generalized "everyone" concerned with animal breeding who will readily recall "instances of this kind." That Darwin could so confidently make this claim and not feel the need to cite confirmatory authorities other than the rather dated observation of the "illustrious Hunter" (whose legendary stature—of which Darwin pointedly reminds us—puts him virtually beyond contradiction) underscores the stereotyping that this interpretation of female choice embodied. This is a very restrictive kind of choice indeed. The female is "comparatively passive" and exerts only "some choice"; she is choosy, not because she is sexually aroused, but because she is not; she "accepts" the "least distasteful" male; she is not the seeker, but the sought. We can only be sure that she exerts even limited choice by a process of circular argument: because of the "results which may fairly be attributed to sexual selection."

Darwin's difficulty was that he was going against entrenched opinion in allowing females any choice at all, particularly in so culturally charged an arena as breeding. Breeders were much more accepting of male combat for possession of the females, taking it virtually for granted. Sebright, discussing animals in a state of nature that are not subject to the degenerative effects of breeding in-and-in because they "are perpetually intermixing," could write in 1809:

The perfections of some correct the imperfections of others, and they go on without any material alteration, except what arises from the effects of food and climate.

The greatest number of females will, of course, fall to the share of the most vigorous males; and the strongest individuals of both sexes, by driving away the weakest, will enjoy the best food, and the most favourable situations for themselves and for their offspring.

A severe winter, or a scarcity of food, by destroying the weak and the unhealthy, has all the good effects of the most skilful selection.[38]

In these passages, which Darwin scored, Sebright not only served up the analogy between natural and artificial selection (which Darwin was to extend way beyond the limitations of natural species change assumed by Sebright), but he also endorsed the principle of male combat and its improving effects—"of course" there was a struggle between males for mates and the most vigorous would mate with more females. This was, of course, hardly news to the grandson of Erasmus Darwin. In the *Descent*, Darwin could confidently reiterate:

It is certain that with almost all animals there is a struggle between the males for the possession of the female. This fact is so notorious that it would be almost superfluous to give instances.[39]

The reason for the certainty of both Darwin and breeders lay in their general conviction of the "stronger passions" and greater sexual eagerness of the male, an equally "notorious" instance of common knowledge: "That the males of all mammals eagerly pursue the females is notorious to every one." This held, Darwin claimed, for birds, fish, reptiles, frogs, insects, spiders, and crustaceans.[40] The classic instance among domesticates was the polygamous and pugnacious barnyard cock, celebrated in the evolutionary poetry of Grandfather Erasmus. A range of other domesticates—pheasant, quail, peacocks, stags, bulls, and rams—bore out the general consensus among breeders that males in general were eager for mating and indiscriminate as to choice.

Of the two known responses to Darwin's 1839 printed questionnaire, one ignored Darwin's query about male preference, while Richard Sutton Ford, agent to the Fitzherbert estate in Staffordshire, held that male domesticates were completely uninterested in the beauty or youth of their sexual partners: "The sexual passion of the males, in cattle and sheep, at least, appears to be wholly indiscriminate, without regard to age, symmetry, or colour." The perceived health of potential mates was beside the point; unless females were in "almost perfect health" they did not come into season and were of no sexual interest to males.[41]

By 1868, Darwin was able to turn all this to account as a rationalization of the predominance of female choice in lower animals. Wallace had objected to Darwin that he had just learned that in all cases the sperm went to the egg, and the male sought the female: "Now if this is so deep seated a law of nature is it not very extraordinary & very improbable that sexual selection should act *always* by the female *choosing* the male, and never by the male choosing the female?" Not so, responded Darwin. This "remarkable law" implied just the opposite: males are indiscriminate searchers and settle for any female they can get; it followed that females had to be the choosy, discriminating ones. Ingeniously dismissive of Wallace's objection as he was, at the same time Darwin indicated some continuing ambivalence over the problem of apparent role reversal in the case of humans and the contravention of the continuum between animals and humans that he otherwise insisted on:

As the male is the searcher he has received and gained more eager passions than the female; and, very differently from you, I look at this as *one* great difficulty in believing that the males select the more attractive females; as far as I can discover they are always ready to seize on any female, and sometimes on many females. Nothing would please me more than to find evidence of males selecting the more attractive females. I have for months been trying to persuade myself of this. There is the case of man in favour of this belief . . . Perhaps I may get more evidence as I wade through my twenty years' mass of notes.[42]

If breeders were unresponsive to Darwin's early attempts to elicit instances of male sexual preference or aesthetic discrimination among potential mates, what of the alternative possibility of female choice? Culturally induced attitudes aside, the general practices of breeders ran counter to recognition of female mating preferences. The studbooks kept by dog, cattle, and horse breeders, with their emphasis on the male line, reinforced attention to male performance in mating. Breeders of thoroughbreds paid considerable attention to the pedigree and performance of the sire and its progeny, but took only minimal notice of that of the mare. From the late eighteenth century, the payment of a fee for the hire of a thoroughbred stallion became customary among farmers and breeders, just as they had traditionally paid for the services of rams and bulls. Farmers paying as high as two guineas to have their mares covered by the best stallions were understandably far more interested in the sexual performance of the stallion and in ensuring that conception was successful. The standard texts of the period advised the use of a "teaser" stallion to test the mating state of the mare, and, if the female proved resistant to a particular mating, a "cheat" might be employed.[43] Female preference for a particular male, if evinced, was a nuisance in controlled mating, something that had to be overcome—a nuisance that males seldom provoked. Females, as every breeder knew, required strict control and no choice in the matter if they were to produce offspring to the requisite standards of the breed.

The necessity for strict control was reinforced by the above-mentioned belief that a female, once mated with an inappropriate or inferior male, thereafter would produce offspring, even to superior sires, with this inferior "taint." The classic case of telegony was associated with the name of Lord Morton, a distinguished horse breeder who reported the phenomenon in an Arabian mare who had been first successfully bred with a quagga (a kind of zebra), and whose subsequent offspring, sired by a purebred Arab, allegedly resembled the striped quagga in coat color and texture. Telegony was a two-edged sword as far as notions of sexuality and inheritance went. Opinions varied as to its cause. Like his grandfather before him, Darwin categorically rejected the traditional doctrine of maternal impressions: "stuff!"[44] He did not, however, adopt Erasmus Darwin's view of the power of the male imagination in conception and development, but endorsed the alternative explanation of the "direct or immediate action of the male element on the mother form," meaning that the female's reproductive organs were materially altered by the initial impregnation, which then influenced subsequent fertilizations even by a different male. As in the case of Yarrell's law, Darwin was predisposed toward telegony because it accorded with his broader views on generation and heredity, and he sought out confirmatory anecdotal accounts by breeders. On publication, he circumspectly abandoned the certification of the young man at the barbershop, and wrote that Lord Morton's law was established in plants and animals by the "authority of several excellent observers."[45]

The prepotency of the male element also of course affirmed contemporary assumptions of male supremacy and the virtues of female chastity—assumptions that Darwin shared and might rationalize according to Malthusian principles. Not surprisingly, telegony was a source of special insight for other theorists besides Darwin. As noted, Paul Strzelecki, the Pacific explorer, applied the phenomenon to the dramatic population decline of Australian Aborigines. Darwin did not share Strzelecki's interpretation of extinction, but he did speculate that the influence of the male on the female form might explain why wives grew to look like their husbands![46]

The notion of female choice was thus obscured or hedged around by conventional breeding practices and beliefs. It did not come readily to the minds of breeders. Darwin had to work hard to elicit confirmatory evidence after he became convinced of the agency of female choice in sexual dimorphism and divergence. Even then, breeders mostly resisted the notion. In the *Descent*, Darwin's carefully compiled list of such instances as he could persuade from trustworthy breeders was small and much qualified.[47]

Darwin had a way of asking questions of breeders that invited the appropriate response. His assumption was that breeders, even "the greatest breeders in England," could have "no theoretical views to support." His correspondence indicates his exasperation with breeders who had views of their own: "When Mr. Brent gets a notion into his head nothing gets it out."[48]

In 1861 Darwin prompted Tegetmeier as follows:

> Though the cock which conquers, naturally gets first choice of wives, yet from analogy with other Birds . . . I imagine the Hen has to be won or charmed to grant her favours.—Can you throw any light on this point, which interests me much?—There can hardly be a doubt that the beauty of male Bird is to charm the female.

In spite of such encouragement, his chief informants on domestic poultry, Tegetmeier and Brent, failed to come up to scratch. Hens, as Brent put it, "do not seem to have any great choice in the matter." The overriding factor was their sexual readiness for breeding, when they would "accept the gallantries of any cock who offers them." Brent delivered what Darwin called "almost an essay" on courtship behavior in domesticates. He offered a perceptive account of a variety of male and female behaviors (including instances of "termagant" hens who would fight strange males and one who took to treading other hens "out of pure wantonness"). He concluded with the jocular point that both hen fowls and hen pigeons, when kept without mates, had "stooped" to him and allowed him to lay his hand on their backs: "which I think can hardly speak much for their ideas of beauty."[49]

Darwin was not amused. To Tegetmeier, who made much the same observation, he replied dismissively—and with a certain distaste at such unnatural conduct—that "it is clear that the instinct of our Hens, (from what you say about touching them) is so vitiated that one cannot judge how they would behave in a state of nature." He repeated this claim in the *Descent*. It was not so much Brent as Darwin who had gotten a notion into his head that nothing could get out, even if it required the expedient of the suspension of his vital domestication analogy.[50]

But it had taken Darwin some time to get the notion *into* his head: first, as an evolutionist, to become convinced that animals, like humans, must exhibit some aesthetic sense; next, to find an explanation through the theory of inherited habit; then, by analogy with artificial selection and through his ethnological readings, to invoke aesthetic choice via sexual selection to explain sexual dimorphism and human racial divergence, shifting from an early emphasis on the dominant male role in the process to the realization that—in animals at least, where males were generally more ornamented or conspicuously colored (particularly among birds)—females logically must play the greater role in any process of aesthetic selection.

The very real difficulties Darwin experienced in eliciting confirmatory evidence from recalcitrant breeders aside, it was Darwin's own thoroughgoing identification of the process of aesthetic choice with the discriminating eye of the breeder that was the greatest stumbling block to the notion of analogous female choice in nature. Artificial selection required human (i.e., male) agency and controlled mating if it was to achieve the desired result. The very essence of good breeding embodied the controlling, artistic, masculine persona of the breeder, who shaped the bodies of domesticates to his fancy.

6.4 Down among the pigeons: "N.B. all pigeon fanciers are little men"

Sebright, who impressed Darwin with his boast with respect to pigeons that he "could produce any given feather in three years" and any "head and beak" within six years, typified the gentlemanly man of breeding who brought his wealth, aesthetic discrimination, and power to bear on the processes of "picking" or "selecting." Sebright wrote the rules and theory of the "art" of mating domesticates and was influential in Darwin's early understanding of heredity and artificial selection. But it was the practices of the humbler fanciers, those whom Darwin disparaged as "little men"—the artisans and tradesmen, the "Spital-field weavers & all sorts of odd specimens of the Human species, who fancy Pigeons"—that had the deeper and more enduring effect on Darwin's interpretation of the art. They gave him the understanding of the "lifetime [of] indomitable perseverance" involved in acquiring the necessary skills and producing the desired results through their discernment of "differences absolutely inappreciable to an uneducated eye—differences which I for one have vainly attempted to appreciate."[51]

Fanciers bred their birds on rooftops and in tiny backyards; lived with their birds almost as members of the family; and, above all, expressed the ideals of their artistry and craft skills in selecting and shaping the forms and plumage of their living birds. Spitalfields was the acknowledged "cradle of the fancy," their specialty the "Improved English Pouter," which Darwin classified in a group of its own and described as the "most distinct of all domesticated pigeons."

The pouter's most distinctive point was its inflatable crop—which, when fully inflated, presented, as Darwin wrote, "a truly astonishing appearance." Pouters were also distinguished from other breeds in that they exhibited a degree of sexual dimorphism related to male display: "The males, especially when excited, pout more than the females, and they glory in exercising this power." This behavior, bred into the pouter, could be stimulated by the artificial inflation of the crop, as Darwin described: if the male bird would not "play," or pout, the fancier would take his bird's beak into his mouth and "blow him up like a balloon; and the bird, then puffed up with wind and pride, struts about, retaining his magnificent size as long as he can." The pouter's other distinguishing points were its upright stance, streamlined body, and decorative fans of feathers over its feet. Its proportions were crucial, standardized by a slide rule calibrated to one-sixteenth of an inch. It was a triumph of the skills of the silk weaver transposed to the living bodies of his birds. The handlooming of the delicate and costly imported silk fiber was a highly exacting skill, not amenable to mechanization, and requiring a "weaver's eye" for its highest expression in the design and execution of the "fancy" and "figured" goods of the craft. The "scrutinizing eye of the pigeon fanciers," so extolled by Darwin, was the "interactive counterpart" of the "weaver's eye" of the silk worker.[52]

The connections that might be drawn among art, breeding, and fashion were well appreciated by Darwin. In a letter to Thomas Henry Huxley in late 1859, he enclosed a few "remarks & Extracts on Pigeons" guaranteed to "make the audience laugh," together with some drawings illustrating the different breeds, that Huxley might use in his forthcoming lecture on the descent of domestic animals at the Royal Institution. The extracts and drawings came from John M. Eaton, a tailor by trade, "an excellent Fancier" and "great winner of prizes," whom Darwin had come to know personally. Eaton authored a series of works on pigeon breeding that Darwin drew on for their explicit discussion of the techniques and practices of fanciers. Eaton devoted a whole treatise to his great enthusiasm, the almond tumbler, described by Darwin to Huxley as "a sub-variety—of the short-faced variety, which is a variety of the Tumbler, as that is of the Rock-pigeon." Darwin was concerned not only with stressing the descent of the almond tumbler from the wild rock pigeon, but also pointing up the "extreme attention & close observation . . . necessary to be a good fancier" in Eaton's meticulous devotion to the minutiae of the art of perfecting the critical "five properties" or characteristics of a single sub-subvariety of pigeon. He cited Eaton's advice that would-be breeders should make themselves "master of one [breed] alone," and even then they should not overambitiously try to breed for all the properties of that breed at once, or they would "have their reward by getting nothing." "In short," chortled Darwin, "it is almost beyond the human intellect to attend to **all** the excellencies of the Almond Tumbler!" To show Huxley "how systematically selection is followed," Darwin quoted Eaton's recommendation that those "particularly partial" to the so-called "Gold-finch beak" in short-faced tumblers should keep the head of a dead goldfinch by them for comparative purposes and select toward this model.[53]

Darwin belittled the artisan pigeon breeders and poked fun at their grammar and enthusiasms. Yet he had a very good appreciation of the intimate connection between the aims of the artist and the breeder. The pigeon drawings that Darwin sent Huxley were, he explained, made by the animal painter Dean Wolstenholme, "an excellent fancier himself," and were "not more exaggerated than the Apollo Belvedere compared with man or the Venus de Medici compared with woman.—They represent the standard of perfection."[54] Breeders were artists who specialized in designing and creating new living forms of beauty or utility, and, like artists, they worked toward an ideal. Darwin's awareness of the aesthetic nature of the breeder's practices was made overt in the *Origin*, where he quoted Lord Somerville on the breeders of merino sheep: "It would seem as if they had chalked out upon a wall a form perfect in itself, and then had given it existence."[55]

Nonetheless, in establishing the analogy between natural and artificial selection, Darwin was forced to downplay the creativity and artistry of the breeder, lest critics claim that natural selection, like artificial selection, implied design. Hence, Darwin emphasized unconscious selection, rather than the methodical or intentional selection of the breeder toward an ideal, as the more important and efficacious form of artificial selection.[56] In distinguishing the two forms in the *Variation of Animals and Plants under Domestication*, Darwin retained the comparison of methodical selection with the vision of the artist: the breeder "has a distinct object in view"—as in Eaton's superposition of a goldfinch beak onto a pigeon—and sets out to preserve and enhance some character that has actually appeared, or "to create some improvement already pictured in his mind." The "even more important" unconscious selection, however, was random and "without method." It was, claimed Darwin, the counterpart of the lesser artifice of fashion, not involving design, subject to the vagaries of caprice, and depending on a "universal principle in human nature, namely, on our rivalry, and desire to outdo our neighbours."

> We see this in every fleeting fashion, even in our dress, and it leads the fancier to endeavour to exaggerate every peculiarity in his breeds. A great authority on pigeons says, "Fanciers do not and will not admire a medium standard . . . but admire extremes." After remarking that the fancier of short-faced beard tumblers

Fig. 23.—Short-faced English Tumbler.

Figure 6.2. The short-faced English tumbler was Darwin's leading instance of the unconscious tendency of fanciers to select to extremes. He made this a universal principle of the practice of artificial selection, which he identified with the exaggerations of fashionable dress and hence with both human and animal sexual selection. *Variation*, 1:152. University of Sydney Library, Rare Books and Special Collections.

wishes for a very short beak, and that the fancier of long-faced beard tumblers wishes for a very long beak, he says with respect to one of intermediate length, "Don't deceive yourself. Do you suppose for a moment the short or the long-faced fancier would accept such a bird as a gift? Certainly not; the short-faced fancier could see no beauty in it; the long-faced fancier would swear there was no use in it, &c." In these comical passages, written seriously, we see the principle which has ever guided fanciers, and has led to such great modifications in all the domestic races which are valued solely for their beauty or curiosity.[57]

Darwin's "great authority" was of course Eaton. Here Darwin's patronizing amusement at Eaton's expression has the effect of trivializing the artistry and creativity of pigeon breeders, demoting them to an "unconscious," artless instinct to "outdo our neighbours." This is a reminder that pigeon breeders are not, for the most part, men of breeding. They are "little men," not constrained by overmuch education or the cultivated tastes of the true connoisseur. They are free to give reign to their fancy, to express their instinctive taste for the bizarre, to produce such unviable abnormalities or "monstrosities" as the pouter, with its grotesquely enlarged esophagus—a bird that can be blown up by its breeder like a balloon—or a short-faced tumbler, whose beak has been so reduced that (as Darwin again quoted from Eaton) "better head and beak birds have perished in the shell than ever were hatched; the reason being that the amazingly short-faced bird cannot reach and break the shell with its beak and so perishes."[58]

Darwin's denigration of the artistry of pigeon selection was achieved in the face of Eaton's own understanding of the aesthetic achievements of the fancier: "To my fancy, I am not aware that there is anything . . . so truly beautiful and elegant . . . as the shape or carriage of the Almond Tumbler approaching perfection, in this property, (save lovely woman)."[59] Eaton's seriousness of purpose and conviction of the artistry of a pursuit associated with costermongers and ratcatchers invites the mockery of real men of taste who do know something of aesthetics. Darwin might effortlessly exert his social superiority in ridiculing Eaton's naive contention that "scarce any nobleman or gentleman would be without their aviaries of Almond Tumblers," were they to appreciate the "amazing amount of solace & pleasure" to be derived from keeping such exemplars of the breeder's art. This belies the pleasure Darwin himself derived from his own forays into pigeon breeding. "I will show you my pigeons!" he had enthused to Charles Lyell, "which are the greatest treat, in my opinion, which can be offered to [a] human being."[60] We may see also, in Darwin's earlier assurance to Huxley that the pigeon drawings he had enclosed represented "the standard of perfection" as the "Venus de Medici compared with beautiful woman," a distinct echo of Eaton's comparison of the beauty and elegance of perfected specimens of his favored breed of pigeon to "lovely woman."

In fact, the differences Darwin insisted on in the two modes of pigeon selection, conscious (or methodical) and unconscious, are not as great as Darwin claimed. In his letter to Huxley, Darwin had conflated them, making no distinction between them. But in the intervening years, Darwin had been made highly aware—even by friendly critics like Lyell—of the problems posed by his apparent attribution of creative powers to lowly breeders and, by analogy, by Darwin's anthropomorphic, voluntarist descriptions of natural selection. Lyell found utterly unphilosophical the "idea of selection ever being allowed to play such pranks as the breeder has played," to "sport with God's creatures & the laws of reproduction so as to perpetuate pouter pigeons & other monstrosities." He could not accept Darwin's comparison of selection with the offices of the Architect:

> The architect who plans beforehand & executes his thoughts & invents . . . must not be confounded in his functions with the humble office of the most sagacious of breeders . . . It is the deification of Natural Selection.[61]

Darwin had to admit that his tactic of defending his domestication analogy from attributions of design by emphasizing and demarcating "unconscious" selection from the artistry of the pigeon breeder was not entirely successful. By the second volume of the *Variation*, he conceded: "Unconscious selection graduates into methodical, and only extreme cases can be distinctly separated." A few pages later this had become: "Unconscious selection so blends into methodical that it is scarcely possible to separate them."[62] By this stage Darwin himself was thoroughly muddled, offering the very same instances culled from his reading of Eaton to illustrate methodical selection that he had earlier used in illustration of unconscious pigeon selection![63]

For all his inconsistencies, Darwin stuck with the fashion metaphor throughout: "Fashions in pigeon-breeding endure for long periods," he wrote.

> We cannot change the structure of a bird as quickly as we can the fashion of
> our dress. . . . Nevertheless, fashions to a certain extent change; first one point
> of structure and then another is attended to; or different breeds are admired at
> different times and in different countries . . . Breeds which at the present time
> are highly valued in India are considered worthless in England.[64]

Even "ancient and semi-civilized people" were subject to the vagaries of fashion in breeding their stock: "Human nature is the same throughout the world: fashion everywhere reigns supreme, and man is apt to value whatever he may chance to possess." In South America, niata cattle, with their disadvantageously shortened faces and upturned nostrils, are preserved, while the Damaras of South Africa value uniformity of color and enormously long horns in cattle, and Mongolians prize only white-tailed yaks: "There is hardly any peculiarity in our most useful animals which, from fashion, superstition, or some other motive, has not been valued, and consequently preserved."[65]

Breeding, through the freaks of fashion, was subject to national, racial, and class differences. But all breeders in all places at all times were united in their admiration of extremes. The "Tendency in Man to carry the practice of Selection to an extreme point" warranted a whole subsection to itself in *Variation*. This "important principle" was evidenced in "useful qualities" in horses and dogs that were made as swift and strong as possible, or in sheep bred for fineness or length of fleece. "With animals bred for amusement, the same principle is even more powerful; for fashion, as we see even in our dress, always runs to extremes." This view, Darwin emphasized, had been "expressly admitted" by pigeon fanciers. It was this universal principle in selection that "necessarily" led to divergence of breeds, and showed how such extreme diversity in domestic breeds such as those of the pigeon had been derived from the one parent stock:

> Selection, whether methodical or unconscious, always tending to an extreme point, together with the neglect and slow extinction of the intermediate and less-valued forms, is the key which unlocks the mystery how man has produced such wonderful results.[66]

6.5 From "short-faced fanciers" to Hottentots

All of this set the stage for the full-blown elaboration of sexual selection in the *Descent*. There, Darwin could confidently write:

> We have seen that each race has its own style of beauty, and we know that it is natural to man to admire each characteristic point in his domestic animals, dress, ornaments, and personal appearance, when carried a little beyond the common standard. If then the several foregoing propositions be admitted . . . it would be an inexplicable circumstance, if the selection of the more attractive women by the more powerful men of each tribe, who would rear on average a greater number of children, did not after the lapse of many generations modify to a certain extent the character of the tribe.[67]

In illustration, Darwin credulously offered his readers the unforgettable picture of the Somali tribesmen, who were "said [on the authority of the explorer Richard Burton] to choose their wives by ranging them in a line, and by picking her out who projects farthest *a tergo*. Nothing can be more hateful to a Negro than the opposite form." This is the human corollary of pigeon selection for a short-faced tumbler. It is instantiated with Darwin's exemplar of a steatopygous Hottentot woman, "considered a beauty" by her tribesmen, who was "so immensely developed behind, that when seated on level ground she could not rise, and had to push herself along until she came to a slope."[68] Here, a certain admired "peculiarity" in women of a particular race allegedly has been selected by male fanciers to an extreme where, as in the case of the short-faced tumbler, it is harmful to women of this race and threatens their viability.

There is, in all of Darwin's descriptions of those differences in non-European races that he attributed to the action of sexual selection, an ascription of the absurd, even the monstrous, just as he considered all domestic productions to have a "somewhat monstrous character" in that they were often injurious or not useful, or simply grotesque ("mere monstrosit[ies] propagated by art"), and could not be perpetuated in the breed without human intervention.[69] At times this becomes overt:

> It seems at first sight a monstrous supposition that the jet-blackness of the negro has been gained through sexual selection; but this view is supported by various analogies, and we know that negroes admire their own blackness. . . . The resemblance of Pithecia satanas [a species of monkey] with his jet black skin, white rolling eyeballs, and hair parted on the top of the head, to a negro in miniature, is almost ludicrous.[70]

No such pejorative attaches to the European preference for a "nearly oval face . . . straight and regular features, and . . . bright colours." However:

> If all our women were to become as beautiful as the Venus de Medici, we should for a time be charmed; but we should soon wish for variety; and as soon as we had obtained variety, we should wish to see certain characters in our women a little exaggerated beyond the then existing common standard.[71]

Look again at Darwin's language. "We" are European males, the fanciers of a certain type of woman idealized in the *Venus de' Medici*, and "we" re-shape "our women" around this ideal to suit the masculine desire for variety and the enhancement of certain sexually desirable physical traits (such as big bottoms, although "we" presumably would not take them to the same extremes as uncivilized Hottentots—fashion in the form of the crinoline or the bustle does this for "us"). Darwin's identification of the sexually se-lecting human male with the manipulative pigeon breeder could hardly be made more explicit. Just as "each breeder has impressed . . . the character of his own mind—his own taste and judgement—on his animals," so the men of different races have shaped and colored "their" women to their divergent tastes—and some tastes are obviously better than others.

In *The Descent of Man*, the beau ideal of the aesthetician, constituted by race, class, and gender, meets the ideal type realized by the pigeon fan-cier. The divergence of pigeon breeds corresponds to the divergence of the human races, both traceable to a common origin, both the product of selec-tive breeding, both determined by fashion or culture, neither offering any survival benefit, both potentially deleterious.

For the same reasons as he had emphasized unconscious over methodical selection in the *Variation*, Darwin claimed that such sexual selection would also be unconscious, "for an effect would be produced, independently of any wish or expectation on the part of the men who preferred certain women to others."[72] The intention of this claim was again to close off any implication of God-driven design through male aesthetic preference, to reduce such masculine choice to an innate, racially distinctive idea of beauty—one, moreover, that Darwin might connect with an instinctive, efficacious appreciation of the beautiful by female animals and birds and so account for nonfunctional male ornamentation or potentially harmful conspicuous coloration.

6.6 Bird-witted females and clever men

And therein lay the rub. When it came to the sexual dimorphism of lower animals, Darwin was forced to turn on its head what he and his contemporaries understood as the norm of courtship and sexual practices: "natural" male sexual preference. To convince his fellow Victorians of the plausibility of male choice (conscious or unconscious) of the perceived prettiest women of their kind (no matter how distorted that perception might be) was one thing. To go against entrenched opinion by attributing not just a sense of beauty to animals, but further, to insist that this aesthetic was primarily exercised through female sexual preference, was, as Darwin himself well understood, thoroughly daunting. This was especially so when the accumulated sexual choices of a drab, sexually coy, bird-brained female supposedly had resulted in such a marvel of beauty and design as the peacock's tail. How could peahens have consistently and exactingly selected to such a refined aesthetic standard and on the basis of such minute differences as Darwin invoked, his critics would demand to know? Little wonder that Darwin, famously given to nausea over much lesser matters, wrote in 1860: "The sight of a feather in a peacock's tail, whenever I gaze at it, makes me sick!"[73]

What sustained Darwin, against all the reasonable and not-so-reasonable argumentation his contemporaries might muster against female choice, was what he knew about breeding and its practices. "I lay great stress," he insisted to Wallace, his most formidable and persistent critic, "on what I know takes place under domestication."[74]

What *did* Darwin know? He had scant evidence of female aesthetic selection under domestication, or in the wild, for that matter. After Darwin's death, Wallace cited Darwin's own breeding authorities to show how "singularly little evidence" Darwin had to show that beauty affects female choice.[75]

In his dispute with Wallace in the critical year of 1868, it emerged that Darwin's major justification of female choice was to rest on complex rules of heredity and development that he had assembled from his collated data on breeding and contemporary physiological theories of inheritance and embryology. Male pouter pigeons came into their own as Darwin's crucial instance of sex-related development and inheritance.[76] Throughout 1868, the letters went back and forth, conceding certain points, backtracking, deploying more and more sophisticated reasons why each must retain his conviction in his own rightness. Birds became the test case on which both parties exercised their ingenuity in trying to outwit the other. By October, having depleted their respective stocks of arguments and exhausted their powers of persuasion, they agreed to disagree with mutual protests of sorrow and esteem.[77]

Darwin's mature views on inheritance and development, their relation to sexual selection and their role in his dispute with Wallace will be discussed in chapters 9 and 12. For the moment, let me simply note how well these views accorded with traditional breeding lore involving prepotency, lines of inheritance, and the conventional differences between male and female sexuality. Darwin's was a thoroughly male-centered view of generation, with males being more highly developed than females and transmitting their masculine superiority—strength, intelligence, weapons, or beauty—primarily to their male offspring. It was males who actively and indiscriminately sought, battled, displayed, and mated. However, for the most part, as breeders affirmed, they did not choose. But neither, breeders insisted, did females.

Darwin's solution was to go against the grain by naturalizing female choice and making human male choice the exception that proved the rule. Darwin's justification of female choice was that it was precisely *because* of their lesser "passion" that females were passive, but discriminating, choosers. Such choice would not be overt or easily discernable. Its demonstration would largely rest, as Darwin was to argue, on the evidence of male display— which "I for one will never admit [to be purposeless]"—and on the "results which may fairly be attributed to sexual selection."[78] Where necessary—and most obviously when it came to humans, where he assumed male choice and female sexual passivity were both routine—Darwin, as I shall show, was forced to some tricky maneuvering around issues of inheritance and his rationale of passive, but efficacious, female choice.

It has puzzled historians who have not seen the force of Darwin's reliance on domestication why Darwin held so tenaciously to the position that, especially among birds, it was male ornamentation and conspicuous coloration that required explanation and not female dullness or camouflage for protection, as Wallace argued. Their other differences aside (and there *were* other significant differences, to which I shall return in ch. 12), we should note that from the outset, in his 1858 paper on natural selection, Wallace was wary of the domestication analogy on which Darwin became so reliant.[79] Although he drew selectively on the phenomena of domestication to support his own views, domestication was never for Wallace, as it was for Darwin, the crucial

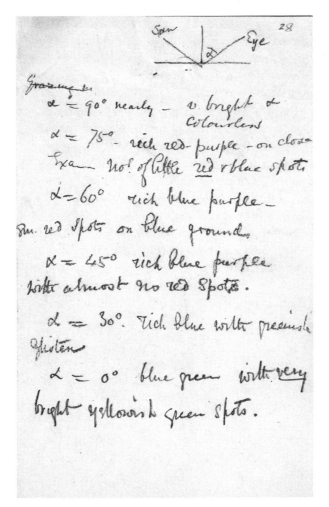

Figure 6.3. Darwin's bird's-eye view: his observations of the details of plumage colors, viewed at different angles to the sun, made while at work on *The Descent of Man.* Darwin Papers, Cambridge University Library, DAR 84.2: 28. By permission of the Syndics of Cambridge University Library.

testing ground for evolutionary processes and the associated problems of inheritance and development.[80] But above all, Wallace had had no such visual reeducation in the art of breeding as Darwin had so laboriously acquired.

It was this insider knowledge as well as the conviction of his belief in the immense modifying powers of the breeder that enabled Darwin to substitute his trained breeder's eye for the bird's-eye view of the peahen, and so analyze the stages by which, he argued, an "ordinary gallinaceous bird with short tail coverts, merely spotted with some colour" had been transformed by the "continued preference of the most beautiful males" into the peacock with its "eyed," shimmering train, the "most splendid of living birds."[81]

All this was to come. In the meantime, the Darwin of the late 1830s and early '40s, in the very early stages of his domestication studies, struggled against the perceptions and prejudices of breeders to transfer the strongly gendered, sexualized conception of aesthetic preference that he had acquired from his readings in aesthetics and ethnology to female birds. But this Darwin had to do if he was to explain nonfunctional male ornamentation and conspicuous colors through aesthetic selection, not divine design.

Ultimately, as we shall see in the next chapter, Darwin's solution of the naturalization of female choice played on the deeper, underlying cultural and social attitudes toward women and their sexuality that breeding lore and its practices embodied. They were attitudes that Darwin and most of his contemporaries shared. These were the very same attitudes that shaped Darwin's unshakable conviction that human evolution and racial divergence could not have been modified or shaped by feminine caprice or by the sexuality and passion of women.

SEVEN

"Better Than a Dog Anyhow"

Children—(if it please God)—Constant companion, (& friend in old age) who will feel interested in one,—object to be beloved & played with.—better than a dog anyhow.—Home, & someone to take care of house—Charms of music & female chit-chat.—These things good for one's health.—*but terrible loss of time.*

—Darwin's notes on marriage, 1838

In *David Copperfield* (1848–50), Charles Dickens made the decorative but irretrievably silly Dora—David's "child-wife"—synonymous with her overbred toy spaniel, Jip. Jip can no more learn new tricks than Dora can balance the household accounts; he shares the dinner table and their inedible meals, and he and Dora are both frightened into hysterics over David's efforts to train them to domestic order and must be soothed with caresses and presents. With his inimitable mix of comedy and pathos, Dickens blurs and confuses the identities of Dora and Jip. Both are ornamental but superfluous—pampered, useless, and ridiculous.

Eventually, Dickens has David abandon his futile attempts to "form" his "little wife's mind," to make her into the ideal companion who will share his literary pursuits and domestic responsibilities. Dora's mind is "already formed." She is an infantile simpleton, the petted product of an improvident father and a Parisian finishing school. In his own susceptible immaturity, David has chosen wrongly. He has succumbed to the attractions of a pretty flirt over the better claims of his proper soul mate, the serenely "modest," "womanly" Agnes, his "better angel," the embodiment of domestic "goodness, peace and truth." Remorsefully, David buys a "pretty pair of earrings" for Dora and a "collar for Jip" and resolves to "make [him]self agreeable" in the hope that "lighter hands" than his would "help to mould" Dora's character; more explicitly, that a "baby-smile upon her breast might change [his] child-wife to a woman." But Dora miscarries. Jip's death under David's caressing hand signals Dora's demise, her fragile doll-like body unable to withstand the rigors of the bedroom. The way is cleared for David's marriage to the faithful Agnes, who gives him the sturdy children and the "natural home" he has earned through the vicissitudes of his own path to mature and responsible middle-class manhood.[1]

There are a number of narratives at work in this semiautobiographical tale of self-made men; but a central motif is the figure of the domesticated dog that does such significant plot duty for Dickens. *David Copperfield* is a veritable Noah's ark of animal-human transpositions, where good people are harmless domestic animals—lambs, horses, little birds, mice, and sheep—while villains are dangerous predatory beasts—snakes, crocodiles, wildcats, bats, and foxes. In others of Dickens's novels, animals are as routinely humanized as evildoers are bestialized.[2]

Like the discourse of animal breeders, transparent to assumptions about human sexuality, morality, and gender relations, the Dickens novel reflected the wider social and cultural resonances of Victorian relations with animals. Discourse about animals captured common value judgments about human behavior and morality; animals represented, explained, and justified social relations.[3]

Even pigeons, Darwin's favored experimental domesticates, might be upheld as "types almost of Christian virtue" for their monogamous "adherence to chastity," shared parenting, and preference for a "settled home" rather than "indulgence in capricious wanderings." The Reverend Edmund Saul Dixon, author of standard works on the dovecote and poultry run, depicted pigeon courtship in blatantly anthropomorphic terms. He transmuted billing into kissing and identified a period of honeymoon before his pigeon pairs settled into chaste domesticity, with the husband taking the lead in house hunting and "with a few sharp pecks if necessary," insisting that his careless mate attend to her domestic duties. He told of an experienced cockbird that took over the responsibility of rearing the young until he had educated a feckless young hen barb into a "steady wife and mother." It quite put Dixon in mind of "those discreet old gentlemen who send their young brides to school before they marry them."[4]

But Dixon also saw a less gentlemanly side to pigeon life, comparable to the domestic relations of the lower orders. Cocks, "being master," do not hesitate to "beat the hen, if refractory, into compliance":

Females in general will put up with any treatment, rather than with indifference and neglect. A woman was one day discoursing with me respecting the virtues of her husband: "'E's a werry good 'usband, Sir, a werry good 'usband indeed. To be sure, he *do* beat me now and then; but you know, Sir, men must have *a little* reckeration!'"

Dixon's heavy-humored slippage from human courtship practices to pigeon mating to wife beating, together with his equally heavy-handed class, gender, and sexual references, was hardly exceptional within the genre of domestication literature. Darwin regarded Dixon's work as "*very* good & amusing," entered into "cordial" correspondence with him, and relied heavily on his observations.[5]

Certain animals offered better models for manners and moral worth than pigeons. Apes, closest to humans in appearance, were endowed with many human qualities, including sympathy and intelligence. Apes in zoos often wore human clothing, like Jenny the orangutan, observed by Darwin in the Royal Zoological Gardens, dressed in a flannel nightgown and drawers. Jenny, for Darwin, settled unequivocally the issue of human kinship with apes. She "kicked & cried, precisely like a naughty child" when her keeper denied her a treat. The expressions and behavior of captive orangutans showed such affinity with human emotions and intelligence that Darwin, as we saw, viewed them as almost more human than the uncivilized Fuegians.[6]

But apes could not, for many naturalists, attain the levels of empathy and wisdom of the domesticated dog. In some respects dogs were considered more human than apes, whose very similarity to humans rendered them disgusting and unseemly and therefore inappropriate for comparative purposes with humans, particularly women. Queen Victoria was not amused but appalled by the "painfully and disagreeably human" sight of Jenny making and drinking a cup of tea.[7] If Victoria was disturbed by so demure and domestic a scene, the response to overt genital displays and the uninhibited sexuality exhibited by various captive monkeys and apes may be imagined. Darwin, uninhibited observer of simian sexual behavior that he was, recognized that such "facts" might be "turned to ridicule" or "thought disgusting" and saw fit to defend his recording of them—even in the privacy of his notebooks (yet with a revealing choice of words)— as "pregnant with interest" to the "philosophic naturalist."[8]

In *The Descent of Man*, Darwin discussed the highly colored rears of some monkeys, like the adult male mandrill (one of Darwin's choicest instances of sexually selected traits in mammals), with some delicacy, leaving their naked "*fesses*" in the original French, or referring to their color patches as "oddly situated." In 1876, with sexual selection under increasing attack,

Darwin wrote more directly of these Interesting "brightly coloured hinder ends and adjoining parts" of such monkeys, which he had discovered had the "indecorous habit" of presenting their rear ends "as a greeting to old friends or new acquaintances." His informant, a German naturalist, had observed that such behavior was connected "with sexual feelings"; but Darwin left the explicit evidence for this in the original German, even in a scientific journal addressed to his peers. The point of his *Nature* article was to use these findings as further corroboration of sexual selection in mammals, of Darwin's claim in the *Descent* that the decorated hinder parts of the mandrill "serve as a sexual ornament and attraction." "I was well aware," he wrote defensively, "that I thus laid myself open to ridicule; though in fact it is not more surprising that a monkey should display his bright-red hinder end than that a peacock should display his magnificent tail."[9]

Darwin anticipated ridicule for his contemplation of the monkey's bottom, and he got it from the influential Victorian aesthetician John Ruskin, who publicly lampooned Darwin's degradation of the "most refined" beauty to sexual selection. Darwin had no sense of the difference between beauty and ugliness, decency and indecency, glory and shame. All alike were reduced to the coarse mechanical sensations and vulgarity of sexual selection. Privately, Darwin affably affirmed Ruskin's taunt: he did feel a "deep and tender interest about the highly coloured hinder half of certain monkeys."[10]

Illustrations and cartoons of naked apes in the popular press were conventionally bowdlerized, although, after the publication of the *Origin* and the *Descent*, *Punch* and other satirical periodicals invoked more bawdy connotations of human and ape interactions, on occasion featuring Darwin himself with hirsute, apelike features. The innuendo of such representations drew on the long tradition of representing anthropoid apes as embodiments of male sexual rapacity, given to carrying off and ravishing women—preferably blond, though their usual victims were African and Asian. The evolutionary argument for the common descent of ape and human provoked renewed satirical and salacious speculation about cross-species sexual relations and the supposed hybrid offspring of such unions. Cartoonists might exploit such an unsettling trope, while pornographers pushed the boundaries of such depictions to explicit and shocking extremes.[11] Bishop Wilberforce's infamous jibe at Thomas Henry Huxley at the Oxford Debate in 1860, where Wilberforce demanded to know whether Huxley claimed his descent from a monkey through his grandfather or grandmother, insinuated the same sexualized, discomforting trope. Legend has Huxley smartly retorting that he would rather an ape for a grandparent than the barbarous Bishop. This took the moral high ground and set the precedent for Darwin's published preference for the ape ancestor over the "savage" who "knows no decency" in the conclusion to the *Descent*.[12]

Darwin ranked the ape as a "noble" animal and praised it for its intelligence and altruism, but he also sprinkled his discussion of human evolution with anecdotes of the perspicacity and faithfulness of his favorite dogs. To an extent, Darwin ranked races on the basis of their domestication of the dog. Such was the superiority of English dogs that they were probably more capable of self-consciousness and forming abstract ideas than the average "savage" woman:

> But can we feel sure that an old dog with an excellent memory and some power of imagination, as shewn by his dreams, never reflects on his past plea-sures in the chase? And this would be a form of self-consciousness. On the other hand . . . how little can the hard-worked wife of a degraded Australian savage, who uses hardly any abstract words and cannot count above four, exert her self-consciousness, or reflect on the nature of her own existence.[13]

The multiplicity of meanings encoded in the dog exemplified many of the tensions in Victorian national, class, and sexual ideologies. Huxley fa-mously had foisted on him the soubriquet of "Darwin's bulldog," a ref-erence to the breed's renowned attributes of courage and tenacity, "coeval with and emblematic of Anglo-Saxon civilization." Middle-class dogs were supposedly as class conscious as they were cognizant of race. George Ro-manes, zoologist and leading Darwinian, credulously recounted the story of a wandering retriever who struck up a friendship with a ratcatcher and his "cur," but who "cut" his lowly acquaintances as soon as his master ap-peared, proving that dogs understood "the idea of caste."[14] Low-caste dogs, like lower-class women, possessed dangerous sexualities that necessitated control and were a potential source of contagion to middle-class humans. Rabies in dogs, like venereal disease in humans, was connected with the undisciplined habits of the curs and street dogs of the poor, the mixed mon-grels that scavenged, fought, and mated without restraint. Like prostitutes, they were a source of deviance and contamination that required vigorous discipline and legislation. At the other end of the scale, the pampered and overfed pets and sporting dogs of the rich were singled out for their corrupt lifestyles and incestuous bloodlines, and associated with the indolence and immorality of their masters—or, more often, mistresses.[15]

The more menacing sado-sexual implications of the dog-human trans-position run more deeply and darkly through *David Copperfield* than the gently humorous or sentimental tone of David's relations with the hapless Dora might suggest. Its sinister undertones are sketched in the early stages of the novel, when David's petted childhood is rent asunder by the black-whiskered Murdstone, who marries David's widowed mother Clara. Murdstone installs a vicious dog, hair as black as his own, and threatens David with the same treatment he metes out to the dog. Goaded beyond endurance, David bites him, whereupon Murdstone mercilessly flogs David into submission: "He ordered me like a dog, and I obeyed like a dog." David is exiled to a kennel-like boarding school, where he is labeled like a savage dog ("Take care . . . He bites") and is harassed by the brutal keeper, Creakle, who whips the "miserable little dogs" in his care. Clara is the precursor of Dora, a pretty, "weak, light, girlish creature" who fawns over the sadistic Murdstone: "I knew . . . that he would mould her pliant nature into any form he chose." She is "trained" and "broken" by Murdstone, fades away, and dies. His mother's property devolves on Murdstone, and the dispossessed David must seek his own fortune. Murdstone symbolically reappears toward the end of the novel, having "subdued" and "entirely broken" his new wife, a once "lively" young woman with a "very good little property," reducing her to abject dependency and near imbecility.

Murdstone is the archetypal wife trainer and breaker of early Victorian fiction, a terrifying example of the domestic and sexual tyranny of the dominant patriarchal male that David cannot bear to emulate in his dealings with Dora and Jip. Yet, while Dickens had David resist his own categorization and treatment as a dog—his dehumanization—neither Clara nor Dora can progress beyond subservient acceptance of their belittling canine status. And, although David is the antithesis of the rock-hard Murdstone, he too subscribes to the pervasive ideology of the controlling hand of the masterful male who molds the doglike pliant nature of the submissive female. It is only through Dora's demeaning lapdog inanity, her desire to remain a petted "silly little thing" rather than be made "uncomfortable" by education—a self-styled "child-wife"—that she eludes David's will to domesticate her. "Lighter hands" than his must do the necessary remolding of Dora's mind, but this means death for Dora and her alter ego Jip. As Dora pathetically recognizes, she is "not fit to be a wife," and she hands on to Agnes the essential womanly tasks of motherhood and reproduction.[16]

There is a fusion of imagery from sentimental fiction, from portraiture, from breeding, naturalist, and field literature that conduces to the most persistent signification of the Victorian dog as the loved but subservient being that thoroughly understood and accepted its inferior position. "Dogs," Darwin insisted in the *Descent*, "have long been accepted as the very type of fidelity and obedience." Such was the extent of their almost pathological love and forgiveness for their human masters that, in the agony of vivisection,

Figure 7.1. Charles Bell's image of devotion from *The Anatomy and Philosophy of Expression* (1806), juxtaposed with a dog "in a humble and affectionate frame of mind" from Darwin's *Expression of the Emotions in Man and Animals* (1872). Bell claimed that uniquely created human muscles were necessary for the expression of the emotions; Darwin challenged this by comparing the facial and bodily attitudes of animals and humans, arguing for the animal ancestry of human emotions and expression. The dog's devotion and love for its master was a state of feeling, Darwin implied, akin to the reverence and gratitude, along with the fear, experienced by the devoted Christian. Bell (1806) 1844, 104 (Courtesy of Wellcome Library, London); *Expression*, 53 (University of Sydney Library, Rare Books and Special Collections).

dogs have been known to lick the hand of the experimenter. This "deep love" of dogs and "complete submission" to their masters was related by Darwin to reverence for a god, as a "distant [but parallel] approach" to the "feeling of religious devotion" experienced by humans.[17]

The dog's very body proclaimed its profound submission to humanity: it was the most malleable of domestic productions, its shape and size responding most readily to the breeder's caprice. Unlike the derogated cat that stubbornly resisted efforts at training or the remolding of its body—thus constituting a threat to household order and public authority—the dog's "plastic body symbolized its desire to serve," to obey.[18]

Within the terms of Victorian domestic ideology, women also were to serve, to obey, to be pliant to masculine whim and will, to stay where men commanded or follow where they led. The young Darwin, reckoning up the advantages of the married state in 1838, made the direct connection: marriage was analogous to pet keeping. A wife was an "object to be beloved & played with.—better than a dog anyhow."[19] Darwin's semifacetious premarital musings were not innocent. The association of women's sexuality and "nature" with dogs and other domestic animals was deeply entrenched in Victorian culture. Talking about animals offered those like Darwin or Dickens who "would have been reluctant or unable to avow a project of domination directly a way to enact it obliquely."[20]

The emotional commitment of the leading feminist Frances Power Cobbe and so many other women to the anti-vivisection campaign of the late Victorian period has been linked with the identification of such women (subject to the domination of men, living in a culture that expected female service in the home, that condoned and even drew entertainment from wife beating) with the suffering of defenseless dogs, cats, and monkeys at the hands of the merciless vivisector. At the same time, and paradoxically, women might exploit such rhetoric and involve themselves in the anti-vivisection cause as a means of testing and extending the limits of their sphere of femininity in ways that did not conflict with their conventional womanly attributes. Darwin, who found himself called to account for his pro-vivisectionist stance in his own household by daughter Henrietta, could only explain the phenomenon by recourse to the same model of femininity: that it was "from the tenderness of their hearts and from their profound ignorance" that women were the "most vehement opponents" of vivisection.[21]

Darwin took great pleasure in the novels of Dickens. They gave in full measure his much-quoted requirement of fiction: a happy ending, and that it should contain "some person whom one can thoroughly love, and if it be a pretty woman all the better." Gillian Beer has suggested that Darwin was "freed from some of the difficulties he experienced in expressing the relation of man to the rest of the natural order" by his reading of Dickens. Their narratives share the "theme of hidden yet all-pervasive kinship." Going beyond this generalization, there is more than a degree of coincidence in the ways in which Darwin and Dickens both characterized animals, especially dogs.[22] The sexual and gendered dimensions of such animal-human transpositions are recognizable in the writings of both men. Darwin's expression may indeed have been inflected by his reading of Dickens as well as other contemporary fiction, but his immersion in the literature of the breeders guaranteed his full exposure to such metaphorical discourse. Darwin and Dickens, breeders, naturalists, feminists, and anti-vivisectionists alike were all drawing on the available rhetoric that collapsed animal into human behavior, and, in particular, on that persistent and powerful trope that transposed the sexuality, submission, and domestication of women and animals.

7.1 Better wives for choosier husbands

Having opted for a wife rather than a dog and well aware of the need for careful choice in the matter, Darwin opted for his cousin Emma, of whose superior Wedgwood pedigree and wealth he could be assured. Still, he brooded on the implications of cousin marriage while in his notebooks he also worried away at the interrelated problems of generation, inheritance, and his newly formed theory of natural selection. In mid-1839, with Emma pregnant with their first child, Darwin was alerted to a promising new work by the *British Foreign and Medical Review*. He laid out fourteen shillings for a copy of *Intermarriage* by the anatomist Alexander Walker.[23] Of all the literature Darwin was to read while he was formulating his theory of sexual selection, none made the link between the breeding of animals and the mating and manipulation of humans more overt than *Intermarriage*; and none so explicitly naturalized and sexualized women's bodies and natures, or so masculinized "choice" in mating and marriage.

Intermarriage was the central text of a popular trilogy on "woman" that ran through several English and American editions. Walker's trilogy offers unique and valuable reading in relation to early Victorian attitudes toward the relations between the sexes, to female sexuality and educability, to issues of class and racial progress, and to the contemporary interest in heredity and interpretations of the laws of selective breeding. His titles virtually speak for themselves: *Beauty: Illustrated Chiefly by an Analysis and Classification of Beauty in Woman* (1836); *Intermarriage; or, The Mode in Which, and the Causes Why, Beauty, Health and Intellect, Result from Certain Unions, and Deformity, Disease and Insanity from Others* (1838); and *Woman Physiologically Considered as to Mind, Morals, Marriage, Matrimonial Slavery, Infidelity and Divorce* (1839).[24]

Walker made the overall aims of his trilogy explicit in the introduction to the first volume, *Beauty*:

> There is perhaps no subject more universally or more deeply interesting than [female beauty]. Yet no book, even pretending to science or accuracy, has hitherto appeared upon it. The forms and proportions of animals—as of the horse and the dog, have been examined in a hundred volumes. Not one has been devoted to woman, on whose physical and moral qualities the happiness of individuals and the perpetual improvement of the human race, are dependent.

What Walker was offering his readers was the promise of human advancement through selective breeding. He deplored the neglect of this socially important science even by those who were skilled in the art of animal breeding.[25] The "great object" of his *Intermarriage* was to bring the "natural laws" and "indisputable facts" of anatomy, physiology, anthropology, and animal breeding to bear on the "transcendently important subject of choice in intermarriage."[26]

Walker's personal history remains obscure. What seems certain is that, having failed as a freelance anatomy lecturer in Edinburgh, Walker pawned his books, begged a more presentable coat and boots to go with his "pan-taloons," and by 1815 was resident in London. There he barely scratched out a living, translating, editing, and writing for literary journals, including *Blackwood's Magazine* and the *Literary Gazette* (Walker was one of its found-ers), and the leading organs of radical medical opinion, the *Lancet* and the *London Medical and Surgical Journal*. During the 1820s, Walker edited an extraordinary multilingual periodical, the *European Review; or, Mind and Its Productions*, a vehicle for a radical materialist "universal science," that claimed continental luminaries like Étienne Geoffroy Saint-Hilaire, Goethe, and Benjamin Constant among its contributors. It foundered after a few issues in spite of Walker's convoluted stratagems for its financing and per-haps because of its grandiose claims and political polemics.[27]

By the 1830s, Walker had turned his hand to more profitable popular monographs, feverishly churning out more than a dozen books, mainly on medical or related topics, including his "woman" trilogy. This became some-thing of a family enterprise, with the recruitment of his son Donald (who contributed a series of works on "manly" exercises and sports as well as *Ex-ercises for Ladies Calculated to Preserve and Improve Beauty*) and the shadowy "Mrs. Alexander Walker" (the alleged author of *Female Beauty*, a ladies' guide to achieving beauty through "Regimen, Cleanliness and Dress"). There was a thriving market for popular physiological and medical works and articles, powered by the new technology of steam printing and fed by the growing numbers of "Grub-Street hacks"—the cash-strapped journalists, naturalists, zoologists, and physiologists like Walker who specialized in selling science to the literate public.[28] Within this context, access to information on sex, gestation, and generation was a marketable commodity, providing "these most delicate inquiries" were appropriately presented. As a measure of its popular appeal and the market for such works, *Intermarriage* was issued in some four English and six American editions.[29]

Reproduction and child-rearing were becoming issues of social and political significance that warranted expert guidance and intervention. Walker made efforts to identify his enterprise with an array of prominent medical men and eminent breeders who shared his concern that the laws of hereditary transmission and interbreeding should be made available for the future benefit of the country and its people. *Intermarriage* featured the endorsement of Sir Anthony Carlisle, conservative councilor and past presi-dent of the Royal College of Surgeons.[30]

According to Foucault's classic *History of Sexuality*, the nineteenth-century bourgeoisie made its own class system not through the aristocratic lineage of blood, but through its sexual reproduction, through the health of its progeny:

the traditional "concern with genealogy became a preoccupation with heredity."[31] The belief in the heritability of disease, insanity, moral insufficiency, and other "hereditary taints" was tightly associated with the need for "prudent" choice in marriage and reproduction by a wide spectrum of medical practitioners, ethnologists, phrenologists, hygienists, social reformers, novelists like Dickens, diarists, and many of the educated upper and middle classes. Marriageable young girls were routinely scrutinized to ascertain that they were free of hereditary disease, particularly "scrofula" (a tuberculous infection of the skin of the neck, thought to be heritable). Families went to considerable lengths to conceal the stigma of familial madness or abnormality. The evils attendant upon consanguineous marriages were regularly aired, and the prevalence of cousin marriages deplored. Life assurance companies began refusing insurance to those whose families had histories of derangement or constitutional disease. By midcentury, there was also a growing middle-class consciousness of the looming problems to be associated with an expanding urban underclass allegedly given to near-intractable alcoholism, criminality, immorality, and pauperism, all of which might be reconfigured as heritable and present opportunities for professional intervention. Fears of national degeneration through reckless "intermarriages" and the unchecked propagation of the socially undesirable began to be articulated.[32]

As a publishing venture touching on these individual concerns and social fears, *Intermarriage* was on to a sure thing, joining several other contemporary works in the same vein.[33] But Walker's interpretation, focused on such a strongly gendered, sexualized, and aesthetic conception of woman, was unique to him. His project for human improvement pivoted on his analysis of female beauty. He set out to demystify traditional aesthetics, to reconfigure and stabilize beauty by applying the methods of "anthropology," a "branch of science . . . strictly founded on anatomy and physiology," to the "external forms of woman." What was previously "mystical and delusive" was to be made scientific and certain through knowledge of the "laws regulating beauty in woman, and taste respecting it in man."[34] A major inspiration for Walker's "anthropological knowledge" was Lavaterian physiognomy, which Walker extended from the typology of faces to "the whole body," giving it a pronounced materialist and utilitarian interpretation.[35]

Walker's work crystallizes, like none other of the period, the close historical ties between physiognomy, anatomy, aesthetics, and anthropology. Furthermore, it was the earliest systematic attempt to extend the new anthropology from racial to gender differences. Walker drew heavily on the writings of James Cowles Prichard and William Lawrence—especially the latter's seditious *Lectures*, large slabs of which were reproduced verbatim in

Intermarriage. Lawrence's materialist anatomy, his location of human intellectual and moral differences in physical organization, and sideswipes at a degenerate and corrupt ruling class were entirely compatible with Walker's own radical anatomy. But Walker drew more direct inspiration from Lawrence's emphasis on the analogue of domestication and his comparison of the influence of ideas of beauty on racial or national differences with the practices of animal breeders; above all, from Lawrence's consideration of the possibilities of regulated human breeding.[36]

Where he differed from Lawrence was in Walker's views on inheritance and in his absolute physiognomic subordination of mind and internal anatomy to external appearance, to beauty. For Walker, beauty was not simply a cause of visual pleasure, but a sign of reproductive fitness, of suitability as a mate: beauty is the "external sign of goodness in organization and function." The internal organization reveals itself on the surface of the body and may be read or decoded by those with the requisite knowledge and applied to mate choice. For Walker it was patently obvious who makes the choice: the form of woman is chosen for examination, "because it is best calculated to ensure attention from men, and because it is men who, exercising the power of selection, have alone the ability thus to ensure individual happiness and to ameliorate the species."[37]

Walker defined three "species" of beauty in woman, the visual representations of his system, founded on his hierarchical division of organs into the locomotive (which control movement); the nutritive, or vital (which control digestion, circulation, and generation); and the highest, the mental (which control sensation, thought, and volition).[38] As beauty is a sign of reproductive fitness, it followed that the most beautiful women are those in whom the vital predominates, those organs controlling generation and parturition. Men should beware those women in whom the mental or locomotive functions have been developed at the expense of the vital. Vital beauties may be recognized by their fair hair and complexions, soft, moderately plump and "voluptuous" forms, and light and graceful movements, "qualities that please, because they announce the good condition of the individual who possesses them, and the greater degree of aptitude for the functions which [she] ought to fulfill."[39] In other words, vital beauties are those best fit to arouse men's desire and ensure their healthy posterity.

This naturalized and sexualized ideal woman was race and class specific. Walker endorsed the traditional aesthetic claim that the "natural" idea of the beautiful had been distorted in almost every nation: the "Negro and the Mongol" have "defective" notions of female beauty. It is only in nations of "very advanced civilization" that the necessary conditions prevail for the production

Figure 7.2. Alexander Walker's three species of facial beauty in women: vital, mental, and locomotive. Walker 1836, 281. State Library of New South Wales, Mitchell Collection.

of both perfect beauty in women and the ability to appreciate it in men; that is, "a brain capable of vigorous thought, sound judgment and exquisite taste." Nor was true beauty to be sought for among working-class women. Walker disparaged the proportions of the lower-class models who posed for the illustrative lithographs. They could not come close to the natural vital ideal, exemplified by that popular icon of female beauty, the *Venus de' Medici*. Walker's description of this marble beauty is lingeringly tactile and highly erotic: "the admirable form of the mammae . . . ; the flexile waist . . . the gradual expansion of the haunches . . . the beautiful elevation of the mons veneris."[40]

Walker's visual and verbal depictions of "beauty" trod a fine line between instruction and offense. *Beauty* (especially its deluxe edition of 1851, issued with tinted plates and gold-embossed leather binding) seems directed to a

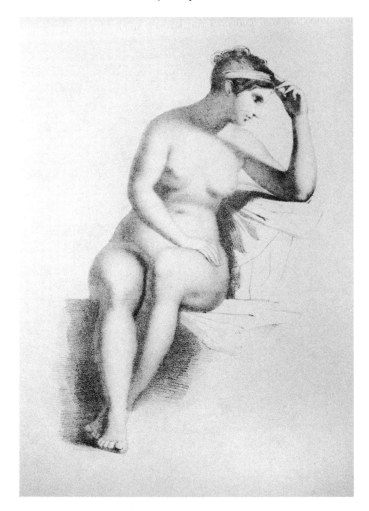

Figure 7.3. A vital beauty. Walker 1836, 226. State Library of New South Wales, Mitchell Collection.

male-only readership who might extract as much titillation as information from its many illustrations of classically posed nude females—the different "species" viewed from various angles to signify the anatomical differences discerned by Walker. *Intermarriage* dwelled on the "erotic orgasm" at some length, going against contemporary opinion to argue that woman was more highly sexed than man and enjoyed "far higher pleasure." Female sexual satisfaction was essential for health and conception, for woman's vital purpose of the reproduction of healthy offspring.[41] *Woman Physiologically Considered* radically championed woman's right to property, divorce, and the custody of

Figure 7.4. A woman whose locomotive system has been developed at the expense of her vital system. Walker 1836, 376. State Library of New South Wales, Mitchell Collection.

her children, going so far as to defend her "natural" right to infidelity when she was deprived of the love that her nature required, deploring the double standard that penalized women for obeying their natural inclination to love while condoning it in men.[42]

Yet Walker's texts were not marginal ones, not part of the radical underground press, but were published by respectable houses and reviewed in fashionable periodicals. *Beauty* was recommended to "mothers as well as fathers of families" by the *Court Magazine and Belle Assemblée*. Other reviewers might cavil at the propriety of entering into the particulars of Walker's arguments, but still gave his works review space.[43] Walker was assured of the backing of the militant *Lancet* (when Walker, ill and broke, retired to Edinburgh in the 1840s, the *Lancet* organized a fund for his support). However, he could command a favorable hearing from the moderate *British Foreign and Medical Review*—also interested in the promotion of a naturalistic science and the expansion of professional power—which commended Walker's efforts to adapt the risky material of *Intermarriage* to the "taste and capacity of the public."[44]

7.2 A science of beauty

Walker's images of naked beauties, with their textual anatomical and physiological validations, were counterparts of the more explicit anatomical models (shown to separate audiences of women and men) on commercial exhibition in London and other major cities.[45] They were edgy representations that Walker sought to legitimate through borrowing their poses from classical art and sculpture and by an appeal to the Western aesthetic tradition of the virtue attached to nakedness and its scrutiny. Walker claimed both the life class and the dissecting room as suitable spaces for the display of "natural purity." But in taking naked bodies from the allegedly disinterested gaze of professionals into the middle-class home, Walker was intruding into a space that was assuming increasing symbolic significance as a sacred site, a domestic sanctuary from the strife and temptations of the public sphere. The propriety of their exhibition under the idealizing force of nude art was complemented by an appeal to science for their further validation and legitimacy. By contrast with the eroticized nude images of *Beauty*, *Intermarriage* confined its visual representations to heads, the scrutiny of which (in the physiognomic tradition, reinterpreted according to Walker's laws of selection) would convey all the information necessary for rational wife choice. They offered a more practical guide to marital choice, familiarized through the proliferation of popular physiognomic and phrenological works. Both sets of images, represented as scientifically valid and as having the high-minded goal of racial improvement, also catered to the visible display that the Victorian public was coming to associate with science.[46]

On another level, Walker's images and text appealed to the traditional association of professional anatomists with artists in pursuit of a common ideal of a more accurate aesthetic rendering of the human body, especially the female body. By the mid-nineteenth century, this was particularly the concern of romantic morphologists, who were considerably interested in the relation of art and aesthetics to gender and race. Robert Knox's *Great Artists and Great Anatomists* (read by Darwin in 1852) and his *Manual of Artistic Anatomy*, both published in 1852, were exemplary of the radical materialist approach to which Walker also belonged (see sec. 10.6).[47] A year earlier, John Goodsir—ex-pupil of Knox, conservative exponent of idealist morphology, and holder of the Edinburgh chair of anatomy—had joined with the painter and art theorist David Ramsay Hay and the mathematician, the Reverend P. Kelland (all founding members of the university "Aesthetic Club"), in the meticulous measurement of the vital statistics of as many artist's models as they could find. This "search after truth," which the professors performed "with extreme care" over a period of some six months, was directed at determining how far ancient Greek sculpture, "the only perfect models of excellence," agreed "in its proportions with ordinary nature." The "leading points" so painstakingly measured included the "full height of the figure . . . the vertical position . . . of the nipples, of the navel, and of the horizontal branch of the pubes" and "the width across the pelvis." The results of their diligent researches were then plotted on a graph to produce the ideal figure, a scientific norm for feminine beauty, published in Hay's *Natural Principles of Beauty* of 1852.[48]

In 1854, Herbert Spencer—ideologue of progress, future member of the Darwinian inner circle, and already a convinced Lamarckian—weighed in with his essay in the *Leader* on "Personal Beauty." This sought to reinterpret the traditional association of beauty with virtue and goodness in developmental and functionalist terms. Invoking physiognomic practice and standard anthropological imagery, Spencer argued for a natural correlation of ugliness with stupidity and beauty with "character" and intelligence. He traced the development of the face from the "projecting jaw, characteristic of negroes" and the "lower classes of animals" (which was "associated . . . with comparative lack of intelligence"), through decreasing jaw size and increasing "protrusion of the upper part of the cranium" in the more civilized races. A parallel progression was traced in individual development, whereby the young of the civilized evinced characteristics of "barbarians," including the "depression of the bridge of the nose," the "forward opening of the nostrils" and a large mouth, all "by general consent called ugly." These ugly features, shared by "infants, savages and apes," were contrasted with those of sculptures of the Greek gods, "admitted to be the most beautiful" and representing "superhuman power and intelligence." It followed, as an "almost irresistible induction," that "aspects that please us are the outward correlatives of inward perfection, while the aspects that displease us are the outward correlatives of inward imperfections." This was substantiated with arguments patently drawn from Walker and Knox, and went to support Spencer's "own conviction": "The saying that beauty is only skin-deep, is but a skin-deep saying."[49]

Around the same time, when pressed by friends to marry as a "cure" for his mental and physical ailments, the persnickety bachelor rejected their "remedy for rationalism," arguing that his "choice was very limited" and he "was not easy to please": "Moral and intellectual beauties do not by themselves suffice to attract me" unless "united to a good physique," and such combinations, said Spencer, were rare.[50] These private aspersions on the physical shortcomings of those intellectual literary women known to Spencer (the closest of them being Marian Evans [George Eliot]—admittedly no beauty by contemporary standards, but then, neither was Spencer) were elaborated on a few years later in the *British Quarterly Review*. There, Spencer claimed that of the elements uniting to produce in a "man's breast that complex emotion which we call love," the strongest were "physical attractions" and the weakest "intellectual attractions." This was not, said Spencer, an assertion "derogatory" of women, but followed from the physiological laws of development that did not permit the cultivation of woman's mind without detriment to her body. Nature's "supreme end is the welfare of posterity." Hence, a woman's most valuable contribution to the evolution of the race was a healthy "physique" that might be cultivated by physical training, while man might contribute a fully developed brain fostered by education.[51]

What these varied attempts at scientific determinations and explanations of beauty held in common was their embodiment in "woman" and their focus on her reproductive organs and function, made explicit by Walker and Spencer and implicit in the "leading points" singled out by Goodsir and his associates. The tight association of beauty with reproduction was made dependent on a devaluation of female intellect and "took the form of a profound sense of the intellectual disparities between men and women."[52] Beauty in women was a sign of reproductive fitness, and was linked, through their sexual choice by men, with the moral and physical development of the human race. A biological narrative thus underpinned notions of the ugly and the beautiful. The ugly was not merely unaesthetic; it was "anti-erotic" and "denied the ability to reproduce."[53]

The other dimension to such midcentury definitions and explanations of beauty and their localization in a sexualized female biology and physiognomy is that they are to be interpreted as early attempts to scientifically stabilize and simplify a multivocal and conflicted social issue that had not yet assumed universal recognition as the "woman question."

7.3 The woman in question: Redeemer or reproducer?

When Darwin read and annotated *Intermarriage*, Chartist agitation was at its height and Parliament had just rejected a national petition of more than a million signatures demanding "universal" suffrage for men. As for women, a particular "woman" was by then well in place. She was the iconic "angel in the house," Dickens's fictional Agnes, the angelic "English lady" or "good wife" of Darwin's letters and notes, the personification of Victorian womanliness, modesty, self-denial, goodness, and purity. She was the product of Evangelical discourse and the growing middle-class accumulation of property and power; her function was to guard the moral and religious values of the propertied. She underpinned the economic importance of marriage and motherhood and was domestic guarantor of middle-class property, propriety, and posterity. Her real-life counterparts were those middle-class women who had no real choice but to marry and little or no choice of marriage partner, whose property devolved to their husbands on marriage, and who had no legal or political existence except through their husbands.[54]

The self-abnegating, asexual "angel" of Victorian domestic ideology did not go unchallenged. There were dissenting voices, notably radical ones, who articulated different conceptions of "woman," ones with economic, political, and even sexual choice—like those of the Owenite socialists, of the radical Unitarians, and the irrepressible Richard Carlile. Carlile offended radicals and respectables alike with his claims that the pleasures of sexual love without fear of unwanted pregnancy were the right of every healthy woman after the age of puberty, and that female disorder "in nine cases out of ten" was the result of unsatisfied sexual desire. Most reprehensibly, he exhorted young women to "assume an equality" by taking the initiative in choosing their lovers.[55]

But even radical opinion was affected by the middle-class construction of womanhood and the ethos of respectability, with its emphasis on the morally regenerative role of the family and on woman's essential contribution to this. The situation was compounded by the failure of most activists and proselytizers to see a plurality of female conditions and roles, of different categories and classes of women with different experiences, needs, and aspirations. Furthermore, there were many "radicalisms" in the early nineteenth century, some of which were in violent opposition to one another, rife with ambivalent rhetoric and diverse and often-conflicting ideas concerning the roles of women.[56]

For all his radicalism, in crucial particulars, the "woman" of Walker's texts did not challenge the conventions. Although she enjoyed sex, her sexuality did not lead her to initiate encounters: she was naturally timid and modest, her passion was "of the gentler kind," and she was "passively voluptuous."[57] She did not select; she was selected. Her reasoning powers were constrained by her reproductive functions. The organ of will in woman was smaller than in man, and confered on her only the will to please, not to achieve or exert herself mentally, physically, or sexually. Her natural place was in the home, where her "natural duties" lay: in the bedroom, nursery, and kitchen. She had no place in the public sphere. "PHILANTHROPY, PATRIOTISM, and POLITICS, not being matters of instinct, but of reason, are unsuited to the mind of woman."[58] There was really no such thing as a "learned woman"; works of genius "exceed the capacity of woman." In any case, Walker warned (in an argument adopted by Spencer), mental exertion in women caused a compensatory "deterioration of their vital system and their natural attractions": a woman "notorious for her mind . . . is in general frightfully ugly" and most probably sterile.[59]

Even Walker's impassioned and sustained defense of woman's right to divorce was structured in terms of the dependent and subordinate position in which her weaker anatomy and intellect placed her in relation to her "natural protector" man, who all too easily might become her oppressor:

> The physical relation of women to men—their beauty, ensures their being beloved; while their feebleness seems to ensure their being oppressed. The fate of women is, indeed, different in different countries; but in all, they are more or less slaves.

In effect, woman must be protected from her own weaknesses, her natural tendency to obey, to be oppressed. Inferiority in body and mind was no excuse for the suspension of woman's "natural rights" in "matrimonial slavery," any more than it was for the "enslavement of the negro."[60]

Walker might draw on contemporary radical rhetoric that likened women's situation to that of a slave or domesticated animal, owned and controlled by men; he might defend their "natural rights" in marriage; but he had no truck with that form of radical or reform-oriented opinion that emphasized women's redeeming effects on society and encouraged their greater participation in the public spheres of the abolitionist movement, adult education, or philanthropic or literary enterprises; still less with the early feminist objectives of female employment outside the home, suffrage, and the intellectual and political parity of men and women. Rather, Walker summoned the household angel—virgo intacta, as it were—sanctimoniously claiming that he deserved the gratitude of the female sex "by showing that nature, for the preservation of the human species, has conferred on woman a sacred character, to which man naturally and irresistibly pays homage, to which he renders true worship."[61]

In the 1843 American edition of Walker's trilogy, the editor, a "Physician," simply reinserted her "soul" along with her more imperative religious, educative, and moral duties into the "natural" procreative, subordinate woman defined by Walker, stripping her of those few radical "rights" Walker had allowed her.[62] Within the complex metonym "woman," the move to her more naturalistic interpretation might thereby be made compatible with evangelical discourse. Reform-oriented professional medical men, or even conservative ones like Sir Anthony Carlisle (who became quite involved in the Walker enterprise, reappearing as "editor" of Mrs. Walker's guide to female health and fashion), might find in Walker's naturalized model an agreeable level of conformity with customary notions of gender difference and the growing emphasis on the essential separation of the public and domestic spheres.

By the 1840s the ideology of separate spheres for men and women had thoroughly infiltrated the assorted middle-class radicalisms, various reform movements, and heterodox religious sects like the radical Unitarians. The negotiation of this ideology was central to all debate on the "woman question," especially for those women active in the redefinition and politicization of the situation of Victorian women. In 1840 the World Antislavery Convention in London ruled that British custom did not permit women to speak at public meetings or act as delegates. Two years later, the very notion of a female franchise was ridiculed at a Chartist convention.[63] Walker's views on women were not so out of step with contemporary popular radicalism as might appear to twenty-first-century perceptions.

Darwin was hardly unaware of contemporary debates on woman's nature and social role. The family, notionally anyhow, advocated education for women, while Wedgwood women, as noted, were much involved in the abolition movement and other humanitarian causes. If his relatives failed to challenge Darwin's conventional preconceptions of women, then the formidable Harriet Martineau was on tap during his London years. At the height of her fame as a political author, she argued for better educational opportunities for women and was a role model for greater female participation in the public life of the nation, if not for Darwin's notion of a "good wife."

Yet, like many other women writers of the period, Martineau was fundamentally conflicted in her own views on women's proper role. She established her early reputation as an "objective" observer of morals and manners through dissociating herself from the pejorative stereotype of the literary "bluestocking," or the excesses and unconventionality of activists such as Mary Wollstonecraft. Throughout her career, Martineau "actively constructed a narrative persona" as a writer that was "detached from her gender as a way of creating 'objectivity'" in her discourses on the woman question, as on other forms of cultural criticism. Erasmus Darwin was right on the mark when he "palliated" his brother Charles's unfavorable reaction to Martineau's unwomanly domination of the conversation by arguing that one "ought not to think of her as a woman." Martineau, in this sense, gained acceptance as a sort of honorary man. Paradoxically, at the same time, this generated depreciations of Martineau and her work as masculine, as an "unfeminine farrago."[64]

Her major sociological study, *Society in America* (1837), and its companion text, *How to Observe: Morals and Manners* (1838), claimed scientific objectivity through the development of the traveler's "powers of observation" and a cross-cultural methodology. She sharply critiqued the disenfranchisement of both women and slaves, making the "degree of the degradation of woman" the test of "the state of domestic morals in any country."[65] Just as for the institution of slavery, the remedy for the oppression of women was education.

Darwin read Martineau's *How to Observe* in 1838, taking notes on her view that the "moral sense" varies from race to race; "no more wonderful," he mused, "than dogs should have different instincts."[66] It may well have been Martineau who directed Darwin to *De l'éducation des mères de famille, ou la civilization du genre humain par les femmes* (1834) by the French feminist Louis Aimé-Martin, a work instrumental in helping to forge the terms of the debate around the woman question in England and America, and also read by Darwin some time in 1838.[67] Aimé-Martin promoted a vision of a world reformed not by the masculine agency of politics but by the feminine agency of motherhood, which, like Martineau's own version of feminism, sought to give women influence but left them still restrained by domesticity.

A mélange of religious pieties, invocations of natural law, and justifications of freedom of the press, private property, and the social and moral centrality of the family unit, *De l'éducation des mères* aimed to liberate women by elevating their cultural roles and social status as mothers and wives (their "natural" vocations as "willed" by God), and thereby advance civilization. Womanhood was to be rehabilitated for this serious purpose from the superficial education that taught only "petty devotional practices . . . mechanical talents [such as piano playing and drawing]," and from the "corruption" that considered "dress as the first essential of life, and beauty as the chief of human qualities."[68]

Aimé-Martin was curiously vague about the appropriate education of mothers (being far more concerned with exalting their "mission"), but was clear that it should be very different from that of men. It was the business of men to "acquire fortune," to "defend the rights of the family in society, and to represent the interests of society in the family circle," while women— "true to the laws of nature"—were to remain at home, plying their feminine qualities of sympathy and maternal love, raising future generations to love God and humanity, and so abating the "fury" of the public masculine sphere of "legislation, the political arena, armies, war." The education of women's "souls" was of much greater value for this powerful moral mission than the education of their "intelligence."[69] Aimé-Martin did suggest that young girls should have "imprinted on their souls an ideal model of all human perfections" and be taught to "subject their inclinations to the guidance of this model." They would then be able to choose husbands who would value and love them for their "innate sense of moral beauty" that would confer on them a "power superior to that of beauty." It was true that in the "present state of affairs," girls were "seldom called upon to make their choice"; the remedy lay in their greater freedom and "enlightenment."[70]

This tenuous conception of the improving social role of educated female choice through the inculcation of a moral "beau ideal" went no further. It certainly made little impact on Darwin. It was not to recur until late in the century, when it was given a socialist twist by, among others, that other male feminist, Alfred Russel Wallace. Darwin, meanwhile, was less interested in Aimé-Martin's moral "beau" than in what he called an "ordinary Beau ideal, Mem[orandum]. Negro, beau"; that is, in those external physiognomies that constituted the different racial ideals of beauty and so influenced male marital choice—an interpretation that he was struggling to transfer to the mating habits of birds. Put bluntly, Darwin was less receptive to the conception of woman as redeemer than to woman as a biological entity, as represented by Walker.[71] Darwin was in the process of turning morality into instinct. By the time of the *Descent*, woman's "mission" had become innate. If women were naturally less selfish and more caring than men ("and this holds good even with savages"), then this was to be attributed to their "maternal instincts."[72]

At this early stage, given his views on heritable habit, Darwin was not insusceptible to contemporary radical Lamarckian opinion on the improving powers of education for both men and women, who would then pass on these acquired traits to their offspring: "Educate all classes," he scribbled amid the welter of notes on generation, breeding, and habit, "avoid the contamination of castes. Improve the women. (double influence) & mankind must improve."[73] Elsewhere in his notes, he put the emphasis a little differently: "Believers [in hereditary character] would only marry good women & pay detail[ed] attention to education & so put their children in way of being happy."[74]

Walker agreed: "Uneducated women are sure to communicate lower mental faculties to children." But such education as he advocated for women was limited to "seconding the purpose of nature, that of making women fit companions for men of sense," a view that meshed nicely with Wedgwood-Darwin family opinion.[75] Walker's system for social reform placed far more emphasis on breeding a better race, and this was to be achieved not by women but by "men of sense," the new men of the middle classes, the makers and achievers, the men on whom the advancement of the species depended. The laboring classes lacked the education and knowledge to breed themselves out. The aristocracy was effeminate, enervated, and degenerate through inbreeding, debauched habits and the disuse of their mental organs: "As to kings in particular, their intellectual faculties are so low, as always to border on fatuity."[76] The real hope for human improvement lay in the heads and hands (or, more precisely, the eyes and genitals) of middle-class men: that of breeding a better race through perfecting the art of rational selection. This art of wife selection was to be informed and illustrated by the practices and knowledge of animal breeders.

7.4 From the horse's mouth: How to judge a woman

In March 1837, two years before Darwin circulated his own questionnaire to breeders, Walker had circulated a privately printed pamphlet to those "who have the means of observing" the "Influence of Natural Beauty, and its Defects, on Offspring; and Law Regulating the Resemblance of Progeny to Parents."[77] *Intermarriage* incorporated many of the resulting letters from "improvers," notably Thomas Andrew Knight, president of the Horticultural Society, to whom *Intermarriage* was dedicated. Their extensive correspondence instantiated Walker's "natural laws of selection" (note, please, the phrase). It was this aspect of *Intermarriage*, Walker's collection "from authentic sources"—especially the "eminent and highly respected" Knight—a "mass of facts sufficient to afford a glimpse of general laws, if not to serve for their establishment," that most impressed the reviewer in the *British Foreign and Medical Review*.[78]

Darwin, primed by this review and embarking on his own extensive researches into domestication and inheritance, thus had high expectations of *Intermarriage*. He read it from cover to cover, marking innumerable passages and making notes. If any doubt remains, there is no way that Darwin now could avoid reading those extracts from Lawrence's *Lectures* on the effects of selection and selective mating in the human species and on the part played by different national ideas of beauty in the formation of race, all served up by Walker in *Intermarriage*.[79]

There was much more that was highly suggestive to an evolutionist, especially one who placed such a high premium on the analogue of domestication and the great modifications achieved by breeders. Darwin collected instances of the effects of crossing, of breeding in-and-in, and of reversion in dogs, horses, cattle, sheep, and humans, paying particular attention to Knight's views as represented by Walker (subsequently cited in the *Descent*). Darwin also noted Walker's views on beauty in women and its three "species." He marked especially Walker's theory that men cannot feel sexual excitement for women who are too like themselves in character, color, and looks, but look for novelty and difference; and Walker's view that it is this "love of difference" that inhibits incest or breeding in-and-in, and leads to those "slight crosses in intermarriage" that are as "essential to the improvement of the races of men" as they are to those of animals.[80] Consequently, women look for strength and passion in their sexual partners, while men's desire is increased by feminine weakness and delicacy, by coquetry and denial before yielding. Strong passion and volition, enhanced by these essential differences, produce offspring strong in mind and body. In support of this, Walker invoked sexual selection among animals: "Nature uses the same means for the preservation of nobleness and beauty among inferior animals: the most vigorous males are always preferred by the females, and the former repel the weaker by force."[81]

But, as he read, Darwin grew increasingly exasperated with Walker's anomalous views on inheritance, aspects of which did not lend themselves to the evolutionary interpretation that Darwin was in the process of formulating. In the face of Walker's many instances of the effects of habit and the inheritance of acquired characters, Darwin was confounded by Walker's assertion that "organization is indestructible": "?? This is the opposite of the case I want—I want new variety," he objected in the margin.[82]

Walker, it must be explained, was utterly opposed to the "common hypothesis of blood," or blending inheritance, where both parents contribute equally to the offspring, which would then have "half the blood" of the father, while the next generation would have only "one quarter of the blood" of this same forebear, and so on. This was the rule of thumb followed by most breeders, including the renowned Sebright, and it also underpinned contemporary terminology denoting human racial admixtures: "quadroons," "octoroons," etc. "Blood," scoffed Walker, "is a groom's term." The true law of inheritance was based not on "blood" but on organization: the whole vital system was invariably contributed by one parent and the locomotive by the other, so that these systems in the offspring retained their distinctive parental conformations and characteristics. The effects of the inheritance and recombination of these systems explained those puzzling cases of reversion where ancestral characters, lost for generations, reappeared, and why it was so difficult to preserve crossbreeds or hybrids. According to Walker, crosses between different human races, although they were often physically or mentally superior to their parent races, sooner or later invariably reverted to the original races. On the eugenic front, this explained why the "faulty organization" of the human race could not easily be eliminated, and why the "excellent Godwin's" radical goal of human perfection was impossible.[83]

Yet much might be achieved by those who followed Walker's rules of selection. Preferably, the male should contribute the locomotive system, along with the back of the head and "cerebel" or organ of will—the physiognomy of the strong masculine man of sense. The female should give the forehead and organs of sensation together with the nutritive organs contained within the trunk—those organs so well developed in the voluptuous "vital" beauties most admired by men. Once these "natural laws" of "selection" were understood, by choosing an "appropriate partner in intermarriage, man . . . has (precisely as the breeder has among lower animals) the power to reproduce and preserve either series of organs—the best, instead of the worst portion of his organization."[84]

Walker's trump card was the late great Knight's endorsement of Walker's "organization" law of inheritance:

> You stated that if in women you were shown merely a face, short and round, full in the region of the forehead, and having what are commonly called chubby cheeks, but contracted and fine in the nose and mouth, you would unhesitatingly predict the trunk to be wide and capacious, and the limbs to taper thence to their extremities; and, so unfailing was this indication also in regard to inferior animals, that if, in adjudging a prize, there was brought before you an apparently well-fed animal of opposite form, or having a long and slender head, you would suspect it to be crammed for show, and . . . should be disposed to reject it.

This was a "practical fact independent of all theory," an "unerring guide in the most important decisions of husbandry," exulted Walker, a "fact of immense extent and bearing" on the even more important business of wife selection. Knight's views on the significance of chubby cheeks as an infallible signifier of good childbearing hips in women were paraded again and again at critical points in the text of *Intermarriage*.[85]

Walker produced further illustrious endorsement of his law with engravings of the instantly recognizable head of the new young queen, Victoria, juxtaposed with those of her parents. These purported to show that Victoria had indeed inherited the forehead of her mother and the nose and mouth of her father. Walker suspended judgment on whatever that might imply for a member of an inbred, degenerate aristocracy, and glossed over the implications of the as-yet-unmarried Victoria's undeniably chubby cheeks.[86]

Darwin, however, was not convinced. His objections mounted up. The real point at issue was that Walker's law of the inheritance of whole systems was not compatible with Darwin's growing emphasis on the heritable small variations or individual differences that would provide the grist for the mill of natural selection—a process that he was developing by close analogy with the breeder's systematic selection of minute variations. How, if Walker was correct, could breeders develop new, stable varieties that would not be sent

Figure 7.5. "The Duke and Duchess of Kent and Queen Victoria, as affording a General Illustration of the Law of Selection." According to Walker, the young queen (center) has inherited her mother's forehead and her father's nose and mouth, while her "chubby cheeks" are indicative of her good childbearing status. Walker 1838, facing page 156. State Library of New South Wales, Mitchell Collection.

back to the ancestral condition? How, indeed, could new species originate? It was not an interpretation Darwin wanted, as he forcefully stated, and he rejected it. Still, doubts remained: "Put the case to Sir J Sebright . . . Ask his opinion of Walkers book."[87]

Darwin remained intrigued by "Mr Walker's law" for more than twenty years. He noted seeming instances of it in his readings in the literature, and he devised crossbreeding experiments to test it against Yarrell's law, enrolling the faithful Tegetmeier. To confuse matters, even Yarrell had been inclined to believe that the male communicated external resemblances more than the female. Walker's views on the inheritance of whole systems or likenesses had some currency among medical men and physiologists; they were taken up and promoted in the 1850s by Reginald Orton, an eye surgeon and fowl breeder whose views Darwin carefully investigated (see ch. 10). Darwin did not dismiss them until he published his own views on pangenesis, or "gemmule" inheritance, in 1868, and even then he allowed for the transmission of whole constitutions, primarily through the male line.[88]

7.5 "The transcendently important subject of choice in intermarriage"

Walker had given Darwin much more to ponder than the specifics of inheritance. To begin with, *Intermarriage* re-presented Lawrence's views to Darwin in a highly suggestive context of animal and human breeding practices and preferences, centered on the problematic of beauty and wife "choice," with the emphasis, semantically and conceptually, on *choice*.[89] Walker made the masculine judgment of woman's beauty, "the critical observation of the external forms of woman," an aesthetically informed process of choice, the key to his notion of human progress or racial improvement. Further, there was the cumulative force of Walker's reiterated analogy between wife "choice" and artificial selection—the analogy Darwin himself was in the process of drawing between the breeder who selects those animals whose traits he wants to preserve and the human male who selects those women whose characteristics most please and attract him. Walker made overt what was mostly implicit: the collapsing of women into breeding animals, as objects of sexual choice whose distinctive points—chubby cheeks, voluptuous curves, and so on—might be evaluated by the discerning, like those of prize cattle or, for that matter, pigeons. This opened up the discursive possibility of Darwin expressing similar, though always more circumspect, views. To this may be added Walker's emphasis on the sexual preference for novelty and difference (to which Darwin also was to accord considerable significance in the *Descent*), and his invocation of male combat and female preference among animals in support of these views.

It may be that Walker's prioritization of external form, of beauty—that is, of choice as a visually and aesthetically determined preference—intimated to Darwin the legitimacy of transferring male aesthetic choice among humans to female aesthetic choice among animals, especially birds, where the looming problematic of useless or potentially harmful male beauty required explanation that could not be made consistent with his newly formed concept of natural selection or with his earlier views on inherited habit. The transposable beauty of women and birds was central to Darwin's argument for the efficacy of female choice among animals. At the same time, Walker's explicit denial of sexual choice to human females reinforced Darwin's own conventional, culturally derived views on women's sexuality and essential passivity in any account he might render of human evolution or racial differentiation.

Intrinsic to Walker's scheme for breeding a better race was the naturalization and sexualization of beauty and so of women as sexual objects, together with his insistence on the primacy of organization, of anatomy, rather than environment or education, in determining women's and men's intellectual and moral characteristics and potential. He offered a scientific rationalization of women's intellectual inferiority and their conventional maternal and domestic roles, traditionally the province of religious discourse. His was essentially a "science of beauty" which "works to secure women's subordination . . . women's identity is bound to and bounded by beauty."[90] A similar discursive practice is at work in *The Descent of Man*. A further point in this connection is that Darwin's focus on such a concept of aesthetic choice, with its embedded conventional objectification of passive female sexuality, lent itself to Darwin's circumvention of any real discussion of human sexuality or erotic desire in the *Descent*. The language of sexual selection in the *Descent* is biological and aesthetic, even romantic in its projection of the language of Victorian courtship and marriage onto the mating practices of animals, but never overtly sexual.

This was an issue of considerable anxiety to Darwin and his publisher John Murray who counseled against including the term *sexual* in the title of a work devoted to sexual selection and scanned the manuscript for any indecencies that might offend potential readers. Darwin dutifully bowdlerized his accounts of female "coyness" and strategically put into Latin one of his prized pieces of evidence for sexual selection—the famous hypertrophied labia that might be referred to male sexual preference of ideals of female beauty peculiar to the Hottentots. At the same time, he elided any implication of female sexual satisfaction or desire or proactive role in sexual selection that his sources might have indicated.[91] Female choice in the *Descent* is represented as essentially a passionless aesthetic exercise of taste and discrimination by an innately modest female who "requires to be courted" and need only decide which of her competing suitors meets her exacting aesthetic standards or, at the very least, settle for the one who is, as Darwin tactfully put it, "least distasteful" to her.

Walker's construction of the "transcendently important subject" of wife choice resonates most strongly in the conclusion to the *Descent*. There, in a much-quoted passage, Darwin wrote:

> Man scans with scrupulous care the character and pedigree of his horses, cattle, and dogs before he matches them; but when he comes to his own marriage he rarely, or never, takes any such care. He is impelled by nearly the same motives as are the lower animals when left to their own free choice . . . Yet he might by selection do something not only for the bodily constitution and frame of his offspring, but for their intellectual and moral qualities. Both sexes ought to refrain from marriage if in any marked degree inferior in body and mind; but such hopes are Utopian and will never be even partially realized until the laws of inheritance are thoroughly known.[92]

Many years ago, regarding this passage, John Durant pointed out that "far from being an idle speculation tacked on to the end of the *Descent*, these ideas were implicit in the very structure of Darwin's thought."[93] This precludes the conventional view that Darwin merely appropriated the lately published eugenic views of his cousin Francis Galton. From his first encounter with wild Fuegians and his experiences with the recidivist Jemmy, Darwin had been preoccupied with issues of civilization and human progress, with questions of inheritance and breeding, which he very early located in the all-purpose idiom of domestication. For Darwin, the process of civilization was akin to the domestication of animals, with the emphasis on breeding and the transmission of innate or gradually acquired qualities and characteristics rather than on training or education, as environmentalists would have it. But where breeders had impressed their own taste and judgment on the shaping of domesticates, humans had shaped themselves through a process of aesthetic self-selection or self-domestication. Sexual selection, as Durant stressed, was for Darwin "simply the most overtly voluntaristic interpretation of a fundamentally anthropomorphic analogy between nature and human artifice."[94]

The projection of this variant of artificial selection into the future suggested possibilities of selective human breeding that had been made overt in the literature many years before Galton was to publish his eugenic views. Walker's *Intermarriage*, picking up where Lawrence's *Lectures* left off, was the most explicit and comprehensive of these early arguments for the extension of the practices of stock breeding to the rational exercise of wife choice for the future benefit of the race.

In this same passage quoted above, Darwin demarcated such rational or selective choice from the "free choice" exercised by the lower animals— that is, from female choice; and implicit in the passage is the assumption that such rational, eugenic choice as he advocated would be exercised by men, presumably informed in the principles of breeding and inheritance, in their choice of marriage partner. (Though he also specified that "both sexes" might exercise negative selection—they should refrain from marriage and reproduction if "inferior in body or mind"— Darwin, as discussed, was an entrenched opponent of contraception.) Walker had specifically denied choice and reason to women on the grounds of their inferior anatomy and intellect, and this chimes nicely with Darwin's argument in the *Descent* that man, being "more powerful in body and mind," had seized the power of selection from woman. By the time Darwin wrote the *Descent*, arguments remarkably similar to Walker's had proliferated in medico-scientific and anthropological texts that sought to displace traditional religious wisdom and authority by drawing naturalistic, scientific limits to the mounting campaign by women and their male supporters for intellectual and social parity. In point of fact, Darwin structured his assertions of women's inferior abilities and nature in direct opposition to John Stuart Mill's countervailing arguments in *The Subjection of Women* (see sec. 13.5).[95]

In the *Descent*, men are attributed with superior energy, intelligence, ambition, perseverance, courage, imagination, and aesthetic creativity, while women surpass men only in tenderness, selflessness, rapid perception, imitativeness, and intuition—these last, Darwin insisted, being "characteristic of the lower races, and therefore of a past and lower state of civilization."[96] Female choice, whether exercised by birds or such humans as Darwin permitted it to, is unconscious, without intention and concerned with trivialities, with useless if beautiful (sometimes even physically impairing or dangerous) ornamentation and the superficialities of appearance; it grades into the artless sexual choices made by the men of "lower races" who have cultivated such absurdities as female steatopygia or hypertrophied labia, or the "monstrous" aberration of a "jet-black" skin; whereas civilized European men admire and select an aesthetically pleasing "oval face", "straight and regular features," and white skin.[97] Ultimately, as Walker had emphasized, only conscious, educated masculine discrimination can result in the rational and intentional moral, intellectual and physical improvement of humanity.

If, on occasion, Darwin attributed an advanced artistry and "almost human degree of taste" to a female bird in explaining, for instance, the exquisitely shaded "ball and socket" ocelli of the argus pheasant (a contention that seemed to undermine his claim that this "most refined beauty" had been achieved without intention, through unconscious selection by a mere female bird), then this was because Darwin, in effect, transferred the eye and selections of the breeder via his own trained breeder's eye to the eyes and choices of the female argus pheasant. It was not so much that Darwin "imagined his way into the animal's mind," as has been argued; it was that he imagined his way into the mind of a putative breeder of the male pheasant.[98] He then had to work hard to dissociate this "not sufficiently guarded" (interestingly, the terminology is Darwin's own) transference from the runaway interpretations of critics and would-be fellow travelers who would take female choice where Darwin did not want it to go, either to intimations of God-given design or of evolutionary change driven by female intention.[99] Hence his backtracking in the second edition of the *Descent* to the "more just comparison" between the taste for the beautiful in animals, and that in the "lowest savages, who admire and deck themselves with any brilliant, glittering, or curious object."[100] In Darwin's own society, as he pointedly reminds his readers, it is the women who—like savages or birds, and presumably lacking the high aesthetic discrimination of cultivated men—"decorate themselves with all sorts of ornaments," borrowing the "plumes of male birds" to charm men.[101]

This was one area of choice that Darwin wholeheartedly might concede to women. Victorian women might be excluded from political choice, from career choice, and (overtly, anyway) from free sexual choice and the choice of family limitation, but in the arena of fashion, they dominated choice. In the course of the nineteenth century, fashion was thoroughly feminized. Men's clothing became duller and uniform, while women's clothing became more ornate, colorful, and expansive, opening up the options for female choice—subject of course to those niceties of discrimination that bore on the class divisions and more subtle distinctions of the Victorian social fabric.

Fashion and its vagaries were a constant point of reference in Darwin's discussions of artificial selection and sexual selection. More than this, I have shown that the fashion metaphor was crucial to his thoroughgoing identification of animal breeder with sexual selector, and that Darwin invoked it at several critical points in his argument for the efficacy of male sexual choice in the divergence of the human races: "In the fashions of our own dress we see exactly the same principle and the same desire to carry every point to an extreme."[102] The next chapter explores the ways in which this same fashion metaphor may have enabled Darwin to make the further critical shift from male choice in humans to female choice among birds and the lower animals.

EIGHT

Flirting with Fashion

The Girl of the Period is a creature who dyes her hair and paints her face . . . whose dress is the chief object of such thought and intellect as she possesses. Her main endeavour is to outvie her neighbours in the extravagance of fashion. . . .

Nothing is too extraordinary and nothing too exaggerated for her vitiated taste; and things which in themselves would be useful reforms if let alone become monstrosities worse than those which they have displaced so soon as she begins to manipulate and improve.

—Eliza Lynn Linton, "The Girl of the Period," 1868

It is no coincidence that the very phrase "female choice" was made familiar to the reading public—and almost certainly to the young Charles Darwin and his sisters—through a popular little fable on the perils of fashionable dress. "The Female Choice" was one of the didactic tales in the much-loved miscellany *Evenings at Home* (1792–96), compiled by the well-known children's writer and poet Anna Laetitia Barbauld, with her brother John Aikin. *Evenings at Home* was intended for the "instruction and amusement of young persons" of both sexes, and attracted a wide Anglo-American audience well into the nineteenth century via continual repackagings by editors, compilers, and publishers on both sides of the Atlantic.[1]

Barbauld was prominent in those radical Dissenting circles frequented by the Wedgwoods and the Darwins, and active in both the antislavery and antiwar movements. Their overlapping networks shared a common interest in the promotion of science and education. Barbauld's texts were used in Wedgwood homeschooling projects; Erasmus Darwin recommended them in his *Plan for the Conduct of Female Education*; and they strongly influenced the educational writings of Thomas Day and Maria and Richard Edgeworth.

Barbauld was attacked alongside Mary Wollstonecraft and Erasmus Darwin by the *Anti-Jacobin*, and counted among those "unsex'd females" who threatened traditional authority and morality.[2]

While it had its reactionary critics, *Evenings at Home* won and retained its middle-class popularity through its emphasis on the family as the site of moral regeneration and political reform as well as its more conventional construction of a meaningful domestic feminine role.[3] "The Female Choice" exemplifies this construction of feminine identity through the choice offered the young protagonist Melissa by two "nymphs" who appear before her as she daydreams in the garden.

The first nymph is "loosely habited in a thin robe of pink with light green trimmings" tied with a flowing "sash of silver gauze," her fair hair dressed in ringlets and decorated with "artificial flowers interwoven with feathers." She reveals her name as "DISSIPATION" and tempts the girl to a "perpetual round" of pleasure with a ball ticket and a "fancy-dress all covered in spangles and knots of gay ribband" that will make Melissa look "quite enchanting."[4]

The second nymph, "HOUSEWIFERY," is clothed in a "close habit of brown stuff," with a symbolic bunch of housekeeping keys at her waist. Her "smooth hair" is confined under a "plain cap." She sternly proffers a distaff and workbasket and summons Melissa to a life of early rising and domestic duties: "You must dress plainly, live mostly at home, and aim at being useful rather than shining." The compensations for this seemingly unattractive female ideal of dullness, domesticity, and self-denial will be "self-approbation, and the esteem of all who thoroughly know you." Housewifery warns that Dissipation's tempting "delights" lead only to "languor and disgust," that she has not shown her real face: "It is time for you to choose whom you will follow, and upon that choice all your happiness depends."[5]

While Melissa hesitates, eyeing the fancy dress, the "bewitching" mask that covers Dissipation's face falls to reveal a "countenance wan and ghastly with sickness, and soured by fretfulness." Horrified, Melissa makes her choice, giving her hand "unreluctantly to her sober and sincere companion."[6]

This little morality tale, familiar to several generations of the reading public, encapsulates the significance of dress to feminine identity. What is articulated as "the female choice" hinges on a choice of dress, which is represented as a life choice on which the future happiness of females depends. Dress choice has moral connotations. Ornate, colorful, conspicuous dress is associated with moral laxity, excess, frivolity, and deception; while functional, plain, dull dress connotes seriousness of purpose, goodness, and integrity. Bound up with this strongly polarized "choice" that allegedly confronts all females is the attribution of an obsession with fashion and manipulative deception to those women who would mask their real natures and physiognomies with dress and cosmetics. The virtuous and healthy need no artifice to reveal their true inner beauty and proper womanly attributes. This was a theme that was to run through the discourse of dress reform throughout the nineteenth century, culminating in the late 1860s—just as Darwin began work on the *Descent*—in the sensationalist attacks launched on the "Girl of the Period" by Eliza Lynn Linton, a journalist, Darwinian, and formidable antifeminist.

For, in spite of the concerted efforts of reformers like Barbauld, of evangelicals who denounced vain, deceptive, and coquetting females, and of an often misogynist press that extracted the maximum of ridicule from the extremes of the crinoline and the bustle, the dress choices of fashionable Victorian females ran more to the bright feathers and flounces of Dissipation than to the drab hues of Housewifery. Through their fashion choices, the visual expression of their identity, Victorian women of all shades and classes found ways of negotiating their ideological relegation to the domestic sphere. Ironically, it was Victorian males who adopted the camouflaged dress and dullness of the nesting bird in the process of fashioning a new masculine identity that had little to do with nesting and more to do with the assertion of social and political authority.

8.1 "They might belong to different species"

It was Thomas Carlyle who, in his *Sartor Resartus* (1836), established the fundamentals of clothes theory: "The first purpose of Clothes . . . was not warmth or decency, but ornament." Clothes do not derive from shame, modesty, or any of the other sexual anxieties associated with the biblical Fall, or from functional protection from the elements: "Decoration" is the "first spiritual want of barbarous man."[7] Carlyle thus directed the comprehension of clothing to anthropologists and social theorists, indicating its evolution from the body painting and tattooing of "barbarians" (comparable to the ragtag ensembles of "Bogtrotters," "White-Negroes," or "Poor Slaves") up to the foppish, peacocking costume of the dandy, who is "heroically consecrated" to the "one object" of dress. Carlyle's satirical attack on the "Dandiacal body" was largely directed against the values and manners of a profligate "effeminate" aristocracy in the process of displacement by an emergent middle-class identity based on a masculinity that rhetorically renounced luxury, vanity, and indolence and formally embraced frugality, industry, and public virtue.[8]

This new middle-class identity fostered modest plain dress for men—a display of manly worth rather than wealth that was "inherently tied to political legitimacy" and that demarcated and excluded women from political power.[9] Men looked powerful, serious, and of singular unified intent. As men's dress became darker, simpler, more uniform, settling into the basic three-piece suit, women's dress became more elaborate, changeable, and invested with erotic meaning.

By the mid-nineteenth century, vivid colors, luxurious fabrics, and decoration were essentially restricted to women's dress, and, with the reinvention of the corset, the desirable feminine shape of sloping shoulders, a long waist, and wide hips was defined. This essential hourglass shape remained as sleeves widened or narrowed, décolletage and hemlines fluctuated, heavy layers of starched petticoats gave way to the innovation of the crinoline (a light cage of whalebone or steel), and skirts underwent competitive inflation. Hats succeeded bonnets and shrank to wisps of straw or feathers, while in the 1870s skirts narrowed again to figure-hugging dimensions and derrieres were plumped out by the bustle.

Nineteenth-century female dress underwent constant change, a sequence of many different styles, each of which "evolved gradually" from its predecessor. It has been suggested that women's fashions "grew up" over the course of the century, from an earlier childlike, high-waisted look to the adult, full-bodied, richly colored, opulent styles of the late Victorian era.[10] This "maturing" was accompanied by a heavy investment in ornamentation and trim—in ribbons, lace, feathers, fringes, flounces, embroidery, knots, braiding, and buttons—their range and availability made possible by the twin forces of empire and industrialization. Women's dress across the classes (although important distinctions remained) became more decorative, more noticeable, as men's clothing assumed a pared-down, de-eroticized, monochromatic homogeneity. The clinging breeches of the Regency dandy and tight cutaway coats that framed the genital area were replaced by looser-cut trousers and jackets that bulked up the figure and concealed the contours of the body (although a variation of the cutaway coat was reserved for formal evening wear until late in the century, and trousers were sometimes narrowed to indicate the muscularity and definition of a manly thigh).[11] A *Punch* cartoon from the crinoline era, depicting a man and woman seen from behind, summed up the pronounced sexual dimorphism of dress with the caption: "They might belong to different species."[12]

The increasingly dimorphic gendering of nineteenth-century dress did not mean that men did not invest considerable money and thought in their attire. For the better-off or those with "dandiacal" pretensions, much care went into the cut of jackets, the quality of suiting, shirts, cravats, and so on. But selections conformed to the predominant masculine creed of "inconspicuous consumption," the significant exception being the vest or waistcoat. Vests shimmered with brilliant silks and gold and silver thread, a flash of luxury and individuality in men's otherwise sober, overtly democratic attire. Their significance lay in the fact that vests were not on open display, being hidden, or partially hidden, under the coat. What was being eschewed was not luxury, but the conspicuous display associated with femininity (and effeminate men). By the latter part of the century, even the vest, as a component of the standard three-piece suit, had assumed the dull conformity of the coat and trousers.[13]

The supposed feminine addiction to fickle fashion and display was adduced as yet another instance of male superiority to feminine frivolity, a natural weakness of women, who lacked the gravitas and reasonableness of the self-made man. "Dress as such, a high pleasure in dress, and indeed a readiness to talk of dress and nothing else" came to be seen "almost as constituting the feminine."[14] The ridiculing of the feminine fixation on fashion was the stock in trade of *Punch* and other satirical magazines that depicted women in crinolines so wide they could not get through doorways or were blown into the air like balloons, or rendered sleepless over the exact shade of a ribbon. Readers were hectored on the horrors and deformities of tight lacing or the pitched-forward gait (the "Grecian bend") enforced by the rising heel and the bustle, while the bustle itself provoked endless coarse hilarity. This satirizing of female fashion as absurd, disfiguring, or disabling was offered as the final proof that women given to such follies and trivialities were incapable of rational thought. Nevertheless, it in no way detracted from the lampooning of those early dress reformers who advocated "rational dress" for women. This reversal of the natural order, the notion that women might wear the pants or "pettiloons," was a never-failing source of amusement, the subject of dozens of cartoons in *Punch*.[15]

There can be no doubt that fashionable display was for many Victorian women, if not a consuming passion, a highly important part of life. Her "proper" career of marriage dictated careful attention to a young woman's wardrobe, presentation, and participation in the courting rituals that would ensure an appropriate partner. Girls prepared for these rites of passage by sewing for and dressing their dolls in the latest modes and by dressing up in their mother's silks and jewelry, as Annie, Henrietta, and their cousins were permitted to do in Emma Darwin's bedroom.[16] The assembly of the trousseau, which included household linen as well as personal items—under-dress, dinner gowns, walking dresses, gloves, bonnets, and so on—

was a serious business. Emma, urged by her fashion-conscious aunt Jessie Sismondi not to skimp on the cost and effort to be "always dressed in good taste," went to London to choose her wedding clothes. She returned with, among other purchases, a "greenish-grey rich silk," a "remarkably lovely white chip bonnet trimmed with blonde and flowers," and a "grand velvet shawl." "No man," Aunt Jessie had warned—even ones such as Charles, supposedly "above such little things"—"was above caring for them, for they feel the effect imperceptibly to themselves. I have seen it even in my half-blind husband."[17]

Darwin was certainly not blind to feminine fashion, as his earlier fascination with the clinging dress of the Spanish ladies of Buenos Aires and his disparagement of Englishwomen, who "can neither walk nor dress," attests.[18] On the death of his daughter Annie, he wrote movingly of her "truly feminine interest in dress," how "such undisguised satisfaction, escaping somehow all tinge of conceit & vanity, beamed from her face, when she had got hold of some ribbon or gay handkerchief of her Mamma's."[19] Nor was he above caring about the social niceties and moral implications of female dress. Early in their marriage, Darwin warned Emma of a "thunder storm" about to break over her head: his father and sisters were "very hot" in their disapproval of Emma's lady's maid, who did not wear the cap proper to her station:

> "Looks dirty," "like grocers maid" & my father with much wrath added "the men will take liberties with her if she is dressed differently from every other lady's maid"!!! Both the girls echoed this—I generously took half the blame & never betrayed that I had beseeched you several times on that score.[20]

For all that they allegedly did not know how to dress, English girls of lesser fortune than Emma invested inordinate time and effort in stitching and embroidering, in selecting fabrics, trimmings, and styles. A proliferation of fashion magazines offered guidance on the latest modes. An elaborate code governed the choice of fabric, color, style, and accessories. Unmarried girls were expected to dress simply and demurely in cottons and muslins. Married women might wear richly colored silks and velvets, and display ornate jewelry or exotic furs and feathers; older women wore darker, heavier clothes. Victorian mourning imposed strict adherence to a complex protocol of dress and observances; women who transgressed these boundaries were subjected to severe social disapproval. Dress was an "index of character" that revealed the inner woman, who was exhorted to eschew affectation and artifice. Powder and paint were associated with "actresses or certain kinds of women," though by the 1870s married women were permitted the subtle use of cosmetics, but "never young girls," where rouge should not counterfeit the blush of innocence.[21]

As cheaper mass-produced clothing, sewing machines, and ready-made trimmings became available and girls and women of the servant and working classes also aspired to fashionable dress for their recreation and churchgoing, the social distinctions and meanings encoded in clothing and accessories assumed increasing significance and subtlety. A cascade of etiquette books and manuals were produced to "mediate class differences and manage lower class dissent."[22] Novels and popular magazines reinforced the message. Details of dress became invested with subtle and ever-changing indicators of status and class rank. It has been argued that "every cap, bow, streamer, ruffle, fringe, bustle, glove or other elaboration symbolized some status category for the female wearer."[23] Women—middle-class women in particular—had to keep up with these fluctuating modes and signifiers of fashion and the niceties of their interpretation if they were to negotiate class boundaries, assert their respectability and status, and not commit some social solecism. Although the demands of fashion and its closely associated etiquette imposed restrictions, social as well as physical, on women, nevertheless, women controlled these indicators. "Ironically, the very signifiers of powerlessness in the gendered frame of reference became eloquent signifiers of power in a class frame."[24]

8.2 Parlor politics

These signifiers of feminine authority extended from personal clothing to the household, its layout, management, and decoration. The inner geography of Victorian houses, especially those "of the better sort," were coded as masculine (dining room, smoking room, billiard room, library, or study) and feminine (drawing room or parlor, sitting room, and boudoir) and furnished accordingly. These areas were demarcated from the domestic workspaces and servants' quarters, which also segregated the sexes. Typically, the wife managed the household, while the husband was responsible for outdoor spaces and concerns. The maintenance of a large middle-class house—with its elaborate furnishings and layout in an era of coal fires, no reticulated hot water, and exacting standards of housekeeping—was possible only with an available, well-regulated, and poorly paid servant class. In 1857, Darwin's butler, Joseph Parslow, was paid £44 per year, the equivalent of the sum Darwin considered appropriate for William's annual allowance at Rugby School. As late as the 1880s, Down House had no bathroom and hot water had to be brought from the kitchen, "but there were plenty of housemaids to run about with the big brown-painted bath-cans."[25]

It is estimated that by 1891, about one-half of all employed British women were in domestic service (around two million). The central task of the Victorian angel was the management of the servants, even if she had only the solitary maid-of-all-work. The mystique of the angel who brought "sweet order" to the house without lifting a finger masked the realities of domestic management, of the organizational skills and decision-making entailed in the running of a house.[26]

The material abundance achieved by the Victorian middle classes was reflected in the furnishings and bric-a-brac that cluttered domestic spaces, especially feminine spaces. New emporia, the outlets of mass production and cheap importation, offered a staggering array of furnishings and decorative items that maximized choice and catered to the growing emphasis on artistic self-expression in home decoration. A high premium was placed on novelty and effect, on lavish decoration and rich ornamentation and color. The accumulation of possessions and ornate, often clashing or brash decor warred with the evangelical emphasis on the redemptive power of the home and its sentimental image as a sanctuary from the cold, commercial world. The conflict was partially resolved by an understanding that sideboards, chintzes, and statuettes had moral as well as monetary value. The wrong choices could lower the moral tone of the household. Bad taste invited moral censure. Darwin dubbed the rented London house in which he and Emma began married life "Macaw Cottage" for its jarring juxtaposition of yellow curtains and bright-blue walls. He surmised to Emma that the "lady who chose" them "was no better then she should be,—certainly she had no better taste, than she ought to have, which, like her character, I presume, is none at all."[27] The family association of bad taste with poor morals is burlesqued in an anecdote of Leonard Darwin's dubious choice of pseudo-Gothic windows for his new house; he forestalled criticism from his brother Horace, whose own house was exquisitely furnished, with the bluff disclaimer: "Well, Horace, I suppose you think these windows are not only ugly but immoral."[28]

That Darwin's sons cared so much for their furnishings and house design might surprise, but Darwin himself was heavily involved in major renovations to Down House in 1845 and again in 1858, writing of making the house "luxurious" for the family and more comfortable for the servants. These were the larger decisions involving "extreme extravagance" that required masculine attention and might not be resolved by the feminine expedient of dyeing the curtains.[29] Until the sweeping reforms of marriage law were enacted in 1882, married women had no control over their property or earnings (if any), and husbands could renounce debts incurred by wives. Husbands assumed authority for furnishing, which (ideally) signified elevated masculine taste and moral and financial probity. Early advisers on household taste (invariably male) largely addressed their advice to the man of the house on the proper tone and furnishing of a "gentleman's residence," while the "views of the ladies" warranted little attention. It was not until late in the century that interior design became the increasing preserve of female advisers for a primarily feminine readership. Victorian women were allocated the "lesser tasks"—of bibelots, furbelows, and draperies. In other words, to women fell the task of turning a "house" into a "home," into the idealized "place of Peace," "shrine," or "sanctuary."[30] Men might furnish rooms; women dressed them.

And dress them they did. Victorian parlors have become a byword for excessive, fussy ornamentation and their embodiment of the superfluous and nonfunctional. They incurred the aesthetic displeasure of notable reformers like John Ruskin and William Morris, spokesmen for the functional or the beautiful. By contemporary accounts and illustrations, every available surface—mantelpiece, sideboard, occasional tables, whatnots, and piano—was ruffled, fringed, and draped. These were layered with decorative runners and doilies featuring the woolwork, beading, and embroidery of the woman of the house, and groaned under an accumulation of wax flowers, china, fans, lusters, silverware, live birds in cages, stuffed ones under glass domes, and other artificial and naturalistic bric-a-brac. Chairs and sofas—already upholstered, brocaded, and tasseled—were further embellished with embroidered cushions, shawls, and crocheted antimacassars. Pictures, hangings, and mirrors jostled for space against patterned wallpapers. Windows were festooned with swags, loops, and drapes and layered with muslin or lace and heavy brocade or velvet curtains.

The parlor was the most visible site of a woman's identity, a public yet private space where her decorative skills and social duty were on display. Its decorative complexity "simultaneously asserted and concealed a relation to the marketplace." It was a "feminized space" where women held sway and men and children were visitors, where women controlled the routines and rituals of domestic life that men had to respect as long as they held to the code of domesticity and separate spheres.[31]

The analogy often drawn between the complex draping, layering, and cluttering of Victorian interiors and women's bodies assumes the Victorian attitude to sexuality—the overt, excessive, repressive prudery and the covert, hypocritical eroticism that manifested itself in pornography and prostitution.

The shibboleth that Victorians primly covered piano legs, because their suggestive curves should be kept decently hidden in respectable company, has enduring currency. If piano legs were indeed covered, it probably had more to do with the passion for decoration than a covering up of sexual passion. In spite of a late-century upswing of moral outrage against the exhibition of nude art and sculpture, representations of female nudes—providing they conformed strictly to the prohibition of body hair and genitalia—generally functioned as images of high culture.[32]

This is not to deny the erotic pull of Victorian dress, both of interiors and bodies—particularly when they were juxtaposed, as they often were, in contemporary paintings of women reclining languidly on sofas or posed in the lush interiors of drawing or dressing rooms. The popular genre of domestic scenes performed "important cultural work" in allowing visual access to the private spaces of middle-class life, showing how a home and woman's place within it should look. Yet, while many of these representations validated Ruskin's conflation of woman and home ("wherever a true wife comes, this home is always round her"), they were not always so unambiguously reassuring.[33]

The latent, disturbing erotics of dress and decor are most evident in those well-known narrative paintings, such as Augustus Leopold Egg's *Past and Present* (1858) and William Holman Hunt's *The Awakening Conscience* (1853), which positioned their figures of adulterous or fallen women in beautifully detailed parlors, playing on the rich visual fields of parlor decoration and women's dress. Their interpretation depended on cultural references and expectations that were not as stable and standardized as has been assumed. Ruskin was so alarmed by potential misreadings of *The Awakening Conscience* (which depicts a kept mistress rising to her feet from her lover's lap as she realizes her wrongdoing, while, unconcerned, he strums on the piano), that he wrote to the *Times* to assert the proper, authoritative reading: every item in the parlor had "tragical" meaning "if rightly read": the furniture was "common, modern, vulgar"; its "terrible lustre" and the embossed, unread books conveyed no "old thoughts of home," only "fatal newness" and portending disaster; even the "very hem of the poor girl's dress, at which the painter has laboured so closely, thread by thread, has a story in it, if we think how soon its pure whiteness may be soiled with dust and rain, her outcast feet failing in the street."[34] Underlying Ruskin's attempt to marginalize readings other than that of a tragic sexual (and decor) victim is his association of women and sexuality with corrupt decorative practice, his identification of "overcharged ornament" with moral "intemperance."[35]

Figure 8.1. *The Awakening Conscience* by William Holman Hunt, 1851–53 (Tate Gallery, London). A well-known Victorian narrative painting by a leading pre-Raphaelite artist, depicting a kept mistress at the moment of her awareness of her wrongdoing. The critic John Ruskin, fearing misreadings of her beautifully detailed dress and parlor decor, wrote to the *Times* to assert their vulgarity and nouveau riche symbolism and the "right" reading of impending social exclusion and misery for the sexually fallen woman. © Tate, London 2015.

Ruskin's publicly expressed concern to clarify and delimit the moral meanings of dress and decor was akin to that expressed in the sensational, anonymous articles in the *Saturday Review*, known collectively by the title of one of them as the "The Girl of the Period." A major criticism by their author, Eliza Lynn Linton, was that the dress of the "modern" or "fast" young woman was modeled on that of the "queens of the *demi-monde*," blurring the lines of respectability. Girls might be mistaken for what they were not; or worse, their dress might lead them to adopt the manners and morals of those whose dress they emulated.[36]

Ruskin's moral concerns and Linton's provocative denunciations need not be interpreted as yet more evidence of Victorian prudery, but rather as recognition of and particular responses to the erotic dimensions of fashionable Victorian dress and decoration. The cultural currency of Linton's "Girl of the Period" owed as much to those who opportunistically parodied and perpetuated her existence in a flood of derivative cartoons, fashions in clothing, and satirical publications. Even to her contemporaries, Linton's concern for the proprieties was excessive. She herself was caricatured as a "prurient prude," a portrait-painting Mrs. Grundy with the legs of her easel primly petticoated, painting an unrecognizable and ugly portrait of her model, a demurely pretty girl.[37]

Victorian dress, as numerous commentators have discerned, connoted both modesty and sensuality. It covered up the body and legs (although a surprising amount of bare flesh was on show in the eveningwear of women of unimpeachable respectability) but, in an undeniably erotic fashion, accentuated the breasts, wasp waist, hips, and buttocks of the wearer. Victorian sexuality was more variable than its negative and overly simplified stereotype suggests. A certain sensuality, historians now agree, was not incompatible with Victorian respectability. The idea was that dress should both express and contain it.[38]

Some critics have seen a parallel between the constrictions and restrictions of Victorian dress and the social situation of women. While there is something in these arguments, female dress was not as constricting, heavy, or immobilizing as might be supposed. Henrietta Darwin, for one, lamented the demise of the crinoline for the loss of its lightness and freedom of movement.[39] The corset was undoubtedly constricting, but it seems that tight lacing was rarely carried to the extremes claimed by detractors and that many women and some medical men saw it as conferring essential postural and back support. It is argued that the century's fascination with the corset (as with all fashion) and the moral debate it occasioned had more to do with legitimating the discussion of sexuality in a period of great sexual reticence.

Figure 8.2. "The Girl of the Period! Or, Painted by a Prurient Prude," a caricature of Eliza Lynn Linton, author of the sensational "Girl of the Period" essays in the *Saturday Review*. She is depicted dipping her brush in gall and venom to paint a distorted "devilish" image of her demure subject. *Tomahawk*, April 4, 1868, 139. University of Sydney Library Collection.

Elegant men's clothing was probably no less demanding and uncomfortable than women's (some men, on occasion, wore stays), but it *looked* more relaxed and certainly more subdued than the "deliberately decorative" and "expressive attire" of the fashion-conscious woman.[40]

And there we have it. What Victorian feminine fashion and decoration were ultimately about was to convey an image, an impression of beauty, according to a particular visual ideal. Dress, Anne Hollander insists, has a "demanding visual dimension." It "must submit not just to mental and behavioral conventions but to visual ones." It is a "form of self-perpetuating visual fiction, like figurative art itself." Fashion is inflected by art, just as art—even nude art—represents the body according to the way clothes have been variously fashioned throughout history to make the female body fit particular ideals of beauty.[41] Similarly, the profusion and excessive lushness of Victorian domestic decoration (along with its fictional and pictorial representations) demonstrated feminine participation in a "new economy of desire," a world of pleasurable imagination and sensual plenitude.[42]

Her dress, like her efforts at home decoration, was one of the few areas of creativity open to the Victorian woman, where she might express not merely her individuality or personality, but her artistry in composing a picture—the visual creation of herself or her parlor. This was more than mere creative image making; it was a moral imperative. Victorian women were constantly exhorted to "look their best" and reminded of their duty to beautify themselves and their surroundings. According to Raverat, the Darwin women—especially Henrietta, who was a notable dresser, on occasion, "magnificent"—saw it as a "duty to that state of life to which it had pleased God to call you" to dress well.[43]

Dress was a "second self," an indicator of the character and mind of the wearer. An 1847 article on the "Art of Dress" in the *Quarterly Review* asserted that while men's dress revealed little about their tastes and character, it was a very different matter for women, for whom dress functioned as a "sort of symbolical language . . . the study of which it would be madness to neglect, [since] to a proficient in the science, every woman walks around with a placard on which her leading qualities are advertised."[44] A whole literature, especially evangelical literature, concerned itself with the morality of women's dress and what it revealed about the wearer. Girls were constantly exhorted to cultivate a beautiful character and expression rather than their persons; to wear only the "moral cosmetics" of good health, modesty, and sincerity; and to let their natural qualities shine truthfully through simple, decorous dress.[45] Such exhortations were in tension with the injunction to be beautiful and to beautify. Many fashion advisers and their readers seem to have inferred that, if a woman's inner self might be read like a book from her appearance, then a degree of subterfuge was justifiable in ensuring that the proper "self" was on display. A woman's "nature" might be pulled into shape or embellished, camouflaged, and presented in ways that conformed more closely to her own vision of herself as a desirable and desiring (albeit modest and respectable) status-conscious woman. In the world of fashionable dress and decor, identity was "increasingly recognized as something which could be deliberately constructed for others to read."[46]

8.3 The creative female eye

Conventionally, discourses of (heterosexual) vision are strongly gendered. Western portraiture, with its historical focus on the female body, specifically the nude body, exemplifies masculine "scopic dominance." The nineteenth-century figure of the flaneur, critically observing, ogling, and exploiting the visual images offered by his social inferiors—especially the "fallen" women of the streets—has come to symbolize the masculine penetrating gaze and its practice of "predatory window shopping." Enough has been said about fashion, however, to challenge the unproblematic rendering of the male gaze directed at female objects of desire. Women did not simply cultivate or create a "look"; they also looked.[47] They could contest or escape from the hegemonic masculinized gaze through the assertion of parlor decor or through an alternative vision of dress.

The question of for whom women dressed was much debated: Did they dress for men, for other women, or for themselves? If women dressed to please men, as was often assumed, why did they persist with such a ridiculous, superfluous, physically alienating, and downright dangerous fashion as the crinoline in the face of the concerted attack on it orchestrated by *Punch*? *Punch* was by far the most popular and influential of Victorian illustrated magazines, "a household word" within a few years of its foundation in 1841, "beginning in the middle class and soon reaching the pinnacle of society, royalty itself."[48] In 1856, Dr. Punch declared war by ridicule on the newly introduced crinoline on behalf of the male sex and "husbands in particular." "Crinolineomania" was a diagnosed "female complaint," for which the cure was to keep its sufferers away from the sources of infection—the milliners and Regent Street—and put them on a strict diet of pin money. If all else failed, then Dr. Punch proposed dispensing weekly doses of ridicule, to laugh women out of their insanity and out of the crinoline.[49]

In her fascinating analysis of this prolonged campaign, Julia Thomas sees *Punch* as actively shaping a debate in which the crinoline was bound up with

CRINOLINE CONVENIENT SOMETIMES. A WARNING TO MOTHERS.

Troublesome Parent. "WHO WAS MAKING THAT NOISE, CLARA?"
Clara. "ONLY ME AND *MOUSTACHE*, MAMMA!"

Figure 8.3. One of Punch's many lampoons on "Crinolineomania." *Punch*, October 4, 1856.
University of Sydney Library Collection.

anxieties about the role of women in Victorian society. The crinoline was a
"metaphor for female emancipation, literally enabling women to occupy a
wider sphere," but it was "also a way of checking this freedom, of emphasizing women's frivolity and irrationality and, by implication, their unsuit-
ability for public life."[50] *Punch* deployed the crinoline as a "political and
paradoxical weapon," depicting it with enormous proportions that it never
actually possessed. It portrayed men dwarfed by grossly expanded women
and confronted by the problems of getting them into carriages, down the
stairs, or across the street. But *Punch* also exhibited disquiet over the sheer
visibility and encroachment of the crinoline into public spaces traditionally
dominated by men, and over the barrier the crinoline imposed against close
contact and access to the wearer's body; over its concealment of her body
and the signs of reproduction that defined the Victorian woman. Beneath
the flow of ridicule lay the insinuation that the crinoline was "unnatural," a
"monstrosity" that concealed a woman's true shape from the defining male
gaze. Husbands were "compelled to keep at arm's length," and crinolines
might become grounds for divorce. In France, where the mania had originated, the crinoline was already a "depopulating influence." On the other

hand, the supposed origin of the disease through French invasion also connoted lax morals and Parisian luxury and excess.[51]

In effect, *Punch* turned the crinoline into spectacle, putting it on display, indeed, creating it as a visual object. So effective was *Punch's* construction of the crinoline that it became cultural currency, accepted even by fashion historians who still draw upon *Punch* cartoons to illustrate the crinoline. The magazine "constructed a knowledge about the crinoline that came to be accepted as truth."[52] Yet, though *Punch* railed against the crinoline for a decade, women as ostentatiously went on wearing it.

We do not have to interpret the crinoline as a conscious act of female resistance in order to agree that there was contemporary recognition of female agency in willfully choosing to persist in wearing a fashion that invited such ridicule.[53] Women might assert themselves through dress, asserting choice in the fashion arena when choice was denied them in so many other arenas. Perhaps the more significant aspect of the "crinolineomania" campaign lies in its crucial visual dimension. By this I do not simply mean the role *Punch* played in the visual construction of the crinoline, but also the assumption that women, as the designated objects of *Punch's* visual treatment, were also observers and assessors of *Punch's* constructions and their sexual and sexist nuances. Victorian women looked, even as they were looked at. And they set their eyes on much more than the pages of *Punch*.

Men too might feel the feminine gaze, whether they (somewhat suspectly) courted it with shoulder padding or fashionable Dundreary side-whiskers, or whether they strove more conventionally through sport and exercise to meet the Victorian masculine ideal of manly endeavor, muscularity, and morality. In fact, the sporting culture gave Victorian females more opportunities for observation of naked male bodies than might be imagined. Swimming costumes were not generally worn until the 1890s. Raverat describes the regular summer crowds of youths sunbathing and skylarking on the banks of the Cam, "as naked as God made them," and the crisis of propriety this occasioned for boating "ladies," who carried parasols in which they might bury their faces to protect their modesty, while the young Raverat, under cover of her own parasol, shamelessly looked her fill of the "noble" sight of well-built masculinity.[54]

Victorian culture has been well described as "relentlessly, explosively visual."[55] It was a culture that promoted visual alertness and acuity. Among the many contributing factors were the new media technologies—wood engraving, lithography, and photography—that revolutionized the quality and multiple reproduction of images, the industrialization of printing and papermaking, the mass marketing of cheap prints, greeting cards, and valentines (staggering numbers of which changed hands—as often initiated by females as males under the convention of anonymity), of illustrated books, periodicals, magazines, and newspapers. The Great Exhibition of 1851 that showcased the technological, economic, and cultural achievements of Great Britain and its colonies and protectorates, that drew more than six million visitors (even luring Darwin out of Down House) to gaze at the thirteen thousand exhibits, whetted the public appetite for spectacle and the expectation of display. Luxuriously appointed emporia and department stores, with their lavishly decorated windows and showrooms, established new commercial standards of elaborate display. "Spectacle and capitalism became indivisible."[56]

The introduction of optical gadgetry like the magic lantern, camera, stereoscope, and kaleidoscope; the popularity of dioramas and panoramas; the accessibility of spectacle offered by theaters, music halls, sideshows, exhibitions, museums, aquaria, and menageries—all added to the vibrant visual experiences of the Victorian public. At the other end of the spectrum stood the Royal Academy and the National Gallery, where high art was exhibited and viewed by the more cultured and refined. This too was widely distributed. Victorian artists often made copies of their own works for multiple buyers or licensed their reproduction. Print runs of popular paintings ran to the thousands. Many well-known artists made their livings by painting and designing plates to illustrate popular literature. A good deal of art, high and low, took women or "woman" as its subject, ranging from representations of the domestic ideal of mothers and wives, to romanticized and sensualized images of feminine beauty and desirability, such as those of the pre-Raphaelites, to the more overtly moral and sexualized genre of narrative paintings or illustrations.[57]

Women were avid viewers and consumers of this same explosive spectacle and display (although there were a few areas from which they were actively excluded, pornography being an obvious one). On the other hand, Victorian gender roles meant that women's opportunities to create display were limited. Artistic creation for women was conventionally restricted to watercolors of flowers and genre pictures, while the recognized higher (and more costly) forms of art—oil paintings of classical, historical, and scriptural subjects—were primarily the preserve of male artists. It was not suitable for women to paint in oils or model in clay, which largely explains the relative dearth of exhibiting Victorian women artists. While there were a few women illustrators, they tended to work within designated feminine areas, such as children's literature. Most women's artwork went largely unseen, even though it was such a culturally encouraged pastime for middle-class women.[58]

It was not until the 1880s that Thomas Henry Huxley's favorite daughter and darling of the Chelsea art set, the highly strung and highly talented Marian—"Mady", who attended Slade School and behaved with the freedom of a "New Woman"—daringly broke new ground in the campaign for women's right to study and represent the nude of high art, imperiling her father's much-vaunted sexual respectability. Her depiction of a nude "lying lasciviously on a beach," which was exhibited at the Grosvenor Gallery, provoked outrage along with the raised eyebrows and disparagement that greeted such encroachment onto a traditional male preserve. The brilliant Mady achieved her limited artistic fame not for her challenging nudes, but for her more conventional portraits of young girls and famous men (including Huxley, Darwin, and her husband, the artist John Collier) before succumbing to depression and hysteria, unable to cope with the care of home or child, and, eventually, dementia.[59]

In a highly visual, consumerist culture that fostered "visual acquisition" yet so restricted women's opportunities for self-expression and creativity, women might channel their creative energies into dress or decoration, creating themselves or their parlors as visual displays in ways that drew on the available visual resources of that culture. These resources were not confined to the more obvious ones of women's magazines, fashion plates, or shop windows, but also included other pictorial and artistic representations. The English "man-milliner" Charles Worth, who dressed European royalty and the rich and famous with his ruinously expensive gowns, well understood that it was the actresses who wore his beautiful designs on stage who had the "great influence over fashion" rather than his more respectable, socially eminent clients. As discussed, other relevant resources, such as the popular hybrid genre of pictures that told stories, required close scrutiny and interpretation of the moralistic and symbolic images they incorporated and might be read in different ways, depending on the assumptions the viewer brought to the artist's representations. Victorian "ways of seeing" were diverse and imaginative. Opinions varied about what to see, how to see it, and what to choose not to see.[60]

What all this adds up to is that the pressures to conform to conventional choices in dress and decoration were not so intense as to leave women no room for creative self-expression. A woman's display of clothing and decoration might be restricted by the prescribed social and moral messages they ideally reflected, but might also project her own aspirations and fantasies, a self-image related to the aesthetic standards of Victorian visual culture. Women could thereby participate in a creative tradition of image making. Many contemporary accounts attest to the aesthetic pleasure—even to the sense of empowerment—that women experienced through their dress and decor. These were socially accepted, even encouraged, practices that allowed women to experience themselves as agents, where the results of their creative endeavors accrued aesthetic value within the terms of the moral exhortation to be beautiful and to beautify.

8.4 Designing women

The continuing need to articulate the cultural prescriptive ideals of feminine dress, manners, and decor suggests the continuing resistance to such pre-scriptions. So long as women were the acknowledged primary consumers and exhibitors of fashion and home decoration, they retained a degree of control over their choices of dress and environment. Even after decades of didactic advice on decoration, the ornamental clutter of middle-class houses increased rather than diminished.[61] Having taken the crinoline as far as it would go, women next went in for trains, pushing their length to fully a yard and a half, provoking renewed ridicule from *Punch*. No sooner had trains left the streets than the crinoline was revived, shrunken and transformed into the "crinolette," a cushion of uncrushable sprung steel or braided wire that conferred the new requisite shape of voluminous roundness on the derriere. There was no end to the transformations of shape and decoration that manipulative women might devise and display. Transitory and fleeting, frivolous and superfluous, deceptive and decorative, absurd and function-less, intimating while inhibiting sexuality, Victorian fashion embodied the feminine as it simultaneously announced female agency and choice.

Those "fast" young women of the 1860s who allegedly adopted demi-mondaine dress and behavior, who devoted "such thought and intellect" as they possessed to dress, who made themselves immodestly conspicuous, who flirted uninhibitedly, who took the initiative in addressing young men and daringly exchanged sexually charged innuendo, were not simply the media creation of Eliza Lynn Linton (although it is to be noted that not even Linton went so far as to allege their literal unchastity). What is intriguing is the extent to which the practices of animal breeding informed her diatribes against the "Girl of the Period." In language remarkably similar to Darwin's contemporaneous description of selective practices by the fanciers of short-faced pigeons, Linton represents this unwomanly young woman as seeking to "outvie her neighbours" by "exaggerat[ing]" existing fashions in dress (by raising her skirts halfway to her knees, cutting her bonnet down to "four straws and a rosebud, or a tag of lace and a bunch of glass beads," and frizz-ing her hair out "like certain savages in Africa"), creating bizarre and useless "monstrosities . . . so soon as she begins to manipulate and improve."[62]

This, as discussed in chapter 6, is the essence of Darwin's reliance on the fashion metaphor in his *Variation of Animals and Plants under Domestication*. Darwin, as we saw, was intent on downplaying the artistry of breeding by relegating it to the lesser artifice of fashion in order to dissociate his domestication analogy from connotations of design. Hence the pigeon fancier is motivated by a "universal principle in human nature, namely . . . [his] desire to outdo his neighbours" into carrying his practice of selection to an "extreme point," even to the production of unviable monstrosities—such as pigeons with beaks so short they cannot break from their shells: "For fashion, as we see even in our dress, always runs to extremes." It is this "important principle" that "necessarily" leads to divergence in pigeon breeds and explains their derivation from the one parent stock. Subsequently, in *The Descent of Man*, Darwin transferred this conception of artificial selection to sexual selection in humans in order to explain racial divergence from an original human stock: each race has its own conception of beauty, and it is "natural to man to admire each characteristic point in his domestic animals, dress, ornaments, and personal appearance, when carried a little beyond the common standard." So, his argument ran, the continuing selection by the "more powerful men of each tribe" of those women considered the "more attractive" according to this "universal" or "natural" principle of exaggeration of characteristic features, would, over the course of generations, "modify . . . the character of the tribe."[63]

Where Alexander Walker (and after him, Herbert Spencer) ascribed a functional value to women's beauty—as an attraction to men in the exercise of their generative duty and as a sign of women's fitness for the bearing and rearing of children—Darwin rejected this utilitarian conception in favor of a relativistic, hierarchical notion of female beauty peculiar to the different human races, arguing that these different ideals were culturally induced, had no utilitarian value, and were subject to the whims of fashion: "In the fashions of our own dress we see exactly the same principle and the same desire to carry every point to an extreme."[64]

What Darwin obviously had in mind was the same contemporary phenomenon of competitive exaggeration of aspects of feminine apparel seized on by Linton and by *Punch* in its visual inflation of the crinoline. It is women who both exhibit and create this excess of dress and decoration, who actively construct their own appearance and their feminine identities. They take pleasure in the frivolous, ornamental, and excessive, those categories traditionally associated with the feminine. They resist the reforming efforts of those who would impose the masculine categories of function, order, restraint, and decorum in dress and decor.

It is highly likely that her reading of Darwin's *Variation* influenced Linton's attribution of the manipulative and improving practices of artificial selection to the designing hussy of her journalistic manufacture. *Variation* was first published in January 1868 and was favorably reviewed in the *Saturday Review* on March 14, the very same issue in which the first and most famous of Linton's articles on the "Girl of the Period" appeared.[65]

Linton was an ardent Darwinian who underpinned her popular attacks on fast girls and the "shrieking sisterhood" by invoking naturalistic limits to their aspirations. Women could not emancipate themselves from the fixed laws of biology and evolution, which Linton, like many Darwinians, grounded within a traditional rendering of Victorian femininity.[66] Whether or not she took her cue directly from Darwin, the notion of the designing, fashionable woman was already familiar through reform discourse and popular satire and caricature. Linton's contribution was to bring the "girl" of her manufacture and denunciation so vividly and vehemently before the sensation-seeking Victorian public in a format that guaranteed attention and notoriety and in a context of growing middle-class antipathy to the threatening economic and political independence of women. The catchphrase and the concept of the willful "Girl of the Period" took fire in the popular imagination. It fueled Linton's rise into "one of Victorian Britain's most influential and powerful journalists," while it popularized the fashion metaphor that Darwin had already applied to the practices of pigeon breeding, turning it back from pigeon breeding to female design and choice of clothing. His readers were well primed for Darwin's subsequent deployment of the same metaphor in the *Descent*. A copy of the review of *Variation* printed in the *Saturday Review* is filed among Darwin's papers; it is most unlikely that Darwin (not to mention the female members of the family) was unaware of Linton's much-talked-about essay in this very same issue of the popular weekly.[67] A degree of reciprocal influence seems inescapable.

The strongly gendered distinctions of Victorian dress, made so much of by *Punch*, invited ready comparison with the sexual dimorphism of birds: bright, decorative female dress and dull, inconspicuous male dress might be represented as artificial secondary sexual characteristics, the reverse of those naturally manifested in the bird kingdom. It is easy to see how readily Darwin might move from humans to birds and back again via these fashion distinctions that manifested his principle of female ornamentation in humans and male ornamentation in birds, with concomitant male choice in humans and female choice in birds. More than this, in the notion of designing women, Darwin had an exemplar both of female ornamentation and of willful female agency or choice—indeed, of female design. It was an exemplar that also in-

THE FAST SMOKING GIRL OF THE
PERIOD.

Figure 8.4. A farcical take on Eliza Lynn Linton's moralizing denunciation of the immodest, unfeminine, masculinized "Girl of the Period." The elaborately dressed "Fast Smoking" girl is partial to "a mild weed on the quiet," expertly chalks her billiard cue, hoists up her skirts to go dancing, and abjures stitching and the household accounts for the perusal of "sensation novels" and studying the "glorious masculine figure . . . all day" at South Kensington art school. *The Girl of the Period Miscellany* (London: Wyman & Sons, 1869), 137. State Library of New South Wales, Mitchell Collection.

sinuated sexuality: these women had designs on men; they designed themselves to attract masculine attention, exaggerating the more erotic aspects of their appearance—hips, breasts, and waist, the recognized female secondary sexual characteristics—and making themselves conspicuous. Like pigeon fanciers, they "manipulate and improve," creating nonfunctional, anesthetic "monstrosities" such as the crinoline. And, like the woman responsible for the clashing decor of Macaw Cottage, they were "no better than [they] should be."

In short, what I am suggesting is that while the feminization of Victorian fashion reinforced Darwin's claim that among humans it was the females who were decorated to attract dull, less ornamented males and the males who exercised choice, it also enabled his transfer of male agency in humans to female agency in birds. I have already stressed the importance in Darwin's theorizing of the transposable beauty of birds and women. Male birds exemplify all that is conspicuous and nonfunctional, even potentially disabling or disadvantageous, in external appearance. They are the bird equivalent of the fashionable woman, with their bright coloring or intricate patterning and decoration, with their long tails or head ornaments. This was an equivalence that Darwin precisely drew.

8.5 Birds of a feather

The section on birds (four full chapters) in the *Descent* is the longest and most detailed of Darwin's discussions of secondary sexual characteristics and the action of sexual selection in animals. He introduced it with the claim that birds are the "most aesthetic of all animals . . . and they have nearly the same taste for the beautiful as we have. This is shewn . . . by our women, both civilized and savage, decking their heads with borrowed plumes, and using gems which are hardly more brilliantly coloured than the naked skin and wattles of certain birds." In the later section on human sexual selection in which Darwin defined and defended the "manner of action" of male choice, he again stressed that it is women who "take more delight in decorating themselves with all sorts of ornaments than do men. They borrow the plumes of male birds, with which nature decked this sex in order to charm the females."[68]

In both instances, Darwin's transposition of women and birds carries with it the sense of female agency in dress and decoration, an agency that is aesthetic in impulse and is designed to attract the sexual attention of human males. Male birds could not be said to design themselves in a way comparable to the dress and bonnet choices exercised by Victorian women. Male bird "dress" (Darwin's terminology) was designed by female birds, who exhibited all the frivolity and capriciousness in dress design of their fashionable human counterparts. Through their cumulative aesthetic choices, female birds ornamented and encumbered the males with, in Darwin's words, "all sorts of combs, wattles, protuberances, horns, air-distended sacs, topknots, naked shafts, plumes and lengthened feathers gracefully springing from all parts of the body."[69] More explicitly, in discussing the many variations in sexual dimorphism within the same group of birds having similar habits of life, Darwin wrote: "It would even appear that mere novelty, or change for the sake of change, has sometimes acted like a charm on female birds, in the same manner as changes of fashion with us."[70]

Darwin deployed this fashion metaphor in specific opposition to the creationist view, recently articulated by the Duke of Argyll, that the great diversity and beauty of bird plumage had been designed by the Creator for his own or human appreciation. "Capriciousness in taste in the birds themselves appears a more fitting explanation," declared Darwin, pointing to parrots, where males and females were both beautifully plumaged "according to our taste," yet still displayed sexual differences in collar and head colors. Or what, he asked, of those strange cases of male birds with shortened tails or crests, in opposition to the general tendency for elongated tails or crests? These "almost seem like one of the many opposite changes of fashion which we admire in our own dresses."[71] The female bird designers of such short-tailed or crested males are the equivalent of those manipulative, frivolous young women conjured up by Linton, who cut their bonnets down to "four straws and a rosebud"; and it may well be that the capricious, nonfunctional choices of female birds disadvantaged or exposed the wearers of their designs to danger—environmental in this case, rather than the social disadvantage or the moral danger that might attend the wrong choice of dress in women. On the other hand, women were also capable of fine discriminations in dress and decor in negotiating the complex class and gender relations of Victorian society. So it was perhaps not too far-fetched to attribute the more subtle and refined, aesthetically pleasing patterning of birds such as the argus pheasant to discriminating female choice.

The female appropriation of natural objects to decoration and dress has been touched on. Flowers, fruit, leaves, moss, seaweeds, shells, feathers, birds, butterflies—dried, stuffed, or reworked in various media—all found their place in the Victorian parlor. Plants in pots, ferns in Wardian cases, live birds in elaborate cages (the symbolism evident even to Victorians), and fish and other marine life in aquaria shared living space with wives and daughters. The female practice of collecting, arranging, and reworking natural artifacts conflated scientific, aesthetic and moral impulses, and was endorsed in a Victorian "fantasy of nature tamed and pressed into the service of a domestic ideal." It complemented the more conventional fancy work—beading, embroidering, crocheting, and woolwork—that middle-class women were expected to produce and display within the home. We may wonder at the aesthetic appropriation of nature that impelled a girl to the skeletonizing of a mouse or frog for entombment under a glass dome in an "exquisite" arrangement, or to the dissection of a pigeon for the manufacture of a

feather screen, involving the removal of eyes and brain and the sewing of
the head and wings together onto a handle to be finished with a ribbon. Yet
instructions were available in the literature.[72]

The incorporation of the wings, heads, tails, and even the more spectacular whole birds—such as hummingbirds or birds of paradise—into an array
of domestic objets d'art was utterly outmatched by their appropriation to
feminine apparel, particularly to the decoration of hats and bonnets. The
highly lucrative plume trade made enormous depredations into bird life.
Wallace, collecting in the East Indies in the late 1850s, estimated that one
shipment of 400 birds of paradise would fetch £500 in London, while a
shipment of 4,000 insects would fetch only £200.[73] In just one year in the
1880s, over 400,000 West Indian and Brazilian birds and some 350,000
East Indian birds sold on the London market. In spite of growing agitation
against the trade, by 1898 in just six days of trading, 34,860 whole birds of
paradise and forty-five packages of their plumes were sold in London. The
devastating inroads into breeding colonies of egrets for the fashionable "ai-
grettes," or sprays of plumes, led to the adoption of the bird as the symbol
of the American bird preservation movement. Although men's hats were
frequently feather trimmed and the "feathered military" were mass consumers of bird plumage, the all-women officers of the Society for the Protection
of Birds, founded in 1889, targeted women and their passion for feathered
decoration as the real offenders. Their campaign played on the maternal
compassion and moral concern for God's creation that all women should
manifest. Those unnatural creatures who "disfigured" their heads with the
"nuptial plumage" torn from parent birds, leaving their babies to starve in
the nests, should be not "feathered" but "tarred." In America, the Audubon
Society offered public lectures on "Woman as a Bird Enemy."[74]

Darwin, who had been no mean destroyer of birds for sport (though he
had long ceased the practice) and for their collection and "preservation" as
scientific specimens, was well removed from the pending conflict among pro-
tectionists, feather traders, and the fashion industry. He took a more benign
view of the fashion and the women who wore the feathers. Intent on collapsing women into birds, he saw only the proliferation of feathered heads
and appropriated them to his theorizing without interrogating the innate
maternal love and tenderness he assumed to be the norm for all women,
savage and civilized alike. "The head," he emphasized in the *Descent*, "is the
chief seat of decoration" in both birds and "savage and civilized" humans.[75]

As if to prefigure his evolutionary transposition of women into birds, from
the late 1860s, *Punch*—indefatigable in commentary on female fashion—took
note of the prevalence of plumes in a series of cartoons. Among the more note-

" I WOULD I WERE A BIRD——"

Figure 8.5. *Punch* prefigures the transposition of women and birds in the *Descent*. *Punch*, April 23, 1870. University of Sydney Library Collection

worthy of these, one, appearing a full year before the *Descent* was published, featured a fashionable young lady adorned head to foot in feathers: "I would I were a bird," she trilled.[76]

The most apposite of "Mr. Punch's Designs after Nature" appeared just a few weeks after the publication of the *Descent*, to which it clearly refers.[77]

This 1871 version depicts a fashionably dressed young woman viewed from behind to show to best advantage the whole peacock she wears on her head in lieu of bonnet. Its stuffed head and crest, perched on the woman's forehead, announce its bird provenance, while its tail feathers cascade down her back to blend with her long, loose tresses, so that her hair appears to have developed the characteristic "eyes," or ocelli, of the peacock. It is the metamorphosis of hair into feathers (or of feathers into hair) that rivets the eye, with its implication that peacock and woman are one and the same. Their common identity is symbolized by the ocelli, which have their underlying evolutionary meaning in Darwin's carefully constructed explanation in the *Descent* of their sexual selection by generations of peahens that have, through their "continued preference of the most beautiful males, rendered the peacock the most splendid of living birds."[78] But ocelli are here transformed into the objects of masculine desire and choice; partly, we may assume, through Darwin's interpretation in the *Descent* of the evolution of European women's "much admired" and erotically significant long tresses through male sexual preference;[79] but mostly through *Punch's* own longstanding lampoons of feminine fashion that depend for their meaning on the Victorian convention of masculine scopic dominance and discrimination. Here they commingle—ocelli and hair—in a naturalization of female ornithological display in the person of a fashionable young woman, in seemingly sophisticated commentary on sexual selection that is prior informed by *Punch's* history of satire on female fashion.[80] The image is a co-construction by *Punch* and the *Descent* that replicates all the contradictions of Darwin's attribution of visual discrimination to a female bird while denying it to human females. The irony is that it is through the same medium of female fashion and decoration, via the notion of the designing woman, that Darwin might attribute an agency and choice to female birds that he conventionally denied to women.

Female birds in the *Descent* do not always behave with the sexual passivity and coyness that Darwin expected of them. In presenting what little direct evidence he had of female choice, he argued: "Not only does the female exert a choice, but in some few cases she courts the male, or even fights for his possession." It so happens that peahens fall into this category: "The first advances are always made by the female." According to Darwin's same authority, some peahens showed such unaccountable preference for "an old pied cock" over his "rival," a "japanned or black-winged peacock, which to our eyes is a more beautiful bird than the common kind," that the pied cock had to be shut up in the stable, and only then had the hens "all courted his rival."[81] Nevertheless, in Darwin's subsequent account of the evolutionary

MR. PUNCH'S DESIGNS AFTER NATURE.

GRAND BACK - HAIR SENSATION FOR THE COMING SEASON.

Figure 8.6. *Punch* commingles feathers and hair in a modishly dressed woman in commentary on Darwin's theory of sexual selection in the just-published *Descent of Man*. *Punch*, April 1, 1871. University of Sydney Library Collection.

development of the peacock ocellus ("one of the most beautiful objects in the world") from a "mere coloured spot," he represented the process of female choice as an exclusively aesthetic practice, divorced from all sexual passion, "unconsciously" exercised by generations of admiring peahens with an eye only for beauty.[82] There is no hint of the aggressive peahen sexuality that had, earlier in the *Descent*, provided corroboration of the existence of female choice, even though it was not of the more beautiful male.

His readers were familiar with descriptions and depictions of bird life that conformed with Victorian conventions of domesticity and sexual morality. For the most part, Darwin worked within the existing conventions of ornithological literature and illustration, appropriating them to his own evolutionary purposes in the *Descent*. Yet, paradoxically, Darwin relied on instances disruptive of these conventions to support his claim of the efficacy of passive—and therefore difficult to observe—female choice. As he told one of his more compliant breeding correspondents, John Jenner Weir, Darwin was otherwise forced to "trust to mere inference from the males displaying their plumage, and other analogous facts." At the same time, he was careful to indicate that such disruptive, overt female sexuality was the exception to the rule, restricted to "some few cases."[83]

While he admitted the difficulty of direct demonstration, Darwin remained utterly committed to and strongly defended the aesthetic sense and discrimination of lowly birds, comparing them with lowly humans: "Low powers of reasoning . . . are compatible, as we see with mankind, with strong affections, acute perception, and a taste for the beautiful."[84] His "best evidence" of an aesthetic sense in birds was exhibited by bowerbirds, made famous by John Gould's description of their "singular habits" and the striking illustrations of the birds and their decorated bowers in his *Birds of Australia*. Gould depicted the bowers as scenes of divinely ordained monogamous domesticity, ambiguously describing them both as nests (with resident female) and "playing grounds" or "sporting-places," avoiding explicit acknowledgment of their courtship and mating functions.[85]

Darwin hijacked Gould's bowers for his own agenda and into the pages of the *Descent*. He specified them as built and decorated for the "sole purpose of courtship" (the females nesting elsewhere). He quoted a description (also provided by Gould) of an aggressively courting male satin bowerbird, with bower ornament in his beak, feathers erect, eyes starting from his head, calling and posturing around the bower "until the female goes gently towards him." This left little doubt of what bowers and male "love-antics" were for, while not doing damage to the convention of "gentle" female sexuality.[86] "But what most concerns us," Darwin emphasized, "[is] that [the

Figure 8.7. The Australian spotted bowerbird with decorated bower. *Descent* 1871, 2:70.
University of Sydney Library, Rare Books and Special Collections.

bowers] are decorated in a different manner by the several species," and this attested to their distinctive, innate aesthetic tastes. Different species built their elaborate display bowers in different ways and decorated them with particular colors and materials peculiar to the species. In an avian equivalent of Victorian naturalistic parlor decoration, males collected shells, bones, leaves, flowers, berries, feathers, or pebbles—often from great distances—and displayed and continually rearranged them "with a decided taste for the beautiful."[87] Whereas beauty was usually invested in the bodies of male birds through female choice, in the case of bowerbirds, beauty was transferred from male bodies to bowers. The tasteful decor of the several kinds of bowers was not preordained by a Creator, as Gould intimated, but, as in middle-class drawing rooms, their diverse and divergent decorations were driven by the eyes and the capricious aesthetic/sexual choices of the females of the different species.

The trouble was that, although Darwin made as much of the fighting, choosy peahens as he could, he also had to argue that such demonstrable instances of female sexual preference were exceptional to the general rule of passive choice among birds. Nor did these observable instances of female preference add up to "direct evidence" of female aesthetic choice, as Darwin had to concede.[88]

In the *Descent*, as he had feared, Darwin was forced back onto "mere inference" from male ornithological display and the "analogous facts" of animal breeding and dress, where he juggled with the sexual inversion requisite to his conviction of the predominance of male choice among humans. It is at the conclusion of the section, euphemistically headed "Preference for Particular Males by the Females," that Darwin—in tacit acknowledgment of the limitations of his evidence of female choice among birds—pulled, as a rabbit out of a hat, his famous thought experiment of the "pretty girl" at the fair choosing from among her strutting, displaying rustic suitors: we may, said Darwin, "infer that she had the power of choice only by observing the eagerness of the wooers to please her, and to display their finery." The antics of these country bumpkins, "courting and quarrelling" over the girl "like birds at one of their places of assemblage," are the human correlate of the ridiculous "love antics" of the male satin bowerbird. Like the female bird, the hypothetical pretty girl makes no overt, discernable choice: "We can judge . . . of choice being exerted, only by the analogy of our own minds; and the mental powers of birds, if reason be excluded, do not fundamentally differ from ours."[89]

Yet this hypothetical "analogy" of human female choice was a barefaced exception to Darwin's normalization of male choice among humans. With sexual selection, Darwin walked a fine line that morphed back and forth, as the need arose, between male and female choice. Whatever the realpolitik of the parlor or fashion, or however matters might be negotiated among "rustics," who, like "savages," had little to do with fashion or urban parlors, the real power of choice in Victorian society lay with men. Even among "savages"—those humans whose "mental powers" and aesthetic sense were most comparable with those of birds—it was, insisted Darwin, men who chose, in spite of the fact that "in most parts . . . of the world," contrary to the feminization of Victorian fashion, "the men are more highly ornamented than the women."[90]

8.6 Back to the painted "savage"

Darwin introduced the crucial section of the *Descent* "On the Influence of Beauty in Determining the Marriages of Mankind" with the claim that even in civilized societies, "man is largely . . . influenced in the choice of his wife by external appearance." But, in order to explain the divergence of the separate races through such aesthetic choice, he argued, "we are chiefly concerned with primeval times, and our only means of forming a judgment on this subject is to study the habits of existing semi-civilized and savage nations." The "notorious" passion for ornament among "savages" was therefore critical to the development of Darwin's argument, and to support it, he invoked Carlyle's thesis that "clothes were first made for ornament and not for warmth."[91]

At one point in *Sartor Resartus*, Carlyle has his hero, Teufelsdröckh, define humans as "two-legged animals without feathers." Carlyle's extended clothes metaphor had multiple layers of meaning, a significant dimension being the ways in which it made tangible the interdependency among the natural, the social, and the spiritual aspects of human existence. The animating spirit of the clothes metaphor is the mind of God.[92] For Darwin, who first read *Sartor Resartus* in 1841 (finding it "excellent"), the metaphor that transposed clothes and feathers had literal face value and eliminated the need for the mind of God by reducing feathers to the same love of novelty and the "most capricious changes of customs and fashions" evinced in the dress choices of Victorian women and the "hideous ornaments" admired by "savages."[93]

Among anthropologists, it was Darwin's own famous encounter with the naked, painted inhabitants of chilly Tierra del Fuego that constituted the key ethnographic evidence for their objection to the conventional clothes for protection or modesty hypothesis. Far from modest intent, "primitive" decoration and clothing were, in large part, intended to make the body more sexually attractive, to accentuate rather than to conceal.[94] Darwin more circumspectly, and typically, elided the overt sexual reference: such body decoration was commonly motivated by "self-adornment, vanity, and the admiration of others." Even so, he went to some lengths to support his claim that this decoration was linked to the varying ideas of beauty and the mating preferences of the "men of different races."[95]

Darwin's own direct experience of "savage" ornamentation extended beyond the body paint of the Fuegians to the tattoos of Tahitians and the Maori and the cicatrices and body decoration of Australian Aborigines. He supplemented his personal experience with gleanings from a wide range of anthropological, ethnographic, and travel writings, collecting further instances of ear, lip, and nose ornamentations, teeth filing, skull manipulations, and the extirpation of hair, especially when he might link such practices with the female sex. He connected the "deep scars on the face with the flesh raised in protuberances, the septum of the nose pierced by sticks or bones, holes in the ears and lips stretched widely open" with the less ornamental, even ugly, "knobs and various fleshy appendages" of some male birds, arguing that it cannot be certain that these appendages "cannot be attractive to the female [birds]" when such "various hideous deformities . . . are all admired as ornamental" by "the savage races of man."[96] When he had to concede, in spite of assiduous searches of the literature, that body ornamentation was more widely practiced by males than females, Darwin explained this discrepancy by reverting to the "characteristic selfishness of man" in denying women the "finest ornaments." Given the opportunity, "savage" women were as vain and as given to finery and frippery as those crinoline-wearing, feather-bonneted women so derided by *Punch*. Darwin could cite the "amusing account of a Bush-woman who used so much grease, red ochre, and shining powder, 'as would have ruined any but a very rich husband.'"[97]

But the main point of Darwin's extended researches into the variable dress and decoration of "semi-civilized and savage nations" was to substantiate the "truth of the principle . . . that man admires and tries to exaggerate whatever characters nature may have given him." So, the "beardless races" extirpated all facial and bodily hair; other peoples flattened the head or nose in order to exaggerate some "natural and admired peculiarity"; the Chinese had naturally small feet, and it was "well known that the women of the upper classes distort their feet to make them still smaller"; while the American Indians preferred "colouring their bodies with red paint in order to exaggerate their natural tint; and until recently European women added to their naturally bright colours by rouge and white cosmetics." This brought the decorative choices of "savages" into satisfactory alignment with those of European women and pigeon fanciers: "We see exactly the same principle and the same desire to carry every point to an extreme" in the "fashions of our own dress" and in the selective practices of fanciers who "ardently desire to see each characteristic feature a little more developed."[98]

So far so good: through his fashion metaphor, Darwin might triumphantly connect the decorative choices of female birds, "primeval" ancestors, contemporary "savages," fashionable women, and pigeon selectors. Yet, through his insistence on the predominance of male choice in humans, the contradictions mounted up. In an attempt to demonstrate continuity of female choice from the "inferior animals" to humans, Darwin suggested that female choice still operated to an extent among "utterly barbarous tribes" (instancing the Fuegians among other practitioners of such primitive behavior).[99] This was in some tension with his earlier claim that "in civilized nations women have free or almost free choice which is not the case with barbarous races."[100] Darwin proceeded to intensify the contradictions by arguing that, because in "the savage state" woman is kept in a "far more abject state of bondage" than among "any other animal," it is "not surprising" that, as is the norm among more highly evolved women and men, it is men who "have gained the power of selection":

> Women are everywhere conscious of the value of their beauty; and when they
> have the means, they take more delight in decorating themselves with all sorts
> of ornaments than do men.

His major evidence for this turned on Darwin's recurrent theme that it is women who "borrow the plumes of male birds, with which nature decked this sex in order to charm the females."[101]

Wherever he went, Darwin circled back to the feminized fashion of his own society and to his fashion metaphor that carried with it the sense of female agency, yet was at the same time charged with the cultural values that allowed only the choice of the trivial, the frivolous and, sometimes, the absurd and disadvantageous to women, and left the important and more serious agency for human progress still with males.

I have argued that Darwin's construction of the sexually selecting, potent human male who molds woman to his fancy was an almost literal transference from the powerful masculine persona of the breeder who selects and shapes his domestic productions to his will, who "has impressed . . . the character of his own mind—his own taste and judgment—on his animals":

> What reason, then can be assigned why similar results should not follow from
> the long-continued selection of the most admired women by those men of
> each tribe, who were able to rear to maturity the greater number of children?[102]

It was his fashion metaphor, I suggest, that paradoxically allowed Darwin to move beyond the masculine persona of the animal breeder to his theorization of the efficacy of female choice among birds. At the same time as Darwin elevated and emphasized the immense modifying power of the breeder, his denigration of the artistry and aesthetics of the pigeon fancier to an artless, unconscious desire to "outdo our neighbours" and his linking of this so strongly and consistently with women's dress that "always runs to extremes," made possible the connection of the selective practices and scrutinizing eyes of the "little men" who designed both pigeons and fancy silks with those of women who chose and exhibited the fashionable exaggerations of dress and decor so denigrated by their critics.

A further connection may be made with the milliners and dressmakers who actually made the fashionable clothes and hats in the proliferating sweatshops or the freezing, ill-lit garrets that featured in popular accounts. From the 1840s, *Punch* and other periodicals made much of the plight of the miserable, exploited seamstresses who stitched, starved, and went blind so that rich conceited women might parade their finery.[103] While there were some influential "man-milliners," such as Worth, clothing for women was predominantly designed and made by women. The very term "man-milliner" that persisted until late in the century denotes the traditional female occupation of milliner (a term in general use for a dressmaker and retailer of feminine apparel as well as for a maker of women's headwear), but it also connotes dandyism and effeminacy. "Man-milliner" as a term of opprobrium came to prominence in the aftermath of the French Revolution. The man-milliner was "not only . . . one of the most despicable members of society," but "one of the most injurious . . . for the employment that degrades this animal might have preserved a woman from prostitution." The man-milliner was as often suspected of cross-dressing as he was accused of subjecting his women customers to "impertinent familiarities." The errant sexuality that impelled men to the expropriation of women's work and clothing was comparable to the "effeminate cast" of the suspect new pleasures of weavers. Weavers, a contemporary lamented, had abandoned their old muscular pursuits: "They are now Pigeon-fanciers, Canary-breeders and Tulip-growers."[104]

Déclassé weavers/pigeon fanciers were not only no gentlemen, they were unmanly, even effeminate; while designing females were not only unfeminine and "bold in bearing" but even, according to Linton, "masculine in mind."[105] The two might meet in their mutual pleasure in the frivolous, ornamental, and excessive, in the head and eyes of the female bird as Darwin set about transferring their choices and practices to the design of the peacock's beautiful but disabling tail or to the macaw's clashing blue and yellow plumage.

Through the metaphor of fashion, Darwin might attribute an agency and choice to female birds that he conventionally denied to women. The Victorian middle-class male was preeminently a self-made man—sober, powerful, and purposeful in attire and intent, disdaining display. He could not have been designed by a woman, by female caprice and sexual preference. Having routed the Great Designer from human evolution and racial divergence, Darwin was not about to cede the Creator's place to a woman—though, in the case of birds, he might be replaced by a female bird, as Darwin explicitly argued.

Wallace entirely missed the point when he attacked Darwin's fashion metaphor:

So, a girl likes to see her lover well and fashionably dressed, and he always dresses as well as he can when he visits her; but we cannot conclude from this that the whole series of male costumes, from the brilliantly coloured, puffed, and slashed doublet and hose of the Elizabethan period, through the gorgeous coats, long waistcoats and pigtails of the early Georgian era, down to the funereal dress-suit of the present day, are the direct result of female preference. In like manner, female birds may be charmed or excited by the fine display of plumage by the males; but there is no proof whatever that slight differences in that display have any effect in determining their choice of a partner.[106]

Wallace had to reach back to the Elizabethans and early Georgians to find the male fashion correlates of male bird plumage, obtusely disregarding the headlong pace of change and the exaggerations of nineteenth-century female fashion that so engaged Darwin's attention and informed his metaphor. But then, Wallace held to an entirely different view of the progressiveness and purposefulness of human female choice.

Darwin's formulation of the threefold analogue of the selective practices of pigeon fanciers, of the frivolous dress choices of fashion-conscious women, and of the aesthetic choices of female birds—all of whom capriciously pushed their selections to nonfunctional extremes—was fundamental to his assertion of the efficacy of female choice among animals. But it was Darwin's conceptions of embryology and inheritance that gave theoretic heft to metaphor, that underpinned their common underlying pattern. It was his faith in this developmental understanding that was absolutely critical to and sustained the making of sexual selection against all the criticisms that Wallace might cast at it. The next chapter examines Darwin's views on the connections among embryos, sex, and sexual selection.

NINE

Development Matters

Another circumstance occurs to us as curiously illustrative of the resemblance which the children of both sexes bear to each other and the female parent. When the plumage of male birds (as usually happens) is more brilliant than that of the female, the young birds of both sexes resemble the latter in their early dress; just as among the part of mankind accustomed to the artificial refinements of clothing, little boys as well as girls are made to wear petticoats before being advanced to the dignity of breeches.

—Review of Alexander Walker's *Intermarriage,* in the *British & Foreign Medical Review,* April 1839 (read by Darwin around that date)[1]

There exists a daguerreotype of Darwin with his oldest son William (aged about four years) in which William is wearing a dress. This was quite conventional. Victorian boys were not "breeched" until around five years of age. This feminized stage of male maturation, familiar to all Victorians, is a fashion referent for the significant embryological theorizing that underpinned the making of sexual selection. It also helps resolve the historical confusion about the major stages and timing of theory construction.

It has been claimed, for instance, that Darwin had all the knowledge he needed to make the distinction between natural selection and both forms (male combat and female choice) of sexual selection around the same time he first formulated his concept of natural selection in 1838.[2] Another suggestion is that it was not until he had developed the domestication analogy in the context of scientific justification for natural selection that Darwin was led to the formulation of female choice through its close analogy with the artificial selection of traits for beauty and pleasure, sometime between 1839 and 1842.[3]

Figure 9.1. Darwin with his firstborn, William, in 1842. William is wearing a dress, as was conventional for Victorian boys who were not "breeched" until around five years of age—a fashion referent for Darwin's emphasis on the resemblance between females and young throughout the animal kingdom. DAR 225:129. By permission of the Syndics of Cambridge University Library.

However, contrary to these interpretations, there is no explicit statement of female choice in Darwin's earliest known formulations of what he later designated sexual selection. In the well-known "Sketch" of his species theory, the short penciled outline he put together just before he and Emma moved to Down House in 1842, Darwin jotted: "Besides selection by death [natural selection] . . . [there is] selection in time of fullest vigour, namely struggle of males." There is a surplus of males at breeding time, "possibly as in man more males produced than females," and thus a "struggle of war or charms." The male with the "fullest vigour, or best armed with arms or orna-ments" would gain "some small advantage," and would mate and "transmit such characters to its offspring." Darwin drew a clear analogy with artificial selection by comparing the process of sexual selection (although he did not use this term) to the practices of breeders: "to man using male alone of good breed."

So in this first version, the "struggle of war or charms" was activated by males who outnumbered the available females, who presumably became the obligatory possession of victorious males. Females apparently had no choice in the matter. Humans were also subsumed under this same prin-ciple of surplus males and consequent sexual struggle for possession of the females.[4]

Over the next two years Darwin "slowly enlarged & improved" his sketch into a comprehensive "Essay" of more than two hundred pages. The es-say of 1844 fleshed out his major evolutionary mechanisms—variation and selection—and his argument based on the achievements of breeders among domesticates, followed by the proofs of descent with modification. It was a carefully crafted document, designed to convince fellow naturalists that his argument for the mutability of species was not guilty of the scientific "sin of speculation," but was based on a "grand body of facts" collated from animal breeding, horticulture, embryology, comparative anatomy, classifica-tion, geographical distribution, geology, and paleontology. Gone were the more extreme opinions and speculations of Darwin's private notebooks. There was no mention of an apelike ancestry for humans. Humans, who had populated the notebooks and were inseparable from his evolutionary views, were scarcely referred to, save for their role as breeders of useful or ornamental plants and animals.[5]

By now Darwin had his term for his principal mechanism: "natural selec-tion," a force in nature analogous to the hand of man in breeding domestic animals. However, as yet, Darwin had no term for what he referred to as a "second agency," the "struggle of the males for the females." He drew a clearer distinction between the more general "law of battle" and "the case of birds," where the contest was waged by males, "apparently by the charms of their song, by their beauty or their power of courtship, as in the danc-ing rock-thrush of Guinea." He also noted that the "sexual struggle" would be severest in polygamous animals, "as in deer, oxen, poultry," where the "males are best formed for mutual war." This kind of selection was "less rig-orous" than natural selection, since the penalty for failure "does not require the death of the less successful, but gives to them fewer descendants." Once more, Darwin drew the analogy with artificial selection in terms of those effects produced by "agriculturalists who pay less attention to the careful selection of all the young animals which they breed and more to the occasional use of a choice male."[6]

Yet again there was no definite statement of female choice.[7]

His earliest known, unambiguous statement of female choice is contained in the first edition of *the Origin of Species* (1859). There, Darwin, in discussing what he now termed "sexual selection" as distinct from "natural selection," distinguished between the "law of battle" between rival males "for possession of the females" and the "more peaceful" contest among birds, which sing or "display their gorgeous plumage and perform strange antics before the females, which standing by as spectators, at last choose the most attractive partner."[8]

There we have it: female choice—though Darwin did not actually use that terminology. He put the concept forward rather defensively: "It may seem childish to attribute any effect to such apparently weak means." But Darwin pinned his faith on the power of his all-purpose analogy with the achievements of breeders:

If man can in a short time give elegant carriage and beauty to his bantams, according to his standard of beauty, I can see no good reason to doubt that female birds, by selecting, during thousands of generations, the most melodious or beautiful males, according to their standard of beauty, might produce a marked effect.

Next follows the vital evidence from embryology and inheritance:

I strongly suspect that some well-known laws with respect to the plumage of male and female birds, in comparison with the plumage of the young, can be explained on the view of plumage having been chiefly modified by sexual selection, acting when the birds have come to the breeding age or during the breeding season; the modifications thus produced being inherited at corresponding ages or seasons, either by the males alone, or by the males and females; but I have not space here to enter on this subject.

On this basis, Darwin affirmed his conviction of the extent of the efficacy of sexual selection: "Thus it is, as I believe, that when the males and females of any animal have the same general habits of life, but differ in structure, colour, or ornament, such differences have been mainly caused by sexual selection."[9]

So, somewhere between 1844 and 1859, a period of fifteen years, Darwin had finally put the pieces together and added the essential element of female choice as a selective agent to the masculine "struggle by war or charms," thus arriving at the definitive theory of sexual selection. As has been emphasized, he also, in 1859, hinted at his explanation of the origin of human racial differences, "chiefly through sexual selection of a particular kind."[10]

Fifteen years is a long time. Can this time frame be narrowed down? I think it can. A major clue here is Darwin's earliest deployment of his closely associated fashion metaphor, which situates the concept of female choice in the late 1850s, when Darwin was at work on what he called his "big species book," which he began drafting on May 14, 1856. This was the unfinished work that he hurriedly pruned and honed into the *Origin of Species* after receiving Wallace's communication on natural selection in June 1858. It was during this period of theoretical consolidation and in connection with his recently formulated principle of divergence and his interrelated breeding, anthropological, and embryological understandings that Darwin naturalized female choice among animals and normalized male choice in humans.

The most important single factor in all of this was embryology. It was not until Darwin could confirm his embryological and closely connected theories of inheritance that he could take the essential final step in the formulation of sexual selection. These significant intellectual themes that came together in Darwin's thinking in the late 1850s are the main focus of this chapter. However, as I have stressed, it was Darwin's already articulated analogy between pigeon breeding and fashion that enabled this process. And there were other important nonscientific themes at play in Darwin's world and work during this critical period of theoretical consolidation that will be discussed more fully in the next two chapters.

9.1 The politics of development

Development really mattered to Darwin. The embryological argument for evolution was "the weightiest of all" to him, his "pet bit" of the *Origin of Species*. Embryology was, he told Asa Gray in 1860, "by far [the] strongest single class of facts in favour of change of form . . . Variations not coming on at a very early age, & being inherited at not very early corresponding period, explains . . . the grandest of all facts in Nat. History, or rather in Zoology. viz the resemblance of embryos."[11]

The changing forms assumed by the embryo during the course of individual development constituted Darwin's crucial piece of evidence for evolution. But there was more to it than this—much more. Throughout his working life, Darwin structured his evolutionary theorizing around the "facts" of "generation": of development and inheritance. These underpinned his three major evolutionary agencies: natural selection, divergence, and sexual selection. In his early transmutation notebooks, as we saw, Darwin took up, extended, and modified Erasmus Darwin's embryological model for evolution. The general principle was substantiated in Darwin's eyes by his readings and interpretations of more recent work in embryology and by his comprehensive study of all known living and fossil barnacles (from 1845 to 1853). Most important, Darwin made it compatible with both a branching taxonomic series and with his conviction that evolution was essentially progressive.[12]

However, at no stage of his writings, published or unpublished, did Darwin present his embryological views and his evolutionary agencies in the one complete, coherent package. They were developed and further tweaked at different stages of his work and writing. His understanding of the facts of generation shifted over time as he worked to pull them into line with his theorizing and make them compatible with the latest literature and his latest interests. The fragmentation and prolonged genesis of Darwin's embryological arguments for evolution have made them difficult to unravel and subject to much disputation among historians.[13] The issue is further complicated by the diversity of embryological interpretations and applications at play in Britain during Darwin's working life. However, through piecing his notes and writings together within these embryological and larger institutional and social contexts, it is possible to show that by the time of his writing of the *Origin*, Darwin had in place his main embryological arguments and their relation to natural and sexual selection and divergence; and, furthermore, that he held consistently to these fundamental understandings for the rest of his working life.

To begin with, it is necessary to understand that at the time Darwin was formulating his species theory in the late 1830s and 1840s, there were two major versions of development current in British embryology and comparative anatomy.

The first and older was embodied in the law of parallelism, sometimes known as the Meckel-Serres law, and had its genesis in the romantic metaphor of gestation. A parallel was drawn between the two series: the ontogenetic sequence of embryonic forms and the taxonomic sequence of adult organisms, both being considered as uniserial or unilinear; consequently the embryo sequentially repeated the adult forms of animals lower in the scale. This was more or less the version adopted by Erasmus Darwin. Associated with this law was the theory of arrests of development, which explained malformations or monstrosities in higher animals as states of organization that were permanent in lower ones. Instead of completing its normal development, the malformed organism stopped short and remained fixed at one of those stages—comparable to the permanent state of lower animals—through which it normally passed during development. The French anatomist Étienne Geoffroy Saint-Hilaire made this theory the backbone of his materialistic evolutionary speculations. In essence, Geoffroy argued by analogy from the production of monsters to the origin of species. He was convinced that in provoking the hatching of monstrosities from hens' eggs by artificially varying the conditions of incubation, he had experimentally illustrated the way new species arose in nature: mechanical and chemical changes in the environment (especially in the respiratory milieu) induced changes in the organism during the embryonic stages that were akin to monstrous development. Through their propagation by inheritance, these embryonic changes brought about the transmutation of species.[14]

From the late 1820s, Geoffroy's transmutationist and materialist interpretations were widely promoted in a British context of a miscellany of radical doctrines that included democratic politics, laissez-faire demands, and doctrines of self-development. This radical version of Geoffroyan anatomy was particularly espoused by London medical reformers, most notably Robert Grant. From his position at the University of London, Grant modeled his progressivist transmutation on Geoffroy's version of embryogenesis. It was this same Grant who awakened Darwin's early interest in generation theory, while the young Darwin was a medical student in Edinburgh.[15]

Just how familiar these embryological comparisons became is indicated by Charles Lyell's detailed refutation of their transmutationist implications in the second volume of his *Principles of Geology* (1832). Lyell admitted parallelism and monstrous arrests of development as a fact, but argued that these "curious phenomena . . . lend no support whatever to the notion of the gradual transmutation of one species into another," let alone to the notion of an ape origin for humans. Darwin was an early convert to Lyell's uniformitarian, gradualist geology, but by the late 1830s, Darwin was reading Lyell's anti-transmutationism against the grain. Lyell's endorsement of the law of parallelism can only have encouraged Darwin's evolutionary interpretation of the same.[16]

By this time, a second, more complex theory of development had been articulated by the Prussian-Estonian Karl Ernst von Baer (1828). This conceived ontogeny as essentially a process of divergent differentiation. As the embryo develops, it becomes increasingly specialized or individualized, and therefore diverges more and more from other animal forms. Like the leading French anatomist and special creationist Georges Cuvier, von Baer held that there were four basic types of vertebrate organization and that the type is manifested in the very earliest stages. Hence the law of divergence is based not on uniserial taxonomy, but on a multitypal or divergent one. The result is that the simple lineal parallel no longer applies, and the embryo repeats not the adult forms of lower animals, but their embryonic ones. However, von Baer's law does allow for some similarity between embryos of "higher," or more differentiated, animals and "lower," or less differentiated, adults of the same type. Such similarities are explicable in terms of lack of differentiation in both cases, the lower animals having a more general, less differentiated organization, comparable in some ways to the embryonic condition of higher animals of the same type.

Von Baer was an unrelenting critic of the earlier unilineal transmutation theories, and his divergent embryological law provided ready ammunition for other anti-transmutationists. These included the foremost British morphologist, Richard Owen, aspirant to the coveted title of the "British Cuvier" and doyen of the Royal College of Surgeons, then under heavy attack by the medical reformers and radicals. Owen worked to reconcile Geoffroy's transcendental anatomy with Cuvier's functional teleology, while secretly developing his own theory of divinely preprogrammed gross embryonic evolutionary change.[17]

The physiologist William Carpenter incorporated Von Baer's law into his influential *Principles of General and Comparative Physiology* (1839). Carpenter's *Physiology*, which went through four editions by 1854, was quickly adopted as the standard physiological text in medical schools. It was a major source of von Baer's embryology in Britain until Huxley's 1853 translation of the relevant section of von Baer's treatise. Yet, until late in the 1840s, British anatomists and physiologists (apart from an informed few) either continued to promote the law of parallelism or, more commonly, to conflate it with von Baer's law.[18] And embryology took on its own, sometimes wildly divergent, politics.

The law of parallelism and the theory of arrests of development proved attractive not only to radical anatomists and political fellow travelers, but also to conservative exponents of British natural theology. The orthodox followers of William Paley absorbed these embryological concepts into a natural theology that demonstrated God's providential design and continuing stabilizing presence in nature and the social order. It was a progressionist philosophy of nature that was both discontinuous and historical: the continuity exists only in the mind of God, who has created a succession of new and higher forms after the extinction of all their predecessors through a series of divinely invoked universal catastrophes. This interpretation was popularized through the writings of the Scots stonemason Hugh Miller and in several of the commissioned volumes of the Bridgewater Treatises, the official compendium of natural theology. The poet Alfred Tennyson even poeticized this notion in one of his most celebrated works, *In Memoriam*.[19]

This progressionist religion of embryology culminated in the paleobiology of the famous Swiss-American naturalist and devout teleologist Louis Agassiz, who cut his teeth on romantic morphology and embryology while a student in Germany. Agassiz made explicit the identification of ontogenetic stages with the fossil sequence, generalized it into a law of nature, and familiarized it through constant reiteration. He applied it obsessively to all the phenomena uncovered by his extensive paleontological and embryological researches, building up an impressive empirical documentation for his law of a threefold parallel between ontogeny, animal taxonomy, and the fossil sequence. Like Owen, Agassiz adopted von Baer's law of divergent development, but the none-too-muted undertones of the romantic gestation of nature impressed a unilinearity on his progressionist conception of nature that persistently neutralized von Baer's divergent schema of development. For Agassiz, the mastodon was an embryonic elephant, and fossils generally were the "embryonic types" of the living organisms of the "present creation," fetal structures that prophetically announced man's own creation.[20]

It was this dedicated lifelong opponent of evolution in all its forms who, with all the force of his powerful reputation, made ontogenetic development the paradigm of all of nature's history and the leading argument for premeditated design. But in the process, Agassiz brought progressionism much too close for comfort to the "development hypothesis" of the sensational *Vestiges of the Natural History of Creation*.

This anonymous evolutionary work, first published in 1844, posited a preordained natural law of progressive gestation by means of which new and higher species were constantly produced; a process that was, its author (the Edinburgh publisher Robert Chambers) claimed, as natural as, and no more to be feared, than the birth of a child into a middle-class Victorian family. It was an immediate sensation, a work of popular science, the product of new publishing technology and new means of communication. Its readily accessible, many revised editions reached hundreds of thousands of readers. *Vestiges* was read and discussed in the most famous household in the land, Buckingham Palace. Many more middle-class readers read it at the family hearth and took comfort from its reassuring domestic imagery. Evangelicals crusaded against its description of the ungodly gestation of nature, while the same provided common ground for reform-oriented phrenologists, radical publishers, middle-class journalists, and freethinkers. The struggling young surveyor Alfred Russel Wallace read it, converted to evolution, and found his vocation as a naturalist.[21]

In *Vestiges*, Chambers argued that the "development" of a new species was the result of an abrupt upward deviation from the normal course of embryonic development. The length of the gestation period was the critical factor, with premature birth resulting in arrest of development and monstrosity, while prolonged gestation led to an advance of development and transmutation of species into a higher form. This close association of his mechanism with the Meckel-Serres law via the theory of arrests of development, together with his invocation of prolonged gestation, predisposed Chambers to the conflation of von Baer's law with the law of parallelism. His leading evidence for this was Agassiz's relation of ontogeny to a progressive paleontological sequence.[22]

The evangelical Christian geologist Adam Sedgwick (Darwin's old Cambridge tutor) belatedly came alive to the pitfalls of progressionist embryology. He poured his scorn and loathing onto *Vestiges*, this "filthy abortion," and held it responsible for the insurrection of the lower orders and the corruption of pure Victorian womanhood—while simultaneously dismissing it as most probably written by a woman. Sedgwick schooled himself in the latest von Baerian embryology and attempted to dissociate progressionism from the insidious and now suspect law of parallelism, made infamous by the *Vestiges*: "false at every step . . . an idle dream of the philosophy of resemblances."[23]

Meanwhile, the continuing pull of the many editions of *Vestiges* for the reading public was largely attributable to its very appeal, so maligned by Sedgwick, to the ordinary processes of generation and reproduction as the key to the evolution of just about everything, from the gestation of the universe to the gestation of species. Chambers even extended his law of progressive gestation to the inevitable extinction of the "lower" races and to support white male supremacy. For, after "completing the animal transformations," the fetal human brain

> passes through the characters in which it appears, in the Negro, Malay, American, and Mongolian nations, and finally is Caucasian. . . . The leading characters in short of the various races of mankind, are simply representations of particular stages in the development of the highest or Caucasian type.

In time, the "best examples of the Caucasian type" would "supersede the imperfect nations already existing." Sex too was "a matter of development," the female being "arrested at a particular stage—that early one at which the female sex is complete."[24]

Vestiges became a sensation in a period during which traditional authority seemed under siege on all fronts—not least in the home itself, where uncertainties about women's place were exacerbated by the increasing movement of women into areas outside the traditional domestic sphere. Not only the pioneering Harriet Martineau, but also, by the 1850s, other women—including the young Marian Evans (the future novelist George Eliot) and Eliza Lynn (subsequently Linton)—were making some headway in the expansive world of metropolitan journalism. Sedgwick might denigrate *Vestiges* as written by a woman, but in progressive Whig and radical circles, the detection of a feminine hand in its writing could be taken to signal that women were at last coming into their own. Its contents, centered on the processes of reproduction and birth, might be viewed as endangering women's morality, or might be seen as proper reading matter for women. On the other hand, *Vestiges* might affirm the reassuring assumption that women and the "lower" races were not as "developed" as white European males. Yet again, read as a "people's science" of progress, the "development hypothesis" might be identified with free trade, national education, and universal suffrage, as offering a program for gradual, progressive reform.[25]

Whatever the meanings and interpretations it accrued as its readers grappled with these shifting political and social tensions, *Vestiges* put embryogenesis center stage and kept it there, opening it up to new possibilities beyond its earlier associations with an irreligious and politically suspect radicalism, or with a conservative, religion-sanctioned doctrine of special creation.

9.2 Darwin, Agassiz, Huxley, and *Vestiges*

A major problem for Darwin was that his own views on embryogenesis and its evolutionary significance came perilously close to the heretical *Vestiges*. Although their mechanisms for evolutionary change were quite different, there was enough similarity between their respective superimpositions of progressive change onto a branching von Baerian embryological substructure to cause Darwin to become seriously worried by the criticism this provoked. *Development* was now a loaded term, whether applied to the individual or to the species, and Darwin became chary of its use. Nevertheless, he clung to the critical piece of the embryological argument that *Vestiges* had made such play of: Agassiz's evidence of a parallel between embryonic stages and the progressive transitions of fossilized antecedents.[26]

This first came to Darwin's attention in 1842, when Agassiz and his young assistant, Carl Vogt, reported that fetal salmon went through a developmental stage in which their tails shared certain characteristics with the earliest fossil fish. In the adult salmon, the tail is symmetrical, or "homocercal," but in an earlier stage of development, the spine arches up into the superior lobe of the tail fin, making it "heterocercal," like the asymmetrical tails of the most ancient fossil fish. Darwin had become increasingly convinced that, while the evidence was fragmentary, Agassiz was fundamentally right, though wrong about the cause of the parallel. It was not divinely preprogrammed, as Agassiz insisted, but was to be explained in evolutionary terms, as Darwin's barnacle researches had further persuaded him. His eight-year immersion in the finer points of barnacle anatomy strongly reinforced Darwin's belief that embryological development not only revealed community of descent or a common ancestry, but also that it displayed the evolutionary history, or "phylogeny," of the organism.[27]

Darwin might withstand the savaging of the embryology of *Vestiges* by Sedgwick and other critics, but its authoritative clubbing by the up-and-coming young naturalist Thomas Henry Huxley forced him to take notice.

Darwin had first met Huxley in 1851 as Huxley rocketed onto the London scientific scene after four long years as assistant surgeon on HMS *Rattlesnake*. Huxley was without money or connections. He compensated (some might say overcompensated) for such lack with his frightening articulacy, daunting intelligence, hard work, and hungry ambition. He had left his fiancée in Sydney and could not bring her to England until he had carved out a secure scientific niche for himself and might support a wife and family. In remarkably quick time, Huxley secured membership of the Royal Society, a Royal Medal (1852), a close friendship with Joseph Hooker, growing enmity with Richard Owen (who was emerging as Huxley's major morphological competitor and bête noire), and an acknowledged expertise in marine invertebrates, morphology, and embryology.[28] Owen might be the "British Cuvier," but Huxley was determined to establish himself as the "British von Baer." Barnacles brought Darwin and Huxley together, but their embryological differences threatened to push them apart.

Huxley at this stage structured his morphological studies within a von Baerian framework that assumed strict lines of demarcation between natural groups, with no transitional forms between groups. Each natural group was based on an embryologically determined general plan or primordial type that precluded transition from one group to another. This offered, as far as it went, an explanation of resemblances between embryos and organisms independent of any theory of common descent. For the militantly anticlerical Huxley, the leading alternative—progressionism—reeked of the cloth and unlawful supernatural intervention in natural affairs. For Huxley in the 1850s, there was no evidence, embryological or paleontological, to support any hypothesis of the "progressive development of animal life in time."[29]

He seized on the opportunity to wield the ax of von Baer against the tenth edition of *Vestiges* of 1853, getting in some good sideswipes at Agassiz's fish-tail claims and Owen's morphology. Huxley was able to exploit the narrowing of the gap between transmutationists and progressionists, capitalizing on the current distrust of the embryological law of parallelism and its association with an ideologically suspect romanticism and a politically dangerous evolutionism. Given Huxley's own struggle to make a living from "pure" professional science, he took particular offense against the populist *Vestiges* notion of a preordained natural law of progressive gestation, which he scornfully condemned as "got up" to appeal to those "ignorant of science."[30]

It was not only a gullible public that was swallowing the offensive development hypothesis. Some experts who should know better were nibbling at its arguments. In fact, the respected William Carpenter had revised this latest edition of *Vestiges*. He tidied up Chambers's conflation of the two versions of embryological development, giving precedence to von Baer's law of divergence. Nevertheless, as Huxley had detected, the "development hypothesis" remained incorrigibly progressive. In the fourth edition of his *Physiology* of 1854, Carpenter began to show the effects of exposure to *Vestiges*, arguing for the Creator's "great scheme" of a "progressive evolution" of organisms in time—though not for their actual transmutation—from the "more general to the more special." He in turn was hammered by Huxley in the *Westminster Review* for this progressionist transgression and for his endorsement of Agassiz's parallel between ontogeny and the fossil record.[31]

Darwin read Huxley's review of *Vestiges* late in 1854. He told the embryological expert that he was "rather hard on the poor author"; Darwin's own views were "almost as unorthodox about species." He was "rather sorry" Huxley did not think more of "Agassiz's embryological stages," for though Darwin saw "how excessively weak the evidence was," he "was led to hope in its truth." Stubbornly, he kept on hoping.[32]

To add to the complexities of Darwin's situation, Agassiz was not only a creationist and staunch antievolutionist, but also emerging as a leading exponent of scientific racism in his adopted American homeland. He brought his great scientific reputation to the promotion of polygenism, the theory of the separate creation of the human races in different geographic areas, or "centers of creation." Much has been made of Darwin's antipathy to Agassiz's endorsement of polygenism—which Darwin condemned as bringing comfort to the Southern slaveholder.[33] Yet it was to Agassiz that Darwin looked for validation of his own views on embryogenesis, not to Huxley, by then recognized as the foremost British expert on comparative embryology.

If we are to understand why, we need to focus on Darwin's above-cited representation to Asa Gray of the great importance of embryology for his evolutionary views: that "variations not coming on at a very early age, & being inherited at not very early corresponding period" explains this "grandest of all facts in Nat. History."

9.3 Darwin, sex, and the embryo

During the twenty-year period from the time of his London notebooks to the publication of the *Origin*, Darwin read across the available embryological literature in some detail. Although at points he wavered, he kept returning to his conviction that embryonic stages revealed a sequence of adult ancestors or more primitive stages of development in the evolution of life.[34]

Darwin did not read von Baer until 1853, when Huxley's translation was published, gleaning what he knew of the embryological law of divergence from secondary sources.[35] Even then, Darwin continued to superimpose his own conviction of progressive embryogenesis and evolution onto the embryological law of divergence.

The reason for this lies in Darwin's view that variation affects "only the more mature periods of life." Natural selection, he reasoned, cannot act on structures that are protected from selection, as in the egg or the womb. This explained both the significance of embryology for determining relationships or homologies among mature organisms and Darwin's predisposition toward reading its ancestors into the embryo. If evolutionary change came about as an additional or terminal step after the completion of normal development, and was inherited at a correspondingly late stage, then the embryo as it ran through its developmental stages would recapitulate "earlier full-grown structures," as Darwin expressed it in his 1844 essay. The variations on which selection would fall would tend to arise at a "not very early period of life" when the animal had to fend for itself, and they would tend to reappear at a "corresponding period of life." This would explain "how the embryos and young of different species might come to remain less changed than their mature parents," and so resemble one another more than their parents. Darwin supported this with measurements he had made of newborn greyhounds and bulldogs, which showed that their noses and legs were the same length, though their adult forms differed so greatly. He particularly distinguished between small heritable variations that occurred after birth and "monstrosity," such as "extra fingers, hare-lip and all sudden and great alterations in structure," which occur before birth and are inherited at a correspondingly early stage of development.[36]

EMBRYONIC DEVELOPMENT.

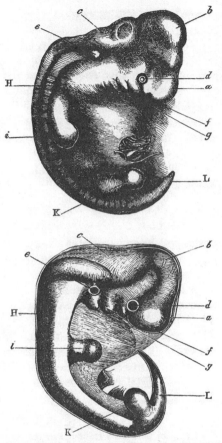

Fig. 1. Upper figure human embryo, from Ecker. Lower figure that of a dog,
from Bischoff.

Figure 9.2. "The grandest of all facts in Natural History": an illustration from *The Descent of Man* showing the similarities between the embryos of human (upper) and dog (lower) at comparable stages of development. *Descent* 1871, 1:15. University of Sydney Library, Rare Books and Special Collections.

This was a view he had held from a very early stage of his evolutionary theorizing, when he rejected the notion that the "production of monsters . . . present[s] an analogy to production of species." Darwin, for a number of interconnected reasons, drew a clear distinction between monstrous change and species change.[37] They may be summarized as his conviction of the conservatism of generation and individual development—the innate tendency for the developing organism to replicate the embedded sequence of ancestral changes—along with his gradualist, Lyellian conception of transformism, which for Darwin meant that any major adaptive change could be acquired only gradually, through the slow accumulation of many small steps over many successive generations.[38] Monstrous change could not lead to or explain species change. Rather, Darwin decided, natural selection was the force that could build new species traits onto the end of progressive embryogenesis by acting on the more mature organism through population pressure and the consequent struggle for existence—the Malthusian factor. This explained why "the birth of the species & individuals in their present forms are closely related . . . each of us, then, have passed through as many changes, as has any species"; in other words, why each of us during embryogenesis passes through the evolutionary history of our species.[39]

After Darwin formulated the "beautiful analogy" between nature's and man's selective breeding, he adopted the view that variation in a state of nature, as under domestication, was somehow randomly environmentally produced (by climatic change or changes in the conditions of life) and then sorted by natural selection, just as the breeder sorts or "picks" those traits he wants. This gave him even stronger reasons for assuming that adaptive species change, having only rarely produced chance natural variations to work on, was very slow and gradual. The "very essence" of his theory was that "little change is produced." A monstrosity, by definition, could play no role in species formation, but was relegated to the domestic domain, there to be dismissed by Darwin as "a mere monstrosity propagated by art," like the famous short-legged Ancon sheep or some varieties of pigeon.

Darwin's mature theory of evolution of 1859 retained this crucial distinction. Domestic races "often have a somewhat monstrous character," while monstrosities in nature are "either injurious to or not useful to the species, and are not generally propagated."[40]

His domestication analogy also gave Darwin further reason to argue that modifications "supervene at a not very early period of life" and tend to reappear "at a corresponding age in the offspring":

It is commonly assumed, perhaps from monstrosities often affecting the embryo at a very early period, that slight variations necessarily appear at an equally early period. But we have little evidence on this head—indeed the evidence rather points the other way; for it is notorious that breeders of cattle, horses, and various fancy animals, cannot positively tell, until some time after the animal has been born, what its merits or form will ultimately turn out.

Fanciers select their horses, dogs, pigeons, and so on, from full-grown animals on the basis of the desired mature characteristics. They are indifferent as to whether the young animal possesses them, as long as they are present in its mature form. This, Darwin argued, supported his analogical claim for the action of natural selection on animals in the wild.[41]

So, from a very early stage of theorizing, Darwin consistently rejected the possibility of monstrous change, or what we would now call "saltation," as an evolutionary mechanism. This meant that he rejected those early materialistic theories of transmutation, such as Geoffroy's, which relied on monstrous or sudden great change. This was an explicit rejection of radical or revolutionary change in the political as well as the biological senses. As we saw, Darwin himself noted that his insistence on explaining major structural changes by the gradual accumulation of minute changes through the constantly acting Malthusian population law on more mature organisms, "baffles idea of revolution" in nature as in "government" and "institutions."[42]

Two further points remain to be made at this stage. First, Darwin never entirely gave up his earlier theory of the effects of inherited habit. In theory, all variations were heritable, so some adaptive variations might be produced and perpetuated in a population through the effects of habit or use; others might be lost through disuse. Acquired characteristics necessarily would be manifested after birth, when they became habitual or useful to the organism and reproduced at a correspondingly late period of development. They could be further sorted by natural selection. In either case, they would become embedded in embryonic change, like those produced solely by natural selection. Secondly, and most important for our purposes, from his very early speculations on transmutation, Darwin explicitly linked embryogenesis with the differences between the sexes, and he saw that difference as both an embryological and an evolutionary *progressive* change from the hermaphroditic condition, via the female, to the definitive masculine primary and secondary sexual characteristics. Furthermore, from the very beginning of these speculations, he related such change to both animal and human sexual selection.

He took his earliest views on sexual development from the anatomist John Hunter's claim that, during the course of individual development, both sexes undergo a development from a common hermaphroditic state. Hunter had argued that the female, tending to show only the primary sexual marks, was the more "hermaphroditic," or less sexually specialized. Mature females remained more like the young of the same species; males when castrated retained more of the "original youthful form" and were therefore more like females. Hunter also described cases where female birds past their reproductive period had taken on male secondary sexual characteristics, including the "wonderful case" of an aged peahen that had displayed the feathers and spurs of a peacock. For Darwin, such instances supported his interpretation that females were less developed than males and were therefore more like the species' ancestor.[43] Furthermore, he explicitly related this interpretation to sexual selection. Here is Darwin, around mid-1838, wondering:

> Is man more hairy than woman[.] because ancestors so, or has he assumed that character—–female & young seem more like mean characters the others assumed . . . Daines Barrington says cock birds attract females by song. Do they by beauty, **analogy of man** if so war not. [bold format indicates a later annotation by Darwin][44]

Tantalizingly, the entry breaks off at this point, and the next four notebook pages are excised—an indication of their importance to Darwin's subsequent theorizing. Presumably they were incorporated into his later writings on sexual selection. But we may unpack this early entry to mean that Darwin is here querying whether men are hairier than women because they have inherited this secondary sexual characteristic from their hairy primate ancestors (the implication being that women somehow have lost this common ancestral hairiness), or because men's extra hairiness is the result of their further development beyond the common hermaphroditic or female stage. In support of the latter, Darwin points to the claim that mature females and young females and males are more alike—that is, they share the common characteristic of less hair (or, as he remarked elsewhere in his notes, in the case of birds, similar plumage)—while the male goes on to assume its distinctive adult hairiness (or mature plumage). He then immediately connects this speculation with the mating habits of cockbirds, which are said to attract females with song, and asks whether male beauty is also an attractant to females. He next embarks on the beginnings of what seems to be an extension of the sexual struggle beyond "war" between the males to one of charms or beauty. Then, sometime later, Darwin inserted the "analogy" of human males being attracted to female beauty.

These speculations linking embryogenesis with sexual difference in both humans and lower organisms and with sexual selection recur throughout the early notebooks, particularly in relation to birds. Darwin also extended the link between sexual juvenility in females from appearance to intelligence: "The female and young of all birds resemble each other in plumage (that is where the female differs from the male?).—children & women = 'women recognized inferior intellectually.'"[45] At this stage (September 1838)

Darwin was still querying the generality of this argument, but the origins of his claim in *The Descent of Man* that women were the intellectual inferiors of men (which he then supported with a similar but more up-to-date embryological argument) are to be found here in the notebooks. Women were more childlike in their appearance and intellect, and this suggested that they were at a lower stage of development (understood as both ontogeny and phylogeny) than men.

The significance of these various early notebook entries on the connections among embryogenesis, sexual difference, and the sexual struggle is encapsulated in a much later note interpolated by Darwin while he was at work on his "big species book": "*Sexual Selection*[.] If masculine character. added to species,. we can see why young & Female alike[.] Good Ch 6 Keep."[46]

This refers to the chapter on natural selection in Darwin's "big species" manuscript, which included a section on "sexual selection." However, most of this section is missing, and sexual selection is alluded to only fragmentarily. By this stage of theory formation, Darwin's comprehensive barnacle researches (for which he received the Royal Medal of the Royal Society in 1853) had vindicated his fundamental thesis that the separate sexes had evolved from ancestral hermaphrodites. The barnacles provided a series in which he could trace this evolutionary development from hermaphrodites proper to those with reduced male organs and tiny "supplemental" males, to females with entirely obliterated male organs and "simple" male companions. If the separation of the sexes were to be fully effected in barnacles, he wrote in his "big species book," males would cease to be complementary and would "assume the full dignity of the male sex."[47]

However, Darwin's researches on the sex life of barnacles (in the course of which he just about turned himself into one—"fixed & half-embedded in the flesh of their wives," as he described the males of one species he studied) were not without complications for his theory of sexual evolution. The complemental males were miniscule rudimentary, embryo-like organisms, living more or less parasitically within the carapace of the females. In some cases they had no mouths or stomachs, being reduced to "mere bags of spermatozoa." Barnacle anatomy and sexual arrangements did not accord entirely with Darwin's insistence that, as a general rule, females were more embryo-like or "primitive" than males.[48]

But for Darwin, analogy had its limits and, where convenient, might be suspended altogether. He held tight to his conviction of the resemblance of females to their young and hence of the females' more primitive or ancestral condition. It suited his general thesis that change was built onto the end of progressive embryogenesis, was inherited accordingly and sequentially embedded in embryonic development, so that the embryo repeated the ancestral stages of its evolutionary history. The significance of all this for sexual selection was briefly summarized by Darwin in the small section on sexual selection in the *Origin of Species* where, as we saw, he invoked the "well-known laws with respect to the plumage of male and female birds, in comparison with the plumage of the young."[49]

Female choice could fall only on variations in male feathering when both sexes are sexually mature, and usually only in the breeding season, when many male birds put on their breeding plumage. This meant that such variations would be selected, according to Darwin's generalization, at a mature stage, and inherited (usually by males) at a correspondingly late stage of development. Consequently, the females and young would resemble one another; hence the "masculine character" was "added to the species."

With this, Darwin brought his evolutionary agencies—sexual selection, natural selection, and inherited habit—satisfactorily into line with his views on embryogenesis, heredity, and ancestry. All agencies acted after birth or on the more mature organism; all were inherited accordingly and were sequentially embedded in embryonic development; so the embryo would recapitulate its adult ancestors. A clear pattern was emerging. Darwin was bringing (indeed by the late 1850s had brought) the complex laws of generation—development and inheritance—under a unified view that accorded with his evolutionary views. Above all, this pattern accorded with the practices of breeders who selected their breeding stock from sexually mature subjects, and so with Darwin's template of artificial selection on which both natural and sexual selection were patterned. The final piece of this pattern remains to be discussed: Darwin's principle of divergence.

9.4 Getting it together: The keystone and the key

As he later recalled, divergence was a "problem of great importance" that Darwin "overlooked" until "long after [he] had come to Down." The trigger seems to have been his 1846 reading of "the most profound paper I have ever seen on Affinities" by the French naturalist Henri Milne Edwards, which he then successfully applied to his classification of the barnacles. Like von Baer, Milne Edwards emphasized the treelike branching or divergence of embryological development and its correlation with the taxonomic hierarchy. The most general characters of a class appeared earliest in development, and these established higher taxonomic affinities. The more characteristics two organisms shared in their development, the more closely they were related.

Of course, none of this was exactly new to Darwin. Embryology aside, Adam Smith's economic principle that more wealth accrues from a division of labor was hardly unknown to the grandson of old Josiah Wedgwood, who had founded the family pottery fortune on precisely that principle. Darwin's application of it to natural history was more an "agronomic version" of the principle, appropriate to a country gentleman familiar with the notion of agricultural improvement and its practices, such as crop rotation. He substantiated it empirically through counts of plant species diversity in small, defined patches of land in his neighbor's field and in the lawns and meadows of Down House. Yet it seems to have been Milne Edwards's physiological application of the principle to embryological development and classification that triggered Darwin's awareness of its generality and significance, reinforcing his own evolutionary interpretation of the same. And Milne Edwards added a corollary that struck Darwin full force. This was his measure of the "highness" of a species within a type: the increasing specialization in embryogenesis illustrated the tendency in higher organisms toward a "division of physiological labour." For Darwin, this provided a standard for determining "highness," which, he told Hooker in 1854, "usually means that form which has undergone more 'morphological differentiation' from the common embryo or archetype of the class": "the division of physiological labour of Milne Edwards" was the "best definition."[50]

Progress was compatible with the divergent pattern of embryogenesis and classification, and—Darwin became more certain in the face of Huxley's criticisms—with the fossil record, imperfect though it was. All he cared about, he told Hooker, was that "very ancient organisms, (*when* different from existing,) sh[oul]d tend to resemble the larval or embryological stages of the existing." The underlying pattern was progressive, "tree-like," branching divergence.[51]

Darwin worked on the principle of divergence in a series of notes dating from around November 1854, first formulating it clearly in January 1855 and then refining it, clarifying its link with natural selection in September 1856 and testing it with his field experiments that continued until mid-1858. His barnacle studies had convinced him of the great abundance of variability in nature, and he moved away from his earlier position that hereditary variation depends on slow geological change. Two critical moves in this direction were, firstly, Darwin's growing understanding that varieties were "incipient species," and secondly, that they could develop by divergence without geographic isolation. This latter realization was of particular significance in the development of Darwin's evolutionary theorizing. It meant that he no longer had to invoke geographical barriers (such as oceans or mountain ranges) or climatic differences in order to account for the isolation of varieties from one another that would prevent their back-breeding and allow them to form separate species. He focused instead on the intensity of competition and selection within and between species and invoked the division of labor. Competition, he reasoned, is most intense among organisms that are most alike. The more organisms differ or diversify, the greater the number that can be supported within a shared territory. There is a selective advantage that accrues to the most different, or extreme, varieties. "It follows from this," Darwin wrote in his "big species book,"

> that the amount of difference which at first may have been very small between any two varieties from the same species, in each successive set of new varieties descended from the first two, will steadily tend to augment as the most divergent or different will generally be preserved.

Divergence "regulates the natural Selection of variations, & causes the Extinction of intermediate & less favoured forms." It bore on, and explained, the "classification or natural affinities during all times of all organic beings," which "seeming to diverge from common stems are yet grouped like families within the same tribes, tribes within the same nations, & nations within the same sections of the human race." Existing beings resembled the "buds & twigs" of a gigantic tree; "all beneath their living extremities may represent extinct forms." This "great Tree of Life . . . fills the crust of the earth with fragments of its dead & broken branches, & covers with its ever living, ever diverging & marvelous ramifications, the face of the earth."[52]

Divergence, as David Kohn makes explicit, was "essential to Darwin's account of how new species are formed because [it] gave him the means to account for evolutionary branching." Divergence "arises from intraspecific competition nested within, or going on simultaneously with, interspecific competition . . . By imposing an interspecific level on top of the competition going on within a species, Darwin effectively introduced the natural equivalent of an external selector." Natural selection resulted from the competition engendered by Malthusian population pressure, while the agency of "divergence selection" regulated or directed natural selection. This external agency, argues Kohn, is "*strongly* analogous" to the selector of artificial selection, whereas natural selection is only weakly analogous to artificial selection.[53]

Darwin introduced his principle of divergence in his "big species book" by making just this point. Domestic productions, he wrote, offer a close analogy to the principle of divergence in that "each new peculiarity either strikes man's eye as curious or may be useful to him; & he goes on slowly

& often unconsciously selecting the most extreme forms" to make, in the case of pigeons, the "several breeds of improved tumblers, carriers, pouters, fantails &c, all as different or divergent as possible from their original parent-stock the rock-pigeon," neglecting the intermediate birds, which become extinct.

> *It is the same with his dress, each new fashion ever fluctuating is carried to an extreme & displaces the last; but living productions will not so readily bend to his inordinate caprice.* Now in nature, I cannot doubt, that an analogous principle, not liable to caprice, is steadily at work, through a widely different agency; & that varieties of the same species, & species of the same genus, family or order are all, more or less, subjected to this influence [of divergence selection]. For in any country, a far greater number of individuals descended from the same parents can be supported, when greatly modified in different ways, in habits constitution & structure, so as to fill as many places as possible, in the polity of nature, than when not at all or only slightly modified. [emphasis added][54]

Like the pigeon fancier, divergence "always favour[s] the most extreme forms." The linking analogy is fashion. And, as I have explained in some detail in the previous chapter, Darwin made the same link in developing the concept of female choice. Divergence and sexual selection both reflect the practices of artificial selection and are epitomized, in both cases, by the practices of fashion. The difference is that female choice more closely mirrors the capricious choices of the breeder and the fashion-conscious woman, while divergence selection is not a matter of choice or caprice, but results from the incessant competition of species with species superimposed on that of variety with variety, as they struggle to fill the diverse available environmental niches.[55]

This is the earliest instance of Darwin's fashion metaphor that I have been able to trace. Significantly, it was around this same time that Dr. Punch began his campaign against "crinolineomania." But of equal significance in this first appearance of Darwin's fashion metaphor is its emergence in Darwin's theorizing coincident with the principle of divergence. It was this crucial principle that was the "keystone" that locked both natural selection and sexual selection into the structure that Darwin built into the *Origin of Species*.[56]

With divergence, everything began to fall into place. Darwin could bring together and explain the diverging domestic pigeon family tree he was in the process of constructing, the branching taxonomic hierarchy, the fossil sequence, and, as he stressed, embryogenesis. If divergence was the "key-

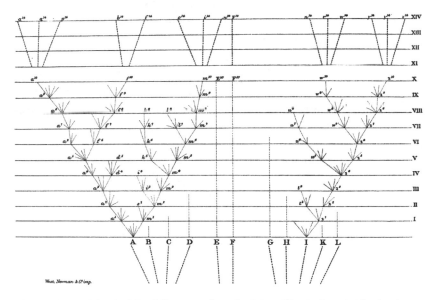

Figure 9.3. Darwin's principle of divergence, from the *Origin of Species* (1859). Like the pigeon fancier, divergence "always favors the most extreme forms." Divergence and sexual selection both reflect the practices of artificial selection, and all are underpinned by embryogenesis, which follows the same progressively divergent path. *Origin*, facing page 117. Courtesy of Wellcome Library, London.

stone," then embryology was the "key" to Darwin's interpretation of the relations among variation, maturation, inheritance and natural and sexual selection. As Lyell, after reading the *Origin* in 1859, appreciatively put it to Darwin, embryology was the "genealogical key to the Natural System."[57] It was the key to the evolutionary past of a species—a key that, in Darwin's eyes, unlocked material evidence of a progressive ancestral history. "We have no written pedigrees," Darwin emphasized in the *Origin*,

> we have to make out community of descent by resemblances of any kind. Therefore we choose those characters which, as far as we can judge, are the least likely to have been modified in relation to the conditions of life to which each species has been recently exposed.

This was why "the leading facts in embryology . . . are second in importance to none in natural history." They were "explained on the principle of slight modifications not appearing, in the many descendants from some one ancient progenitor, at a very early period in the life of each . . . and being inherited at a corresponding not very early period."[58]

As Darwin clarified in chapter 10 of the *Origin*, "This process, whilst it leaves the embryo almost unaltered, continually adds, in the course of successive generations, more and more difference to the adult." This is why embryos resemble ancestors: "The embryo comes to be left as a sort of picture, preserved by nature, of the ancient and less modified condition of each animal." Darwin also made explicit his belief that this picture was of an adult, not an embryonic ancestor. While he scrupulously reported Huxley's criticism of it, Darwin expressed his support for Agassiz's "law" of the parallel between the "geological succession of extinct forms" and "the embryological development of recent forms." He "fully expected" to see it confirmed: "For this doctrine of Agassiz accords well with the theory of natural selection." Indeed, so strongly predisposed toward Agassiz's law was Darwin that, he argued, even if it could never be verified because of the great difficulty of securing fossil evidence from the very early stages of vertebrate history, it still "may be true." If it were to be proven, then it would make "intelligible" the "vague yet ill-defined sentiment, felt by many palaeontologists, that organisation on the whole has progressed"; and, furthermore, that it had done so through natural selection, which ensured that the "inhabitants of each successive period in the world's history have beaten their predecessors in the race for life, and are, in so far, higher in the scale of nature."[59]

As Darwin represented it, Agassiz's law was both a consequence of and evidence for his theory of progressive evolution by natural selection and divergence. Agassiz's law thus was absolutely critical to Darwin's evolutionary theorizing, and Darwin's continuing resistance to Huxley's determined demolition of the same is explained.[60]

Darwin's principles of embryology and inheritance underpinned and explained the resemblance of embryos to other embryos and to ancestral adults through the agencies of natural selection and divergence. In the course of evolution, new variations would appear toward the end of individual development (ontogeny); natural selection, directed by divergence selection, would then tend to cause divergent variations to be selected. As these new variations were added on to the end of individual ontogenies, new divergent varieties would be produced. Over the course of time, under the continuous, simultaneous action of natural selection and divergence, these varieties (being incipient species) might result in, say, two new modified species. Each species would be able to trace back through the course of its individual development, past the point at which they diverged, their common progressive evolutionary history, or phylogeny. Divergence was not incompatible with progress.[61]

As I have stressed, a similar embryological argument underlay Darwin's theory of sexual selection. As he wrote, the development of the separate sexes followed from the same "division of labour, to use Milne Edwards's expression, [which] tends to perfection of every function."[62] Some characters developed in only one sex, usually the male, and they too usually developed late in life and were inherited at a correspondingly late stage. In those cases where the adult male was more conspicuous or colorful than the adult female, the young of both sexes closely resembled the adult female. New variations in color or decoration appeared at the end of individual male development, at sexual maturity. Through sexual selection—in this case, female choice—these new variations were selected and added on to the end of individual male development, and over time, the males and females would diverge more and more. A similar embryological explanation accounted for those secondary sexual characters—such as larger size, or weapons such as horns or spurs—that gave individual males the advantage in male combat for possession of the females. In both cases, the divergent sexual history was preserved in the different stages of development from the young to the adult male. It followed from this that the female, as Darwin had stressed in his early notebooks, was less modified and more "generic," or closer to the ancestral form of the species, than the male.

These are the embryological arguments fundamental to Darwin's three evolutionary agencies of natural selection, divergence, and sexual selection. They are patterned on the practices of breeders, female choice being the most closely aligned with the picking or selective practices of breeders, and further interconnected and explained by embryogenesis. Put in this way, they reveal their common origins, processes of theory building, and theoretical interconnections. Although Darwin explained them all in this way at one stage or another, at some points he equivocated; and, as I have remarked, he did not, at any stage, put and present them coherently together, developing and fine-tuning them at different stages of his writing. A related and perhaps more significant reason was that, as Darwin recognized, his embryological arguments were to be regarded as simplified, "default" positions or generalizations—generalizations that were fundamental to his evolutionary agencies, but that did not do full justice to their complexities and explanatory power.[63]

As Darwin fully appreciated, natural selection plus divergence had to explain not only those instances that conformed to this fundamental pattern of embryos resembling ancestors and other embryos, but also those instances that did not. As his barnacles had impressed on him, some embryos or larvae were free-living, and so subject to the pressures of natural selection and divergence at this early stage of development. Adults some-times were less complex than their larval stages. Ancestral stages might be omitted or drop out of development. Some embryos closely resembled their adult forms. All these exceptions to the rule required explanation through the same agencies of natural selection and divergence. And, over time, Dar-win and his embryologically oriented allies (notably Fritz Müller and Ernst Haeckel) did explain them. The ability of natural selection and divergence to explain both similarity and dissimilarity in embryonic structure became tri-umphant proof of its great superiority over Agassiz's dominant explanation of the Creator's preordained plan, which could not account for exceptions to the plan (see sec. 12.3).

Similarly, in the case of sexual selection, Darwin was aware that not all instances followed his basic pattern of similarity between the females and young and of male inheritance of secondary sexual characters. He also had to concede a measure of female inheritance, and by the time of the writing of the Descent, he identified not one, but six embryological rules that governed the complexities of resemblances between adults and young birds, and the timing and sequence of changes in individual development.[64] Neverthe-less, the dominant rule remained that of resemblance between females and young.

By 1871 Darwin could confirm this dominant pattern of resemblance throughout the animal kingdom. As he then acknowledged, he had "not elsewhere discussed this subject, and as it has an important bearing on sex-ual selection, I must here enter into lengthy and somewhat intricate details." The laws of inheritance were "extremely complex," as were the "causes which induce and govern variability." The variations thus caused were "preserved and accumulated by sexual selection," which was in itself an "extremely complex affair, depending, as it does, on ardour in love, courage, and the rivalry of the males, and on the powers of perception, taste, and will of the female." Sexual selection was also "dominated by natural selection for the general welfare of the species." Hence the ways in which individuals of either sex were affected "cannot fail to be complex in the highest degree." Yet all this complexity was made intelligible by the "general and remarkable coincidence between the period of variability and sexual selection."[65]

Put more succinctly than Darwin's rather circuitous discussion in the De-scent, by this time Darwin had identified two major "rules" of inheritance ef-fecting sexual difference: "Variations which first appear in either sex at a late period of life, tend to be developed in the same sex alone; whilst variations which appear early in life in either sex tend to be developed in both sexes." He was "first led to infer that a relation of this kind existed," Darwin wrote, from the fact that whenever and in whatever the adult male has come to differ from the adult female, he differs in the same manner from the young of both sexes. The generality of this fact is quite remarkable: it holds good with almost all mammals, bird, amphibians, and fishes; also with many crustaceans, spiders and some few insects . . . In all these cases the variations, through the accumulation of which the male acquires his proper masculine characters, must have occurred at a somewhat late period of life; otherwise the young males would have been similarly characterised; and conformably with our rule, they are transmitted to and developed in the adult males alone. When, on the other hand, the adult male closely resembles the young of both sexes (these, with rare exceptions, being alike), he generally resembles the adult female; and in most of these cases the variations through which the young and old acquired their present characters, probably occurred in conformity with our rule during youth.[66]

It was these understandings of inheritance and development—his utter, unshakable faith in them—and their fundamental relation to sexual selection that, for Darwin, confirmed the agency of female choice in birds and sorted the issue of the role of sexual selection versus natural selection (through resistance to disease) in the divergence of the human races. It was primarily the embryological argument that persuaded Darwin to introduce the significant sentence on sexual selection "of a particular kind" that could throw "some little light" on the contentious issue of the origin of the different human races in the *Origin of Species* (see sec. 11.7). It subsequently powered Darwin's dispute with Wallace over sexual selection (see ch.12), and above all, it instilled the necessary confidence and self-belief to write and publish *The Descent of Man, and Selection in Relation to Sex*.

Given these complexities—which he was still unraveling—Darwin's anxiety in the late 1850s to obtain corroborating evidence for his basic embryological generalization of resemblance between embryos and of embryo-ancestor resemblance through the inheritance of late-appearing selected variations becomes understandable, as does his predilection toward Agassiz's "law" and his vulnerability to Huxley's criticisms of the same.

9.5 Back to fashion and the family

It remains only to add that Darwin's penchant toward reading its ancestors into the embryo and young individuals carried over to his own children. As soon as William was born in 1839, Darwin commenced taking notes on his firstborn's physiognomy and development, observing his "animalcule of a son" like a new species of barnacle or some newly hatched pigeon. As more children arrived, they were added to the record, and Darwin, with Emma's assistance, built up a file of observations that he later published as "A Biographical Sketch of an Infant" (1877).[67]

The underlying theme of Darwin's observations was that his son's early developmental stages recapitulated the behavior and intelligence of lower animals, advancing into those of his savage ancestors while still very young. The infant William was afraid of the large caged carnivores in the Zoological Gardens, although he had never seen them before and had no knowledge of them, making Darwin suspect that "the vague but very real fears of children, which are quite independent of experience, are the inherited effects of real dangers and abject superstitions during ancient savage times." Never one to waste an observation, Darwin added that this was consistent with "what we know of the transmission of formerly well-developed characters, that they should appear at an early period of life, and afterwards disappear." The fears of his ancient adult forebears had been inherited by William and recapitulated in his development, but pushed back to an earlier juvenile stage by subsequent and more advanced inherited ancestral beliefs and behaviors, just as Darwin expected they should be.[68]

Darwin also took careful note of William's "observation on dress," which was "very curious." He recorded that William instantly noticed Emma's different boots and bonnets, her new hairstyle, and the artificial flowers that came from her cap. Gender differences in his children's development were of considerable significance. William was a precociously jealous male who flew into a passion when Darwin pretended affection for a doll in his presence; he also exhibited aggression from an early stage and became adept at throwing things at anyone who offended him. "I shall be curious," added Darwin when making this note, "to observe whether our little girls take so kindly to throwing things when so very young. If they do not, I shall believe it is hereditary in male sex, in the same manner as the S. American colts naturally amble from their parents having been trained."[69] Predictably, Darwin's daughters showed no signs of recapitulating such instinctive masculine traits, being more passive, gentle, and compassionate. Annie, by Darwin's account, was slower to exhibit reason than William and expressed her inherent feminine nature in her neatness and dexterity in manipulating small objects and in her love of feminine finery.

Boys and girls were inherently different; even highly civilized boys exhibited the habits and fears of their savage male forebears at an early developmental stage—behaving like so many miniature Fuegians, aggressive, throwing things, a prey to "abject superstitions," yet highly observant and liking bright, garish colors—before they advanced to a manlier, more civilized state and put on their breeches. Boys, inevitably, were more developmentally advanced than girls, who remained in petticoats all their lives; women were less intelligent and innovative than men, but compensated through their caring and more conformable natures. Women also, Darwin conceded, were more perceptive and intuitive than men; but this was, as Darwin could embryologically validate by the time of the *Descent*, "characteristic of the lower races, and therefore of a past and lower state of civilization."[70]

PART TWO

"For Beauty's Sake": The Making of Sexual Selection

TEN

Critical Years: From Pigeons to People

What effect w[oul]d idea of beauty have on races and selection. [I]t w[oul]d tend to add to each peculiarity. V. our aristocracy.

—Darwin's notes on *Types of Mankind*, around Christmas 1855

The period from 1856 to 1858 was critical. During these years Darwin moved to the definitive articulation of sexual selection. This only emerged after he actually set to work in mid-1856 drafting the chapters of his "big species book," the work that became the *Origin of Species*. By this time, Darwin had been pondering the question of species for nearly twenty years, marshaling enormous quantities of information from across the spectrum of natural history, breeding literature, and beyond, doggedly pursuing his experiments and inquiries, testing his arguments with a few trusted confidants. His scientific reputation was established, his intensive eight-year study of barnacles behind him, and the Royal Medal of the Royal Society—the "Philosophic Order of Knighthood"—secured. The time had come to publish.

Nevertheless, Darwin was cautious of putting his heterodox ideas before a large audience, anxious to make his theorizing and evidence as comprehensive and watertight as possible. He was confident of natural selection; less sure of his "secondary" principle of sexual selection, as yet incompletely formulated, the concept of female choice inconclusive, awaiting definition, yet to be linked with other aspects of his theorizing. But the need for it became more imperative as the act of writing focused his attention on the problem of how to account for those aspects of the living world that could not be explained by his primary agencies of natural selection aided by use-inheritance: sexual dimorphism and the more general but pressing issue of beauty. The beauty of the natural world lay within the largely unchallenged

realm of natural theology. Darwin, committed to its naturalistic interpretation, was compelled to find and promote a persuasive alternative to the
established creationist view that God had created the beauty of living things
for his own and human delight. To transfer such creative power to the eyes
and brain of a female bird was a hugely intimidating task. Little wonder that
the very sight of a peacock feather made Darwin sick.

Chapters 10 and 11 chart Darwin's moves as he worked to pull the composite strands of sexual selection together, to comprehend and formulate
the concept of female choice, and to integrate both female choice and male
combat with his views on embryology, inheritance, and his newly formed
principle of divergence. He had a number of strands to work with, some
semi-choate, some almost mature. His barnacle work had confirmed his
early notebook thesis that the separate sexes had evolved from ancestral
hermaphrodites and had further convinced Darwin that Agassiz was right
about the parallel between ontogeny and the fossil sequence, that—put in
evolutionary terms, rather than in Agassiz's creationist ones—embryological
development displayed the evolutionary history, or "phylogeny," of the
organism. Early in 1855 Darwin had taken up pigeon breeding with the
intention of confirming this conviction, while his close association with
breeders deepened his understanding of their art and his confidence in the
validity of transferring the practices of artificial selection to sexual selection.
In mid-1856, *Punch* initiated its campaign against "crinolineomania," ridiculing the expanding dimensions of the crinoline, bringing the excesses of
female fashion forcefully to public attention. The conditions were ripe for
Darwin's determination of female choice, which was in place by the time
the *Origin* was published.

He left no detailed record of his path to the *Origin* version of sexual selection. It has to be inferred from his extant notes and annotations. As previously indicated, the section on sexual selection in Darwin's draft of chapter
6 of his "big species" book (the vital chapter on his evolutionary agencies
of natural and sexual selection) is missing. Darwin completed the first draft
of this chapter on March 31, 1857, and revised it between April 14 and June
12, 1858. His manuscript table of contents of this revised draft of chapter 6
lists a section on "Sexual selection" from folios 7b to 7k.[1] We may assume
that these ten missing pages were inserted along with the new section on divergence (which is extant) during Darwin's revision. I have identified two of
these pages on sexual selection, thought lost, on the verso of two of a series
of pages of calculations of the sex ratios of lambs, puppies, and calves, made
while Darwin was at work on the *Descent*. Darwin's thrift is both a source of
frustration and a boon to the historian. As I shall show, these retrieved pages

resolve all doubt that, by this stage, Darwin finally had decided on female choice as the efficacious agency in sexual selection among birds.[2]

The extant pages of this same revised chapter 6 contain the first expression of Darwin's fashion metaphor, which introduces the section on divergence and which, as I emphasized in the previous chapter, links divergence with artificial selection and so with aesthetic selection, and ultimately with female choice. All this evidence goes to suggest that Darwin had the essentials of his theory of sexual selection, including female choice, in place by, at the latest, June 1858.[3]

But before this coalesced in his thinking, Darwin went back to where it all began: to those issues of race and aesthetics that had first triggered his notion of the role of different ideas of beauty in racial divergence. We know that Darwin initially had intended to include a section on "Theory Applied to Races of Man" in chapter 6 of his "big species" book. Sometime before his revision of June 1856, Darwin decided against this. All that remained in the *Origin* was his hint that the human races had originated through "sexual selection of a particular kind." Darwin's earliest known explicit reference to the notion of the cumulative action of male aesthetic choice in modifying human races was made around the end of 1855 on a slip of paper inserted into the back of a work notorious for its racist, proslavery agenda. This was *Types of Mankind* (1854) by Josiah Nott and George Gliddon. And Darwin's first datable use of his neologism "sexual selection" occurred in the notes he made early in March 1856 on another work that has accrued infamy for its racist interpretations, Robert Knox's *Races of Men* (1850). This was no coincidence.

The 1850s was a time of heightened racial consciousness, fomented by the expansion of empire, a series of historic conflicts, and the forging of a new sense of national identity that challenged and displaced the more humane views associated with evangelism and progressive liberalism.

10.1 The convergence of gender, ethnicity, and "race"

Thomas Carlyle's notorious "Occasional Discourse on the Negro Question" of 1849, republished in 1853 with the contemptuous substitution of "Nigger" for the more respectful "Negro," is symbolic of the incipient racism of this period.[4] It also symbolizes the convergence of gender and race that was as critical to Darwin's construction of sexual selection as his interlinked pigeon breeding and embryological understandings.

The "Discourse" pandered to the influential West Indian planter lobby that claimed to have been ruined by abolition. Carlyle satirized a "Black

Quashee" given to "ugliness, idleness, rebellion," a derogatory portrayal of West Indian blacks, sitting "with their beautiful muzzles up to their ears in pumpkins," who must be goaded to work by the "beneficent whip" if the sugar islands were not to be reduced to a squalid "tropical dog kennel." While British evangelicals were lauding the principle that black slaves were "men and brothers," Carlyle was intent on elevating the notion of "natural masters" and categorizing those races that refused to work as subhuman, a different species born to be mastered.

Carlyle's racist, proslavery diatribe revolted many, including Darwin. But it was the political economist and progressive liberal John Stuart Mill, fierce antagonist of slavery, who was provoked to scathing rebuttal. Mill feared the celebrated Carlyle's influence in the looming clash with the "owners of human flesh" across the Atlantic. He articulated the counterview that "one kind of human beings" is not "born servants to another kind." The differences between whites and blacks were not "original differences of nature," as Carlyle's "vulgar error" would have it. With education, blacks could rise to the civilized level of whites, just as, Mill subsequently argued, the differences between men and women were culturally induced and might be remedied by education.[5]

Mill's liberal environmentalist views were under siege in the general drift toward new definitions of race and gender as biologically based and innate. Carlyle's targets were less "poor Quashee" than those philanthropists and Utilitarians like Mill who promoted abolition. They were "windy sentimentalists" and "effeminate types," whose "unhappy wedlock" would "give birth to progenies and prodigies; dark extensive mooncalves, unnamable abortions, wide-coiled monstrosities, such as the world has not seen hitherto!"[6]

Carlyle's sexually laden invective bears on the wider construction of national identity. Historians agree that the racial theory of the mid- and late nineteenth century was largely a form of cultural self-definition. For Carlyle, men were self-made, manly, and independent, masters in their own house, necessarily dominant over their dependents—women, children, and servants (including black servants or, indeed, slaves). Men were born to command, women to obey, just as blacks had to be mastered by "red-blooded *men*" who knew how to dominate inferiors. Philanthropists and abolitionists were weak and feeble, too effeminate to wield the whip. "Quashee" was feminized, "a pretty kind of man" (Carlyle veers from the stereotype of the ugly black to his contemptuous, aestheticized, feminized dismissal), lazy but amenable, best compared, like women, with dogs.[7]

The analogy between slavery and women's subordination was a commonplace for liberal progressive thinkers, ranging from Mary Wollstonecraft to Mill. It informed Darwin's play on it in his relationship with Emma: in an inversion of the usual image, she was the master and he was her "nigger" or her "slave." Carlyle subverted this understanding. The comparison of woman's bondage with that of slaves fostered the derogation of both as dependent and childlike. In the interlinked social, racial, and anthropological hierarchies of the later Victorian period, the equation of woman-as-child-as-primitive supplanted earlier emancipatory rhetoric.

The 1850s saw the beginnings of the "first wave" of feminism as a small but vigorous circle of women, including those politicized through the radical Unitarians and the closely associated abolition movement, laid claim to rights and opportunities previously reserved for men. These predominantly middle-class women campaigned for access to higher education and the professions, for married women's property rights, for the reform of (male) sexual conduct, and eventually, for the vote. The tensions generated by these campaigns within the milieu of Victorian domestic ideology constituted the "woman question," which became the subject of intense interest and debate in all areas of society, including science and medicine. In the course of these concurrent debates, the woman question became indissolubly linked with the "negro question."[8]

The orthodox background to the convergence of gender and race was symbolized in the famous emancipationist and evangelical antislavery slogan "Am I not a man and a brother?"—extended by abolitionist women to encompass woman: "Am I not a woman and a sister?" The paradoxical conviction that slaves were brothers and sisters, yet not equal to their would-be liberators, was underpinned by the "cultural racism" that equated slaves and black people with children who required education to achieve civilized, Christianized adulthood. Women in the antislavery movement were particularly engaged with the moral issue of the family and, implicitly, with questions of sexuality and miscegenation. The separation of slave families, the breaking of maternal ties, and the sexual abuse of female slaves were the themes of the pietistic and phenomenally successful *Uncle Tom's Cabin* (1852), which sold over a million copies in Britain and reinvigorated the antislavery campaign. The patience, submissive spirit, and natural Christianity of the original Uncle Tom was the evangelical ideal with which actual black rebellion conflicted.[9]

As always, the imbrications are more complex. Beneath Carlyle's dismissive rhetoric of 1849 ran his deeper fear of black sexuality and his disgust at miscegenation that drew on the vocabulary of embryology and monstrous births made widely available by *Vestiges*, but reinvigorated by a new interpretation that played directly into the "negro question," one that also deployed embryos and monsters in a radically different view of racial origins. This was the race science of the above-mentioned Robert Knox. It was Knox who, in 1850, coined the notorious slogan that has come, more than any other, to symbolize the racism of the mid-nineteenth century: "Race is everything: literature, science, art—in a word, civilization, depends on it."[10]

In 1849 Carlyle's denunciation of the dominant abolitionist and emancipatory discourses was in the minority. By 1859 the popular perception of race and racial difference was undergoing significant change. This was the "Golden Age" of capitalism, when Britain consolidated her economic and political rule over various non-European peoples whose supposed backwardness and inferiority became justification for their suppression and subordination by white colonists and administrators. The claim of universal brotherhood and sisterhood increasingly gave way before the growing notion of imperial hierarchy. The assertion of empire was celebrated in the Great Exhibition of 1851 and its relocation and permanent establishment as the Crystal Palace Exhibition at Sydenham in 1854. These events brought unprecedented crowds to the Metropolis and, through their representation of British technological achievement and colonial possession, heightened public consciousness of racial and cultural difference and the superiority of British national identity. Even the prosaic Emma Darwin sobbed with emotion when Clara Novello sang "God Save the Queen" at Queen Victoria's opening of the Crystal Palace.[11]

The Crystal Palace was the setting for a new, popular display of ethnological tableaux. These were organized by Robert Gordon Latham, vice president of the Ethnological Society, to show the different peoples of the world and their cultures and customs in relation to the animals and plants of their particular geographic habitations. The human models were supposedly remarkably "life-like," great attention being paid to skin color, hair, features, and ornaments (though the young Alfred Russel Wallace, returned from his South American travels, thought them "ludicrous," having a "shabby and dilapidated" appearance, "so utterly unlike the clear, glossy, living skins of all savage peoples"). They were intended to instruct viewers in orthodox monogenetic theory—being billed as "different varieties of the human species"— but visitors and critics were free to make their own interpretations. Tellingly, the tableaux did not include models of British or European ethnic groups (the study of which was an acknowledged subdivision of British ethnology and the major focus of Knox's *Races of Men*), but the more exotic peoples of Asia, Africa, and the New World, who might be understood as lagging behind British moral and technological advances.[12]

Contemporary London ethnological showmanship also featured sideshows of living exotics—Zulus, San people, the so-called "Aztecque children," and Australian Aborigines. Just how readily prejudices might be con-

CRYSTAL PALACE—SOME VARIETIES OF THE HUMAN RACE.

Figure 10.1. Two fair, fashionable visitors to the Crystal Palace—pointedly more advanced in dress and manners—take their tea against a backdrop of "life-like" naked "savages" sporting nasal piercings and protruding lip ornaments. *Punch's Almanack for 1855*, viii. University of Sydney Library Collection.

firmed as much as wonder aroused or curiosity satisfied, is encapsulated in the varying reactions to the troupe of performing "Bushmen" [San people], brought to London from the Cape around the end of 1846. Thousands visited to gawp and marvel. Knox introduced them to a "crowded scientific and philosophical audience" at Exeter Hall. They were represented in the *Pictorial Times* as a "little above the monkey tribe, and scarcely better than the mere brutes of the field." For others, responding to the *Vestiges*, they "narrowed the gap" between monkey and man "amazingly." Charles Dickens, champion of the poor, registered his reaction in his 1853 satire on the "Noble Savage" in *Household Words*:

> I call him a savage, and I call a savage something desirable to be civilized off the face of the earth. . . . Think of the Bushmen . . . exhibited in England for some years. Are the majority of persons—who remember the horrid little leader of that party, in his festering bundle of hides, with his filth and his antipathy to water . . . and his odious eyes shadowed by his brutal hand, and his cry of "Qu-u-u-u-aa!" (Bosjeman for something desperately insulting I have no doubt)—conscious of an affectionate yearning towards that noble savage, or is it idiosyncratic in me to abhor, detest, abominate, and abjure him! . . . the world will be all the better when his place knows him no more.[13]

Dickens's facile mix of comedy and disgust is underpinned by his certainty that such odious beings would disappear through "civilization"; whether by improvement to the nonsavage standard or through extermination by their betters is unclear. Either way, they would cause no further offense to the eyes and ears or threaten the persons of their white superiors.

This complacency was challenged by the great rising of 1857–58 on the North Indian plains, known to British history as the "Indian Mutiny," which brought the issues of racial and colonial conflict to public attention as never before. The horrors and insults inflicted and imagined on British women and children shocked and enraged. Dickens on that occasion privately made clear what he meant by "civilization": were he commander in chief of India, he would do his "utmost to exterminate the Race upon whom the stain of the late cruelties rested . . . and raze it off the face of the earth."[14]

The cavalier references of Dickens to racial genocide, facetious or not, may be seen as the hyperbolic expression of a growing tendency to view race itself as the cause not just of physical difference, but of innate moral and intellectual differences that could not be "improved" by education and environment. This was not simply a Dickensian idiosyncrasy. Both Dickens and Carlyle were commenting in a context where "Bushmen" were so little regarded as human that "their heads were sometimes taken as trophies, stuffed and mounted, and their skins were collected and sold at auctions in Europe along with the skins of rabbits, hippopotamuses, and quaggas." The transformation of the skulls, heads, and skins into curiosities or collector's items (or, as was increasingly the case, into scientific objects in a museum) detached indigenous people from their contexts of conflict and resistance against white colonial incursions, from the subtleties and complexities of indigenous culture, and effectively dehumanized them.[15]

As a sign of the times, the "stuffed skin of a Hottentot woman" was said to be on exhibition in London in the 1850s; "a great curiosity, no doubt," scoffed Knox. This was the prized acquisition of Sir Andrew Smith, KCB (knighted in 1859 in the face of his thorough incompetence as director general of medical service in the Crimean War), physician, explorer, zoologist, ethnologist, and founder of the South African Museum, where this "Hottentot Woman of Smith's" was initially exhibited. On Smith's widely published authority, "Bushmen" were "deeply versed in deceit," masters of "cruelty . . . in its most shocking forms," repulsive in their eating habits, and prone to "laziness and . . . thieving." They were "pure Hottentots," not diluted by intercourse with other races, but were "approaching extinction" through an inability to change their "habits and conduct" and "the gradual extension of civilized life." Darwin had met with Smith at the Cape (where he possibly viewed the "Hottentot Woman") during the voyage of the Beagle and kept in touch with him after Smith's return to London. In 1857, indicative of his esteem, Darwin proposed Smith as a fellow of the Royal Society. It was to this "excellent authority" that Darwin, while at work on the Descent, looked for information on the much-debated issue of the conformation of the genitalia of Hottentot women.[16]

In these critical years, Darwin's sympathies were not engaged by colonial conflict or indigenous extinction, but by the issue of slavery that was polarizing opinion in the years before the American Civil War. As the friction among abolitionists, slave-owning Southerners, and the industrialized North veered toward open conflict, its implications for British identity and the British economy were significant issues in debate. Concurrently, in the new science of physical anthropology, a strengthening pluralist line challenged the conventional religion-dominated unitary theory of human racial origins. Pluralism hardened into "polygenism" (the neologisms polygenist and monogenist were coined in 1857) and scientific racism, in a context of increasing anticlericalism and secularism, allied in a number of leading instances with racist and proslavery apologetics.[17]

There was a strong class basis to Victorian racism: a hierarchical social structure fostered the view that reinforced the racial and gender superiority of the white English middle-class male. It was also a period when English workers still considered themselves in need of emancipation, and this affected working-class attitudes to the emancipation of blacks and middle-class women. But at base, racism was a means of rationalizing inequalities and justifying privileges in a society that—notionally, anyhow—subscribed to an egalitarian ideal: "Liberalism had no logical defense against equality and democracy, so the illogical barrier of race was erected: science itself, liberalism's trump card, could prove that men were *not* equal."[18]

If we add a gender dimension to this analysis, and consider the well-documented role of science and medicine in attempting to prove that women were not the equals of men in a period when women were beginning to lay claim to masculine prerogatives of liberty and opportunity, its explanatory power is significantly extended. "Race" and "sex" were "flash-points" for many social questions in mid-Victorian England. They were, above all, strategies by which the social and ideological contradictions at the core of English culture might be managed.[19]

They were strategies with implications for the liberal, humane, and gentlemanly Darwin's construction of sexual selection as he worked to pull it together and integrate it with his related conceptions of natural selection and divergence, with contemporary theories of embryology and inheritance, and with his pigeon observations and experiments.

10.2 Pigeons: "Watching them outside & . . . watch[ing] their insides"

Darwin began his collection of pigeons early in 1855 and entered the world of dealers, breeders, and fanciers. It was at this stage that he met the invaluable William Tegetmeier and that "very queer little fish" Bernard P. Brent. Within a year, Darwin had "almost every breed known in England." His daughters were enchanted, detecting different pigeon personalities to go with the extravagant fans, puffed-up bodices, or headdresses of the diverse breeds.[20] Darwin too enjoyed his foray into pigeon-keeping, but he had a grim scientific purpose: to pursue an intensive program of embryological research, aimed at showing that each of the common breeds of pigeon revealed traces of common descent from the wild rock dove. This required the study of their embryology and maturation from squab to adult.

In order for their bones to be measured and compared, newly hatched pigeons had to be killed and skeletonized, a prospect from which Darwin shrank. He recruited his cousin William Fox, a "goodnatured" fellow who would "kill Babies" and who kept a veritable "noah's ark" of domesticates:

> The chief point which I am & have been for years very curious about is to ascertain, whether the *young* of our domestic breeds differ as much from each other as do their parents, & I have no faith in anything short of actual measurement.[21]

Before long Darwin had overcome his scruples and set up his own "chamber of horrors."[22]

He also pestered a wide array of colonial officials, professional collectors, skinners, and medical men for pigeons from "each chief quarter of the world," specifying his precise requirements. Parcels of pigeon bones, feathers, and live birds converged on Down House—testimony to Darwin's growing scientific stature and networking skills—to add to those he was breeding and, reluctantly, slaughtering. It was a smelly, messy task, but essential to Darwin's aim of proving what he already believed to be the case: that all the fancy pigeons of the world had descended from the common rock dove. This was a controversial issue among fanciers; many, like "Mr. Dixon of Poultry notoriety," argued "stoutly for every variety being an aboriginal creation." Slowly and methodically, primarily on the basis of his accruing embryological evidence, Darwin began to construct a pigeon family tree.[23]

His pigeons, dead and alive, were beginning to give him the evidence he needed to withstand Huxley's onslaught on Agassiz's embryology. As Darwin explained in the section on the embryological evidence for evolution in the *Origin*, he "compared young pigeons of various breeds, within twelve hours after being hatched." He made careful measurements of the proportions of the beak (a major diagnostic feature for classification in birds), "width of mouth, length of nostril and of eyelid, size of feet and length of leg, in the wild stock, in pouters, fantails, runts, barbs, dragons, carriers, and tumblers." Some of these birds, when mature, "differ so extraordinarily in length and form of beak" that they would "be ranked in distinct genera, had they been natural productions." But the proportional differences in pigeon neonates "were incomparably less than in the full-grown birds." Some "characteristic points of difference . . . could hardly be detected in the young." There was, however, "one remarkable exception to this rule."[24] This was the young of the short-faced tumbler, which emerged from the egg with its adult proportions—including its extraordinarily short beak—intact. This was, in Darwin's terms, a definite monstrosity, where the peculiarity appeared early in development and was inherited at a correspondingly early stage, was deleterious to the animal, and could only be perpetuated with the assistance of the fancier.

Darwin also crossbred his pigeon breeds in violation of accepted practice, in order to show that the different breeds could interbreed and produce viable offspring that were "perfectly fertile," and therefore were all members of the same species. By back-breeding his pigeon crosses, Darwin produced individuals that showed some of the plumage characteristics of the ancestral rock pigeon. Furthermore, he was able to show that in some pigeons, the males and females had developed small but discernible differences—such as the more crenulated heads of the males of the English carrier, and the male pouter with its more pronounced pout.[25]

But it was Darwin's association with practicing pigeon breeders that gave him his most valuable insights into artificial selection in action. These fanciers might be "little" and "odd," but they could "pinch, bustle and crown birds, creating forms scarcely less marvelous than the changing female fashions" that Darwin encountered in London streets or chuckled over in the pages of *Punch*.[26] And they did so, not by crossing (as some assumed), nor by picking and breeding from sudden great mutations or "monstrosities" (though the short-faced tumbler may have originated in this way), but by selecting miniscule variations in mature birds—variations of feathering or beak or body shape and size so minute that only the long-trained eye could see and appreciate them—and subsequent breeding in-and-in. Fanciers gradually pushed the bodies of their living birds to extremes, inflating breasts and fans, shortening already shortened beaks, just as, Darwin began to see, fashion created and cultivated the extremes and excesses of female dress.

Pigeons, in effect, became Darwin's outstanding evidence for both natural and sexual selection. He constructed his divergent "genealogy" for domesticated pigeons primarily on beak size, but he shifted this diagnostic feature to the embryonic forms of the beaks.[27] This was yet more evidence of the classificatory significance of embryology. Above all, his pigeon experiments bolstered Darwin's confidence in his thesis that species-altering variations generally appeared late in development and were inherited at a correspondingly late stage of development, and so strengthened his resistance to Huxley's criticisms of Agassiz's law.

His pigeon family tree was still a work in progress when Darwin invited Charles Lyell to Down in early April 1856. But enough had been established for Lyell to be simultaneously impressed and horrified. He recoiled from a theory that "brings man into the same system of progressive evolution on which developed the orang out of an oyster." Evidently Darwin went the "whole orang," revealing his theory of natural selection and elatedly presenting his pigeon evidence—living, bottled, and skeletonized. "The young pigeons are more of the normal type than the old of each variety," Lyell noted in his journal. "Embryology, therefore leads to the opinion that you get nearer the type in going nearer to the foetal archetype & in like manner in Time we may get back nearer to the archetype of each genus & family & class." Had they been classified by an ornithologist in the wild, Darwin's pigeons would represent "three good genera and about fifteen good species." It seemed as though Darwin had a good case, and embryology was the clincher. Lyell was sufficiently up to the mark to appreciate the importance of the arch-progressionist Agassiz's views for Darwin's evolutionary interpretation of pigeon embryology: "Darwin thinks that Agassiz's embryology has something in it, or that the order of development in individuals & of similar types in time may be connected."[28]

10.3 Darwin, Agassiz, and the "American school" of race theorists

Lyell and Agassiz were old acquaintances. It was Lyell who had sponsored Agassiz's entrée into the American scientific and social worlds. The uniformitarian geologist disagreed profoundly with the global catastrophes invoked by the creationist Agassiz, with his progressionism and theory of multiple centers of creation. But the spiritual (and somewhat racist) Lyell did not dismiss Agassiz's views on the separate creation of the human races so cavalierly. He kept tabs on Agassiz's views and fed him tidbits of information: "As to Agassiz saying the negro's brain is like a child's fourteen or sixteen years old . . . Owen says the same of the adult male stolid and uneducated agricultural labourer. Tell Agassiz this, and see if it is new to him."[29]

Toward the end of 1855, Lyell, whose American travels kept him alert to the intensifying racial polemics, sent Darwin some pamphlets by the South Carolina parson and naturalist John Bachman, and these led Darwin to a key text in the maturing of his views on sexual selection: Nott and Gliddon's *Types of Mankind*. Bachman was a dedicated opponent of those blasphemous pluralists who denied the descent of the human races from Adam and Eve. He was also, as Darwin knew, an expert on breeding and domesticates. Following in the ethnological tradition of Prichard, Bachman drew on the analogy of domestication to argue that the human races were one species, just as the different breeds of fancy pigeons were capable of interbreeding and so demonstrably came from one ancestral type, the rock pigeon.[30]

Bachman was also a slave owner and anti-abolitionist, finding enough in the scriptures to endorse the special relationship of "masters" to their black "servants" within "our domestic institutions." He represents the difficulties of categorizing the respective positions of what were shortly to become known as "polygenists" and "monogenists" on the burning issue of Southern slavery. Theological apologists for slavery had no need of secular science to justify an institution they believed to have biblical warrant. Nor was polygenism invariably the scientific underwriter of slavery. For Bachman, what was at stake in his confrontation with those who outrageously categorized the human races as separate species with separate origins was neither their racism nor the issue of slavery, but biblical authority. Pluralists were promulgating doctrines contrary to "the laws of nature" and thus "injurious to morals" and "subversive of Christianity."

Darwin was not impressed by Bachman's theologically bound, waffling defenses of racial unity: "Shows what a good man will write when involved in controversy," he cautioned himself. He was well ahead of Bachman. Darwin could explain how fancy pigeons had come to diverge from the rock pigeon, and he was tracking an analogous explanation for the human races. His interest piqued, he sought out and purchased his own copy of the target of Bachman's pamphlets: *Types of Mankind*, which included an introductory "Sketch" by Agassiz.[31]

By the 1850s, the biblical doctrine of unity of descent and the Quaker philanthropy that underpinned Prichardian ethnology was beginning to seem more than a little shaky. A "new science" of anthropology was in the making, which formally eschewed biblical authority. It opposed the theological concern of the ethnologists to derive all human races from the one stock and advocated a strict anatomical methodology that placed great emphasis on describing, measuring, and classifying the physical "types" of humanity, forming rigid categories that maximized racial differences. Its more extreme exponents linked the doctrine that the human races were separate species with a racist defense of black inferiority and slavery. Among the more extreme were the authors of *Types of Mankind*. Nott and Gliddon were disciples of Samuel George Morton, the famous Philadelphia craniologist who took up where Lavater and phrenologists left off in "objectifying" a hierarchy of human races through physical characteristics of the brain, notably by its size.

Morton's measurements of cranial capacity established the hard "facts" that credentialed the "American school" of polygenists and fueled its anti-Prichardian rhetoric. According to Morton, there was no connection between climate or environment and race. American Indians were patently of one race (or rather, Morton thought, one separate, created species, or "primordial type"), from the Arctic fringes to Tierra del Fuego. Morton's brain measurements and subsequent ranking met "prejudiced expectations"—whites on top, Indians in the middle, and blacks on the bottom of the hierarchy. With this, the "Superiority of the Caucasian Race" and the "inferiority" of the "Negro" were mathematically demonstrable.[32]

Morton also entered the already heated dispute over hybridity. If the human races sprang from the one stock or species, their union would produce fertile "mongrels," "half-breeds," or "mixed-breeds," comparable with the mongrel offspring of the different breeds of dog; if, however, they represented separate species, their offspring should be sterile "hybrids." The interfertility of all human races was the strongest claim of the opponents of pluralism. But, countered pluralists, the intermixture of species was "unnatural" and might be perpetuated only under artificial conditions, although the "mulatto" (offspring of the union between white and black) was obviously fertile (even if only to a limited extent, many pluralists argued). Morton's tactic was to dissociate the standard definition of a species from its inability to produce fertile hybrids. Fertile "mulattos" were no proof against their parents being separate species; fertile animal hybrids were not as unusual as they were generally considered to be. The old definition must go: a species had to be redefined as a "primordial organic form." Thus Morton regarded certain breeds of dog as separate species on the grounds that their bones preserved in Egyptian tombs showed them to be as distinct from other breeds as they now are. The same might be claimed for those specimens of Caucasian and black humans mummified in the same tombs or depicted on their walls. This muddying of the waters was a boon to pluralists, and it back-footed the unitarists. If, as Prichardians like Bachman insisted, an analogy was to be drawn between domestic breeds of animals and the human races, then, as reinterpreted by Morton and his followers, it offered little support for human unity.[33]

Types of Mankind was a posthumous paean to Morton's work (he died in 1851), giving it a thoroughly racist spin and extra political heft in the growing antebellum conflict over slavery. Nott, a Southern slave owner and physician, was an active proselytizer for slavery; but he was also a committed secularist intent on proving "that the diversity of races must be accepted by Science as a *fact*, independently of theology." His targets were more often religionists like Bachman than abolitionists. Gliddon, an English-born Egyptologist and adventurer, had been United States consul at Cairo, using his position to plunder tombs and ship off skulls for Morton's cranial collection. These crania, together with the evidence of Gliddon's copies of hieroglyphs and ancient monuments, showed that "the Caucasian and Negro races were as perfectly distinct in that country upwards of three thousand years ago as they are now." Negroes were, and would remain, servants and slaves.[34] Together, Nott and Gliddon mounted a caustic and damaging attack on the dominant Prichardian unitarist doctrine. The effectiveness of this was greatly enhanced by Agassiz's endorsing "Sketch."

Agassiz's "Sketch" outlined his thesis that the different "types of man" had originated in different geographic areas of the earth in conjunction with "natural combinations of animals" circumscribed within the same "definite boundaries." Furthermore, it was *"beyond all question* that individuals of *distinct* species may, in certain cases, be productive with one another as well as with their own kind." Morton's was the only acceptable definition. According to Agassiz, human racial differences were "of the same kind and even greater than those upon which the anthropoid monkeys are considered as distinct species." He concluded with an endorsement of the pluralist position that "what are called human races, down to their specialization as nations, are distinct primordial forms of the type of man." However, although there was no genealogical connection among the races, there was a spiritual one. This was the higher moral stance denied to those who, "contrary to all the modern results of science," assumed the unity of the human races and thus promoted the godless "Lamarckian development theory, so well known in this country through the work entitled 'Vestiges of Creation.'"[35]

The notion that, in lending his support to polygenism, Agassiz was innocent of any intent of endorsing slavery or innate black inferiority, has been exploded. Agassiz had visited slave ports and plantations where he had women and men of different tribes degradingly stripped and daguerreotyped for anatomical study of their differences. He shared Nott's visceral horror of miscegenation and defended him against all criticism: "[Nott] is a man after my own heart." His scientific eminence gave the views of Nott and Gliddon credibility and played into contemporary politics of race and slavery. Equally well documented by Desmond and Moore is Darwin's passionate hatred of slavery and its apologists.[36] But Darwin's response to Agassiz's pluralism and to Nott and Gliddon's *Types of Mankind* was more complex than this might suggest.

Darwin was coming to identify Agassiz as the most formidable opponent of his as-yet-unpublished species theory. In America, Agassiz's stature stood as high in society as in science. He was a consummate networker of high ambition, talent, and determination who captured unprecedented public support and funding from private benefactors and government bodies. His theory of multiple centers of creation might be regarded as "bosh" by Darwin, but it was becoming the dominant view among American zoologists: "I seldom see a Zoological paper from N. America, without observing the impress of Agassiz's doctrines," Darwin moaned to Joseph Hooker. He urged Hooker and Asa Gray (Hooker's penfriend and Agassiz's colleague at Harvard, currently being cultivated by Darwin as a likely ally) to combat Agassiz's creationist views: "How awfully flat I shall feel, if when I get my notes together on species &c &c, the whole thing explodes like an empty puff-ball."[37]

Yet there were aspects of Agassiz's work that Darwin's own views might accommodate. He was more than happy to lean on Agassiz's cognitive and social authority in the matter of embryology to defend his theory of evolution by natural and sexual selection; but when it came to the issue of human and animal divergence, the same primary motivation of defending his own interests accounts for Darwin's eagerness to disparage and undermine Agassiz's threatening polygenist views. There was more to it than this. Darwin may have been utterly unsympathetic to its proslavery platform, but he was not altogether opposed to certain of the themes and claims in *Types of Mankind*.

10.4 Reading *Types of Mankind*: "What effect wd idea of beauty have on races in selection"

Darwin read *Types of Mankind* from cover to cover a few days before Christmas 1855, leaving a trail of notes and marginalia. If he was offended by its "white supremacist" content, Darwin's annotations do not show this. Nor did he register any disgust at its atrocious caricatures of Hottentots and other Africans. These were reproduced from woodcuts made by that "accomplished lady," Mrs. Gliddon, under the supervision of the authors, who were intent on making them as grotesquely like chimpanzees and orangutans as possible. Darwin's more forceful criticisms were reserved for Agassiz's "Sketch": "How false," "what forced reasoning!" He made marginal jabs at Agassiz's endorsement of Morton's "primordial organic forms" and a final expostulation at the outrageous inconsistencies of his arguments: "Look at same race in United States & S. America oh pro pudor Agassiz!"— meaning that the same race of Indians inhabited two of Agassiz's separately defined geographic zones, the temperate areas of North America and the tropical areas of Brazil. "Oh for shame Agassiz!"[38]

Darwin pasted an extended note into *Types of Mankind*. It is this note, when read in conjunction with Darwin's marginalia, that arrests attention. "As mere naturalist," Darwin noted, "*excepting* from blending of races . . . I sh[oul]d look at races of man as deserving to be called distinct species, yet I consider as descended from common stock, so come back at common belief; only difference is name whether to be called species or variations." On the basis of their crossbreeding and blending, humans constituted races or varieties that had descended from a common ancestor; but it made no difference to Darwin's theoretical views whether on the basis of their differences of color, physiognomy, and so on, they were called species or varieties. For Darwin, varieties were, after all, incipient species; what was at stake was the issue of their common descent.

It was just at this point, immediately after this note, that Darwin made the following riveting entry: "What effect w[oul]d idea of beauty have on races and selection. [I]t w[oul]d tend to add to each peculiarity. V. our aristocracy."

What, in this particular work, had provoked this sudden critical statement of the essence of racial divergence through sexual selection? Well, on the flip side of his inserted slip of paper, Darwin helpfully jotted a few clues:

> It will be quite necessary for me to state most strongly how impossible it is to guess the steps by which even var[ietie]s, as of human race (*or of Pigeons*) have attained their characteristics. . . . I am beginning to conclude that it is more difficult to account for small variations, as of man, when there is *no* adaptation than greater differences, when adaptation. . . . Nothing is more odd than similarity of Fuegian & Brazilian . . . I may contrast Man with Monkeys, for on my theory, the Monkeys have varied. [Darwin's emphases][39]

Pigeon varieties were the analogue of the human races; both were formed in the same way, through nonadaptive selection—that is, by Darwin's template of the fancier's selection and addition of tiny variations, though the intermediate steps were difficult to ascertain. Natural selection might account for greater adaptive differences, as in the different species of South American monkeys; but natural selection could not account for the sameness of the Fuegian and the Brazilian Indians, who occupied such radically different climates and whose habits of life also were so different. Darwin, in line with his developing theory of divergence, was by now beginning to appreciate that climate was not an essential factor in species change. The ranges of species were limited not by climate, but by other species.[40]

This meant that Darwin was in fundamental agreement with Nott and Gliddon's critique of the traditional environmentalist view that climate was the defining factor in the formation of race. Pencil in hand, he followed their arguments. Prichard, "the grand orthodox authority," scoffed Nott, had published no fewer than three editions of his *Physical History of Mankind*. "Alas! for his fame, Dr. Prichard had continued to change his costume with the fashion." Neither climate nor the inheritance of acquired characters—both invoked by Prichard at different stages—could account for any change in the human races. "Jews, Persians, Hindoos, Arabs" who had migrated to different climates and lived among different races had remained unchanged for more than a thousand years. The children of white-skinned Europeans in New Orleans and the West Indies remained as white as their parents. The only option left for advocates of the unity of the human species was to fall back on Prichard's earliest notion of "'*congenital*' varieties or peculiarities, which are said to spring up, and be transmitted from parent to child," like the often cited "porcupine" or six-fingered families, "so as to form new races." But did anyone ever hear of a race with porcupine skin or six fingers? To the contrary, such congenital varieties were "always swallowed up and lost." "Well argued," wrote Darwin beside this passage.[41]

Darwin had long dismissed the role of gross or deleterious mutations in species formation. He also agreed that there was no adaptive value in differences such as skin color when the same race of people occupied such different regions and climates. There had to be another explanation for the preservation of such nonadaptive racial differences, one that would account both for their selection and preservation, so that they should not be "swallowed up and lost" in the population at large. What Darwin had in hand was his pigeon analogy, along with his old notebook aesthetics and, of course, his well-worn copy of Lawrence's *Lectures*: the different races, like pigeon breeders, had different ideals of beauty, and it was this that directed and *added up* their sexual choices to form and preserve racial variation. This was an explanation consistent with Darwin's embryology, as I have been at pains to explain: "What effect w[oul]d idea of beauty have on races and selection. [I]t w[oul]d tend to add to each peculiarity."

But why had beauty again suggested itself to Darwin at just this juncture; and why so forcefully as to warrant its assertion in this particular context?

The part answer is that *Types of Mankind* preserved the venerable aesthetic/physiognomic hierarchy of racial beauty and, moreover, illustrated it, vigorously and persuasively. For instance, on the one page, the head of the *Apollo Belvedere* is juxtaposed to a "Greek" skull exhibiting the aesthetically endorsed facial angle of Camper; below them are depicted the markedly prognathous head and skull of a "Negro" and "Creole Negro"; below these are the head and skull of a decidedly less prognathous "Young Chimpanzee" (see fig. 10.2). Over the page, Mrs. Gliddon's woodcuts degenerate into the egregious caricatures of "Hottentots," "Negroes," chimpanzees, and orangutans. For comparative purposes, at the bottom of this page are two line drawings of a similarly prognathous "Negro" and a "Nubian," taken from Gliddon's sketches from the wall of an Egyptian tomb estimated to be "3200 years old."

These representations of a hierarchy of racial beauty that had persisted for as long as recorded time were supplemented with textual descriptions of racial "beauty" and "ugliness." Readers were invited to contrast the "manly and beautiful countenance" of the Egyptian pharaoh Rameses II, which would "not suffer by comparison with the finest Caucasian models," with the features of the defeated African Negros taken from the same tomb wall and to draw their own conclusions. The "repugnance" of the

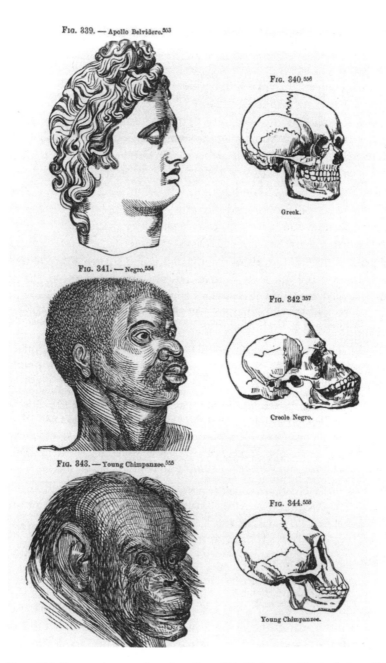

Figure 10.2. The notorious woodcuts from *Types of Mankind* executed by the "talented" Mrs. Gliddon under the supervision of the authors, Josiah Nott and George Gliddon. The profiles and skulls of a chimpanzee and a Negro lead up to those of the classical ideal of male beauty, the *Apollo Belvedere*, in illustration of Camper's facial angle and the representation of a visual hierarchy of racial beauty. Nott and Gliddon 1854, 141. University of Sydney Library Collection.

Egyptians toward "Negritian races" was a "solid fact in primeval history." Circassian crania show their "full share of manly beauty." "Western Africans, from Guinea to Congo . . . are ordinarily very ugly, and represent the purest Negro type," while the "lowest and most beastly specimens of mankind" are the "*Hottentot* and the *Bushman*," which are "but little removed, both in moral and physical characters, from the orang-outan."[42]

It was not only civilized men who were repelled by such racial ugliness; all "barbarous" or "primitive stocks" of humans, past and present, were characterized by their innate "repugnance towards other people," and it was this that inhibited their crossbreeding and preserved their "fixed" or "permanent type." Among the more civilized types, the Jews were "genealogically, perhaps, the purest race living," because of their religious proscription on intermarriage. Their instantly recognizable "type" or "indelible image" had remained fixed since the time of the "Egyptian Empire," traceable in their monuments and portraits, illustrating that "no causes are in operation which can transmute one type of man into another."[43]

Darwin marked this passage and another immediately preceding it where "Prof. Agassiz's researches in embryology" were invoked as having "most important bearings on the natural history of mankind." These had shown that in the fetal state, the anatomical structure of the Negro could not be distinguished even from that of the Teuton (the highest type of man in Morton's scale), but "after birth, they develop their respective characteristics in diverging lines, irrespective of climatic influences." This was a "most important law; and it points strongly to *specific* differences." Why should Negroes, Spaniards, and Anglo-Saxons, though the same in embryo, all diverge at birth and develop their specific differences even after ten generations in America, and the Jews, moreover, follow the same law even after 1,500 years?[44]

So, packaged within the one tome (albeit with an offensive proslavery agenda), Darwin might mark the old visual hierarchy of racial beauty, the assertion of racial repugnance to intermarriage, the historical persistence of the Jewish "type" through its prohibition on intermarriage, Agassiz's views on embryological resemblance and specific divergence, and the irrelevance of climate to racial divergence. More than this, in the course of its sustained attack on Prichardian environmentalism, *Types of Mankind* often called on the work of Robert Knox and the "very able book of Mr. Lawrence." William Lawrence was the obvious source of a particularly pertinent digression—in the midst of a discussion of racial interbreeding (deemed to be generally pernicious in its effects)—on the improving effects of intermarriages between the "impoverished nobles of the Norman stock" in England with "wealthy commoners of the homogeneous Saxon." Such intermarriages "reinvigorate the breed" and give it "more manly beauty," whereas the inbred European aristocracy includes "some of the most abject specimens of humanity anywhere to be found."[45]

In short, there was more than enough here to prompt Darwin's revealing allusion to the cumulative effects of the continuous selection of pretty commoners by "our aristocracy" in his mesmerizing note on racial aesthetic selection. Moreover, Darwin came well primed to *Types of Mankind*. A few months earlier he had reestablished contact with that "very clever, odd, wild fellow," Edward Blyth.

10.5 Reenter Blyth, the go-between

Darwin reopened his correspondence with Blyth with a clearly stated objective: to milk this expert on domesticates and their wild relatives for evidence and opinion on "the races of domestic animals . . . with reference to Ethnology." Surely Darwin had in mind Blyth's suggestive paper of 1835, where Blyth had drawn just such explicit comparisons. Blyth obligingly agreed that the study of domesticates could "throw some important lights" on the issue of the origin of the human races. From Calcutta, along with pigeon skins for Darwin's growing collection, poured forth an inexhaustible stream of information and references, in the course of which Blyth gave Darwin the assurance he wanted of his pigeon analogy: "The more I see of domestic Pigeons, the more obvious becomes *the conclusion* . . . of their having all descended from the *livia* [the common rock pigeon]."[46]

It was Blyth too who inspired Darwin to ransack old breeding manuals in order to trace the history of fancy breeds. Pigeon breeds had not remained constant but had undergone some remarkable changes over the previous few hundred years. The short-faced tumbler had not appeared until the early eighteenth century, its beak becoming even shorter since then, while the beak of the English carrier had lengthened over the same period. Pouters had developed more pout, while the hoods of jacobins had become yet more exaggerated. "Fashion goes in extremes with Fanciers," mused Darwin, leafing through one dusty old tome—an observation that was to be given increasing weight as he pulled together the threads of his transference of the breeders' art to natural and sexual selection.[47]

Just before Darwin read *Types of Mankind*, he took delivery of a stack of "Notes for Mr. Darwin" from Blyth that triggered the requisite cascade of memory, reading, rereading, and reasoning. In the sleepy heat of a Calcutta Sunday, Blyth had attempted a "sort of summary or recapitulation" of his

views on domestic breeds and the human races. The result was more a gar-
rulous ramble around the issues than a straightforward argument. Yet Blyth
scratched away industriously and assembled and displayed an absolute treasure trove for one with an eye to see and appreciate it. And Darwin most
certainly had that eye.

Blyth was especially interested in those strongly marked "abnormal" varieties or "breeds" that had been produced by *"direct human interference"*
from a *"casual monstrosity,"* such as the famous short-legged Ancon sheep.
It was Blyth's view that "extremes of climatal and local influence" had been
"immensely over-rated" in the production of the breeds of domestic fowls.
India had none peculiar to it because Hindus had "some absurd prejudice
against common poultry," whereas the "Chinese & Indo-Chinese are just the
very people to be taken with any outlandish monstrosity, & to try & propagate it as a curiosity, & to keep such a breed up by the selection of parents,
the China-men evincing a most especial taste for monsters of all kinds (witness their carvings)." Diminutive bantams were "much cherished" by the
Burmese, who went in for extremes by also favoring the "largest (& ugliest)
of all fowls, the Cochin Chinese." Europeans demonstrated their more cultivated taste by breeding beautiful gold- and silver-spangled birds: "We have
now the sum (pretty nearly) of the variation . . . fostered & carefully 'bred'
by a succession of 'fanciers' during so many centuries."[48]

If different cultural and ethnic preferences and breeding practices explained the origins and perpetuation of such different breeds of fowl or
pigeons, what of the human races? Blyth was doubtful that the "races of
mankind," which exhibited "aberrations of merely trivial import," fell into
this category. However, some human varieties did, such as the famous "Porcupine family & 6-digital folks" and "certain Hottentot peculiarities." "Suppose," speculated Blyth, "we were to select from the Bojesman [Bushman]
race, & continue the propagation from *the choicest specimens only*, for a few
generations! Could the result be doubted? Anymore than with mere albi-
nos." Beyond his reiteration of his old Lawrence-inspired views, Blyth was
now drawing on a new source of inspiration. Suddenly, he asked Darwin,
"Have you seen Knox's curious volume on the races of mankind?"[49]

Darwin had not. He had, however, read Knox's *Great Artists and Great
Anatomists* in October 1852. He now made a note to read the *Races of Men*—"a
curious book (Blyth)," but did not read it until March 1856. In the meantime, Blyth's suggestive notes touched off a train of thinking that began to
coalesce with Darwin's reading of *Types of Mankind*.[50] It was likely around
this time (December 1855) that Darwin reread Blyth's original 1835 paper,
which contained his views on the improving powers of male combat and on

the role of aesthetic preference in maintaining the separate human races along with his adjoining references to Lawrence's *Lectures*.[51]

Nevertheless, as stressed, Darwin's earliest datable use of his neologism "sexual selection" was made in his notes on Robert Knox's *Races of Men*. In a follow-up letter, Blyth again referred Darwin to "Dr. Knox's remarks on species": "Knox is an original thinker, and amusingly indisposed to hide his light under a bushel; moreover not a little opinionated on some matters, & few have better grounds on which to form opinions."[52]

Blyth was not the only one to commend Knox's views to Darwin. Darwin's interest in racial affairs was accelerating, and his reading lists for this period reference an accumulation of publications and articles by ethnologists, travelers, and missionaries on the diversity, distribution, and "natural history" of the human races. Just two weeks after he went through *Types of Mankind*, with its numerous references to Knox, Darwin read the Reverend Baden Powell's "excellent Essay on the Philosophy of Creation" (1855). Baden Powell positively reviewed "theories of the physical evolution, or origination, of new species in past epochs"; he made much of Knox as "one of the most zealous supporters of the principle of transmutation in this country" and offered an extract from the *Races of Men* in support of this claim.[53]

Finally, Darwin read the *Races of Men*, most likely a copy borrowed from the London Library. He made a series of notes on it and then went back and forth from these notes to his own copy of *Types of Mankind*.

10.6 "An original thinker": Robert Knox

Knox was a highly controversial figure, tarnished forever by his association in the public mind with the notorious Burke and Hare, body snatchers turned murderers, who in 1828 had sold their victims to Knox's then-thriving anatomy school in Edinburgh. He was, moreover, a "savage radical" whose materialistic, anti-theological polemics won him few but the wrong kinds of hangers-on. After his death, his idiosyncratic views were turned to the service of a vehemently racist and proslavery agenda that Knox had never endorsed in his own lifetime.

Knox graduated in medicine from Edinburgh University in 1814. He saw service as an army surgeon at Waterloo and then at the Cape of Good Hope, where he experienced at first hand the bloody frontier conflicts involving the British, the Boers, and the Xhosa. His interest in human variety stemmed from this period, as did his entrenched anticolonialism. Knox next studied pathological anatomy in Paris from 1821 to 1822, where he came under the sway of the new transformist, transcendental anatomy practiced by Étienne Geoffroy Saint-Hilaire. Knox's Parisian acquaintances included William Frederick Edwards (brother of Henri, the zoologist), who was to become a leading proponent of the permanence of human "types" and founder of the Paris Ethnological Society; Samuel George Morton, future pluralist; and the Quaker abolitionist Thomas Hodgkin, founder of the Aborigines Protection Society. All were to play primary roles in the coming debates on racial theory.[54]

By 1826, when the young Darwin was in Edinburgh studying medicine, Knox was on his meteoric rise, the best anatomist in Edinburgh, if not in the whole of Britain. He cut an extraordinary figure—he had only one eye and a pox-ravaged face—with a dandy's taste for frilled shirts, enormously high collars, diamond rings, gold chains, and puce coats. The dazzling Knox snatched paying students from the moribund university anatomy classes to his flourishing extramural school and enthralled audiences with his Saturday morning public lectures on "Comparative and General Anatomy and Ethnology." At this stage Knox seems to have been a Prichardian environmentalist, a position he later renounced.[55]

Knox's views underwent radical change as his fortunes shifted after the Burke and Hare affair. He was forced to give up his dwindling and unprofitable anatomy school, leaving Edinburgh in 1842. Embittered and nihilistic, he scratched a precarious living from hack journalism and public lecturing, dying in London in 1862. It was during this latter period that Knox elaborated and published his "moral anatomy," or race science, which incorporated his radical anticolonial politics, his materialist version of transcendental anatomy, and his related antiprogressive evolutionary views.[56] The Knox that Darwin read in early 1856 was very different from the one he well may have heard lecture at the height of his rhetorical powers thirty years earlier in Edinburgh.

Knox's later writings are fragmentary and not readily integrated into a coherent body of work. His publications ranged over topics from fishing to prostitution, which a fickle public might find of interest; and where Knox wrote within his own expertise, he subordinated theory to reaching a wider popular audience. With his public lectures on race in Newcastle, Manchester, and Birmingham in the late 1840s, Knox recognized that he had struck a major chord, and he repeatedly played on such a marketable theme. His lectures were published verbatim in the *Medical Times*, then collected into his major ethnological work, *The Races of Men: A Fragment* (a title conceding its fractured structure). Nevertheless, Knox's *Races* (which went through two authorized editions in his lifetime, 1850 and 1862, plus a cheap transatlantic

Figure 10.3. Robert Knox, onetime dandy and celebrated anatomist, newly down-at-heel race theorist and hack journalist, lecturing on the Bushmen Family in London in 1847. Contrary to accepted opinion, Knox was an abolitionist, anticolonialist, evolutionist, and monogenist. After his death, Knox's idiosyncratic theories of racial origins and inevitable antagonism and conflict were pressed into the service of the proslavery and imperialist Anthropological Society of London and became identified with its anti-Darwinian polygenist platform. *Pictorial Times*, June 12, 1847, 376. National Library of Australia Collection.

spinoff of the 1850 edition) represents a particularly vivid dramatization of racial difference, which Knox grounded in material organic form and presented as central to the understanding of history, art, literature, religion, laws, and morality. It also brings to the fore the essential aesthetic dimension to this understanding, as well as its unequivocal location in the body and form of woman.

Where Nott and Gliddon's racial hierarchy held to the traditional aesthetic of masculine beauty, Knox was a primary exponent of the shift to the centrality of feminine beauty in the determination of racial aesthetics, a shift from the embodiment of classic ideal beauty in the *Apollo Belvedere* to that of the *Venus de' Medici*—the same embodiment of ideal beauty in Darwin's writing. In Knox's case, this feminine ideal was given greater cogency and immediacy through its realization in the faces and forms of certain contemporary "beauties" to be found on the streets of London. In short,

Races of Men brought before the British public and into Darwin's study the pivotal recognition of female form in the analysis of racial difference, the recognition fundamental to Darwin's human-centered theory of sexual selection.

The generalized congruence between Darwinian and Knoxian conceptions of race, remarked on by a number of historians, is attributable less to the direct influence of Knox's *Races of Men* on Darwin than to the common context of the genesis and institutionalization of their racial views. Both eschewed theological and environmental explanations of racial difference; both drew on the notion of the human races as discrete biological units with distinct moral and mental traits; and both played heavily on the themes of intraracial sexual attraction and interracial repulsion and conflict.[57] When we add to this known mix the extra dimension of a woman-centered aesthetics, their points of agreement (as well as disagreement) are brought into sharper focus.

Knox was greatly influenced by the views of the radical anatomist Alexander Walker, whose *Intermarriage* had so interested Darwin in 1839. Knox's former pupil and earliest biographer, Henry Lonsdale, viewed Walker's writings as a major inspiration for Knox's race science. Not surprisingly, Walker and Knox were acquainted. When Walker, old and ill, appealed to Sir Robert Peel in 1849 for a government pension in recognition of his contributions to physiology, he enclosed a letter of the highest esteem from Knox (much good it did him): "No one has thought more clearly on the great physiological questions than you have." Just before Knox died, he was in the process of reediting Walker's *Intermarriage* and was making notes for this project.[58] He adapted Walker's views on racial hybridity and his antiprogressionism to his own theory of race. Knox was as much influenced by physiognomy, with its emphasis on external form, and like Walker, he authored several works on art and anatomy. Aesthetics thus played a central role in Knox's racial theory, and the concluding section of *Races of Men* is devoted to the "theory of the beautiful." There, Knox insisted on the necessary connection between the aesthetic evaluation of the beautiful and the scientific analysis of racial difference and human history, and (what is of particular interest to Darwin's construction of sexual selection) the centrality of female form to this.

In essence, Knox was intent on reducing all social and biological phenomena to the basic biological category of race. In this sense, he was profoundly racist. His infamous and much misconstrued principle—"Race is everything"—summed up his conviction that human history could only be studied through the application of biological method: "The basis of the view I take of man is his Physical structure . . . his Zoological history."

According to Knox, the human races were so different and distinct that they were entitled to the name of species: "call them Species, if you will; call them permanent Varieties; it matters not." Any anthropological or political theory that did not take this fundamental principle of innate and ineradicable racial difference into account was "unscientific": "Wild, visionary and pitiable theories have been offered respecting the colour of the black man, as if he differed only in colour from the white races; but he differs in everything as much as in colour. He is no more a white man than an ass is a horse or a zebra."[59]

For Knox, the materialist, the "mind of the race, instinctive and reasoning, naturally differs in correspondence with the organization." Human character, intellect, and morality were neither divinely induced nor environmentally produced, but rooted in the "all-pervading, unalterable, physical character of race." By maximizing these supposedly fixed racial differences, Knox was able to construct an elaborate racial history of Europe and her colonies. What was previously construed as national conflict was better understood as racial conflict, as each race sought to dominate its own geographic locale and erect its own government and civilization in keeping with its own distinctive nature.[60]

If contemporary racism was a factor in the construction of national identity, of white male British superiority, as is generally agreed, then Knox's racism does not fit easily within this framework. In placing anticolonialism on a biological footing, Knox's race science was antithetical to the mounting jingoism of the period. Each race was suited to its particular original environment and climate—each was "perfect in its own way"—so the attempts by Europeans to colonize the tropical world were futile. They could not long survive in a place for which the Saxon or Celt was not physically or mentally fit. Nor, being the equivalent of a species, could their biology be altered, either by environmentally induced change or through hybridization with an indigenous race. To this end, Knox invoked both the "innate dislike of race to race, preventing a renewal of such intermarriages" and Walker's law of reversion: human hybrids were a "monstrosity of nature" unable to "hold their ground" for more than a few generations, a claim eagerly snapped up by Nott, who cited Knox in the opening pages of *Types of Mankind*. But Knox applied this in a way utterly opposed to Nott's white supremacist views: sooner or later, Knox argued, natural law inevitably must assert its effects and the colonizers would be exterminated, either through their inability to adapt to their "unnatural" environment or through "natural" and inevitable racial conflict. Europeans might enslave indigenous peoples, take their lands, or massacre them wholesale (unlike Darwin, Knox was in no doubt as to who was responsible for the widely predicted extinction of indigenes), but they could only temporarily repress the inevitable struggle of race against race. Ultimately they would be defeated by the tropically adapted, energetic, and fierce Negro: "From St. Domingo he drove out the Celt; from Jamaica he will expel the Saxon; and the expulsion of the Lusitanian from Brazil, by the Negro, is merely a matter of time."[61]

Race was, in Knox's hands, an insurmountable barrier to nation building and imperialist might, to the "progress" of the so-called civilized. However, he took an equally pessimistic view of indigenous prospects for improvement. If Europeans could not colonize the tropics, neither could the "dark races" be civilized. The Christian principle "love thy neighbor" was at odds with the intractability of racial aggression. For all the efforts of misguided philanthropists, the "dark races of men" could not change their anatomy and interrelated moral and mental capacities, which were inferior to the European. Philanthropic protectionism was mere pious cant, compromised by the hypocritical denial of the innate antipathy of the European toward men of color, exemplified by the refusal to recognize their rights as men.[62]

Much historical disapprobation has focused on Knox's "offensive" and "defamatory" pronouncements on the "dark races," especially his reiterated claim that the black races could never be civilized (an assertion capitalized on by the anti-emancipation lobby in the face of Knox's consistent anti-slavery/anticolonial stance).[63] Knox was hardly immune from the racial stereotyping of the period. Yet he took a more sympathetic view of indigenous Africans than, for instance, did Andrew Smith (who had succeeded Knox as army surgeon at the Cape), or for that matter than Darwin did of the Fuegians. Left to themselves, the "Hottentot" and "Caffre" lived "in harmony with all around" them. Knox abhorred slavery as greatly as Darwin and was outspoken in his condemnation of its contemporary defenders and practitioners. The Americans, with their vaunted cries of liberty, were the greater hypocrites: "The rights of men is a phrase forever in their mouths; by men we now know they mean white men." Moreover, Knox's "abuse," as his contemporaries complained, was directed more often at those European "races" that crowded the metropolis and whose racial distinctions he represented as the proper subject matter for anthropological inquiry and political policy.[64] No one escaped: Celts were "lazy, worthless"; Gypsies were "without a redeeming quality"; the Jew had "no ear for music . . . no love of science or literature . . . invents nothing." But nor did "low and boorish" Saxons have any musical ear, talent for art, invention, or "genius." They were hard working and lovers of liberty, but given to great tyranny in their dealings with other races. Only "Slavonians," the South German originators of the highest transcendental science, were spared Knox's scalpel. Little wonder that the *Christian Examiner* labeled Knox a "boastful dogmatist, eaten up with German nonsense and his own conceit, uttering in snatches and riddles a theory of materialism and despair."[65]

To add to the confusion, Knox commonly has been characterized as a pluralist or polygenist who denied the fundamental unity of races. This is yet another misinterpretation, stemming from Knox's insistence on the human races—or rather, "species"—as immutable biological entities and his anti-transmutation polemics.[66] Paradoxical as it may seem, Knox subscribed to a theory of organic development, or "evolution," as both Blyth and Baden Powell recognized. Darwin also initially read Knox as an evolutionist, but had him tagged as one who considered Geoffroy Saint-Hilaire as the "great solver of the problem" of the origin of species.[67]

Darwin was right, up to a point. Knox certainly was inspired by his "illustrious" French mentor, arguing by analogy from the production of monsters to the production of species, but there were important differences. Knox repeatedly denied any progressive, unilineal "transmutation of species, the one into the other," as Geoffroy and his British followers had assumed. For Knox, new species were not produced by gradual, progressive change in the mature organism (i.e., by transmutation, or by Lamarckian inheritance of acquired characters), but by gross embryonic (or "generic") change.[68] Species, as in von Baer's law, diverged from a common embryo. In this sense they were immutable, being "fixed" for all time in the embryo, and they did not emerge in any necessarily progressive way. Knox's theory of "generic descent" supposed the embryos of all members of the same natural family or genus to contain within themselves all the "characters," or incipient structures, of all possible different species of that genus. New species arose through the action of the "law of deformation," which opposed the "law of specialization [or speciation or individualization]" and therefore suppressed "some parts of the organ or apparatus already existing in the generic being [or embryo]." These "deformations," or monstrosities, were constantly generated. Those that were not "viable" perished; those that were compatible with existing geographic and geologic conditions reproduced and increased in number, and so a new species was established. The human races were the result of such monstrous change.[69] Although Knox usually confined his evolutionary theorizing to a particular genus or family, he assumed a common hereditary descent for the different genera, including humanity. The human embryo was a microcosm of all life, "worm, mollusk and fish," and humanity was consanguineous with all other animals that have lived or may live; most explicitly: "Mankind is of one family, one origin. In every embryo is the type of all the races of men."[70]

Knox was indisputably a monogenist, a member of the school that Gliddon was shortly to denounce as "professing to sustain dogmatically the unity . . . of human races."[71] However, his theory of evolution (which rejected gradual, continuous organic change in favor of radical nondirectional change—the abrupt nonlinear embryogenesis of new species—with unchanging persistence of species over long steady-state intervals), was not a theory with which Darwin, the committed gradualist, could concur. Nor had Darwin any patience with Knox's affirmation of inherent transcendental laws of deformation and specialization that determined the course of embryological development and the origin of a new species; though he took note of Knox's claim that the human embryo contained within itself all the species or races—extinct, extant, and future.[72] Darwin was in the process of articulating his own explanation of embryological development consistent with his theories of natural and sexual selection and divergence, patterned on his analogy of artificial selection—an analogy that Knox, intent on materializing the inner transcendental laws of development, totally rejected.

Rather, it was Knox's instances of racial stability and his aesthetically inflected views on race and gender that caught, and held, Darwin's attention.

10.7 Knoxian aesthetics and Darwinian "sexual selection"

Aesthetics enters into the Knoxian scheme through his embryological law of specialization, which is ever striving to achieve perfection. This leads to the beautiful particularity of form that individual living women sometimes display and which reached its apotheosis in the people of classical Greece. Their sculptors had produced the most beautiful form on earth, "the Venus, a real, not an ideal form." The "mysterious and wonderful secret" of the voluptuous Venus was that it hides the ugly interior and exemplifies the beauty of form or exteriority: "In the exterior, beauty resides—that alone she decorates—all within is frightful and appalling to human sense." It is the ugly interior that reveals the unity of organization that links humanity to all living beings and is manifested in the sequence of embryonic development. The "numerous metamorphoses" of the human embryo "embrace the entire range of organic life, from the beginning to the end of time," including the forms of fossil animals, lower animals, and "inferior races of men." This is why children for all their youthful appeal cannot be called beautiful: their forms and features tend toward brute or embryonic forms. The old and infirm display the "unseemly" and ugly skeletal interior through their scant flesh. The non-white races "exhibit the outline of the interior" more strongly than the "fair races generally." Landscape and the picturesque do not satisfy the human *eye for form*." It is only in those humans where the law of specialization, which gives exterior form, overcomes the deforming tendencies of unity of organization that true beauty results and can be perceived and appreciated by men of taste: "The human *form* alone satisfies the *human mind*"; it connects the "history of race with the perfect."[73]

It is an "aesthetics of surface and appearance" that imbues all Knox's transcendental-inspired, yet thoroughly materialistic, race science. If the human mind is to comprehend human history and evolution, then it is to outward form—the visual signs of nature's perfection—that humans must look. Biological form is the lens through which the origin of races and species must be viewed: their outward form is where "we must look for the more remarkable characteristics of animals; it is it alone which nature loves to decorate and to vary." Knox thereby linked his race theory—through the concept of form and the importance of exteriority—to a universal standard of beauty, which he located in the body and face of "woman." Thus it is that woman's external form is the "only absolutely beautiful object on earth," the "highest manifestation of abstract life, clothed in physical form, adapted to the corresponding minds of her race and species." While this may seem to suggest that female beauty is racially relative and an abstraction or idealization, Knox was adamant that perfect beauty is both universal and materially manifested: "The correct mind rejects everything which is ideal, or what never had an existence. The monstrous creations of the disordered Hindoo, Chinese and Saxon minds; these are ideal, fictitious, false; the Venus is real." Perfect beauty was not only manifested in the classical Venus, but also in the ancient Greeks who perfected her and on whom she was modeled. Yet this universal female beauty was not specific to any particular race: the classical Greeks constituted "a *mixed race*, an anomaly on earth, a thing repudiated by man and animals,"—persons, not a race, "matchless and perfectly beautiful," since disappeared. Indeed, according to Knox, perfect beauty, displaying the classical Grecian facial angle, was to be encountered on the streets of London, where mixed-race females abounded.[74]

So Knox, who argued that a hybrid population could not be designated a race, that it was unviable, a "monstrosity of nature," yet localized perfect universal beauty in a hybrid or mixed race. Similarly, Walker had claimed that crosses between the different human races were often physically or mentally superior to their parent races, yet over time invariably reverted to the original races. Hence all utopian visions of human perfection were impossible. Here Walker and Knox agreed. Both assigned a positive aesthetic value to racial amalgamation, but viewed it as an ultimately unattainable ideal. Both located perfect beauty in woman's form, but Knox dissociated form from function. For Knox, as for Darwin, beauty was functionless, pure form that was materially realized in the exterior shapes and characteristics of humans and animals.[75]

Nevertheless, a racial hierarchy of beauty underpinned Knox's race science. While he poured scorn on the notion of Saxon moral and physical superiority, he was equally certain that among existing races it was by and large the Saxon race that exhibited superior beauty—it was the fairest of all races, demonstrating the combination of flaxen hair and blue eyes that constituted the best beauty—and it was probably the admixture of Saxon blood that was responsible for the perfect beauty of the ancient Greeks and those beauties walking the London streets. Women of other races were scaled to lower levels of the hierarchy: not even the admirable Slavonians might be called beautiful; "Gypsies" (Roma) "have not the elements of beauty"; the Jewish face, while it sometimes appeared of "transcendent beauty . . . will not stand a long and searching glance," because "then the want of proportion becomes more apparent."[76]

These two latter "races" particularly interested Knox (and Darwin as note-taker), as they had, according to Knox, persisted unchanged over long intervals of time and in very different climates, simply because women of pure Gypsy or Jewish blood would not intermarry with or take lovers from other races. This was the "grand secret" of racial permanency. It was an innate, natural aversion, similar to the "same horror that the Saxon has for the Negro." It had nothing to do with cultural or religious prohibitions, as Gypsy and Jewish men "had no such aversion to the Saxon fair"; as a result, the Jewish race in Britain was much intermixed. The pure strain of Jewish blood stretched from the time of the "Young Memnon" (the Egyptian pharaoh whose bust in the British Museum Knox identified—and illustrated several times over—as having the distinguishing Jewish physiognomy) to the Jewish quarter of Rotterdam—where Knox had gone "almost on purpose," Darwin noted, "& found perfect likeness of Young Memnon—but Nott & Gliddon . . . say a hybrid race, see to this. see to this" (a double reminder, stressing its significance to Darwin).[77]

See to this Darwin did, taking down his copy of *Types of Mankind* to crosscheck the reference, and even visiting the British Museum to check out the profile of the Young Memnon for himself.[78] But the more significant and more immediate impact on Darwin is there in the extended rumination he appended to his notes on Knox's *Races of Men*.

Darwin had to admit that "on direct evidence it is doubtful whether there is any good evidence of change in races of men within historical period." But it "can be deduced from indirect evidence, as with the races of domestic animals & plants, & species generally." The breeds of pigeons *had* changed over the last several hundred years, and Darwin supposed that "this analogy must in man hold good" with all that followed from it as to variation, preservation of some, and extinction of others. "Ask Andrew Smith," Darwin's expert on race, about Knox's surprising claims about the unchanging Gypsies; yet Smith himself, by Knox's account, had documented no change in the complexion of the descendants of white settlers after three hundred years at the Cape, and other such instances had been recorded by Nott and Gliddon. Knox did claim, and Darwin pounced on it, that tall men in Britain tended to die early of pulmonary consumption—hence the greater mortality of the Foot Guard—but when sent to fine climates, "as the Cape & Australia such persons live readily," and their descendants "seem to become a tall race in these latter countries." Darwin transcribed, word for word, Knox's explanation for this: "But this arises merely from the circumstance that the tall children, who wd. die in Europe survive at the Cape & in Australia."[79]

Pondering all this, Darwin concluded his notes on the *Races of Men*: "Fuegians & Brazil, climate & habits of life so different good instance of how fixed races are, in face of very different external conditions. The slowness of any changes explained by constitutions selection & sexual selection."[80] This meant that human racial evolution was brought under the same natural agencies as animal evolution, with the proviso that racial divergence was not only independent of climate but also of habits of life: it was an extremely slow process governed by natural selection (through resistance to disease, e.g., Knox's instance of the Foot Guard), and, above all, by "sexual selection" (Darwin's new-coined terminology, here denoting the aesthetically determined choice of partners of the same race and the innate antipathy to interracial mating).

There were enough sparks in Knox's *Races of Men* to jump at Darwin from the center of a blaze that was already well lit, both in Darwin's long-term theorizing and in the literature at large.

10.8 The intersection of beauty, sexuality, and race

Others around this time, besides Knox and Darwin, were drawing on the intersection of aestheticized categories of gender and race in the construction of a hierarchy of beauty, in which white European women were elevated to special status as the living embodiment of classical beauty, while women of other races were scaled to lower levels of the hierarchy.

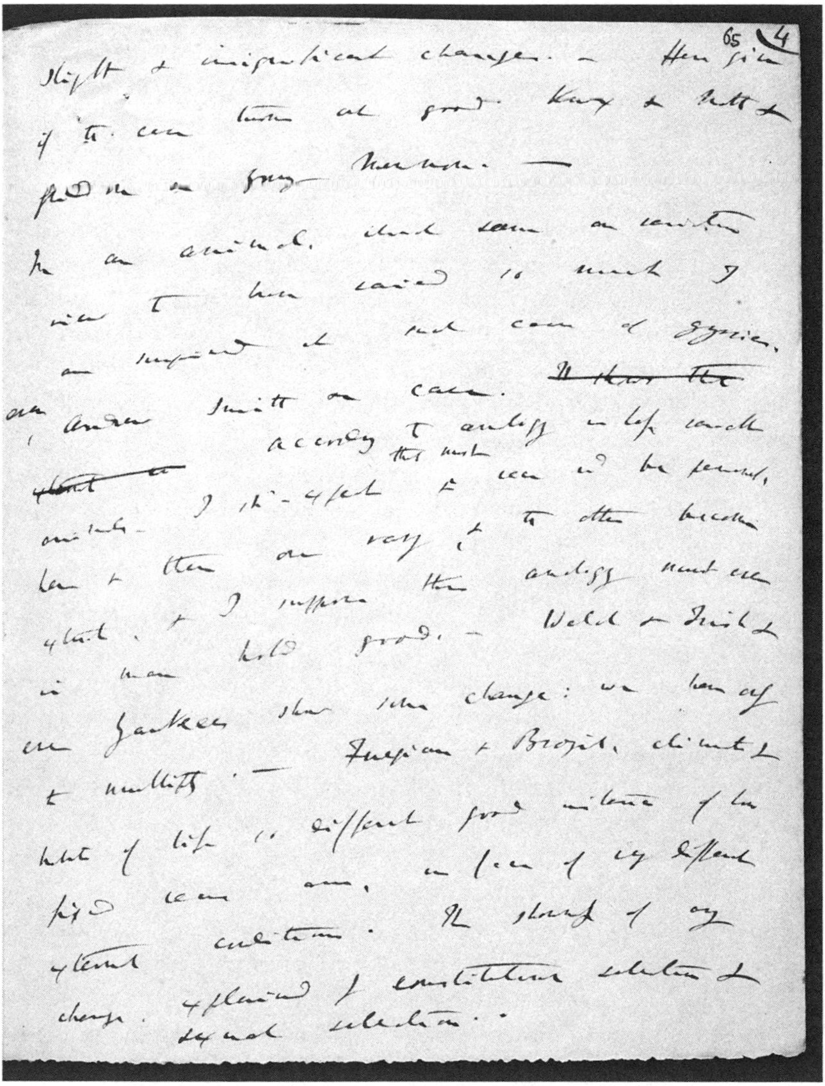

Figure 10.4. Darwin's earliest dateable use of his neologism *sexual selection* on the final page of his notes on Knox's *Races of Men*, made early in 1856. His conclusion reads: "Fuegians & Brazil, climate & habits of life so different good instance of how fixed races are, in face of very different external conditions. The slowness of any changes explained by constitutions selection [through resistance to disease] & sexual selection." DAR 71:65. By permission of the Syndics of Cambridge University Library.

In 1848, William Frederick Van Amringe, a New York lawyer, had produced a bloated tome with a bloated title: *An Investigation of the Theories of
the Natural History of Man, by Lawrence, Prichard, and Others, Founded upon
Animal Analogies: And an Outline of a New Natural History of Man, Founded
upon History, Anatomy, Physiology, and Human Analogies*. His intention was to
undermine the conventional analogical method of Lawrence and Prichard
that founded the unity of the human races on the basis of animal comparisons. This was a dangerous practice that could lead to victory only for
the "progressive developists" who would evolve humans from animals.
The natural history of the human races must be placed on "entirely a new
basis," the "psychical character of mankind," which focused on moral, tem-
peramental, aesthetic, and intellectual differences. This new methodology
proved that, from an "original single center of distribution or creation,"
there were "at least four distinct species of men in the world." These distinct
human species had not originated through "climate, mode of living, or any
natural causes now in operation"; nor could they be accounted for by "accidental, or congenital varieties springing up in the human family." Rather,
they had originated through the "great natural law" of a "difference of taste
for sexual beauty in the several races of men . . . which has been instrumental in separating, and keeping distinct, the different species of men, more
effectually than mountains, deserts, or oceans." Sexual love was directed by
the perception of beauty (which included morality and intelligence) and
was peculiar to each human species. Hence the "taste for personal beauty,
in each species, was incompatible with the perception of beauty out of the
species." This enduring natural "prejudice" against interracial unions was
both God-given and entirely in harmony with natural law.[81]

Van Amringe described his thesis as the "joint offspring of Lawrence and
Prichard, although it bears no resemblance to either parent." Its parentage
is, in fact, directly traceable to the "Anthropological Works" of "Mr. Walker,"
whose frequently cited views on beauty, marital selection, and inheritance underpinned Van Amringe's central creative "law of sexual love"; these ensured
that even if the human races did transgress against nature to crossbreed, their
offspring could not long endure before reverting to the parent type.[82]

Van Amringe's views were taken up by Nott's associate Henry Hotze, who,
as a Confederate agent, was to achieve some influence in London anthropological circles during the American Civil War. Hotze approvingly appended
Van Amringe's view—that each race had its own aesthetic standard and that
this was critical to separating and keeping them distinct—to his 1856 translation of the foundation text of racial purism, Joseph Arthur de Gobineau's
Essai sur l'inégalité des races humaines (1853–55). Contrary to Gobineau's

assertion of an "inequality in point of beauty among the different races of men" according to the universal standard of beauty, the separatist, proslavery Hotze argued that the relativist supposition that each race had its own notion of beauty was a "sign of a more *radical* difference among races."[83]

Van Amringe's and Hotze's interpretations of racial aesthetic selection show the currency of the idea (kept alive through the many pirated versions of Lawrence's *Lectures*, still rolling off the radical press) and how it might be adapted to changing social and political interests. Meanwhile, Walker's views on beauty and inheritance were enjoying something of a renaissance, and not only via Knox. The deluxe edition of Walker's *Beauty* was issued in 1851 and excited the attention of collectors—it was just the kind of work that might find a place on the shelves of that dilettante aesthete, Erasmus Darwin. As previously discussed, in 1854 Herbert Spencer's articles on "Beauty" in the *Leader* melded Walker's and Knox's views into the Spencerian embryological-based aesthetic thesis of the evolution of beauty as a development from the unintelligent, embryonic ugliness of brutes and savages to the godlike beauty and intelligence of the civilized races. Spencer's discussion of the physiognomic resemblance and common ugliness of "infants, savages and apes" is clearly derivative from Knox's aesthetics, along with his corollary claim that mixed races, while sometimes deemed beautiful, are not fit to survive and will die out unless there is intermarriage with the original race (see sec. 7.2).

The eye surgeon and fowl breeder Reginald Orton revived Walker's "pioneer[ing]" views on the inheritance of male and female characters and set out to corroborate them in his pamphlet *On the Physiology of Breeding* (1855), owned and annotated by Darwin. Orton's claims set Darwin experimenting with pigeon crosses in an effort to determine matters for himself. Throughout 1856 and 1857 Darwin enrolled Tegetmeier in attempts at testing the theory promoted by Walker and Knox of continual reversion to an ancestral state. In July 1856, in an article on "Hereditary Influence, Animal and Human" in the *Westminster Review* (also read by Darwin), George Henry Lewes remarked on Orton's "curious facts" on inheritance, which Lewes crossbred with those of the French physician Prosper Lucas, Spencer, Knox, and others to arrive at the conclusion that what we call "National Character" is the outcome of the history of "hereditary transmission" of both physical and psychical characters in the different races:

> The Jew . . . never altogether merges his original peculiarities in that of the
> people among whom he dwells. He can do so only by intermarriage, which
> would be a mingling of his transmitted organization with that of . . . another

race. This is the mystery of what is called the 'permanence of races' . . . That the Jew should preserve his Judaic character while living among Austrians or English, is little more remarkable than that the Englishman should preserve his Anglo-Saxon type while living among oxen and sheep; so long as no intermarriage takes place, no important change in the race can take place, because a race is the continual transmission of organisms.[84]

10.9 The innate repugnance to intermarriage

Racism and radicalism do not sit easily together. As Knox's views began to filter through a popular press already well saturated with racial opinion, his race science inevitably was dissociated from its radical roots. Carlyle might play on the monstrous consequences of interracial couplings in deriding the abolitionists; Benjamin Disraeli argued in Knoxian terms against the emancipation of West Indian slaves, "All is race"; a reviewer in the liberal *Westminster Review* welcomed Knox's original and clarifying pronouncements on the "limitation of application [of civilization] which the benevolent spirit is disposed to brook ill" and identified it with the proslavery agenda of Nott and Gliddon's *Types of Mankind*. Knox himself reinforced such conflations by regularly accusing the press—notably the *Times*, Darwin's favored newspaper—of plagiarism.[85]

Knox's audience was less interested in the intricacies of his system than in his emphasis on the intractability of race and racial conflict and the (usually conservative) political implications they might draw from it. And if such a racial determinist as Knox equivocated over the alleged advantages of pure against the disadvantages of mixed blood (the affirmation of which "stood at the heart of the foundation texts of racism"), then the emphasis was placed squarely on the repulsion of one race for another. Prichard had asserted the axioms of interspecific repugnance and intraspecific fertility among species in the wild, arguing by analogical reasoning that since there was no "invincible" repugnance between men and women of different races and since human "mixed breeds" were invariably prolific, hence "the several tribes of man are but varieties of the same species." Lawrence had invoked the domestication analogy to draw the same conclusion. As the status of racial crossbreedings became the "key site for conflict" and monogenists and polygenists argued interminably over the fertility of hybrids, the issue of the natural repugnance of one race for another assumed crucial significance. Polygenists might subvert the analogical reasoning of monogenists "by appropriating the axiom of interspecific repugnance and continuous intraspecific fecundity to their own agenda." If races were species, all was explained. Contra Prichard, the claim of interspecific repugnance was proof that the human races were separate species.[86] Yet, for those monogenists who theorized that aesthetic discrimination influenced human matings and had a creative role in the formation of race (Lawrence, the early Prichard, and more recently Van Amringe), the assumption of its aesthetic opposite—aversion or repulsion, which kept the races distinct—was implicit in their theorizing. The prominence Knox gave to innate racial antagonism and antipathy in his *Races of Men*, endorsed and widely promoted by Nott and Gliddon in *Types of Mankind*, only could add fuel to the flames.

In Darwin's case it seems to have fired his conviction of the essential role of aesthetic selection in the formation of race. It was the culmination of more than twenty years of thinking, reading, and note-taking since his primal encounter with the "naked, painted . . . hideous savage" of Tierra del Fuego. For all their differences, both Darwin and Knox were drawing on a shared aesthetic tradition of the priority of external, visible, nonfunctional form in the making of aesthetic judgments. Although Darwin's aesthetics was relativist and Knox's universal, both subscribed to the notion of a human racial hierarchy of beauty centered on woman's face and form. Darwin could accommodate Knox's assertion of racial persistence within the duration of recorded human history, and as far as Darwin was concerned, the distinction of race or species was purely nominal when races were to be considered as incipient species. From his earliest evolutionary speculations, Darwin had assumed the natural antipathy of the human races and invoked the notion of interracial or intervarietal repugnance to prevent back-breeding in both humans and animals and so allow the formation of new species: "There is in nature a real repulsion amounting to impossibility [of interbreeding between different forms] . . . we see it even in men; thus possibility of Caffers [Kaffirs] & Hottentots coexisting proves this." This was Andrew Smith's influential view of the "pure" Hottentot lineage, early imbibed by Darwin and perpetuated by the ethnologist Robert Latham. Despite his monogenism, Latham made much of the physical differences between the races, interpreting Prichard's views on racial intermixture to mean that "purity of blood" was the "rule rather than the exception." Darwin pounced on Latham's "excellent remark" in his *Man and His Migrations* (1851) to the effect that the physically distinct "Hottentot" and "Kaffir" occupied overlapping territories, having displaced other and intermediate types to perpetuate their own distinctive kinds.[87]

Nor did his pigeons let him down. Darwin's crossbreeding experiments had encouraged him to think that his pigeons had shown "plainly a liking each for its own kind." Early in 1856 Darwin had visited Matthew Wicking, "a jolly old Brewer," to view his extensive pigeon collection. Wicking, whom Darwin had met at the London Philoperisteron Society, had "more experience than any other person in England in breeding pigeons of various colours." Yet, like other expert breeders whom Darwin consulted, Wicking was oddly "without having any theory" of his own on the issue. He "unhesitatingly" had confirmed that pigeons, even when given plenty of choice in the matter, "would prefer to match together." That other authority on pigeon marital relations, the Reverend Edmund Saul Dixon—who did have theories of his own to promote—had asserted, in support of his creationist views, the dovecote pigeon's "actual aversion to the several fancy breeds." It was all grist to Darwin's mill.[88]

Darwin had Blyth's assurance of the common descent of all domestic pigeons from the wild rock dove, and Blyth was feeding him descriptions of the diverse pigeons bred by different nationalities: Indian, Chinese, Persian, Egyptian.[89] Blyth had earlier pointed to the role of racial appreciation of different kinds of beauty (or, according to Blyth, the Chinese aesthetic preference for the monstrous—a clear echo of Blyth's reading of Knox) for selecting and breeding the different varieties of fancy pigeons and fowls; now he was accumulating instances (witting or not) that Darwin might use to support an analogous explanation for the role of different notions of beauty in forming and keeping separate the different human races. But while Darwin was ready enough to transpose the agency of the pigeon selector to the human male—even to males of the most "inferior" races—he still was some way from assigning it to any kinds of females.

ELEVEN

Putting Female Choice in (Proper) Place

We come to this that all plumage is related to sexual selection, & laws all depend on degree to which they have happened to be confined to adult males. A Hen [bird] by selecting fine colours, makes her own sex beautiful. . . .

All this applies to theory of Human races = It makes sexual selection far more important than I thought.

—Darwin's notes on Yarrell's *British Birds*, written November 1858

In April 1856, not long after writing his notes on Knox's *Races of Men*, Darwin was galvanized into unaccustomed hospitality at Down House. First Lyell, then Huxley, Hooker, and Thomas Wollaston (a "first-rate" entomologist, thought by Darwin to be open to speculation on species variation) came and "grew more & more unorthodox." Their reaction to Darwin's cautious disclosure of his evolutionary views and guided tours of his pigeon houses was sufficiently encouraging for Darwin to begin the monumental task of writing up his species theory. He rejected Lyell's suggestion that he should publish a short abstract of his views to stake his priority: "It is dreadfully unphilosophical to publish without full details."[1]

He set to work on his "big book," mapping its structure, ransacking the accumulation of twenty years of reading, experimentation, and note-taking that covered so many diverse fields: geology, plant and animal distribution, animal husbandry, behavior and instincts, comparative anatomy, embryology, ornithology, ethnology, and the rest. All had to be pulled together into the "one long argument" for evolution by natural and sexual selection. Darwin was now in hot pursuit of the issue of sexual selection in general and its relation to human racial divergence in particular for his proposed "note on Man."

However, he was yet to put female choice in theoretical place. He had not been able to firm up his fundamental embryological argument in the face of Huxley's continuing opposition. And, while he had accumulated many instances of sexual dimorphism and variable sexual characters in animals—especially birds—which seemed explicable in terms of aesthetic choice, Darwin was challenged by the difficulties of extending an aesthetic sense to lower animals. He investigated male combat in mantids and crickets, and puzzled over insect musicality. The chirping of male crickets seemed an obvious attractant to females, but the "musical instrument of the male Cicada" was "hard to understand by Selection."[2] Early in 1856, he had seized on a promising report that the hen of a Japanese breed of poultry with extremely long tails: "disowns cock when robbed of [its tail]." "Very good," he crowed. This instance of female aversion toward the mutilated male seemed understandable in terms of its obverse: the hen's preference for the more beautiful male. But, could the greater beauty of male butterflies be explained in the same way? It was "more difficult" to attribute an aesthetic sense to insects. And what might be the "cause of beauty of snakes?"[3]

The major obstacle in Darwin's way to female choice was less a lack of information than his acculturated presumption of the predominance of male sexual preference in sexual selection.

11.1 Not much choice for females

In his *Races of Men*, Knox had attributed the racial purity of Gypsies and Jews to female aversion to males of different races; in essence, to female choice. Unlike Walker and Darwin, Knox held no strongly entrenched views on woman's sexual passivity. To the contrary, his radicalism led him to argue against the legislation of prostitution on the grounds that such legislation would infringe on the rights of men "amongst whom we are bound to include women." Every woman, proclaimed Knox, had an "innate right . . . in her own person," and "to use her person as she may think fit, in so long as she commits no outrage on society."[4]

Darwin was immune to such outrageous reasoning. The very notion of women's sexuality lay beyond the bounds of gentlemanly discourse. Darwin and his circle, however, might exchange sexual innuendo about the reproductive practices of lower organisms. In this, they were typical masculine exponents of Victorian "classic moralism" which assumed that, though sex was not something for serious discussion, it might be a subject for jokes.[5] Darwin could "pity the Hen-Cock-Roach from your description of the male apparatus" when Huxley described the anatomy of the common cockroach.

Huxley responded with ribaldry to Darwin's question whether "the cilio-grade acalephes [jellyfish] could not take in spermatozoa by the mouth": "The indecency of the process is to a certain extent in favour of its proba-bility, nature becoming very *low* in all senses amongst these creatures." Darwin appreciatively repeated this witticism to Hooker, following it with the famous outburst: "What a book a Devil's chaplain might write on the clumsy, wasteful, blundering low & horridly cruel works of nature!" The "Devil's chaplain" was a soubriquet Darwin associated with the radical agi-tator Robert Taylor, whose even more notorious fellow freethinker, Richard Carlile, had not only advocated birth control but also preached women's right to sexual choice outside the bonds of marriage—a choice long enjoyed by ostensibly respectable men. Such disreputables, dedicated to destroying the family and society, well might bring jellyfish reproduction to bear on the evolution of indecent sexual practices. Darwin would never make overt such unmentionable connections, even while he was convinced that humans in-deed had evolved from a headless hermaphrodite like the jellyfish.[6]

Meanwhile at Down House, contraception continued unmentioned. At the end of 1856, at the age of forty-eight, Emma gave birth to their tenth child, Charles Waring, after two earlier miscarriages and months of "wretched" ill-ness. Women might be inferior in intelligence, but greatly exceeded men in endurance. Earlier in the year, Darwin had marveled at the "pedestrian feats" of Frances Hooker, who with true Victorian stamina trudged up and down the Swiss Alps at her husband's side: "How I wish I was one half or one quar-ter as strong as Mrs. Hooker." His brother Erasmus—who had a fondness for strong women—favored the expression "as strong as a woman," and this, Darwin affirmed, was "quite right." In the event, there having been recent press coverage of the dangers of chloroform administered during childbirth, Darwin withheld it from Emma "till she skriked [*sic*] out for it, & yet she never suffered at all."[7]

There was quite a population explosion among Darwin's unorthodox circle: Henrietta Huxley (Huxley and his Australian fiancée finally had been able to marry in 1855, when he secured the directorship of the School of Mines) gave birth to a son on New Year's Eve 1856, the first of their eight children. Frances, also pregnant by this time—her "state (you are as bad as I am) is to be pitied," Darwin nudged Hooker on hearing the news—delivered the Hookers' fourth child in August 1857.[8] Pregnant women, like the female cockroach, were to be pitied, while men might bond over such evidence of male carnality and reproductive success.

Strong women going about their proper business as good wives and mothers were one thing; Snow Wedgwood (daughter of the redoubtable

Fanny), an opinionated young woman beginning to question her place in a society that denied women choice of little except their dress and draperies, was less admirable. "Snow came here today," Darwin complained to William early in 1857, "& this, between ourselves, is rather a bore." William, in the sixth form at Rugby, had attained the manly age of shaving, had purchased a cane "to whip the [younger] Boys" in time-honored fashion, and was considering his proper business of training for a profession, probably law. Henrietta at thirteen showed few strong woman tendencies, either of her mother's or of her cousin Snow's persuasion. Diagnosed with a persistent debility of unknown etiology, she was prescribed breakfast in bed—a badge of invalidity to which Henrietta clung for the rest of her long life.[9]

While Darwin was having difficulty in naturalizing female choice, he was well on his way to normalizing male aesthetic selection as a major determinant in the divergence of the human races.

11.2 The beauty of the Hottentot "bump"

Within weeks of his reading of *Races of Men*, Darwin had fired off what he prefaced as "some very peculiar questions" to the African explorer Charles John Andersson and the New Zealand geologist and naturalist Walter B. D. Mantell. Darwin's "ridiculous" questions concerned the sexual preferences of African and Maori men: Was their "idea of beauty, either in face or whole person, in their women . . . like ours?" And "do the chief men generally succeed in getting for their wives the handsomest women, or do they care more for a good work-woman?" Andersson vouched for "savage" preference for "a good figure" over facial beauty—his implication being "good" by European standards—while Mantell claimed the Maori preference for a woman who appeared handsome to European eyes. However, more promisingly, Mantell also thought that, while Maori "aristos" preferred wives skilled in making mats or roasting sweet potatoes, "in the old times almost every girl pretty or promising to be so . . . was sure to be taken to some chief." He dampened the effect of this by asserting that "rank" was "another charm" in a wife, and that a chief had thereby "greatly increased" his and his children's status. Darwin did not allow this qualification to stand in the way of one of his most cherished beliefs: that the aristocracy's privileged selection of the racial beau ideal accounted for the good looks of the English upper class—"V. our aristocracy." Mantell's verification of the savage practice of droit du seigneur duly appeared in the *Descent*, without qualification.[10]

It was most likely around this time, when Darwin actively was pursuing this very issue, that he turned up his old copy of Lawrence's *Lectures*:

"found," he wrote triumphantly in the margin, double-scoring Lawrence's passage on the power the "great and noble" have always possessed of selecting the most beautiful women for mates, whereby they have "distinguished their order, as much by elegant proportions of person, and beautiful features, as by its prerogatives in society."[11] He also double-scored those passages in which Lawrence cited the explorer Chardin's account of the means by which the Persian nobility reputedly had "completely succeeded in washing out the stain of their Mongolian origin" by "frequent intermixtures with the Georgians and Circassians, two nations which surpass all the world in personal beauty." However, Darwin wrote "poor" against Chardin's further remark, quoted by Lawrence, that without this continuing interracial injection of beauty, "the men of rank in Persia, who are descendants of the Tartars, would be extremely ugly and deformed": the Tartars might be ugly, but they were not deformed. Darwin was increasingly certain that what one race considered a deformity might be an "essential point in *Beauty*" in another.[12]

This applied especially to the "peculiar" anatomy of Hottentot women, discussed at some length by Lawrence. Darwin noted his description of the supposedly characteristic elongated nymphae—these, said Lawrence, were not remarkable in size; they were natural, not artificially produced; and there was no connection between their length and Hottentot notions of beauty. Lawrence was "consciously high-minded" in his efforts to rebut the tendency to exaggerate differences between Hottentots and the rest of the human species (Darwin was not deterred: in the *Descent* he relied on the authority of that trusted informant Andrew Smith, who gave a grossly exaggerated account of the famous nymphae and their function as a sexual attractant). Darwin found more noteworthy Lawrence's discussion of "another striking peculiarity" of Hottentot women: the "vast masses of fat accumulated on their buttocks," which Cuvier had referred "to those which appear in the female mandrills, baboons, &c; and which assume, at certain epochs of their life, a truly monstrous development." This was an analogy from which Lawrence dissented; nevertheless, "Hottentot women Baboons steatopyg[i]a," Darwin jotted tersely inside the back cover of the *Lectures*. Pondering his annotation on some later occasion, Darwin inked over his penciled reference to "steatopyg[i]a," etching it into his memory and his theorizing on sexual selection.[13]

If, on Cuvier's authority, Hottentot steatopygia was analogous to the sexual swellings of the female baboon that appeared at estrous and were presented in open invitation to the male (a spectacle not unknown to the zoo-going public), and if such "monstrous" swellings were beautiful to a baboon, why should they not be beautiful to an African, who also liked black skins, flat noses, and thick lips? Darwin, however, would elide the unmistakable sexuality of such display (especially as it seemed to relate to active female participation—no sign of coyness here) with his catchall of aesthetic appreciation.

The Hottentot bottom, cliché of racial and sexual difference, recently had received the attention of Darwin's cousin, Francis Galton. Galton, who had adventured through Africa and the Middle East (where he was rumored to have had more than a passing interest in pretty slave girls and concubines), fancied himself a connoisseur of beauty. He later was to compile a "Beauty Map" of Britain, whereby, in Knoxian fashion, the ugly Scottish Celts came off badly by comparison with the Saxon beauties of South England. In his popular narrative of his African travels, Galton gave a burlesqued account of his trigonometric calculation, from a decorous distance, of the allegedly vast bottom of a "Venus among Hottentots":

> I was perfectly aghast at her development. . . . I profess to be a scientific man, and was exceedingly anxious to obtain accurate measurements of her shape; but there was a difficulty . . . I did not know a word of Hottentot, and could never therefore have explained to the lady what the object of my foot-rule could be . . . I therefore felt in a dilemma as I gazed at her form, that gift of bounteous nature to this favoured race, which no mantua-maker, with all her crinoline and stuffing, can do otherwise than humbly imitate. . . . Of a sudden my eye fell upon my sextant.

Galton's "scientific" calculation of the dimensions of the Hottentot buttocks, whose possessors could "afford to scoff at crinoline," reached a wide audience already well attuned to the accretion of the grotesque, the lewd, and the comical around the overexposed image of the Hottentot Venus in the popular imagination.[14] Darwin was among them: "Good," he chortled. In London in early May 1856, he sought out this experienced cousin's views on racial beauty: Galton "thinks savages & ourselves have different ideas of Beauty—two very pretty girls in one tribe were not admired by natives—too slim & light."[15]

William Lawrence's *Lectures* had alerted Darwin to the possibility of aesthetic selection as a formative factor in race as early as 1838. He read his well-worn copy a number of times over the years, assimilating aspects of it to his theory building. On this reading, Darwin was intent on accumulating instances of racial differences in secondary sexual characteristics that he might attribute to the action of his neologism "sexual selection," such as fatty buttocks, or beards and body hair. He also collated those practices and adornments that Lawrence related to the "fantastic ideas of beauty" of the different races, especially those of females rather than males: "Ears," "Lips," "Tattoo females." His annotation "exaggerate form of head," however, was a reference to Lawrence's discussion of the "Carib custom" of flattening the heads of both sexes in conformity with their notion of beauty, which, according to Lawrence, followed the "great Humboldt's" general rule:

> Nations attach the idea of beauty to everything which particularly characterises their own physical conformation, their natural physiognomy. Hence it results, that if nature has bestowed very little beard, a narrow forehead, or a brownish red skin, every individual thinks himself beautiful in proportion as his body is destitute of hair, his head flattened, and his skin covered with cannotto [red pigment].

This notion of exaggerating features already naturally present in the different human races was, of course, entirely compatible with Darwin's understanding of the pigeon fancier's art of selecting and adding up those small variations that took his fancy, taking them to extremes of useless feathers and frills or excessively short beaks. Fatty buttocks and artificially flattened heads were of a piece: both were the product of a self-selecting process in conformity with different aesthetic ideals. The Hottentot bottom, with its sexual and fashion referents, was the outstanding example of how such "extraordinary and unnatural appendages"[16] had evolved through male agency, and Darwin again singled it out in the latest edition of Prichard's *Researches into the Physical History of Mankind*.

Prichard's *Researches* had grown into a veritable encyclopedia of multiple volumes, so packed with information as to make it difficult to follow any coherent argument. It was still *the* monogenist text; but by contrast with Lawrence's *Lectures*, with its major shifts and revisions and its gathering contradictions, it had lost the essentials of a unified argument for racial unity. It was more "a work of reference" that defied abstraction as Darwin now turned to it, plundering its volumes for as many instances of different racial notions of "beauty" as he might assign to the effects of sexual selection: "Bump in Hottentot, & . . . many other particulars which I have *omitted* to mark, w[oul]d require selection to separate." Eagerly he hunted them down: "Beauty . . . Beauty . . . Beauty"; the "Chinese admire Chinese beauty"; differences of "Skin & hair"; all were down to the effects of sexual selection, not to climate or isolation, as Prichard seemed inclined to think. Riffling his way through Prichard's tomes, the constant recurrence to domesticates— the priority Prichard gave to his strict "analogical method" of reasoning his way from domesticates to the unity of the human races—suddenly struck Darwin afresh: "How like my Book all this will be."[17]

11.3 Truth and the embryo

Darwin's "big book" seemed set to be bigger even than Prichard's, stuffed with all the supporting evidence and detail Darwin could throw into it, moving analogically from the breeder's selection of domesticates to nature's selection of species. He was pushing the analogy of artificial selection as far as it would go, beyond the "eminently domesticated" man, ever deeper into nature: "Man's Sexual characters like tufted Ducks," he noted on one of his trawls through Prichard's *Researches*. He collected relevant information into a portfolio labeled "Ch 6 Sexual Selection."[18]

He researched and experimented as he wrote, pushing Tegetmeier, Fox, anyone he might enroll or cajole, near and far, on to the provision of birds, eggs, puppies, geese, and rabbits—alive or skeletonized—and to undertake experimental crossings; filling in the branches on his pigeon family tree through his study of neonates; floating seeds in saline solutions and testing their viability; measuring out his plots of turf and counting species; recruiting the family and servants to his experiments; and badgering his circle of confidants for more and more information and support. Support, above all, was what he needed. Darwin was convinced of the "truth" of his theory, but others must be convinced of it too.

Wollaston was not as unorthodox about species change as Darwin had hoped, though Darwin kept up a persuasive correspondence. Lyell, for all his lawyerly caution and abiding concern over "Man's" status, Darwin optimistically thought to be "coming round at a Railway pace on the mutability of species" (Darwin had just sunk £20,000 into the Great Northern Railway in confident expectation of its future profitability). Hooker had his reservations, but was endlessly supportive. Huxley was a worry. Although impressed by the pigeons, he persisted in whacking the embryological evidence for progressive development along with his assaults on Owen and Agassiz. His "tone [was] very much too vehement" Darwin told Hooker, and he had told Huxley so too—a most un-Darwin-like reprimand. Darwin declined to back him for admission to that bastion of gentlemanly intellectual opinion, the *Athenaeum*, for fear (he said) that Huxley would be blackballed by the unctuous Owen.[19]

Darwin no longer feared being branded a radical, subversive of religion and respectability, and might make his views known. Chartism had evaporated as a mass nationwide movement, and England had emerged—shaken but intact—from the revolutionary events of 1848 into a period of calm and prosperity. It was a context receptive to a theory compatible with an ideology of progress through competition and the entrepreneurial opening-up of new markets through innovation and the division of labor. The new makers and shakers were those younger men like Huxley: active in the metropolitan press, forging a new language of politics, inspired not by older doctrines of socialism and republicanism but by a militant secularism and anticlericalism that helped shape the course of popular liberalism in the 1850s and '60s. Huxley was a most desirable convert and Darwin made every effort to win him over, but Huxley must tone down the aggression and guard his tongue if he was to project the responsible, respectable image Darwin wanted of supporters. Women's sensibilities were particularly susceptible to offense by any hint of irreligion, as Darwin well knew. Not all wives were as tolerant of unorthodox opinion as Henrietta Huxley. Huxley recently had so alarmed Louisa Ramsay, the young wife of the geologist Andrew Ramsay, by his "want of faith" that she "worked herself into a fever" and Ramsay was forced to cancel their projected tour of the Swiss Alps with the Huxleys.[20]

Nevertheless, Huxley was the acknowledged expert on the development of lower organisms. Darwin pumped him for information to support his fundamental theory of sexes: the evolution of separate male and female individuals from a common hermaphrodite ancestry, with the males being more developed than the females. In spite of Huxley's recalcitrance, Darwin was pushing ahead with the embryological argument for natural and sexual selection. Agassiz had published little of late, being preoccupied with his exhaustive collection and classification of North American fauna, living and extinct. Rather than approach the éminence grise himself, Darwin sought information on Agassiz's latest researches from one of his earlier barnacle correspondents, James Dwight Dana, the Yale geologist and zoologist. Dana obligingly reported that at the recent meeting of the American Association for the Advancement of Science, Agassiz had exhibited some young garpikes that "had the tail of the Ancient Ganoids. That is the vertebrae actually continued to the extremity of the upper lobe. This upper lobe . . . drops off as the animal grows & the fish then is of the modern type of form."[21]

Darwin, excited by this promising new evidence for Agassiz's claim of living embryological resemblance to the heterocercal fossil fish tail, since minced by Huxley, wrote back: "What a striking case of vertebrae in tail of young Gar-pike; I wish with all my heart that Agassiz would publish in detail on his theory of parallelism of geological & embryological development; I *wish* to believe, but have not seen nearly enough as yet to make me a disciple." He took the uncharacteristically bold step of telling Dana that he had become "sceptical on the permanent immutability of species." Agassiz, Darwin predicted, "will throw a boulder at me . . . ; but magna est veritas [great is truth] . . . & those who write against the truth often, I think, do as much service as those who have divined the truth." For all his proper caution, Darwin was convinced that truth was on his side, and he could put the embryological theory of parallelism of "the great Agassiz" to good service in its support. To help his cause along, Darwin appealed to Dana's known abolitionist sympathies. The *Times*, avidly read by the Darwins, was covering the course of the current American presidential campaign, centered on the issue of states' rights on the slavery question, upheld by the Democratic nominee, James Buchanan (subsequently elected president), versus the pro-union, antislavery Republican candidate, John C. Frémont. "You will think us **very impertinent**," hinted Darwin, "when I say how fervently we wish you in the North to be free."[22]

This political impertinence did no harm: "I believe there is real truth in the results of your labours, and the best of foundations for general laws or principles," Dana reassured Darwin. Agassiz's embryological principle "is subordinate to a more general law of progress—a law which involves the expression of a type-idea in forms of groups of increasing diversity, and generally of higher elevation: always resulting in a purer & fuller exhibition of the type." It followed from this "grand law" that it was "natural that the history of an individual in its particulars should sometimes run parallel with that of the paleontological history of the tribe to which it pertains." Dana was no evolutionist; but Darwin was pleased enough with his encouragement of the search for "real truth" and Dana's endorsement, contra Huxley, of the progressive, divergent nature of the parallel between individual development and the fossil history of its tribe to record his satisfaction on the back of Dana's letter: "This note contains . . . Dana's belief that in Embryonic changes & geological expression there is a certain parallelism from the unfolding of the type idea to its full display."[23]

Richard Owen, Huxley's bête noire, was the leading morphological exponent of the archetype as the ideal plan on which all organisms were patterned and of its progressive unfolding in time. Huxley used Owen's "metaphorical mystifications" as a stick to beat him with, but Darwin was more receptive to the "type idea," readily transposing Owen's archetype into ancestor. In March 1856, Darwin had worked his way through the revised edition of Owen's *Lectures on Invertebrata*, noting his "laws of embryological development" (which followed Von Baer, given a progressionist tweak by Owen). Then, in November, at the Geological Society, as Hooker agog reported to Darwin, Owen launched a "cutting telling & flaying alive assault" on Huxley's antiprogressionist paleontological views. Huxley for once "did not defend himself well." What is more, William Carpenter— possibly still smarting from Huxley's critical review of the fourth edition of his *Physiology*—had seized his opportunity and "barbed" the wounded Huxley. Darwin, who had a "very high" estimation of Carpenter's *Physiology* (in which Carpenter, having just revised the tenth edition of *Vestiges*, had yielded to progressive tendencies), may have taken more embryological courage from Hooker's account of Huxley's professional flaying. His copy of the 1854 edition of the *Physiology* was heavy with marginalia, and he dug out some embryological nuggets, among them an endorsement of Agassiz's law: "Fossils approach nearer to Archetypal form & to embryos of recent forms." Carpenter's authority might be set against Huxley's. Hooker, moreover, assured Darwin that his manuscript on geographical distribution (a section of Darwin's "big species book" largely directed against Agassiz's theory of multiple centers of creation) had "very much delighted & instructed him." Hooker had "never felt so shaky about species before."[24] Darwin might forge ahead with renewed confidence.

He had another ace up his sleeve. After reading Lewes's article on heredity in the *Westminster Review*, Darwin had acquired a copy of Prosper Lucas's hefty two-volume work on inheritance, a major source for Lewes. Darwin read through Lucas's *Traité* in early September 1856, annotating it copiously and referring a number of his notes to "Ch. 6 Sexual Selection." These notes concerned sexual dimorphism in plants and animals, the law whereby "sexes transmit commonly to own sexes," and animal mating behavior. Darwin also drew on Lucas to combat Walker's laws of male and female inheritance and continual reversion, lately reinvigorated by Orton (see sec. 10.8). But the *Traité*'s major significance lay in its corroboration of Darwin's view that variations occurring in the adult recurred at a "corresponding age of the offspring." Lucas gave many instances of the age-related recurrence of hereditary conditions and illnesses, such as gout that appears at the same age in grandfathers, fathers and sons. "This important to me," Darwin noted. "Heriditariness [*sic*] at same age."[25]

This trumped the evidence his pigeons were giving of the same pattern of inheritance: species-altering variations, for the most part, occurred in the mature organism and were inherited at a correspondingly late stage of development—they were, as Darwin constantly stressed, "added" on to the end stages of development. It followed that earlier embryonic stages must represent ancient, ancestral adults, the progenitors of current species. The most compelling case for the terminal-stage addition of variations to embryonic development was that of sexually dimorphic species (particularly birds), where the resemblance between the young and females meant that the "masculine character" was "added" to species. Sexual selection could act only on sexually mature organisms, just as the fancier selected from mature pigeons. The young breeds were too much alike and too little indicative of how they would appear at maturity to warrant their selection at this stage. Darwin by now had clarified the link of his new-formed principle of divergence with natural selection, and he was ready to add sexual selection to the mix. The common underlying pattern of all three evolutionary agencies was that of the breeder who favored the "most extreme forms" to make the divergent breeds of pigeon. Embryology was the key to this pattern.

Embryology had always carried a political charge; now it was accruing new meaning in the context of the developing conflict over slavery and the forging of polygenist race science. Since *Vestiges* had popularized the idea that the fetal human brain repeated not only the forms of lower animals but also those of the "lower" races, from the "Negro" to the "Caucasian," race theorists, both monogenists and polygenists, had embellished it.[26] Darwin had marked the passage in *Types of Mankind* in which Nott drew on Agassiz's embryological schema to claim that the fact that all races were alike in embryo before diverging at birth to arrive at their permanent forms—even after hundreds of years outside their original climatic zones—attested to the specific distinctness of races. On the other hand, a divergent embryology was fundamental to Knox's monogenetic race theory. Darwin had scooped up Knox's leading instance of the "cuticular fold" at the inner angle of the eye: "so common with the Esquimaux and Bosjeman or Hottentot . . . so rare in the European, but existing in every foetus of every race." And there was Lawrence's description of the development of color in the "Negro" baby: eyes blue, lighter complexion, becoming darker after birth. Did this mean that black skins had diverged from an initially white or intermediate form? If he thought this (as he subsequently was to imply in the *Descent*), Darwin's terse annotation did not say so: "Eyes of Negro[e]s at Birth."[27]

But was skin color more connected with resistance to disease after all, as many, including Prichard and Knox, thought? And if so, might not constitution, rather than sexual selection, better explain racial divergence? Were both in play, as Darwin's note on Knox's *Races of Men* indicated, and to what extent? A dark skin, long assumed to confer protection against sun and disease in the tropics, might be more useful than it was sexually desirable. Darwin, despite his relativist aesthetics, always had trouble with the notion that a black skin might be considered alluring, even to those with black skins: "monstrous" as it may seem, he was to parlay the claim in the *Descent*. Toward the end of 1856, a letter came from a veteran West African army surgeon who claimed that light-skinned, "sanguineous" soldiers lived "*twice* as well and *as long again*" than their darker complexioned fellows in the tropics. How to interpret all this? The Indian subcontinent, with its rigid hierarchical caste system and religious subdivisions that prohibited intermingling, offered unparalleled opportunities for inquiry into complexion, sexual selection, and disease resistance; and Darwin apparently instituted these via Blyth, his man on the ground. He later told Lyell that he had "some rays of light" on the issue of racial origins, but "mutiny in India had stopped some important enquiries." From January 1857 India was in the throes of the First National War of Independence, initiated by "mutiny" of the sepoy troops in the East India Company, and Blyth's voluminous letters dried up. Darwin did not hear from him again until Blyth wrote to announce, along with some obligatory pigeon news, the firing of a "royal salute" to the "arrival of the glorious garrison of Lucknow," a triumph of civilization and science over "barbarism."[28]

Race, however it was to be accounted for, was becoming a minefield. Among Darwin's correspondents, feelings ran almost as high at the hint of a black ancestry for man or the "contamination of negro blood" as they did over the issue of bestial descent from the ape. If an ape ancestry was inevitable, Lyell agonized, "where [would] Man commence & in the form of what race, Red, White or Black? savage or civilized, superior in stature & form or inferior or of average beauty"? Lyell was reading the literature and struggling to reconcile it with his religious beliefs and assumption of black inferiority. Huxley, the hard-nosed professional and scourge of the clergy, had little doubt of black inferiority. He praised *Types of Mankind* for its assertion of the independence of science from theological bonds: it was "remarkable" for its scientific presentation of the "Diversitarian" case for racial permanency. Nevertheless, its "incessant" reference to the "slavery question" was in bad taste and, moreover, illogical. Were "the negro . . . a metamorphosed orang," it still would not justify slavery. Slavery was evil, not because of the "hypothetical cousinhood" of white and black, but because its "brutality . . . degrades the man who practises it." Huxley's representation of science as the apolitical, neutral arbiter of "truth" was the central plank in his platform for the advancement of science in a clergy-ridden, superstitious, and hypocritical society. He was no philanthropist caught up in "Uncle-Tom-Mania," but determined to look on the issue of race with a chilling, scientific objectivity.[29]

Was Darwin, with his abolitionist sympathies and with the fraught issue of color by no means settled in his own mind, to become yet another "good man" mired in controversy and made to look foolish?[30]

11 4 "Overwhelmed with my riches in facts"

Birds, however, were another matter. His pigeon experiments on target, Darwin was rushing ahead with his accumulation of embryological evidence for natural and sexual selection. Eaton's "curious Book" on pigeons affirmed the fancier's difficulty of judging the forms of young tumblers and the practice of taking their characteristic short beaks to such extremes that "many of the shortest beak birds perish in egg Q[uote] (Ch 6)." "Poor excellent Mr. Yarrell," Darwin's long-term ornithological authority, died in September 1856. A life's work, Yarrell's extensive library and collection, went under the auctioneer's hammer. His *British Birds* yielded up some "Curious & important rules of colour in Birds compared to young Ch. 6 Sexual Selection."[31] Darwin's old notebook jottings on the sexual dimorphism of magpies, jays and starlings were retrieved and, along with his early statements of the resemblance between young and female birds, were scissored out and went into his delegated portfolio; consistent with the practices of pigeon fanciers, the "masculine character," with all that it implied, was "added to species": "Good Ch 6 Keep."[32] No controversy here. As every Victorian paterfamilias knew, females were simply not the equals of males. Darwin's own sons and daughters, under the benign scrutiny of this self-trained gentleman fancier, were conforming to expectations.

He was hard at work, he told Fox in February 1857, "overwhelmed with my riches in facts."[33] He had to work even harder—against all tradition, against his own acculturated, conventional assumptions—to transfer the breeding practice of exaggerating natural characteristics beyond the pigeon fancier; beyond the savage male's fancy for outrageous, even "monstrous" anatomies and skin colors; beyond the high-ranking cultured Englishman's preference for pretty faces (naturally, of the best English kind); on to the female's passive (could Darwin possibly countenance the notion of active sexual choice by females?) but necessarily efficacious "choice" of those characteristics of *male* beauty that could not be explained by natural selection, by male aesthetic selection, or by his old fallback mechanism of habit and the inheritance of acquired characters. His embryological argument for sexual selection was forcing him to the only possible conclusion: in those cases where females and young were alike, where males were more ornamented than females, where such ornamentation had no connection with male combat and no utility in the struggle for existence, then, logically, it had to be the females who were doing the choosing and the "adding up" of mature masculine characteristics, a process repeated or "condensed" in embryonic development.

Still Darwin hesitated. The very idea of female choice went so against the grain. He pondered counterinstances: "On animals as Bull & Stallion, having much more choice than w[oul]d think. Ch 6." He turned his notes inside out, he ransacked old and new volumes on birds and insects, and he asked pointed but fruitless questions of various authorities in an effort to resolve the issue.[34]

With William Yarrell gone and Blyth silenced, he became more demanding than ever of Tegetmeier. Tegetmeier's recently revised edition of an earlier work on poultry was the source of an important new insight. Darwin singled out a passage on cockfighting to the effect that the "comb is a fearful advantage for the foe" in combat, so gamecocks were routinely "dubbed" (their comb and wattles—those dangling adornments that distinguished the cock from the hen—were removed). Ornamentation was not simply of no use in the larger adaptive struggle for existence; it could be a fearful disadvantage in male combat for the females. Here was a well-documented instance where the male was exposed to injury "for beauty's sake!"[35]

If, as seemed more and more the case, females were doing the choosing among the lower animals, then they were pushing fanciful variations to disabling extremes. Females in such instances, rather than males, were the equivalent of pigeon fanciers. It was all down to fashion. Fashion dominated the pigeon fancy, and fashion determined the ballooning dimensions of the egregious crinoline. Dr. Punch, now at the height of his campaign for the extinction of crinoline, was not interested in any advantages it might possess in female eyes, but in lampooning its dangers and general disabling ridiculousness—yet more evidence of feminine folly and extravagance.

By the end of March 1857, Darwin finished writing his first draft of chapter 6 "On Natural Selection." What he wrote on sexual selection at this stage we do not know. Curiously, there is no entry for sexual selection in the table of contents Darwin then drew up, before his later revisions of 1858. He also left dangling his proposed "Theory applied to Races of Man." Perhaps he had intended to elaborate the concept more fully in this as-yet-unwritten section where it was uppermost yet not entirely resolved in his thinking.

11.5 "Man" avoided

Darwin was now seriously ill, staggering under his huge burden of writing and research, reeling before the implications of his postponed section on "man" and beset with domestic worries. Henrietta seemed no better, and the baby, though of a "sweet, placid & joyful disposition," was without his "full share of intelligence." Two stints at a luxurious new hydrotherapy establishment at Moor Park, Surrey, revived him enough to finish his huge chapter on variations. Doubt assailed him afresh over the precise relation of embryological divergence to classification, and he posted this section to Huxley for his opinion. Darwin had sought to reconcile the view of the French embryologist Gaspard Brullé—that the more specialized or complex organs appeared first in the embryo—with that of Henri Milne Edwards. Huxley, with some relish, "*brûler'd* Brullé": Milne Edwards was correct that the more distinct the adults, the earlier their embryological resemblance ceased; Brullé's notion was ridiculously akin to that of a house builder who began with the "cornices, cupboards & grand piano" instead of the basic rafters. Darwin, who recently had embarked on some "very extravagant" house renovations of his own, was "*extremely* obliged" for this expert correction and duly crossed this section out of his manuscript. Nevertheless, he held fast to his evolutionary interpretation of Agassiz's embryological law.[36]

He worked on the iconoclast Huxley, pressuring him to convert to a view of a natural system of classification as "simply genealogical." Huxley, busily asserting his professional presence in the leading scientific societies and clashing with Owen on all fronts, was not interested in genealogies: Darwin's "pedigree business," scoffed Huxley, had as much to do with pure zoological classification as "human pedigree has with the Census." Back came Darwin with a pointed question for the expert embryologist: "It might be asked why is development so all-potent in classification . . . I believe it is, because it depends on, & best betrays, genealogical descent." How could Huxley be so blind to the glaringly obvious relationship between embryology and evolution? The outstanding instance of a genealogical classification was that of the human family, and the human embryo was the exemplar of fetal development; but Darwin placed this critical case for race relations under Huxley's determinedly agnostic gaze without effect. Huxley next declared against the "doctrine that every part of every organic being is of use to it," singling out the beautiful luster of the hummingbird and the forms and colors of flowers for particular mention. Darwin was far from thinking these had direct utility in the struggle for existence, but they were of indirect utility for reproductive success: "The doctrine that structure is developed for variety or beauty sake would, if proved, be fatal to our theory." Wearily, Darwin covered this criticism too in the "big book"—bigger than ever, now entitled *Natural Selection*.[37]

Fortunately, Darwin had Asa Gray, "a very loveable man," in correspondence; the anti-Agassiz patter that went round the tight threesome of Gray, Hooker, and Darwin was a source of strength and amusement. Agassiz's defense of his creationist views—"nature never lies"—was a running joke among the trio. Darwin, convinced of the accumulation of evidence in favor of his principle of divergence (largely on the basis of plant data provided by Hooker and Gray), took the momentous step of letting Gray know of his "heterodox conclusion that there are no such things as independently created species." He sent Gray an abstract of his *Natural Selection* on September 5, 1857. Plants presented continuing difficulties, "but in animals, embryology leads me to an enormous & frightful range." Gray was sworn to secrecy for fear that, Darwin said, the "Author of Vestiges . . . might easily work [Darwin's views] in." Embryology, while so suggestive, brought with it potentially damaging associations. Darwin's abstract included a discussion of the "accumulative power" of natural selection, "by far the most important element in the production of new forms," and of the principle of divergence and its significance for a branching classificatory tree of all life, living and extinct; but it did not mention sexual selection or humans.[38]

Darwin was still pursuing the issue among the *"many* horrid puzzles" thrown up by his writing with none other than Alfred Russel Wallace. Darwin had exchanged several letters with Wallace, who had been collecting in the Malay Archipelago since 1854 and had dispatched domestic bird skins to Darwin. Both Blyth and Lyell had alerted Darwin to Wallace's published paper of 1855 on geographic distribution, in which Wallace elaborated the "law which has regulated the introduction of new species": "Every species has come into existence coincident both in time and space with a preexisting closely allied species." But Darwin, while he noted proprietarily that Wallace "uses my simile of tree," casually dismissed Wallace's views as creationist: "It seems all creation with him." Nevertheless, Darwin had written to Wallace in May 1857, sprinkling a little spice of encouragement while staking his own priority: "I can plainly see that we have thought much alike & to a certain extent have come to similar conclusions . . . this summer will make the 20th year (!) since I opened my first-note-book, on the question how & in what way do species & varieties differ from each other. I am now preparing my work for publication." Though he hinted that he had "learned much" from his study of domestic pigeons, Darwin did not give his own views away. Instead he fired off more requests and questions to this interloper, a mere paid collector who could have no significant theoretical insights to share: "Can you tell me positively that Black Jaguars or Leopards are believed generally or always to pair with black?"[39]

If Darwin could not resolve this particular puzzle among humans, it might be referred to the mating behavior of the big black cats. Wallace obligingly reported back that the "black & the spotted [Jaguars] are generally confined to separate localities" and that, of the thousands of skins traded, he had never heard of a "particoloured" one (the only portion of this letter that Darwin retained, snipping it out for inclusion in his portfolio on sexual selection). Then Wallace evidently asked a leading question of Darwin: Was he planning to discuss "man" in his book?[40]

"Man" and his origins were of intense interest to Wallace, whose conviction of evolution had been sparked by reading Lawrence's *Lectures*, then confirmed by *Vestiges* and his experiences of cultural and organic diversity and geographic distribution in the more exotic parts of the world. Wallace was of relatively humble origins, a surveyor by training and early occupation, with progressive social convictions. He believed in the betterment of humanity not through the Victorian mantra of the self-made man, but through social reform—an anathema to Darwin, who believed as profoundly in the necessity of competition. Wallace had not been a gentleman observer of indigenous peoples, but, as a professional collector, he actually had lived among indigenes along the Amazon in South America and on the islands of the Malay Archipelago and depended on their local knowledge and expertise. He was not shocked and horrified by savage encounter, but recorded his "unexpected sensation of surprise and delight" at his "first meeting and living with man in a state of nature." Wallace now was piecing together a theory of species change that paralleled Darwin's in many respects, but was founded directly on the divisions and boundary lines separating the many human tribes he encountered, which he also might relate to the spatial arrangements and competitive struggles among indigenous and intruding plants and animals.[41]

Just before Christmas 1857, Darwin sent a rather patronizing response to the would-be theoretician, in which he again emphasized his twenty years of work on the species problem: "I go much further than you," said Darwin complacently. On the vexed issue of "man," he intended "to avoid [the] whole subject, as so surrounded with prejudices, though I fully admit that it is the highest & most important problem for the naturalist."[42]

So there went "man"; banished from the book. He would bring nothing but trouble, as Lyell had repeatedly warned, and Darwin was well rid of him. Darwin would pull the threads of his argument tight around the gaping hole of his absence and hope for the best.

11.6 "Interrupted"

The year 1858 began well enough. Huxley might be recalcitrant on some issues, but Darwin had his respect and that of the young Turks who clustered around the fiery professional. They included the physicist John Tyndall, the naturalist and comparative anatomist George Busk, and Herbert Spencer, the wordy, tendentious philosopher-aesthetician, now intent on systematizing all knowledge under "the evolution point of view." William Carpenter, now registrar of the University of London, also was on tap for information and experimental work. Darwin was hopeful that the views of this "great physiologist" on embryology "surely" would push him into acceptance of evolution. Hooker was *corrupted* and Gray more rather than less in favor of species change. Blyth was back on track, dispatching pigeons and information on peacocks and macaws.[43]

In February, Darwin took delivery of a complimentary copy of the first two parts of Agassiz's *Contributions to the Natural History of the United States*. Mendaciously, he thanked Agassiz for his "magnificent present": Darwin had "eagerly turned over the pages & can plainly see that there will be much of the highest interest to me." Indeed there was. A few days later he prodded Huxley, "I shall be very curious to hear what you think of Agassiz's Contributions." What Darwin thought may be gauged from his annotations. "All rubbish," he scribbled next to Agassiz's extension of his embryological parallel to geographic distribution. However, there were a surprising number of nondissenting marks. Perhaps the biggest surprise was Agassiz's assertion of the necessity for "study of the habits of animals and a comparison between them and the earlier stages of the development of man" as the "one road" to a proper investigation of the relation of human and animal instincts and faculties. Agassiz confessed he "could not say in what the mental faculties of a child differ from those of a young Chimpanzee." This was hardly news to Darwin, making comparable observations of his own children; but with Huxley and Owen currently at one another's throats over this very issue of ape and human similarities, the extent to which Agassiz could take his embryological speculations, while obstinately opposing any evolutionary explanation of them, was astounding.[44]

It was on his embryology that Darwin focused, particularly on the details of Agassiz's "Parallelism between the geological succession of animals and the embryonic growth of their living representatives." In evidence of his intention of incorporating it into his own writing, Darwin put Agassiz's summation of this relationship in quotation marks and partly underscored it:

["]It may therefore be considered as a general fact, *very likely to be more fully illustrated as investigations cover a wider ground*, that the phases of development of all living animals correspond to the order of succession of their extinct representatives in past geological times.["] [Darwin's emphasis]

This was the essence of what he wanted from Agassiz, and it was this, together with Agassiz's high authority, that Darwin brought to bear on his discussion of the relation of embryos to ancestors in the *Origin*. He also paid close attention to the section where Agassiz discussed von Baer's views on the relation of embryology to classification, disregarding Agassiz's strenuous efforts to keep the particular progressive "type plans" of development separate and fixed. Darwin took from this section an affirmation of his view that the development of parts proceeded in order of their "importance"—importance, as Huxley had brought home to him, being understood as "important for classification; if so[,] simple case[,] as might be expected."[45] The sequence of embryological stages was the key to both classification and ancestry, and all were progressively divergent—Darwin was absolutely sure of it.

Darwin now stood ready to appropriate the central explanatory motif of the romantic morphologists, Agassiz's prized embryological law, to his theories of natural and sexual selection. Agassiz's work on classification was "all utterly impracticable rubbish," he told Huxley. "But, alas, when you have seen what I have written on this subject, you will be just as savage with me."[46]

Lyell, wrestling endlessly with the issue of an ape ancestry and tying himself in knots over how the inferior races best were to be accommodated in any evolutionary scheme, had by far the best appreciation of the dangerous path trod by Agassiz. Lyell saw the writing on the wall, and it was embryogenesis writ large:

> There are none who so much aid the transmutationist as they who push the doctrine of progressive development farthest—since their assimilation of the successive appearance of species and genera to embryonic develop[ment] in an individual is the very opposite of the arbitrary fiat which at other times they invoke, it is creation working by law, according to a prescribed pattern & by a force analogous to that displayed in an individual from the embryo to the adult.

Darwin was proposing something far more radical than Lyell's lawful creative development, and his prescribed pattern was that of the pigeon fancier—a notion Lyell had difficulty comprehending and was never comfortable with. But Lyell was alive to the larger implications; and so too, in his own way, was Darwin. He was again ill, capable of only a few hours writing a day, as he faced up to "some very difficult" issues concerning "palaeontology, classification & embryology."[47]

On March 9, 1858, Darwin recorded that he began work on "Divergence & correcting Ch. 6." It was seemingly at this stage—with all the pieces of his argument finally in place, with his embryological evidence secure—that Darwin inserted his ten new pages on sexual selection and the new extended section on divergence, replete with fashion metaphor, finishing his revision on June 12. The two recovered pages, thought lost, show conclusively that Darwin finally had decided on the agency of female choice among birds; however, they also indicate that he was wary of pushing the concept too far. He could find little evidence of actual female choice among domesticates, always Darwin's favored test case. And he was troubled by the extent to which structures gained through sexual selection for combat or ornament might prove a "great encumbrance" to their possessors, such as the horns or antlers of some deer, or the dangling wattles and combs of some birds (the problem to which Tegetmeier had alerted him). Nor could he assume the cock's comb and wattles beautiful to the hen, as this scrawled fragment states:

> I presume the hen has no choice left her in selecting her partner, even if we could stretch our belief so far as to suppose a fine rosy comb was attractive to her eyes, like the brilliant colours and strange antics of the Rock manakins seem to be to the hens of that bird. I suppose we must attribute many of these characters, as the brush of hair of cock turkey's breast, to those same laws of growth, which have produced the horn of the Creve coeur fowl, or first commencement of the wattle in the carrier pigeon.

The rock manakin, otherwise known as the "Rock thrush" (the terminology Darwin used in the *Origin* in referring to this bird), belongs to a large family of birds endemic to the tropical regions of America, notable for the striking color patches and spectacular courtship dances of the assembled males. The female manakin— it seems beyond doubt from this extract—chooses her mate from among brightly colored, performing males who compete for her attention; but the domestic hen has "no choice" in escaping the sexual attentions of the rapacious cock. Some structures, in Darwin's eyes, defied interpretation as aesthetic and were to be attributed to laws of correlation of growth, which meant that any changes in the growth or development of one structure might bring about corresponding changes in other structures. This same caution is expressed in the *Origin*, where Darwin invoked the same instance of the tuft of hair on the turkey cock's breast as attributable not to female choice, but to the action of the principle of correlation.[48]

Female choice was finally in theoretical place. What is more, Darwin was even prepared to consider extending it to humans. Toward the end of 1857, he had read a review in the *Athenaeum* of the Reverend J. Shooter's account

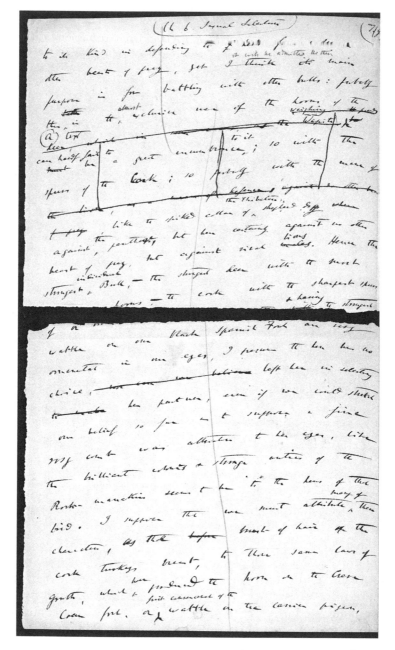

Figure 11.1. The two recovered pages on sexual selection from chapter 6 of Darwin's "big book," originally inserted around April to May 1858. The top fragment concerns structures peculiar to male combat or defense against rival males, such as the spurs of the cock and the mane of the lion. The more interesting bottom fragment indicates Darwin's attribution of female choice to the rock manakin, whose "brilliant colours and strange antics" are attractive "to the hens of that bird." DAR 85B:34r, 25r. By permission of the Syndics of Cambridge University Library.

of *The Kaffirs of Natal* (a work Darwin subsequently obtained and cited in the *Descent*), and made a note for his portfolio on sexual selection: "'The Kaffirs though generally not black admire that complexion: there has been a man among them so fair, that no girl w[oul]d marry him' (So bird-like selection in man)." Darwin, girding up for the act of writing on color and racial divergence in humans, was looking interestedly for opportunities to transfer what seemed to be an authenticated biologically based, capricious process of female choice from birds to humans. However, the great majority of travel and ethnographic accounts, including the venerable Lawrence's and Prichard's, signified conventional male choice in any description of marital customs among non-Europeans. Nor had Darwin at this stage determined whether skin color was due entirely to sexual choice or if resistance to disease played some part in racial divergence.[49]

Although "man" evidently had been dropped from his book, as he had informed Wallace, Darwin continued to collect material on the human races and sexual selection. The more famous *African Travels* by the Reverend David Livingstone (1857), read by Darwin in April 1858 ("the best Travels I ever read"), described the "Banyai," a "Negro" tribe of the African interior, as admiring a "light coffee and milk colour[ed]" skin. Darwin wondered "if sexual selection might come into play in altering black colour.—Chapter 6 Man." Darwin also recorded a conversation with Hugh Falconer, for many years superintendent of the Calcutta Botanic Garden, to the effect that "Hindoos have same idea of Beauty with us, as shown by their choosing Cashmere women, who are fair, for their concubines (but then same type of structures in Hindoos & us)."[50] Hindus were more like white Europeans than black Africans, so it might be expected that their choices (indubitably masculine in this case) would match those of Europeans.

Darwin's very conventional views on the female sex are captured in his letters to Emma during yet another stint at Moor Park in April 1858. He was reading a novel "which is feminine, virtuous, clerical, philanthropical & all that sort of thing. . . . I say feminine, for the author is ignorant about money matters, & not much of a Lady—for she make[s] her men say 'My Lady.'" A little later, Darwin exchanged gossip with Hooker about the "dreadful case" of Mrs. Forbes, widow of the zoologist Edward Forbes (a close friend of Hooker), whose second marriage was revealed as bigamous. This merry widow who got her just desserts also was judged no lady by Darwin: "I did not like what little I saw of her . . . so I was amused & I confess rather pleased to hear of [Hooker's] outburst of 'cold, unsympathetic &c.'" Meanwhile, Dr. Lane of Moor Park, Darwin's favorite hydrotherapist, was up for adultery, accused by a lady patient, whose sensational allegations of their affair Darwin dismissed as delusional. In all this, Emma represented all that a man with a stomach in a "liquid state" could want in a woman: "Ah Mammy   what an inexpressible blessing it is to have one whom one can always trust—one always the same, always ready to give comfort, sympathy & the best advice— God bless you my dear, you are too good for me."[51]

Hydrotherapy having "made a man" of him once more, and with divergence and sexual selection finally behind him, Darwin was just two days into his chapter on pigeons when his writing, he recorded, was "interrupted" by receipt of Wallace's famous letter, posted months earlier from Ternate, containing Wallace's version of the theory of natural selection.

"Interrupted" is an utterly inadequate term for Darwin's total devastation. He had completely underestimated and misread Wallace: he had been "forestalled," and his life's work was "smashed." He was shell-shocked, still staggering from the explosion, when acute illness erupted at Down House. Henrietta came down with the dread diphtheria; the children's nurse and "our poor Baby" followed a few days later with scarlet fever. Darwin was beside himself, his work and carefully ordered domesticity in tatters, his children's lives in danger. Desperate, he put the matter of priority entirely into the hands of Lyell and Hooker. These stalwarts, as told, arranged for the joint reading of extracts from Darwin's essay of 1844 and his 1857 letter to Gray, along with Wallace's manuscript, at the Linnean Society meeting of July 1, 1858. It was a calculated strategy: the Linnean was one scientific society where Owen had little influence, where Lyell, Hooker, and Darwin were all council members; Huxley, not yet a member, worked behind the scenes, vetting the documents. The papers were read without a ripple of dissent and Darwin's priority was secured, without Darwin and without consultation with Wallace—a world away from this "gentleman's agreement" in London.[52]

Darwin was almost, but not quite, beyond caring; little Charles Waring was buried on the same day, and the family fled from further contagion to the Isle of Wight, where they stayed until mid-August. There, Darwin pondered his next move. What now to do about the book? He had completed eleven chapters of a projected fourteen, about a quarter of a million words. Hooker urged the need for a quickly published "abstract" in some appropriate journal, which, once Darwin got going, grew into a book-length manuscript. Back home, safe in his study, once more the cosseted center of his orderly, chosen world, he worked like one possessed, so nearly "forestalled," quelling his demons and his rebellious stomach with brief hydropathic sessions, driven on by the need for speedy publication. He chopped here and expanded there, telling Hooker that the enforced process of "abstraction" had "clarified my brains much, by making me weigh relative importance of the several elements." He was almost glad of Wallace's preemptive letter now, for the public support he had received from Lyell and Hooker and for forcing him to publish.[53]

11.7 Publication: "Sexual selection far more important than I thought"

Darwin's main objective was to get his major principle of natural selection across to a mixed scientific and popular audience. Much would have to go. In his published Linnean Society extract, his theory of sexual selection received only brief mention as a "second agency" producing secondary sexual characteristics: "namely, the struggle of the males for the females."[54] The section on sexual selection was cut to the bone in the *Origin*, but it included the vital embryological argument with respect to the "plumage of male and female birds in comparison with the young." Ultimately, for Darwin, it was this that confirmed the agency of female choice in birds and sorted the issue of the role of sexual selection in the divergence of the human races. It was primarily the embryological argument that decided Darwin to interpolate the significant sentence on sexual selection "of a particular kind," which could throw "some little light" on the origin of the different human races.[55]

Fortunately, Darwin (with some help from his amanuensis Emma) left a hitherto unnoticed record of the steps of this reasoning. Toward the end of 1858, while working on the *Origin*, Darwin read (or, more properly, reread) two major ornithological works. The first and most important was Yarrell's multivolume *British Birds* (1843). Darwin's lengthy abstract of this work—some fourteen pages of scratchy notes—gives revealing insight into his theoretical concerns at this stage and how he resolved them. His major focus was again on Yarrell's "Curious Law[s] of *colouring* in Birds" and their relation to sexual choice. He extracted three distinct patterns of plumage resemblances in young and adult birds from Yarrell's account: the first and most significant was the rule that when the male and female were unlike, the young were like the female; the second covered those cases where male and females were alike and the young differed from both; the third rule applied to those exceptional cases where development "added to female"; that is, where the female was brighter. But there were yet other cases where the breeding and winter plumage of the adults was different and the young were "intermediate" in plumage; others, such as crows, in which all were alike; and the puzzling case of plovers, which nest in the open and so are exposed to predators, in which the males and females put on "equally or very equally" breeding plumage: "Looks as if sexual & [natural] selection were not compatible—I do not understand Law!!! In which case the sexual selection affects & is transferred in different way equally to young & Hen?"[56]

In an effort to resolve the problem, Darwin drew up a "Summary on Colouring of Birds": "Nat[ural] Sel[ection] might make adults different from old," so the young would retain the older or ancestral appearance while in the nest or suckling, as in striped lion cubs or the spotted fawns of deer. But in birds "after young have left nest or can fly & feed themselves (The plumage in nests [purely] embryonic) there seems no cause for thus having, as far as adaptation is concerned a different plumage. Again it is very doubtful how far plumage in any case is for adaptation," particularly when the separate sexes are so different. In summary:

> When male differs so greatly from hen, it is clearly sexual selection, partly in relation to males fighting together or for charming hens . . . As sexual characters are completed in adult age they would tend to be transmitted to corresponding age—they will not [necessarily?] affect young; & when they are confined to male, [characters of] young will resemble female." [57]

This was the essence of it, conforming to Darwin's long-held embryological understandings. However, he had to allow for those exceptions embodied in Yarrell's other laws: in some cases the sexual characteristics would not be confined to the male, but transmitted in some degree to the female, so both sexes would be alike; in other cases the masculine characters might be transmitted to the young, so that, as in pheasants, young males can be distinguished from young and adult females. In still other cases the transference was so great that the young and adults are all alike, as in crows and rock pigeons. "We come to this that all plumage is related to sexual selection, & laws all depend on degree [to] which they have happened to be confined to adult males." Males dominate inheritance, but "A Hen [bird] by selecting fine colours, makes her own sex beautiful."[58]

The transformative power of female choice was at last acknowledged, even to the extent of (indirectly) transforming her own sex through inheritance of male color. And generation (embryology plus inheritance) was the master key that unlocked this understanding and forced this acknowledgment.

Finally, Darwin wrote a set of directions for writing up his section on sexual selection in the *Origin*: "Begin [with] I suspect that all plumage of Birds is due directly or indirectly to sexual selection," not to "adaptation" or "conditions of life," though "no doubt" natural selection "will determine

Figure 11.2. Hummingbirds, exemplars of sexual dimorphism and spectacular beauty, from *Descent* 1871. In his *Monograph on Humming-birds* (1849–61), the ornithologist John Gould cataloged the extraordinary variety and gorgeous coloring of the many different species of hummingbirds, arguing that their beauty was a sign of divine craftsmanship and served no purpose other than that of ornament. Darwin co-opted Gould's hummingbirds in support of sexual selection: their modifications were often carried, "as Mr. Gould showed me, to a wonderful extreme," and were comparable to the modifications wrought by pigeon fanciers, their "sole difference" being "that in the one, the result is due to man's selection, whilst in the other . . . it is due to the selection by the females of the more beautiful males." *Descent* 1871, 2:77. University of Sydney Library, Rare Books and Special Collections.

that no characters injurious should have been [formed] by sexual selection" (yet it was "wonderful that it will permit" structures such as the tail of the peacock, which is, however, restricted to males, the peahen having a "lesser tail," which makes for safer incubation of the young). While male characters may be transmitted to females and the young, the general rule where "we know male sexual characters [have] long been confined to male" was for these "not [to] be transferred to the female," so the females and young would be alike.[59]

Then, directly after this, at the very end of Darwin's notes on Yarrell's *British Birds*, came this long premeditated conclusion: "All this applies to theory of Human races = It makes sexual selection far more important [for racial divergence] than I thought." This has been crossed out and a more authoritative summation scrawled across the bottom of his notes: "Applies to Human Races."[60] With this, Darwin made the definitive decision, primarily on the basis of his embryological argument, that skin color in humans was, like the color of bird plumage, predominantly if not solely due to sexual selection, rather than through natural selection via resistance to disease.[61] The way was cleared for his *Origin* declaration on the determining role of sexual selection in human racial divergence.

The other significant ornithological source for the writing of the *Origin* was Darwin's old copy of John James Audubon's authoritative five-volume *Ornithological Biography* (1831–39), the earlier volumes of which had traveled on the *Beagle* with Darwin. This was searched anew just after Christmas 1858 for its many instances of mating behavior and plumage characteristics. Darwin marked many passages describing ornithological courtship practices that attributed mating choice to the female of the species. At some stage, Emma, following Darwin's marked references (and using her own judgment in her selections), copied out relevant passages: "Picus auratus [the golden-winged woodpecker]. The males do not fight but the hens are followed by even half a dozen of these gay suitors 'until a marked preference is shewn to one, when the others leave her'"; "Icterus phoeniceus [the red-winged starling] when the hens arrive they are pursued by the males until becoming fatigued, she soon makes a choice." But not all females were passive participants: in *Caprimulgus Virginianus* [the common nighthawk], "several males court in highly entertaining manner one female, but 'no sooner has she made her choice than hen approved gives chase to all intruders & drives them beyond his dominions.'" Emma could also query instances that did not fit the pattern of the dominant male obtaining a female, or young male turkeys fighting among themselves for no apparent reason: "this seems useless." Audubon's account of the courting practices of *Tetrao cupido* [the greater prairie chicken], where "a mated pair is sometimes attacked by another male & they fight," met with wifely approval: "I can understand this."[62]

Enlightening as Emma's notes are on the Victorian resonances of ornithological mating practices, they also show that Emma had a very clear notion of what Darwin wanted with respect to plumage characteristics of young and mature birds: in the case of the American goldfinch, she reported, "Old males lose their beauty in winter; at that season young & old of both sexes resemble each other . . . N.B. The extreme difference of plumage shews most directly not due to *external* influences, tho' *light* has some effect."[63] Plumage differences in bright males and dowdy females and young were not adaptations to environmental conditions, as Emma knowingly stressed, but were due to sexual choice, and it was the females, by and large, who were making the choices.

Darwin, as author, moaning and groaning over his self-imposed task, had to present the concept in a way accessible to his audience and consistent with his defining pattern of pigeon selection. The efficacy of male combat was conveyed with energetic descriptions of battling cocks, deer, and alligators "fighting, bellowing, and whirling round like Indians in a war dance, for the possession of the females"; Darwin was even prepared to extend his "law of battle" to male salmon and stag beetles. But the very notion of an efficacious female choice, which was evidently confined to birds, was hedged about with preemptive, disarming admission of its "difficulties": "[It] may appear childish to attribute any effect to such apparently weak means." Next came the indication that Darwin could, had he the space, "enter on the details necessary to support this view; but"—and here followed the essence of Darwin's glib transference from the selective practices of the masculine fancy to the aestheticized sexual choices of female birds—"if man can in a short time give elegant carriage and beauty to his bantams, according to his standard of beauty, I can see no good reason to doubt that female birds, by selecting, during thousands of generations, the most melodious or beautiful males, according to their standard of beauty, might produce a marked effect."[64]

11.8 Analogical issues

From all that has been said, it is evident that Darwin's famous analogy between artificial selection, or "breeding," and nature's power of selection, or natural selection, is not a true analogy as analogy is usually understood. It is more a spectrum or continuum of instances that Darwin carefully built up in the opening chapters of the *Origin*. It extends from the "barbarians" of Tierra del Fuego, who preferentially and uncaringly ate their old women rather than their dogs in times of need; up through the "savages" of Africa and Australia, who foster their semi-wild dogs and stock; to the semi-civilized, who breed from their "best" animals without any conscious thought of modifying the breed; on through those gardeners and sheep farmers who preserve their choice stock and seed for propagation, though still with no aim of varying the breed; then to those "hybrid" pigeon fanciers like the "jolly old brewer" Wicking or the "funny" Eaton, who practiced both unconscious and methodical selection; and finally, at the pinnacle of methodical selection, stand the gentlemen breeders like Lord Somerville of merino sheep fame and Sir John Sebright, breeder of the beautiful Sebright bantam who can produce any feather to order. It is a socioeconomic/ intellectual hierarchy stretching from the Fuegians to the aristocratic Sebright; it is complicit with Victorian values; and it functions to collapse the distinctions between the ignorant, ultra-unconscious practices of Fuegians (living in a state of nature at the far ends of the earth, soon to disappear from it) into natural selection (which happens without a selector at all). It is a "spectrum of more and less volitional circumstances" that provides a powerful rhetorical means of dealing with the problem of agency in Darwin's theorizing, primarily of solving the looming conceptual problem of selection without a selecting agent—that is, without an intelligent supernatural being who actively selects or directs the process of selection.

Yet, as I have been at pains to show, Darwin's "collapsing" analogy[65] works hardest and most effectively at the level of the pigeon fancier, who unconsciously, artlessly selects toward an extreme. Like the fancier, divergence "always favour[s] the most extreme forms." It links with another collapsing analogy: that of fashion, which also always runs to extremes. In the case of humans, the capricious practices of the fancier are transposed directly to those more powerful, manipulative males who have the pick of the women and so unconsciously design women to their fancy. As becomes is: human males act not as pigeon fanciers (as in the classic analogical relationship), they are fanciers—not of pigeons, but of women. They are identified with breeders and are ranked accordingly, from the "lowest" savage, whose aesthetic discrimination is comparable with that of birds—being attracted by the gaudy and valueless, even the monstrous—to the aristocratic connoisseur, who takes his pick of the most beautiful women. Human beings are subsumed into nature through evolution, and their selective practices, whether of pigeons or mates, cannot be divorced from nature. In the case of female choice, the analogies of fancier and fashion function together to elide the gap between the capricious choices of the male pigeon fancier and the feminine agency of the fashion-conscious woman, the victim of "crinolineo-mania," who, in the *Descent*, is resolved back into nature as she decks herself with the plumes of birds. In the hierarchy of intelligence, women explicitly are ranked with savages—their reasoning powers are embryonic compared with those of civilized men, being of a "past" and "lower" kind—and so with birds. How far Darwin sensed or was aware of these analogical slippages in his reasoning remains unclear; there can be no doubt, however, of their rhetorical power in Darwin's published writings, or of the resonances they evoked in his Victorian readers.

This analogical heavy lifting was realized only partly in the *Origin*. It became clearer with the publication of *Variation of Animals and Plants under Domestication* in 1868, followed by the *Descent of Man* in 1871, when Darwin filled in the details and gaps he left in squeezing his big book into his "little book," the *Origin*. The gaping hole of course was "man," deliberately excluded, save for the famous "light" to be thrown on his origin and history. Yet his determining role in Darwin's construction of sexual selection is there in Darwin's deliberate hint that sexual selection "of a particular kind" could throw "some little light" on the origin of the different human races, "but without here entering on copious details my reasoning would appear frivolous."

With this hint, Darwin "potentially" may have "solved a problem polarizing science, fuelling racial antagonism and dividing the human family," as claimed; but his solution was fueled by that very same racial antagonism.[66] By stressing the frivolity of his reasoning, Darwin not only deployed his favorite tactic of disarming criticism by admitting difficulties up front, but he also glossed the deeper resonances of a theory that at its core depended on innate racial antipathy and an aesthetic hierarchy that endorsed the lesser beauty (put bluntly, the ugliness) of races lower in Darwin's scale. The act of selecting according to some standard necessarily implies the rejection of those who do not meet that standard. In the case of pigeons, the many rejects ended up in "puptons" and other pigeon-based dishes that kept Emma and the Down House cook on their culinary toes during Darwin's pigeon-breeding years. In the case of humans, where the standard was a racially based notion of beauty, if the purity of the standard (in other words, the race) was to be maintained, the rejects must be denied reproduction—an inevitability that is less "frivolous" than it is a reification of racial prejudice. Put simply and clearly, the race-making power of sexual selection in Darwin's theorizing depended on maintaining and/or explaining the antipathy to interracial mating. In this sense it did not challenge racial antagonism but reinforced it, as demonstrated by its earlier promotion by van Amringe and its eager adoption by Hotze as a prime indicator of innate racial differ-ence. By making sexual selection (which acts through human agency) and not natural selection (which has no selecting agent) responsible for human racial divergence, Darwin's "solution" to the racist apologetics for the slavery he so abhorred did not resolve but, rather, attenuated the problem of racial antipathy by shining a new, if softer, "light" on it.

The more contentious issue, as perceived by Darwin, was: Which human agency—masculine or feminine? By overtly bringing the human races within the jurisdiction of sexual selection, he challenged another deep-seated prej-udice, that of giving sexual choice to females, however much he glossed their agency as primarily aesthetic. If Darwin was forced to transfer sexual choice to female animals through theoretical exigency (as we know he was), then, in all theoretical consistency, he must reimport this same female agency back into the human races. Humans, after all, were part of nature, and there was a necessary continuity from animals to humans. This had implications for human sexual selection that troubled Darwin as much as it could be expected to bother his Victorian readers. We know the answer he gave in the *Descent*: human males, through their greater physical and intellectual strength, had seized the power of choice from females. Yet at the time of the publication of the *Origin*, Darwin still was dithering over the issue. The decision having been made that human skin color differences were due to sexual choice, the application of "bird-like selection" to man—which, as we saw, he considered in at least one instance—made for good theory. But the vexed question remained: In humans, who made the sexual/aesthetic choice, females as in birds, or males as tradition and most ethnological in-terpretations asserted? His answer was to leave it unanswered.

Without proper elaboration of his argument and without a mass of sup-porting evidence, which frankly Darwin did not have, he ran the risk of inviting ridicule for a theory that relied on the vagaries of pigeon fanci-ers and the freaks of fashion for its rationale. Further than this tantalizing hint he was not prepared to go; not at this stage, that is. The vital issue at stake was natural selection. As he later told Lyell, "with respect to the Races [of men]," Darwin had "one good speculative line, but a man must have entire credence in N[atural] Selection before he will even listen to it." Sexual selection—aesthetic choice in particular—was "secondary" or subsidiary to natural selection; it explained, above all, what the primary principle could not: the persistence of the frivolous and the non-useful, of the beautiful and superfluous, in the struggle for existence. Above all, caution prevailed. Darwin had "guarded [him]self against going too far [on sexual selection]" in the *Origin*, he later told Henry Bates (expert on Amazonian butterflies and insect mimicry) in 1860; hence his cautious attribution of the ugly tuft of hair on the turkey breast to the principle of correlation. "If I had to cut myself up in a Review, I would have worked & quizzed Sexual selection."[67]

11.9 The "good woman" triumphant

In the event, the *Origin*, once published, stirred little immediate comment or inquiry on sexual selection among Darwin's reviewers or correspondents; almost all attention was riveted on natural selection and the shocking, polarizing implications of an ape ancestry for humans. One of the few to remark on sexual selection was that connoisseur of Hottentot backsides, Francis Galton. So circumspect had Darwin's expression been about this issue of sexual choice in birds, so teasing his allusion to its role in racial divergence, that Galton called for clarification. Darwin properly professed scientific caution, although he indicated where his ideas were tending: "Till I compare all my notes, I feel very doubtful about the share males & females play in sexual selection; I suspect that the male will pair with any female, & that the females select the most victorious or most beautiful cock, or him with beauty & courage combined."[68]

This explanation, though sexually reticent in Darwin's usual manner, was couched in terms congenial to most Victorians. This is Darwin's earliest expression (late 1859) of an argument that was to assume increasing prominence in his justification of female choice. All complexity of theory and practices, all the arduous elements of theory construction and coherence (the interwoven practices of pigeon fanciers, fashion, divergence, embryology, and inheritance) aside, it was all down to women who were "too good" for their demanding, selfish spouses; to the "good woman" of Victorian domestic ideology, cliché for the moral superiority and lesser sexual passion of females. It was on this very Victorian notion of feminine sexual passivity that Darwin's ultimate defense of female choice rested. Aesthetic discrimination displaced sexual passion as a primary motivation for mating in females and the rationale for female choice: it was precisely *because* of their lesser "passion" that females were passive but discriminating choosers. If they were to mate indiscriminately with any and every male, as males were presumed to do with females, there could be no female preference among contending suitors—no female choice. It was as simple as that.

The differentiation between active male sexuality and passive female sexuality was contained within the Victorian "double standard," the code of sexual mores that normalized sexual activity in men as an expression of "masculinity," while condemning it in women as deviant or pathological. This code had particular implications for middle-class women, where feminine virtue, together with moral superiority, was rigorously maintained as crucial to Victorian domestic ideology. The distinction between the "good woman" and the "fallen woman" was "defined and redefined in an attempt to create clear moral boundaries and prevent any possibility of confusion." The Victorian dress code was designed to obviate such confusion. Prostitutes, whose increasingly overt presence on the streets was causing widespread alarm and hostility, were a conspicuous source of contamination and sexual deviancy; but so too were the women of the "residual" poor, who also represented a health hazard and public danger through their reckless sexuality and prolific breeding. The gross overcrowding of slum dwellings, the resulting promiscuity, incest, filth, and depravity invoked by public health and evangelical reformers, all conduced to this Malthusian vision of unbridled carnality, excess population, and disease. Workingwomen who labored long hours in factories and mines (often in a state of minimal dress) in close and indecent association with men, and unmarried mill girls who flaunted their finery and independence, also were sexually suspect. These class dimensions to female sexuality made it easier to construct and maintain a coherent image of respectable middle-class femininity. Despite evidence of mutually pleasurable sexual relations between consenting middle-class husbands and wives and conflicted views on the nature of female desire, consensus on the generalized image of passionless, respectable womanhood was reached through opposing it to the uncontrolled sexuality of the excessively passionate and undeserving poor, who, like savages, did not know the meaning of moral restraint.[69]

At the same time, the construction of middle-class masculinity was shifting away from older notions of aggressive, unregulated carnality, associated with a decadent and displaced aristocracy (which has its latent echo in Darwin's views on aristocratic "pick of the women"), toward a more domesticated definition in which the marriage bed was perceived as the sole proper place for male sexual activity.

This concern is articulated in Darwin's letter to Fox at the height of the scandal over Darwin's hydrotherapist Dr. Lane, where the issue of deviant female sexuality featured prominently in the *Times* accounts of the case of Lane's alleged affair with a female patient. This came to light when Lane was cited as co-respondent in the divorce case brought by his patient's husband under the new divorce act of 1857. The trial hinged on whether the patient, Isabella Robinson, driven to hysterical erotomania by a conveniently diagnosed uterine disease, had imagined the whole affair, as Lane and his supporters contended. Extracts from her diary were cited as evidence and repeated in the popular press. In his letter to Fox, enclosing a copy of the

Times report, Darwin expressed amazement at "the unparalleled fact of a woman detailing her own adultery, which seems to me more improbable than inventing a story promoted by extreme sexuality or hallucination." It "altogether make[s] me think Dr. Lane innocent . . . I fear it will ruin him. I never heard a sensual expression from him." Darwin's attitude reflects the Victorian belief that there should be "discipline and unobtrusiveness" in all sexual activity: sensuality in females was deviant or hysterical—not even to be expressed in the privacy of a diary—while respectable middle-class men not only would refrain from such extramarital activity, but also from any improper "sensual" suggestion.[70] Darwin's own solution to the temptations of sociable watering places was to cultivate an older spinster patient, Miss Mary Butler, who amused the rational naturalist with anecdotes of ghosts, which he passed on to the family at Down House, along with "other plumes borrowed from you," as he flattered her. So highly did he regard Butler's "kindness" that he timed at least two of his visits to coincide with her schedule. Toward the end of 1859, he especially urged this surrogate "Mammy" to accompany him to his new spa at Ilkley, where, a refugee from the expected storm, he proposed to sit out the publication of the *Origin*: "If you were there I should feel safe & homelike."[71]

By the late 1850s, the organized women's movement was under way. The *English Woman's Journal*, the early mouthpiece of feminism, was founded in 1858. The home—that sanctuary from a competitive, cruel world—seemed under siege by women who were able to petition for divorce under the 1857 divorce act, who were demanding rights to property and custody of children in matrimonial disputes and paid professional work outside the home. In point of fact, the 1857 Matrimonial Causes Act was committed to upholding the double standard by legislating that a man could divorce his wife for adultery (as alleged in the case of Isabella Robinson), while a wife required additional grounds such as desertion or cruelty (the plaintiff, Henry Robinson, had a sequence of mistresses and two illegitimate children—issues deemed irrelevant to his suit for divorce). The act's emphasis on the sexual fidelity and trustworthiness of a wife who stood to lose all through adultery while it was condoned in men (who were seen to transgress only if they went to far as to install the mistress in the matrimonial bed) was designed to protect the paternity of a child when it came to vital issues of inheritance, property, and class reproduction.[72]

Nevertheless, these perceived threats to domesticity generated new insecurities and "provoked a deep sexual antagonism." It was within this context that social pundits like William R. Greg, Darwin's old Edinburgh friend, and the physician William Acton maintained that sensual inclinations were dor-

Figure 11.3. A woman contemptuously says no, exciting Victorian unease; from Darwin's
Expression of the Emotions. According to Darwin, this photograph "represents a young lady,
who is supposed to be tearing up the photograph of a despised lover"; her expression of
disdain seems to declare that "the despised person is not worth looking at, or is disagreeable to
behold." *Expression*, 255. Courtesy of Wellcome Library, London.

mant in girls and women until thoroughly aroused (Greg), or that women
were little troubled by sexual feelings (Acton). If women were aroused, said
Greg, the consequences for society would be "frightful." In Acton's case, his
notorious and much-misunderstood claim was meant to reassure young
men, made nervous by contemporary notions of the spermatic economy
and the evils of masturbation, that they should not "be deterred from marriage by any exaggerated notion of the duties required of [them]." Not coincidentally, both Acton and Greg were leading agitators for the regulation
of prostitution.[73]

In Darwin's case, there is more than a little contradiction between his theoretical insistence on female sexual passivity and his fear of social ruin, were unmarried women to get their hands on contraception. There was, it has emerged, another dimension to the Robinson divorce case that greatly concerned Lane and his defenders, who included, besides Darwin, the phrenologist George Combe and Robert Chambers, the secretive author of the anonymous *Vestiges*.

Isabella Robinson was an intelligent, well-read woman with an interest in literature and science as well as many acquaintances among the Edinburgh intelligentsia, where she first met Lane and became infatuated. She knew Chambers (who published some of her poems), and she corresponded with Combe, who, with many of the Edinburgh set, also came to Moor Park for treatment for that indefinite malaise that beset so many Victorian intellectuals. Isabella earlier had alarmed Combe by taking phrenology to its radical conclusions of materialism and atheism, proposing to publish her view that belief in an afterlife obstructed scientific and social progress. Combe praised her reasoning ("far beyond the average even of educated women"), while discouraging her from publishing views that he largely shared but saw as endangering his own and phrenology's standing. Isabella desisted, but asked pertinently, "Why should human life differ so materially from animal existence?" She also confided her belief that the institution of marriage was arbitrary and unjust; she was possibly the author of an article to that effect in *Chambers' Edinburgh Journal* in 1853.[74]

However, when the sexual scandal engulfed her and threatened Lane's reputation and livelihood, both Combe and Chambers rallied to his side in branding her an irrational and deluded erotomaniac. Combe was shocked to learn that he figured in Isabella's diary in relation to phrenology and the disputed immortality of the soul. He and Chambers had a vested interest in suppressing Isabella's exultant and unrepentant account of adultery that apparently promoted the view that men and women were but animals, with no immortal souls and nothing to fear by expressing their sexual appetites—charges that their more excitable critics had leveled against both phrenology and *Vestiges*. Their united strategy was to discredit the diary and pronounce Isabella insane, the sufferer of some hitherto undiagnosed uterine malady that had affected her mind. Combe persuaded the deeply shamed Isabella to cooperate with this interpretation, while Chambers arranged for an editorial in defense of Lane in the *Times*. This was published on July 6, 1858, shortly after Darwin's letter to Fox, and argued that Isabella was "either . . . as foul and abandoned a creature as ever wore a woman's shape, or she is mad. In either case her testimony is worthless."[75]

By writing of her adultery, even in a private journal, Isabella had put herself beyond the realm of reason; in any case, the word of a self-confessed adulteress was not to be trusted, not even her confession of adultery. The press, given an unrivaled opportunity to talk about sex in acceptably moralistic and allusive terms, rushed to follow suit. Isabella Robinson was either mad or bad, and most certainly dangerous. No man was safe; especially not clergymen or physicians, whose occupations routinely exposed them to such deluded oversexed women of a certain age and put them particularly at risk of false accusations. The orthodox medical press, usually intent on marginalizing hydropaths like Lane, lined up behind him in the face of this perceived professional threat. Medical men too had an interest in the medicalization of the putative female disorders of sensuality and hysteria. So great was the media feeding frenzy that the Matrimonial Causes Act was modified so that Lane, as co-respondent, might appear in court to testify in his own defense. The case was dismissed; the naturalistic premises of phrenology and the *Vestiges* emerged uncontaminated by sexual scandal; Lane's good name was secured ("I am glad to say that not one of Dr. Lane's patients has given him up & he gets a few fresh ones pretty regularly," Darwin reported from Moor Park early in 1859); Henry Robinson was refused his divorce but might wreak vengeance on his wife by removing her children and impoverishing her; Isabella Robinson—the woman who had dared to express her interwoven evolutionary naturalism, her marital unhappiness, and her illicit, unwomanly sexual passion in the privacy of her diary—was exposed and ruined, her supposed mental and reproductive disorders laid bare to public scrutiny.[76]

Darwin need not have concerned himself; there were many powerful forces at play in guaranteeing the essential social myth of the passionless, submissive Victorian wife and mother. Not least was the need to dissociate respectable scientific naturalism, espoused variously by Combe, Chambers, and Darwin, from any suspicion of sexual and moral irregularity.

The sexually passive, morally superior, accommodating "good woman" so often invoked by Darwin was little more than a comfortable fiction, perpetuated within the sexual politics of Down House and the wider society, made necessary by theoretical contingency, and nourished by the ambiguities concerning women's sexuality aired in contemporary debates on the Matrimonial Causes Act, the Married Women's Property Act, and the Contagious Diseases Acts—debates that followed thick and fast on one another from the late fifties on.

In the event, the women's movement proved well able to accommodate the view that men had carnal sexual appetites and women did not and, what is more, capitalize on it. The great crusade against the Contagious Diseases Acts of the 1860s, spearheaded by Josephine Butler, directly challenged the sexual double standard that condoned licentiousness in men while penalizing it in those victimized women who were forced into the "slave class" of prostitution through necessity and for the satisfaction of male lust. The aim of the crusaders was to enforce male conformity with the same standards of moral restraint and sexual purity that were enjoined upon respectable middle-class girls and women. The "social evil" of prostitution was to wither away before the reforming power of morally superior and sexually restrained womanhood and through the regeneration of family life based on a single standard of chastity.[77]

These social and political conflicts—centered on constructions of femininity and female sexuality and the strains and contradictions they engendered within Victorian domestic ideology (and no less within feminism and the women's movement as a whole)—played no small part in the post-*Origin* history of sexual selection. Wallace's views on women and their pivotal role in the socialist society of the future were poles apart from Darwin's socially subordinate "good" women. Yet Wallace's ultimate vision of responsible female choice in shaping the future society was as dependent on this self-same Victorian integration of feminine moral superiority with lesser sexual passion. The trajectories of their conflicting positions on sexual selection are the subject of the next chapter.

Early in 1859, Wallace, when the news finally reached him, graciously ceded to Darwin his priority in the matter of natural selection. He did not have much choice about it; the matter was decided for him without his knowledge or consent. But Darwin, reading his own theory into Wallace's communication, was mistaken in the "striking coincidence" of their views. Buoyed up by the success of the *Origin* and the increasing institutional and social power of the Huxley-led "young Guard" Darwinians, Darwin began the more leisurely elaboration of those aspects of his theorizing that had been compressed into the *Origin*.[78] Slowly it began to emerge that he and the cofounder of natural selection had radically different views on a number of issues relevant to both natural and sexual selection. Darwin found that he must do battle with Wallace if he was to preserve his cherished principle of sexual selection.

TWELVE

The Battle for Beauty: Wallace versus Darwin

Sexual selection . . . is *altogether* your own subject.

—Wallace to Darwin, May 1, 1867

After the publication of the *Origin*, Darwin did not return to serious re-search on sexual selection until 1867, when he recommenced work on his long-deliberated "essay on man." His "sole reason for taking [man] up," he then told Wallace, was because he was "pretty well convinced that sexual selection has played an important part in the formation of races, & sexual selection has always been a subject which has interested me much."[1]

Over the course of its long conception, from Darwin's defining first con-tact with the "hideous" Fuegian other, sexual selection was inextricable from the politics of race, sexuality, and gender. By the late 1860s, these had taken on new dimensions in the larger contexts of Britain's rise to imperial power and an increasingly urbanized, industrialized, and secular society. It was a society concerned with issues of national identity, made uneasy by grow-ing liberal demands for expansion of the suffrage and better educational opportunities for women and workers, and faced with the looming threats of industrial unrest, trade unionism, and socialism. These concerns were refracted more immediately for Darwin and his circle in the Huxley-led drive for the professionalization of science and the popularization of "Darwin-ism." Darwinism was Huxley's lever for shifting power from a privileged, ecclesiastical elite to a new technocratic elite of professional scientists whose authority to guide the progress and organization of society rested in right reasoning and reliable natural knowledge. From the epicenter of the influen-tial X Club, his coterie of activists founded in 1864 (an exclusively masculine enterprise, though the select "X's" occasionally brought their "Yv's" to more frivolous social events), Huxley intensified his campaign for Darwinian cognitive and social authority.[2] This campaign became particularly acute in relation to the politically sensitive "science of man." The context of Darwin's writing of the *Descent* thus included a crucial institutional component in the ongoing struggle for Darwinian control of the "new science" of anthropology, a struggle that left its indelible impress on Darwin's elaboration of sexual selection.

At the end of 1866, after years of corrosive ill health, Darwin finally delivered the manuscript of his major work on the *Variation of Animals and Plants under Domestication* to his publisher John Murray. This was the reworked portion of his unpublished "big book" on natural selection that pulled together his accumulated pigeon and other evidence from domesticates. His intention to include a chapter on "man," that "eminently *domesticated* animal," was prevented by the size of his manuscript. With *Variation* in press (though Darwin tinkered with its editing throughout 1867), he resolved to make this into a "little essay" because, he told the Brazilian-based naturalist and ardent Darwinian Fritz Müller, "I have been taunted with concealing my opinions."[3]

Since he publicly had committed to an explanation of human racial divergence through the action of sexual selection in the first edition of the *Origin*, Darwin's hope that, where he had left off, others might take up the challenge of a thoroughgoing analysis of human descent had been frustrated. After years of encouragement, Lyell's best-selling *Antiquity of Man* of 1863 endorsed the great age of humanity in the light of recent fossil discoveries, the unity of the human races, and an animal ancestry, but, to Darwin's crushing disappointment, balked at the last step: Lyell reverted to a "creational law" to account for human moral and intellectual faculties. Darwin welcomed Huxley's bravura *Evidence as to Man's Place in Nature* of the same year, which tackled head on the contentious issue of a common ancestry of ape and human: "Hurrah the Monkey Book has come." Based on Huxley's earlier lectures to workingmen, it was an immensely popular book, with its "punchy street prose" and its inspired (quickly iconic) frontispiece of a row of skeletons, progressing through the slouching great apes to an upright *Homo sapiens*. Darwin's bulldog, however, was less interested in detailing human racial origins than in sinking his teeth into his inveterate anatomical enemy Richard Owen. Their public conflict was subject matter for cartoons and parody, at some risk of descending into farce. And, while Huxley was an aggressive advocate of "Mr. Darwin's hypothesis" of natural selection, he had shown no interest whatever in sexual selection. As told, Wallace raised high hopes with his promising 1864 "Man paper," but caused major offense by rejecting Darwin's offer of his accumulated notes and dissenting from his cherished beauty-based argument that "a sort of sexual selection has been the most powerful means of changing the races of man."[4]

12.1 The institutionalization of race and the "irrepressible" woman question

To compound matters, Wallace had delivered his paper to a meeting of the notorious Anthropological Society of London, recently seceded from the Quaker-dominated Ethnological Society. In line with the American school of Nott and Gliddon, the Anthropologicals—led by the charismatic James Hunt—favored an anatomically based interpretation of human racial diversity that maximized racial differences and defined the races as separate species. Race was, for Hunt—as it had been for his avowed mentor, Robert Knox—the key to "scientific" political legislation and social procedure. But Hunt turned Knox's race science into a monstrous parody of itself, justifying a virulent racism that endorsed slavery and, none too covertly, the Confederate cause in the American Civil War.[5]

The war tore America apart, divided opinion in Britain, and was a source of gut-wrenching moral tension for Darwin. Two highly divisive issues were the Northern blockade of Southern ports, which caused a "cotton famine," throwing hundreds of thousands out of work and threatening the British cotton industry; and, an issue that deeply concerned an emotionally febrile and patriotic Darwin, the *Trent* affair (when a British mail steamer was forcibly boarded on the high seas and two Confederate envoys taken prisoner by a Union naval captain), which almost provoked war between Great Britain and the Union at the end of 1861. But it was above all the Northern failure to abolish slavery that most agitated Darwin. Lincoln's Emancipation Proclamation did not come until the end of 1862; even then, the horrific toll of the ongoing war fired pacifist calls and a petition in England for an end to the bloodshed—sentiments reiterated by Darwin in his letters to Asa Gray, a fervid Union supporter.[6]

The Anthropological Society was founded at the beginning of 1863, at the height of the war. Henry Hotze, a Nott disciple and paid Confederate agent, was in London with the mission of promoting British recognition of the Confederacy. Hotze saw the propaganda potential of the new society, and he and Hunt made common cause in the hope that the "negro's place in nature" might be "scientifically ascertained and fearlessly explained." It was Hunt's paper by that name, in parody of Huxley's *Man's Place in Nature*, that set the tone for the new society. Hunt's inner "Cannibal Clique" cultivated an unsavory image: they exhibited the skeleton of a "savage" in their window and went out of their way to flout the conventions. Theirs was a male-only stronghold from which women were rigidly excluded—in calculated opposition to the more genteel Ethnological Society, which admitted women as "visitors"—and they prided themselves on their free and open discussion of such essential topics as female circumcision, phallic symbolism, and inevitably, the anatomy of the Hottentot Venus. They attracted a large, enthusiastic membership devoted to the dissemination of ultraracist, antiliberal, antifeminist propaganda in the guise of physical anthropology. Initially they sought to enroll the leading Darwinians, including Darwin and Huxley, whose naturalistic, secular stance seemed more congruent with their own than with the rival Ethnologicals. However, Hunt's offensive tactics speedily alienated the Darwinians. Led by Huxley, the Darwinians took over the Ethnological Society. By the time of Wallace's paper, the two institutions were locked in a no-holds-barred conflict over the control of Victorian anthropology, a conflict that was not resolved in the Darwinians' favor until 1871. Wallace—perversely, in Huxley's view—continued to attend Anthropological meetings. The future feminist defended his presence in the enemy camp by pointing to the "absurdity of making the *Ethnological* a *ladies' Society*" where "many important and interesting subjects cannot possibly be discussed."[7]

The Anthropological Society held the further attraction of being the only London "learned" society to give time to lengthy debates on Darwinian evolution, usually in opposition to its "unity of man" thesis. However, close analysis shows that its members were not opposed to the Darwinian thesis per se, but rather to Huxley's deployment of it. Huxley, the leading Darwinian spokesman on "man's place" in both the natural and social worlds, was as intent as Hunt on drawing sociopolitical lessons from the science of "man." In an increasingly secular and scientifically minded age, the Anthropologicals and the Darwinians offered two competing versions of a legitimating scientific naturalism. While Darwinism remained a controversial doctrine, particularly in its application to humanity, it was ideologically and professionally expedient for the Anthropologicals to maintain their independence of it, generally by defaulting to a bastardized "Knoxian" race science. From their flourishing institutional fortress, Hunt and his followers were able to resist incorporation into the Darwinian anthropological model proffered by Huxley and to oppose the takeover of London science by the Darwinians. Both parties were well aware that Darwinism was not incongruent with polygenism, and several leading Darwinians were instrumental in bringing this forcefully to Anthropological notice.[8]

Wallace's 1864 paper was in fact an attempt to "harmonise" polygenism and natural selection. He argued that all races were derived by natural selection from a single originally homogenous stock, but that racial traits, once developed, were fixed and very ancient. Their common ancestry lay so far back in time that it might fairly be said that "there were many originally distinct races of men." Once racial physical distinctions had evolved through natural selection, this process was checked by the development of the mind: changes in external form were then subordinate to advances in mental and moral characteristics. His audience might have been more receptive to this compromise had Wallace not gone on to draw a splendid utopian vision of an earthly "paradise" peopled by a superior "homogeneous race" of "perfect beauty" and perfect equality—an "eloquent dream" from the incipient socialist that outraged the Cannibal Clique, who reserved their most vitriolic attacks not for the rival Darwinians, but for those liberals like John Stuart Mill who suffered from such "rights-of-man mania."[9]

Darwin had as little time as the Anthropologicals for the socialist dream; his science was rooted in competitive individualism. Nor—although he abhorred slavery as the ultimate evil—was he theoretically or ideologically opposed to a naturalistic anthropology that acknowledged the "negro" and "woman" as inferior to the white European male. He subscribed to the *Anthropological Review* (the major journal of the Anthropological Society), owned and mined its translations of major anthropological works for information on race and sexual selection, and engaged several of its stalwarts in correspondence. He agreed with Hooker that Huxley's "vicious undignified" public spats with Hunt were harmful to the Darwinian cause.[10]

Nevertheless, the sensitive and humane Darwin was in no way receptive to the political extremism and excessive bad taste, the sheer vulgarity, of the ungentlemanly Anthropologicals: they were a "bête noir to scientific men." Hunt's racial slurs and proslavery rhetoric could not but deeply offend a committed abolitionist. Hardly had Darwin assimilated the "grand, magnificent fact" that slavery in the United States was formally abolished with the ratification of the Thirteenth Amendment to the Constitution at the end of 1865, than the Eyre affair erupted in Britain, reigniting all Darwin's emancipationist, humanitarian fervor. Edward John Eyre, governor of Jamaica, brutally suppressed the Jamaica insurrection of 1865 by declaring martial law, summarily executing its black and mixed-race leaders, and unleashing a violently repressive military force that killed many hundreds of revolutionaries.[11]

The 1850 clash between Carlyle and Mill over Carlyle's proposal to return Jamaican blacks to bondage prefigured their much more momentous one over Eyre's controversial actions. The Eyre affair deeply infiltrated British domestic politics. It divided the public, intellectuals, and Darwinians. Darwin subscribed—along with Huxley, Wallace, Spencer, and Lyell—to Mill's committee for the prosecution of Eyre, while Carlyle led the opposition pro-Eyre defense committee, which included Dickens, Kingsley, and Tyndall, with Darwin's closest friend Hooker providing covert support. So passionate was Darwin about the issue that, in an unprecedented display of temper, he "instantly turned on" William in a "fury" of indignation when his son disparaged Mill's committee.[12]

Eyre was a celebrated fellow of the Anthropological Society, which kept up a militant drumbeat of pro-Eyre agitation. Armed with Knox's *Races of Men*, Hunt invoked the "inexorable" Knoxian laws of race antagonism and subordination in support of Eyre's bloody repression of black revolt: Knox, "this great practical anthropologist," had proved beyond doubt that "we English can only successfully rule either Jamaica, New Zealand, the Cape, China, or India, by men such as Governor Eyre." Over the years, the erstwhile "savage radical" (who died just two weeks before the society's inaugural meeting) was regularly exhumed by Hunt. The dead Knox, institutionalized by Hunt, came to represent much of what the living Knox had set himself to oppose; thus he posthumously supported the infamous Eyre, rejected home rule for Ireland, endorsed British imperialism, became an apologist for slavery, and opposed the extension of the franchise to women and blacks. Little wonder that Darwin, like many others, came to identify Knox with all that Hunt and the Anthropological Society stood for.[13]

Little wonder also that Wallace's continuing presence in the ranks of the Anthropologicals—together with his spurning of the high honor of the offer of Darwin's notes on man and sexual selection—rankled. Added to these offenses was Wallace's gratuitous criticism of those "men of science" who were "so dreadfully afraid to say what they think & believe." Hooker and Darwin shared commiseration: it was "all very easy" for an unencumbered bachelor like Wallace. They had vulnerable children and "many kind & good relations" who would be "grieved & pained" if they were to say all they thought. Darwin, seriously affronted, began to "fear [that Wallace] will not do what he ought in science."[14]

The Anthropologicals, untrammeled by theological or social restraints, were flourishing. Grown even more assertive, they launched a formal assault on the scientific establishment, demanding official recognition of their "new science." Wallace chaired the first anthropological subsection at the British Association meeting of 1866, welcoming "all students of man, by whatever name they might call themselves . . . to meet harmoniously to state their views and opinions." Hunt, however, had the numbers and was not to be appeased. With their continuing hostility toward the "Darwinian club" in the Ethnological Society, the all-too-successful Anthropologicals represented a major threat at a stage when the Darwinians were still seeking to establish themselves scientifically and, even more crucially, socially. Huxley, the master strategist, committed himself to reconciling and uniting the two societies under the "proper direction" of the Darwinians.[15]

His 1865 paper "On the Methods and Results of Ethnology" made a number of concessions to the Anthropologicals: Darwinism had the potential for "reconciling and combining all that is good in the Monogenistic and Polygenistic schools." Huxley identified eleven semipermanent racial types on the basis of physical and anatomical distinctions, including skin color, hair, and skull shape. Arguing that belief in the diversity of human species did not necessitate diversity of origin, Huxley postulated that these "persistent modifications" had diverged in some remote epoch from an original monophyletic origin and had evolved by natural selection, perhaps having evolved so far as to prevent fully fertile crosses.[16]

In the same year, Huxley spelled out the racial and gender implications of Darwinism in his well-known address, "Emancipation—Black and White." Nothing could better symbolize the convergence of the touch points of gender and race than Huxley's choosing to address the "negro" and "woman" questions as one; and nothing could better express the ambiguities at the core of Huxley's enduring reputation as a "progressive" in these areas. Just as Huxley earlier had declared against the "fanatical abolitionists," he now dissociated himself from the slavish adherents of the "new woman worship." While he supported the abolition of slavery and votes and education for women on liberal grounds, Huxley could reassuringly assert that the innate inferiority of women and blacks would never endanger white male supremacy. It was "simply incredible" that "when all his disabilities are removed, and our prognathous relative has a fair field and no favour . . . he will be able to compete successfully with his bigger-brained and smaller-jawed rival, in a contest that is carried on by thoughts and not by bites."[17]

So much for black emancipation. Most of Huxley's ammunition for his Darwinian "Whitworth gun" was expended on confronting the "irrepressible" woman question. He held fast to the unshakable Victorian conviction that "in every excellent character, whether mental or physical, the average woman is inferior to the average man, in the sense of having that character less in quantity and lower in quality." But although nature had not made men and women equal, these "facts" did not afford the "smallest ground" for refusing to give women the same educational, political, and civil rights as men. Law and custom should not add to the biological burdens that weighed women down in the "race of life." Let us have "sweet girl graduates," Huxley said—let women even become merchants, barristers, or politicians—it would make no difference to the status quo:

> Nature's old salique law will not be repealed, and no change of dynasty will be effected. . . . The most Darwinian of theorists will not venture to propound the doctrine, that the physical disabilities under which women have hitherto laboured in the struggle for existence with men are likely to be removed by even the most skillfully conducted process of educational selection.[18]

Huxley's 1865 drawing of Darwinian barriers against the complete equality of men and women was a winning strategy. It was ratified by Darwin in the *Descent*, and by other prominent evolutionists who joined forces with anthropologists, psychologists, and gynecologists to forge a formidable body of biological theory which purported to show that women were inherently different from men in their anatomy, physiology, temperament, and intellect—that women, like the "lower" races, could never expect to match the intellectual or cultural achievements of men or obtain an equal share of power or authority.

But not even such newly erected biological barriers were strong enough to keep women in their place. Feminist agitators and pioneers like Emily Davies, Lydia Becker, Josephine Butler, Elizabeth Garrett, Sophia Jex-Blake, and Frances Power Cobbe were on the go, inciting social, legal, and cultural change. Women were infiltrating places of education, science, professions, and politics, not as observers, but as would-be participants. The all-rounder Becker—sometime botanist, biologist, and astronomer; founder of the feminist Manchester Ladies' Literary Society; and subsequent leading suffragist—even corresponded with the increasingly famous Darwin, though on the overtly innocuous subject of botany. Darwin (slyly?) responded to her request for suitable scientific reading material for her "little society" by sending his recent paper on adaptations to prevent self-fertilization in the purple loosestrife: "On the Sexual Relations of the Three Forms of *Lythrum salicaria*."

Becker and her sister science enthusiasts seemingly took this analysis of plant sex life in their stride. Even in an age of great sexual reticence, botany was a fashionable female hobby and still the preferred subject for scientifically minded girls and women. Biology and medicine were more problematic; it took strong motivation indeed for women to flout this conventional sensibility. Women's rights activists were not slow to turn Victorian prudery to advantage by arguing that women would make more fitting obstetricians and gynecologists than men. Another tactic was the establishment of women-only colleges, or for female students to be taught anatomy and reproductive biology in separate classes (a practice advocated by Huxley, though by 1874 he bowed to the inevitable and even employed a woman demonstrator for mixed classes in physiology). Such apartheid stratagems, based on full acceptance of Victorian sexual mores, were not without their ideological hazards; but they were all part of the long struggle to break down the strongly gendered barriers of medicine and science.[19]

In 1867, as Darwin embarked anew on sexual selection, the issue of women's suffrage came to the fore with Mill's magnificent championing of the cause in the House of Commons. Mill was the most prominent advocate in England of civil and political equality of the sexes, an ardent proponent of women's rights in marriage, employment, inheritance, and the ownership of property, education, and the vote. Mill's fundamental egalitarian argument during the Eyre affair—that blacks were members of the same human family as whites and were entitled to the same legal rights—was the same one he extended to women. Given the same education and liberal opportunities, women would prove the equal of men. But there Darwin would not follow him; what is more, in this case Darwin opposed Mill by adopting an anatomical argument promoted by the Anthropologicals.

Spurning Huxley's overtures, the Anthropologicals had taken up the German-born Carl Vogt, one of the most prominent European exponents of Darwinism, a militant materialist, as their preferred evolutionary authority on black and female inferiority. Vogt's Lectures on Man (translated and published by Hunt in 1864) offered the more congenial "anthropological" picture of multiple lines of descent: the different races most likely had evolved from separate ape ancestries. Vogt's advocacy of polygenism within a Darwinian framework made his Lectures "especially sound and philosophical." Added to this was Vogt's explanation for the anatomical unfitness of women for serious scientific debate: recent work in comparative anatomy had demonstrated that the crania of adult women were more childlike than those of men, thereby referring them to the inferior development of the "lower" races. This anatomical difference increased with the development of the race, "so that the male European excels much more the female than the Negro the Negress." This meant, as Vogt interpreted it, that there could be no possibility of sexual equality among "progressive" civilizations:

> Just as, in respect of morals, woman is the conservator of old customs and usages, of traditions . . . and religion; so in the material world she preserves primitive forms, which but slowly yield to the influence of civilization . . . In

MISS MILL JOINS THE LADIES.

[See Page 43.

Figure 12.1. In triumphal comment on his failed championing of female suffrage, a scowling Mill exits Westminster in feminine attire, helped out of his seat by his successful conservative opponent, while a fellow liberal parliamentarian enjoys his port in exclusively masculine company beneath a portrait of Governor Edward John Eyre, target of Mill's campaign to bring him to justice for Eyre's brutal suppression of the Jamaica Insurrection of 1865. *Judy*, November 2, 1868. Courtesy of National Portrait Gallery, London, NPG D22624.

the same manner woman preserves, in the formation of the head, the earlier stage from which the race or tribe has been developed, or into which it has relapsed. Hence, then, is partly explained the fact, that the inequality of the sexes increases with the progress of civilization.[20]

Among "lower" races, the occupations of the sexes were similar—Bushmen and women shared the same tasks—but among civilized nations there was a sexual division of labor, both physical and mental, which could be bridged only at the cost of a common degeneracy.

Conventional ethnological wisdom held that the level of civilization of a society was to be judged by the status of its women: "savages" treated women as slaves and beasts of burden; "Turks" made them sexual slaves and chattels; only in advanced "civilized" societies were women treated with respect and consideration. Martineau had sought to extend this ethnological understanding by making the "degree of the degradation of woman" the test of "the state of domestic morals in any country" and urging the emancipation and equality of women as the highest measure of social progress. Vogt stood this understanding on its head by making the degree of civilization of a society dependent on sexual inequality.[21]

Vogt's "well-ascertained facts" proved as acceptable to Darwinians as they were to Hunt. His *Lectures* were commended for their "Huxleyan outspokenness."[22] Though he dissented from Vogt's "origin of Man from distinct Ape-families," Darwin pounced on his anatomical evidence for the innate and continuing intellectual inferiority of women. He marked this up in his copy, and he put it to use in the *Descent*. More than this, Darwin was proud to claim Vogt as an adherent on its opening page. Vogt earlier had worked with Agassiz on fossil fish and helped pioneer the parallel between embryological development and the fossil sequence. The embryological argument, enlightened by Darwinism, permeated Vogt's *Lectures*; it meshed with Darwin's near obsession with the same, with his long-term emphasis on the similarities between women and children, with the resemblances between the females and young in all animals in which the adult sexes differ; in short, with Darwin's views on racial and sexual divergence through the primary action of sexual selection.[23]

If Hunt had thought to outmaneuver Huxley and the British Darwinians by championing Vogt, in the longer term he was beaten at his own game. For the moment, he might exploit the proliferation of competing versions of human origins offered by the Darwinians and resist Wallace's and Huxley's compromise gambits. Provocatively, from the official platform of the British Association for the Advancement of Science, Hunt invited the reclusive "Mr.

Darwin himself . . . to come forward" and apply his own theory to the origin and future of mankind.[24]

Darwin was at last ready to do so. He would answer Hunt's taunt and take up Wallace's challenge to say what he thought in a "little essay"[25] that grew into the two-volume *Descent of Man, and Selection in Relation to Sex*.

12.2 "Sexual selection has always been a subject which has interested me much"

We know what Darwin had thought over many years. This was given greater immediacy by these institutional, cognitive, and ideological dissonances as they played out in the larger contexts of Victorian culture and society. In particular, they clarify why "the differences between the so-called races of man" loomed so large for Darwin in the writing of the *Descent*. Sexual selection was Darwin's ready-made answer to the problems posed by all interested parties: divergence from an original common ancestry through male combat and sexual choice, determined by specific racial notions of beauty, overcame the disjuncture between climate and race (so harped on by the polygenists) while it solved the issue of the long-term persistence of those nonfunctional external differences like skin color, hairiness, and face shape (also given such exaggerated significance by polygenists) that Wallace and Huxley were seeking to explain through the action of natural selection. Huxley was prepared to invoke a degree of infertility between races which would stabilize these "persistent modifications," while Wallace was suggesting that natural selection had ceased to act on these physical differences once they had evolved "at the very infancy of the human race," confining its subsequent effects to the evolution of mind. Darwin's long-term conviction that these differences were purely aesthetic and had no "beneficial" role to play in the evolution of race eliminated the need for such adjunctive explanation. The different racial ideals of beauty would both lead to racial divergence and (by inhibiting intermarriage) sustain racial differences. In a marginal note on Wallace's 1864 paper, Darwin cursorily reduced "all on this head" to "S[exual] Selection."[26]

By 1867 Darwin had further grounds for his increasing investment in sexual selection. In 1860, as Huxley, Hooker, and others did battle on behalf of the *Origin* in the public arena, Darwin had begun a study of orchids. Pottering around the greenhouse took him out of the glare of unwelcome controversy; but, of greater import, orchids were Darwin's "'flank movement' on the enemy." Creationists were invoking the traditional view that the final cause of plant sexual organs was "to minister to the gratification of the senses of man by the beauty of their forms and colours." Darwin's intensive study of orchid flowers was to demonstrate that their beauty and bizarreness had evolved by natural selection as an attractant to insects to ensure their cross-fertilization. They were "beautifully adapted" to insect visitation, not divinely designed for human appreciation.[27]

This had implications not just for natural selection, but also for sexual selection. All organic beauty was referable to sex and reproduction, with the exception of the lowliest of asexually reproducing organisms. Yet *their* beauty too was explicable in noncreationist terms: "Do you really suppose that . . . Diatomacea were created beautiful that man after millions of generations [should] admire them through the microscope?" Darwin challenged a correspondent. "I [should] attribute most of such structures to quite unknown laws of growth."

Female choice, however, was still sticking in his gorge: "[The] sight of a feather in a peacock's tail," he told Gray around the same time (April 1860), "makes me sick!" The Fuegians were on his mind. Earlier in the year he had fired off a list of queries to the Falklands, among them one on the Fuegian idea of feminine beauty: "do they admire women with strong American cast of countenance, or such as at all approach Europeans in appearance?" He had to wait years for a response from Thomas Bridges, and then it was not the one he wanted. He hoped to find corroboration of female choice among domesticates, but his main informants, Tegetmeier and Brent, declined to give him any encouragement whatever on that score.[28] It was Henry Bates, Wallace's friend and early traveling companion, recently returned from his eleven-year collecting expedition in the Amazonian basin, who gave Darwin "admirable" support: "I thoroughly believe in your theory of sexual selection."[29]

Bates, like Wallace, made only a bare profit from the sale of his extensive collection of different tropical species—mostly insects, thousands of which were new to Western science. Back in London he found the going hard. A committed evolutionist (early converted by *Vestiges*), the unassuming Bates put his observations at Darwin's disposal, providing the first real evidence for natural selection at work. Darwin enthused over his discovery of a series of intermediate forms between two distinct species of Amazonian butterflies living in separate localities. These were not hybrids, but different races or incipient species, on their way to becoming separate species. Bates had "got a glimpse into the laboratory where Nature manufactures her new species." Equally valuable was his concept of insect mimicry (the superficial resemblance between unrelated species). Bates supposed that mimicry functioned as a protective device—if, for instance, an insect was unpalatable to birds, then if another species came to resemble it, it too would benefit from not being eaten. Bates collated an "immense number" of instances of such "deceptive dress" (as Darwin dubbed it, evoking resonances of his fashion metaphor), especially among butterflies. Protective mimicry was of selective advantage, enabling the mimicking species to escape predation; the better the mimicry, the better the survival rate. It was a subtle but "rational" explanation, which undercut one of the "strongest arguments" of those who believed in "the miraculous in creation." Batesian mimicry, as it became known, was natural selection's "greatest triumph in explaining adaptations."[30]

Even more timely, though, was Bates's endorsement of sexual selection, exemplified by the gradation he traced from species of butterfly in which male and female were "exactly alike" up to those in which "the [sexual] divergence reached its highest point." Although Bates, when pressed, could not give any exact observations of "female butterflies pointedly selecting their male partners," he had seen females being pursued by males. Moreover, he was certain that in a number of species where the males were "distinguished by rich colours," the males greatly outnumbered the females (thus increasing the competitive edge for the efficacy of female choice). Significantly, in another species where the males and females were alike, both sexes were found in almost equal numbers. Darwin was ecstatic:

> Hardly anything in your letter has pleased me more than about sexual selection . . . though I am fully convinced that it is largely true, you may imagine how pleased I am at what you say on your belief.—This part of your letter to me is a quintessence of richness.

Even better, Bates followed this up with an account of "double dress" in some butterflies, "answering the two purposes in the welfare of species. The high colour [of the inner wings] being explicable by sexual selection & the plain colour [of the underside] by the check caused by insectivorous animals."[31]

Butterflies might rival birds as outstanding, elegant instances of female choice: fragile butterflies were built sometimes for deception or protection, but primarily for beauty. Darwin took Bates under his wing, finding his new protégé employment (as assistant secretary to the Geographical Society) and negotiating publication of Bates's popular book on the Amazon that included his butterfly observations.[32]

When Lyell in 1865 repeated to Darwin the creationist arguments for the existence of beauty in the natural world—recently revived by George

Douglas Campbell, the Duke of Argyll—Darwin recommended that Lyell "tell him about Sexual Selection."[33] Darwin was possessive of his cherished theory too. He offered minimal encouragement to James Shaw, a Scottish schoolmaster who tried to engage Darwin in discussing beauty and sexual selection and had taken it upon himself to answer the duke.[34] Darwin publicly combated the duke (though without naming him) in the fourth edition of the *Origin*, published in December 1866, considerably expanding his discussion of beauty in unseen marine organisms, in flowers, fruits, butterflies, birds, and humans. For the first time, he put in print his view that the human ideal of feminine beauty was not innate and fixed, but racially relative: "We see this in men of different races admiring an entirely different standard of beauty in their women; neither the Negro nor the Chinese admires the Caucasian beau-ideal." Although Darwin did not spell out its implications for the origin of the human races, discerning readers might draw their own conclusions from his earlier standing reference to "sexual selection of a particular kind" in determining human racial differences.[35]

12.3 Embryos, ancestors, and "my poor dear child" pangenesis

Recent research in crustacean embryology and sexual dimorphism had powerfully reinforced Darwin's theoretical position on sexual selection. In 1864 Darwin had taken delivery of Fritz Müller's *Für Darwin* (1863), in which, inspired by Darwin's interpretation of embryology in the *Origin*, Müller confronted Agassiz's view of development as the manifestation of the Creator's preordained plan. Müller's extensive researches into crustacean development demonstrated the total absence of a "general plan or typical mode of development." He ridiculed the notion of establishing a "true and natural system" according to the rigid embryological rules of the "Old School" (that of von Baer and Agassiz), and gave a ridiculous example of the result if the classificationist were such a "zealous embryomaniac." Müller's point was that Agassiz could not account for the many exceptions to the plan, but Darwin could.[36]

Darwin's barnacle researches had impressed on him the realization that allied embryos sometimes departed markedly from that resemblance on which he placed such emphasis. A degree of adaptation of active free-living embryos to their conditions of existence was to be expected, with subsequent obscuring of the embryonic similarity between allied animals. But until reading Müller's work, Darwin failed to perceive the force of the argument that the incidence of exceptions to this general rule could urge against Agassiz and the special creationists and in favor of natural selection. Müller also showed that some higher crustacea pass through an embryonic stage akin to that known as the Nauplius form of the free-living larvae of many of the lower orders of crustacea; he suggested that the common ancestor of the crustacea had a Nauplius form. Müller's "Nauplius theory" went into the fourth edition of the *Origin*. Darwin used it to advantage to claim along with Müller (overriding Huxley), that the "earlier larval stages do show us, more or less completely, the form of the ancient and *adult* progenitor of the whole group" [emphasis added].[37]

Für Darwin enabled Darwin to firm up his earlier conviction that individual development was the key to ancestral history. The real beauty of Müller's work was that the basic generalization established by Agassiz remained as a central argument for evolution by natural selection, but now the many demonstrated ontogenetic deviations from the presumed ancestral pattern also could be turned to good account. Natural selection could always be invoked to account for the deletion or suppression of a useless ontogenetic stage or the unrecognizable modification of another through adaptation to new conditions of existence.[38]

With this, Darwin's triumph over Agassiz—now slowly submerging under the rising tide of Darwinism, his once-numerous acolytes drifting from his side—was complete. And Darwin's earlier conclusion that variations generally supervened at maturity and were inherited at a corresponding age, being "added on" to embryonic development, also was affirmed— the exceptions being those free-living larval variations that were inherited at a correspondingly early stage of development.

This had implications too for sexual selection, which acted only on mature organisms, with the masculine characters being "added" to the males. Müller described several cases of sexual dimorphism among crustacea, including a peculiar instance where males of a particular species of isopod (*Tanais*), which until maturity resemble the females, lose their feeding appendages and under two separate (or dimorphic) forms develop either extended "smelling" or clasping structures and "appear only to live for love." Müller interpreted these as the result of natural selection, but for Darwin the appendages of the divergent forms of males had "probably been acquired through sexual selection": the one being "best able to find the female" and the other "best able to hold her when found," both forms leaving the "greater number of progeny to inherit their respective advantages" over less advantaged intermediate forms. Müller's finding was yet more evidence for his views on sex-limited inheritance and for sexual selection.[39]

Darwin heaped praise on *Für Darwin*, arranging for its English translation and publication as *Facts and Arguments for Darwin* (1869). Müller, "one

Figure 12.2. Diagram of the two forms of males of the isopod, *Tanais*, from Fritz Müller's *Facts and Arguments for Darwin* (first published as *Für Darwin*, 1863). Müller showed that until maturity, the males resemble the females; they then lose their feeding appendages and develop either "smelling" (depicted on the left of the diagram) or "clasping" structures (on the right) and "appear only to live for love." Darwin interpreted these divergent male structures as acquired through sexual selection, either for finding the female or for holding her when found—yet more evidence for his views on sex-limited inheritance and for the efficacy of sexual selection among the lower crustaceans. Müller 1869, 21. University of Sydney Library, Rare Books and Special Collections.

of the most able naturalists living," became a favored confidant: "You have done admirable service in the cause in which we both believe. Many of your arguments seem to me excellent, & many of your facts wonderful . . . nothing has surprised me so much as the two forms of males."[40]

In 1866, the work on ascidians (sea squirts) by the Russian zoologist Alexander Kovalevsky alerted Darwin to embryological evidence of an ancestral link between invertebrates and vertebrates.[41] In the same year, another new recruit, the formidable Ernst Haeckel—the leading German propagandist of the new creed of *"Darwinismus"*—consolidated the Darwinian takeover of Agassiz's transcendental law. Haeckel's "fundamental biogenetic law" (much indebted to Müller's *Für Darwin* and first put forward in Haeckel's *Generelle Morphologie der Organismen* of 1866), popularized as the slogan "Ontogeny recapitulates phylogeny," was to prove enormously influential. It went on to dominate embryological research and the search for evolutionary genealogies until the end of the century—even Huxley eventually succumbed to its lure.[42]

Inspired by these developments, Darwin brought together all the phenomena of variation, reproduction (both sexual and asexual), development, and heredity under his "Provisional Hypothesis of Pangenesis" (1865–68). Darwin's pangenesis (which originated in the early 1840s and had some familial inspiration in the speculations of Grandfather Erasmus) was that every bodily cell buds off a representative part of itself—a minute, maturing, impressionable "gemmule"; these gemmules aggregate through mutual attraction, when conditions are right, to form the "true germ" (or female element) and the male element; when fertilization takes place (or even if it does not, as in parthenogenesis), these combine to determine the character of the new organism. Pangenesis had the supreme attraction for Darwin in that he made it explain and connect so many phenomena, some otherwise inexplicable, that he had researched over the years: sexual and asexual reproduction, healing, regeneration, reversion, the inheritance of acquired characters, parthenogenesis and alternation of generations, hybrid sterility, prepotency, the case of Lord Morton's mare, the effects of castration, the stages of development and the repetition of ancestral forms, and the "important principle of inheritance at corresponding ages." All were made intelligible by pangenesis—at least, as far as Darwin was concerned. His closest allies, Huxley and Hooker, were not so sure. Darwin defied Huxley's cautionary dissent from his cherished theory of invisible gemmules and went ahead anyway, publishing his "provisional hypothesis" as the penultimate chapter of *Variation* in 1868.[43]

One factor in Darwin's persistence in his "foolish parental affection" for his "poor dear child" pangenesis (despite the criticism it aroused and in the face of his own fear that it would be "still-born") was that pangenesis had particular explanatory value for sexual selection and especially female choice—issues with which he was currently much concerned. He devoted a special section of *Variation* to "Inheritance as Limited by Sex," giving evidence of such variation among domesticates, and stressing its relevance to the origin of secondary sexual characteristics:

> New characters often appear in one sex, and are afterwards transmitted to the same sex, either exclusively or in a much greater degree than to the other. This subject is important, because with animals of many kinds in a state of nature . . . secondary sexual characters . . . are often conspicuously present . . . The principle of inheritance as limited by sex shows how such characters might have been first acquired and subsequently modified.[44]

The test case, as ever, was pigeons, which, Darwin stressed, rarely show sexual difference "throughout the whole great family." The males and females of the parent form, the rock pigeon, are indistinguishable; yet with pouters "the male has the characteristic quality of pouting more strongly developed than the female." English carrier pigeons also exhibited conspicuous sexual differences in the greater development of the wattle over the beak and round the eyes in adult males. "So that here we have instances of the appearance of secondary sexual characters in the domesticated races of a species in which such differences are naturally quite absent."[45]

In Darwin's hands, pangenesis could be made to explain all those "laws" governing sexual plumage in young and adult birds that he had sought to disentangle while he was at work on the *Origin*: why "all plumage is related to sexual selection & laws all depend on degree to which they have happened to be confined to adult males." According to pangenesis, as a general rule, "both sexual elements [male and female] agree in power," and the offspring was a blend of, or intermediate between, both parents. However, if the gemmules derived from one parent "have some advantage in number, affinity, or vigour," then the characters of that parent would predominate in the offspring. Thus "one sex [usually male] sometimes has stronger power of transmission than the other." Most secondary sexual characters "which appertain to one sex, lie dormant in the other sex"; thus a grandfather's red beard or tendency to baldness would not be expressed in his daughter, but would be transmitted through her to his grandson. Pangenesis thereby could explain how males dominated inheritance, especially in those cases of female choice that particularly interested Darwin, where males were more brightly colored or ornamented than females and the young resembled the females. Pangenesis also could explain those other cases where both male and female offspring inherited some share of masculine color or patterning through its general principle of equal inheritance, how, as Darwin had put it in 1858 in his notes on Yarrell's *British Birds*, a hen bird "by selecting fine colours, makes her own sex beautiful."[46]

He brought these arguments to bear on female choice in the 1866 edition of the *Origin*. The preponderance of conspicuous coloring and ornamentation in male butterflies and birds was due to female choice and male sex-limited inheritance. When the female was "as beautifully coloured as the male" as in certain birds and butterflies, "the cause simply lies in the colours acquired through [female choice] having been inherited by both sexes, instead of by the males alone." At this stage Darwin did concede some role to natural selection in inhibiting equal inheritance by the sex in greater need of protection—or, as it came to be known in his dispute with Wallace, the "protective principle": the "proximate cause" of the transmission of some ornaments to the male alone could be seen in such instances as the shorter-tailed peahen—if she had the long tail of the male, she would be "badly fitted to sit on her eggs," or in the "modest attire" of the female capercailzie, who, if she were coal-black like the male, would be "far more conspicuous on her nest" and more exposed to danger.[47]

These critical understandings of development and inheritance were central to Darwin's famous dispute with Wallace over sexual selection.[48]

12.4 "Sexual selection . . . is *altogether* your own subject"

Early in 1867, Darwin began his muster of key informants on sexual selection. Blyth was the first: "*Sexual Selection.*—too many questions to ask." What Darwin particularly wanted was a more precise understanding of the vital laws governing "sexual plumage in birds" with respect to the differences in adult and juvenile plumage in various species at the breeding and winter seasons. He compiled a set of "*Blyths Laws*," which grouped the various relations of immature and adult bird plumage. This was subsequently reworked into the six laws governing such relations that figured in the *Descent*.[49]

Fritz Müller was contacted to help determine "how low down in the animal scale" sexual selection extended. Then Darwin turned to those two experts, Bates and Wallace, for assistance in resolving the finer and more problematic points relative to sexual selection. Wallace had recently taken up the subject of "adaptive mimicry" and sexual characters in butterflies and other insects in earnest, rivaling Bates. This work was complicating the interpretation of color, patterning, and sexual dimorphism. Bates modestly referred Darwin to Wallace to answer his query about the origin of the beautiful, artistic colors of caterpillars, which, being sexually immature, were not susceptible to sexual selection. In his letter to Wallace, wittingly or not, Darwin drew up the lines of battle:

> If anyone objected to male butterflies having been made beautiful by sexual selection, and asked why should they not have been made beautiful as well as their caterpillars, what would you answer? I could not answer, but should maintain my ground.[50]

Wallace replied, suggesting that the colors of caterpillars had a protective function. Many mimicked foliage, bark, or twigs; others had developed hairs and prickles that made them distasteful to birds; yet others with a disagreeable taste or odor probably displayed warning colors to inhibit their being mistaken by predators for the more palatable kinds. He proposed an experiment to test this: insectivorous birds should reject gaudy colored caterpillars, while readily devouring duller edible ones. Wallace then fired the first salvo in the battle for beauty:

> I sometimes doubt whether sexual selection has acted to produce the colours of *male butterflies*. I have thought that it was merely that it was advantageous for the females to have less brilliant colours, & that colour has been produced merely because in the process of infinite variation *all colours* in turn were produced. Undoubtedly two or three male butterflies do often follow a female, but whether *she* chooses between them or whether the strongest & most active gets her is the question. Cannot this also be decided by experiment?[51]

Prior to this, although Wallace had dissented from Darwin's beauty-based view of human racial divergence through masculine "pick of the women," he had not queried the efficacy of female choice in birds and insects. In an 1865 paper on the Malay Archipelago, he had affirmed that the "wonderful train of the Peacock and the gorgeous plumage of the Bird of Paradise" and many male insects "rejoicing in rich colours or sparkling luster" were the result of female choice. However, in the same paper he did suggest that some instances of sexual dimorphism in butterflies were due to natural selection. In some species of swallowtail butterflies, the female alone was mimetic—probably, Wallace thought, because it was of advantage to females, who moved more slowly when carrying eggs and were more exposed to predation when depositing them. In these cases, the female was more conspicuous than the male. Wallace then had extended this explanation to cases in which female butterflies had "colours better adapted to concealment."[52]

Now Wallace was tending to explain all forms of butterfly sexual dimorphism as due to the acquirement of protective coloration by the female by means of natural selection, undercutting one of Darwin's prized instances of the production of insect beauty by female choice and throwing into question the whole basis of the action of sexual selection in the production of color in sexually dimorphic species. Darwin reacted tactfully, praising Wallace's "ingenious" suggestion about the caterpillars while reaffirming his view that the beauty of male butterflies, as of male birds (and the origin of human racial differences), was all "due to sexual selection." He did not think the experiment suggested by Wallace (of rubbing the wings of some male butterflies to discolor them and then seeing whether the females would on average choose the unrubbed, best-colored ones) was feasible. Darwin once had a dragonfly painted with gorgeous colors, he told Wallace, but had "never had an opportunity of fairly trying it."[53]

Darwin was disturbed enough by Wallace's dissent to take up his idea of a defining experiment, focusing on birds rather than butterflies. As usual, he outsourced the work to William Tegetmeier, who was roped in to carry out the definitive trial of female choice in the fowl run. In spite of earlier discouragement, Darwin had not given up hope, and now he had a new twist, thanks to Wallace's notion of disfiguring the male and putting female choice to the test. Tegetmeier was to "cut & mutilate" the tail and saddle feathers of a cock—even better, a peacock—to see whether "his beauty spoiled . . . he would continue to be successful in getting wives." Tegetmeier, however, stuck to his view that among polygamous birds there was no such thing as female choice: gamecocks, with their combs and hackles (erectile neck feathers) closely trimmed, were as well received by the hens as untrimmed ones. He suggested an alternative experiment in monogamous birds of "dy[e]ing a white male pigeon magenta . . . and seeing whether his wife knows him." Darwin seized on this notion: he was "still inclined from many facts strongly to believe that the beauty of the male bird determines choice of female with wild Birds, however it may be under domestication." He "would be most particularly obliged" if Tegetmeier were to dye "a pigeon or two" and try the experiment. But Tegetmeier proved recalcitrant, evidently disinclined to put Darwin's views to the test, although reminded at intervals by his demanding patron.[54]

The need for a book on sexual selection was hammered home in early 1867 by the Duke of Argyll, who brought out his earlier criticisms of the

Origin and his views on God-created beauty in his popular work, *The Reign of Law*. The duke's subordination of natural selection to divine design was very appealing to those (including certain Darwinians like Asa Gray) who wanted some accommodation with natural theology. Natural selection, conceded the duke, could explain the "success, establishment and spread of new forms," but it could not explain the origin of beneficial variations. Only divine law could. What really provoked was the Duke's refusal to take any account at all of the beauty-making power of sexual selection, which he dismissed as "beside the question." Argyll focused instead on the limitations of natural selection: How could it possibly account for the marvelous "eyes" on the tail feathers of the argus pheasant, so exquisitely shaded to represent the "ball and socket" ornamentation of human art? What of the four hundred different species of hummingbirds, the males of which were all distinctively colored and ornamented—with crests, spangles, frills and tippets, or tail plumes? There was "no connection" between such splendid ornamentation and "any function essential to their life." A topaz crest was "no better in the struggle for existence" than one of sapphire. Why was beauty exclusively confined to one sex? The only answer to the great variety of form and ornamentation "which Creative Power seems to have worked with these wonderful and beautiful birds" is that their beauty was created "for its own sake."[55]

Huxley refused even to read the work, demoting its author to a "Dukelet." Hooker disgustedly reduced him to a "little man," even "lower than [Richard] Owen"; this was "too severe," protested Darwin, only half-facetiously, "for a Duke in my eyes is no common mortal." Only Charles Kingsley—ardent theist, author of the evolutionary fairy tale *The Water Babies* (1863), and a major popularizer of Darwinian evolution—demonstrated his good appreciation of sexual selection. Kingsley, with a flourish of masculinity, found the duke's book "very fair & manly," telling Darwin, "He writhes about under you as one who feels himself likely to be beat."

> What [Argyll] says about the humming birds is his weakest part. He utterly overlooks *sexual* selection by the females . . . he has overlooked that beauty in *males* alone, is a broad hint that the females are meant to be charmed thereby—& once allow that any striking new colour w[oul]d attract any single female, you have an opening for endless variation.[56]

Darwin, impressed by this affirmation of solidarity with superior Darwinian muscle, vented his feelings on the duke's book as "clever & very arrogant." He had "not a word to say against" the view of "the Deity having created objects beautiful for his own pleasure . . . but such a view c[oul]d hardly come into a scientific book." He was "glad" that Kingsley was "inclined to admit sexual selection"; Darwin was "more than ever convinced of [its] truth." He referred Kingsley to the fourth edition of the *Origin*, though there were "other explanations," including the "enclosed ingenious letter by Wallace." Kingsley should note that Darwin also had alluded to the "cause of female birds not being beautiful [through the need for protection while nesting]; but Mr. Wallace is going to generalize the same view to a grand extent."[57]

Indeed he was. Wallace was preparing a major article on "Mimicry, and Other Protective Resemblances among Animals." He was "more than ever convinced of the powerful effect of *'protective resemblances'* in determining and regulating the development of colour,—more especially in the *females* of insects." Wallace was now inclined to "impute the absence of brilliant or conspicuous tints in the female of Birds (when it exists in the male) almost entirely to this *protective* adaptation because in Birds, the female while *sitting* is much more exposed to attack than the male." The test case was the "wonderful" instance of the red phalarope, in which the female takes on the more gaily colored plumage in the breeding season: "Now strange to say the sexual habits are exactly reversed also, the *male alone* sitting on the eggs!!" This had led Wallace to "consider why it is that in a number of groups of conspicuously coloured birds, the sexes are *alike*, or at least *equally conspicuous*, contrary to the more general rule." He was "immediately led to the very simple reason that in these cases *sexual selection* had acted unchecked on both sexes, because the habits of the species were such that the female was not more exposed during incubation than the male." Wallace gave an impressive list of such instances: kingfishers, rollers, woodpeckers, parrots, tits, and wagtails. On this basis Wallace proposed a "law,—that where birds nidificate [nest] in *holes in the ground*, or *in holes in trees*, or build *covered nests*, the females will generally be as gaily coloured as the males."

The real sting came in the tail of his letter, where Wallace went on to claim, against Darwin, that the "primary action of sexual selection is to produce colour pretty equally in both sexes, but that it is checked in the females by the immense importance of protection, and the danger of conspicuous colouring." The cases of females being more brightly colored than males when they did not incubate, and "almost always" as brightly colored when they did incubate in *"perfect concealment,"* "proves . . . that the male admires gay colours in the female as well as the female in the male,—and that the *direct cause* of the prevalent dull colours in the female, is solely their *danger*; & does not at all shew that the males have no *taste in colour*."[58]

Wallace was not just delimiting the action of sexual selection in birds and insects and giving the major role to natural selection ("The case of the birds is exactly parallel to that of insects"[59]); he was also explicitly delimiting the explanatory power of female choice by claiming mutual selection by males and females. Implicit in Wallace's increasing emphasis on the protective principle was his belief that variations first appearing in one sex would be equally inherited by both sexes; where the female was exposed to greater danger, natural selection would prevent male choice from acting on the female and also convert equal inheritance into sex-limited inheritance, so that only the male would inherit the brilliant coloration produced by female choice.

In summary, for Wallace at this stage, mutual selection and equal inheritance accounted for those cases in which males and females were equally conspicuous; where one sex (usually female) was endangered and required protection, natural selection checked the action of sexual selection and converted equal inheritance into sex-limited inheritance, producing duller females. Darwin, by contrast, believed that although the most common form of inheritance was equal inheritance by both sexes, variations appearing in one sex (usually male) were fairly often sex-limited *from the first*. Thus female choice alone, in conjunction with sex-limited inheritance, could explain *both* conspicuous males and duller females; hence there was no need to invoke natural selection.[60]

Their differences were not always so clear-cut. In fact, Darwin initially reacted to Wallace's argument that the female had been rendered dull for protection by asserting his own priority: "Your view is not new to me." He pointed Wallace to the fourth edition of the *Origin* where, as he told Kingsley, he had invoked just such an explanation. He was about to discuss the "whole subject" of sexual selection in his forthcoming "essay on man": "I have collected all my old notes & partly written my discussion & it w[oul]d be flat work for me to give the leading idea as exclusively from you." In any case, he could not say he was "fully satisfied" with Wallace's arguments. Then, belatedly feeling that he had not given Wallace his due for the "value & beauty" of his generalization on the relation of nidification to color, he added a postscript to this effect.[61]

But the damage had been done. Wallace responded by assuring Darwin that sexual selection "is *altogether* your own subject" and dispatching his notes to Darwin, which Darwin might make "what use of you like" in his proposed essay on man. He had thought of writing a short paper on *"the connexion between the colours of female birds & their mode of nidification,"* but would leave it to Darwin to treat "as part of the really great subject of *'sexual selection.'"*[62]

Darwin was severely embarrassed; hoist with his own petard. Wallace once again had out-gentlemanned Darwin over the touchy issue of priority.

He returned Wallace's notes forthwith, apologizing for "a touch of illiberality." Wallace could work up the subject of protective coloration very much better than he could: "I earnestly & without any reservation hope that you will proceed with your paper." In this same letter, Darwin discussed the relevant "laws of inheritance," conceding that in those cases where characters arise in either sex and are transmitted to both sexes, "the survival of the fittest has come into play with female birds & kept the female dull-coloured." There was then, it seemed, substantial agreement with Wallace: "It is curious, how we hit on the same ideas."[63]

However, this carefully negotiated consensus was fragile. Wallace attributed to Darwin a much greater concession to the role of protection in the production of sexual dimorphism than Darwin was about to concede; while Darwin was already seeking to reassert female choice plus sex-limited inheritance as playing the dominant role. He continued to collect information on sexual dimorphism in crustaceans, spiders, annelids, butterflies, birds, and mammals from Blyth, Bates, and Müller, all of whom were receptive to his notion of sexual selection. He fought back with fishes. Albert Günther, assistant keeper of the zoology department in the British Museum, assured him of "many cases in which the male presents structural differences from female, solely for ornament." These many fish cases "make one doubt Wallace's view of importance of nests," Darwin noted. In the stickleback, "it is the male which is gaudily attired & he . . . guards the nest."[64]

There was, above all, the critical issue of the human races. Darwin's "essay on man," as he persistently reminded Wallace, was structured around "the whole subject of sexual selection, explaining as I believe it does much with respect to man."[65] On this issue Wallace was equally adamant: natural selection was quite equal to the task through "*correlation* with *constitutional adaptations* to climate, soil, food & other *external conditions.*" Wallace moreover could pit his extensive personal experiences of indigenous life against the notion of a beauty-based process of wife selection or of savage female choice:

> Stealing wives from other tribes for instance is a *very common* practice, & it would . . . tend to check any selective action. Youth is almost the only thing a savage cares about, and the handsomest & finest women very often become prostitutes & leave few or no offspring. The women certainly don't choose the men, & the men want chiefly in a wife, a *servant*. Beauty is . . . a very small consideration with most savages, as it is rare to find a woman so *plain* as not to leave as many or more offspring than the most beautiful. This is of course a delicate subject to go into . . . You must have facts of which I am quite ignorant.[66]

Man, "delicate" as the issue might be, was the ultimate sticking point. The long-awaited response from Thomas Bridges had finally arrived in January 1867, with its disappointing assertion of Fuegian preference for beauty in a woman "the nearer [she] approaches to the Caucasian race." Darwin did better in his search for sexually sensitive facts, when during a visit to London in February he sought out that old authority on Hottentots, Sir Andrew Smith. Smith, too ill to see him, later wrote a response to Darwin's written queries. He could not give Darwin "anything satisfactory as to what kind of women savage men prefer." But "so far as the Hottentot is concerned," Smith, who had not set foot in Africa for some thirty years, could say with certainty that "he regards a woman with huge *posteriors* as first rate." In the past, Hottentot men had "valued highly" the famous lengthened nymphae, "but now [presumably made more fastidious by civilization] he rather views these ugly developments as undesirable, if not as deformities." Smith had heard of women in whom the nymphae were so elongated that they were able to "encircle the man's loins and fix him by them until the appetite of both was thoroughly satisfied," though, admittedly, Smith himself had never seen any quite so long. The elongated nymphae were not artificially produced, but were a naturally occurring characteristic that first appeared at puberty. It was up to Darwin "to reconcile this speciality in the Hottentot's formation," to explain "what brought it into existence."[67]

The Hottentot backside and genitalia, on Smith's authority, became Darwin's prized pieces of evidence for human sexual selection. He filed Smith's salacious sexual fantasy away in his portfolio on racial selective practices, selectively edited it, and cited it in the *Descent*. He also owed to Smith the further embellishment of the Hottentot woman, "esteemed a beauty" for her "more than ordinarily" developed posterior, who could not rise to her feet unless seated against a slope, also recounted in the *Descent*. The African penchant for fatty bottoms, down to the deliberate selection of the one "who projects farthest *a tergo*," was endorsed by that other noted authority on African women, the notorious adventurer Richard Burton, prominent member of Hunt's inner Cannibal Clique. Darwin had this marked up in his copy of the 1864 volume of the *Anthropological Review*, also to be cited in the *Descent*, though he conveniently overlooked this expert on erotica's further claim that he had "practically found that a woman who would be called pretty in Europe, would also be admired in Asia, Africa, and America."[68]

Darwin's unreflecting trust in Eurocentric perceptions and interpretations of other ethnic practices and attitudes has been remarked on. This was especially the case with his other concurrent "hobby-horse," the physical nature of the expression of the emotions. He collected information willy-nilly on both beauty and expression, but more systematically and on a larger scale on expression, having an extensive list of queries privately printed and circulated around the globe—especially to more remote regions where valuable "observations on natives who have had little communication with Europeans" might be had. The effects of frontier conflicts, repressive colonial policies, missionary activities, or diverse cultural practices on the perceptions and behaviors of observers and observed otherwise went unremarked. A few observers did comment on the difficulties of observing expression in "natives" who masked their emotions in the presence of foreigners or whose culture inhibited the expression of the emotions.[69]

Meanwhile, Wallace, given carte blanche by Darwin, took the bit between his teeth and ran wild with his protection thesis. His two major productions of late 1867, on mimicry and birds' nests, were directed toward demonstrating that almost all coloration in animals was adaptive (that is, protective) and hence the product of natural selection, not sexual selection (though he still retained a limited role for sexual selection in birds). It was Wallace's stated view that "none of the definite facts of organic nature, no special organ, no characteristic form or marking, no peculiarities of instinct or habit, no relations between species or between groups of species—can exist but which must now be or once have been *useful* to the individuals or races which possess them." This was a "necessary deduction" from natural selection. The cofounder of the theory of natural selection was hardening into an extreme utilitarian.[70]

Darwin had hoped that Kingsley, sympathetic to sexual selection, might publicly combat the Duke of Argyll; but it was Wallace who "smashed" him in a review in the September issue of the *Quarterly Journal of Science*. Wallace's attack on the duke amounted to an extended defense of the explanatory power of natural selection with only minimal reference to sexual selection. True, male hummingbirds were made splendid through sexual selection, but the females were made dull through the need for protection. Most forms of beauty, including orchids—and just about everything else raised by the duke—might be explained by adaptation through natural selection.[71]

Darwin "admire[d] every word," though he could "not but think that you push protection too far in some cases." However, he said, he had been "fully converted" by Wallace's paper on the mode of nesting and sexual dimorphism in birds: it was a "capital generalization." Wallace could be generous in return: he "may perhaps" push protection too far, it was his "hobby just now"; he supposed Darwin was getting on with his book on "Sexual selection & Man, by way of relaxation."[72]

Darwin was not. He was wrestling with the proofing of *Variation*. He could not resist incorporating new material, especially in support of pangenesis. He also was busy negotiating its foreign translations.

Darwin had become wary of his translators after his experiences with Clémence Royer, who translated the first French edition of the *Origin* in 1862. Royer was a clever, headstrong woman who rejected the royalist politics and Catholic beliefs in which she was raised to "recreate" herself as republican, atheist, and feminist. As an antidote to the repressive "myths" of religion, she immersed herself in the liberating "positive truths" of science, emerging as a committed Darwinist, a biological determinist who sought to explain all of human morality and society in evolutionary terms. She was a force for the popularization of Darwinism in France through her many evolutionary publications, which reinterpreted those aspects of Darwinism that were antithetical to her idiosyncratic blend of feminist and elitist views. She added a vehemently anticlerical preface to her translation of the *Origin* and extensive footnotes, giving her own spin on Darwin's text. A bemused Darwin initially pronounced Royer "one of the cleverest and oddest women in Europe"; but by the time of the second Royer edition of 1866, she had become "the verdammte Mlle Royer."[73]

He was concerned that her aggressive anticlericalism (her "enormously long & blasphemous preface") was affecting the acceptance of his ideas in France, and he was annoyed by her insistence on interpolating her controversial views into her translations. Royer was one of the earliest to suggest eugenic remedies for what she saw as the lamentable tendency of Christian piety to promote the preservation of the weak and unfit at the expense of the strong and the healthy. This tendency ran counter to the beneficial effects of natural selection and social progress; moreover, through their selection of partners on the basis of passivity and beauty, men had weakened human development:

> One must conclude that in order to hasten the progress of the race, it will become necessary to ask for woman a part of what has been asked for man, that is, strength united to beauty, intelligence united to gentleness, and for man, a little idealism united to vigor of mind and body.[74]

Darwin could accommodate eugenic suggestions when they came from accredited Darwinian players like Galton or Greg, but Royer did not play by the rules. She had no respect for his authority, and Darwin had his own views on marriage selection and human evolution. Her sins were compounded by the fact that she was a woman—a Frenchwoman, and an immoral one at that.[75] He arranged an alternative translator for *Variation*, angering Royer, who promptly brought out a third and unauthorized French edition of the *Origin* in which she savaged his cherished pangenesis—"abusing me like a pick-pocket," Darwin complained to Hooker. "How *nasty* women are when spiteful," Hooker affirmed in masculine solidarity, "& French women perhaps the worst in the world."[76]

Mercifully, "my Ladies" were disinclined to obtrude their own opinions into his work. Henrietta (now a young woman of twenty-four and coming into her own in her proper role of amanuensis) read and corrected the proof sheets of *Variation*, earning nothing but praise: "All your remarks, criticisms doubts & corrections are excellent, excellent, excellent."[77]

Royer was not the only would-be exponent of Darwinism to cause concern. It was difficult enough to keep control of British Darwinism, with its various adherents running off in all directions; in Germany the "terrible 'Darwinismus'" was provoking just that counterreaction Darwin was anxious to avoid. The radical Carl Vogt reportedly was touring the country delivering lectures "like a travelling preacher," but not in the true faith—rather, in favor of "materialism in the absurdest form." Darwin was not about to chide Vogt (they maintained a friendly correspondence, though Vogt was rejected as German translator of *Variation*[78]), but he was prevailed on to intercede with his most devoted disciple. Haeckel's "quite unnecessary remarks and immoderate sharpness" were provoking alarm among more moderate German Darwinians who had "tried in vain to mitigate his fury"; but Haeckel would listen only to Darwin. Darwin did his best, telling Haeckel, "It is easy to preach & if I had the power of writing with severity I dare say I sh[oul]d triumph in turning poor devils inside out & exposing all their imbecility." Nevertheless, Darwin was "convinced this power does no good, only causes pain." He added, "As we daily see men arriving at opposite conclusions from the same premises it seems to me doubtful policy to speak too positively on any complex subject however much a man may feel convinced of the truth of his own conclusions."[79]

It was advice that Darwin struggled to bring to bear on his own accumulating differences with Wallace.

12.5 A "metamorphosed (in retrograde direction) naturalist"

There could be no doubt that Wallace was a misfit. He was coauthor of the theory of natural selection, and his modesty in ceding priority to Darwin won great admiration and respect. He had returned to London in 1862 as something of a celebrity, a naturalist of repute. His string of subsequent publications had enhanced his scientific standing, though they had not secured him a major teaching or research post. Darwin and Hooker, despite earlier reservations, did not doubt his "extraordinary talents"; he had established friendly relations with the leading Darwinians, but he was perceived as not quite of their circle. He was socially diffident, and he had a background in "trade," selling butterflies and beetles to finance his expeditions; Lady Lyell snobbishly thought Wallace "shy, awkward, and quite unused to good society." He also was far too inclined to go his own way. He did not necessarily toe the Huxleyan line (witness his persistence in maintaining his membership of the Anthropological Society); he had no formal scientific training and remained outside the professional community; he earned a precarious living from occasional teaching, school examining, and publishing; he was not a member of the exclusive and influential X Club (though this included other nonprofessionals, such as Spencer); most embarrassingly, he recently had converted to the increasingly popular "parlor religion," spiritualism. Darwin had received a copy of his pamphlet *The Scientific Aspect of the Supernatural* toward the end of 1866, in which Wallace declared his personal belief in the authenticity of séances, clairvoyance, and otherworldly phenomena he had witnessed. This public avowal ran directly counter to the Huxleyan-fostered ideology of scientific naturalism—the pursuit of a professional program of objective research, free of political, religious, and all extrascientific taints. Huxley was a scathing critic of spiritualism's "disembodied gossip" as well as of orthodox supernatural belief, especially as they obtruded on his ideology of the neutrality of science. He was shortly to introduce the useful neologism *agnosticism* for his secular form of unbelief.[80]

Yet Wallace was hardly unique in his belief in the existence of a spirit world that could communicate with and influence the material one. Recent scholarship has documented the intense sense of religious doubt and questioning that drew many Victorians into spiritualism, which seemed to offer a vehicle for mediating between the competing claims of traditional religion and materialist science. It attracted thousands of men and women from diverse socioeconomic, political, and religious backgrounds, including some of Britain's foremost intellectuals and scientists. Wallace's complex commitment to spiritualism included the assumption that spiritualist phenomena might be brought within the scope of legitimate scientific inquiry and validation; but his spiritualism also was structured by its relation to Wallace's belief in human equality and social reform. Many old radicals in the wake of the collapse of Chartism and the failure of the 1848 revolutions took to spiritualism—including Robert Owen, whose utopian communitarian ideals preached in the London Hall of Science had inspired Wallace's early socialist, humanitarian awareness. If political agitation and organization could not undermine individualistic competition and bring about the earthly paradise, spiritual forces might drive humanity to cooperative ends.

Owenite socialists and communitarians had promoted the equality of the sexes, including equal education for boys and girls. Wallace's spiritualism reinforced the Owenite foundations of his gathering feminism. Spiritualism was one of the few arenas in which Victorian women could play influential, even dominant, roles. This interpretation helps make sense of what has been seen as Wallace's credulity in tenaciously defending mediums (most of whom were female) against continuing allegations of fraud. His immersion in spiritualist circles also gave the shy and unassuming Wallace the opportunity of socializing with women (and of course men) who shared not only his spiritualist beliefs, but also his egalitarian and sociopolitical commitments. He earlier had suffered pain and humiliation when his engagement to a "Miss L" was abruptly terminated; but in 1866 Wallace, at forty-three, married the twenty-year-old Annie Mitten (daughter of the botanist William Mitten), also a spiritualist. Their marriage was enduring and mutually satisfying, and no doubt contributed to Wallace's sense of what he thought gender relations ought to be.[81]

It seems easy enough to relate Wallace's insistence on mutual selection and equal inheritance in the early stages of his dispute with Darwin over female choice to Wallace's growing radical egalitarianism of that same period. Yet it is important to realize that Wallace retained throughout this dispute and beyond it a fairly traditional rendering of femininity. For all his feminism, he did not escape Victorian assumptions of an essentially passive female sexuality and of the central significance of the institution of marriage in any society, including the society of the future. Wallace never was susceptible to the antimarriage stance adopted by certain Owenites, and he thought the advocacy of free love by radical contemporaries like Grant Allen "detestable." Not unlike Darwin, Wallace feared free love would destroy the family and undermine parental affection—the mainstays of Wallace's personal existence. His previously discussed 1868 exchange with Darwin over the "deep-seated law of nature," by which the male invariably seeks out the female and never the reverse, is a case in point. Wallace accepted Darwin's counterclaim that it was precisely because the male was the "searcher" with "more eager passions than the female" that males were not selective in their mating choices, but females were: Darwin's argument was "good . . . as far as it goes."[82]

Wallace's conventional view of female sexuality is most evident in his famous volte-face of 1890, when, after more than twenty years of forceful criticism of sexual selection, he advocated free female choice in marriage (uncoerced by economic need) as critical to biological and social progress in a postsocialist society. Even so, his "woman of the future" was not the seeker but the sought; her sexual and social power was the result of her socially secured ability to say no to those suitors who did not come up to her exacting moral standards. In Wallace's hands, sexual selection had less to do with sexuality or aesthetic choice than with the exercise of woman's spiritually endowed superior moral judgment (though he also thought that the higher education of women would enhance this). Nevertheless, his insistence on situating females at the center of any scheme for future social improvement echoes his earlier insistence in his dispute with Darwin that dull-colored egg-laying and nesting females were more essential for the future of the race than flashy evanescent males, and that therefore the protective principle was of far greater import in biological evolution than any unsubstantiated frivolous choice of spontaneously produced colors and frills: "The prolonged existence of the males is in most cases quite unnecessary for the continuance of the race." It was a notion that Darwin as forcefully resisted.[83]

The better-known reason adduced for Wallace's rejection of the notion of female choice is "the firmness of his conviction that the aesthetic faculties were a part of the 'spiritual nature' conferred upon mankind alone by a supernatural act."[84] However, this requires considerable qualification. There can be no doubt that spiritualism—validated, as he was convinced, by his own intensive investigations of psychical phenomena—powerfully reinforced Wallace's growing theistic convictions. By the late 1860s, he began to integrate these views into his evolutionary writings, particularly his writings on human evolution. Rather than being incompatible with evolution, for Wallace, spiritualism completed biological theory, and he ascribed to it those human attributes that he considered inexplicable by natural selection, including the human appreciation of natural beauty.[85] But Wallace's increasing emphasis on the limitations of natural selection in the history of human evolution was as much a correlate of his reawakened Owenite political views as of his conversion to spiritualism.

When Wallace had presented his 1864 paper on "man" to the Anthropological Society, he had been very much under the sway of the leading Social Darwinian, Herbert Spencer. Although the outlook of his paper was utopian, the mechanism Wallace proposed for social progress was Spencerian. Improvement was to be achieved through the survival of the fittest (the Spencerian terminology that Wallace urged on Darwin, as less open to misinterpretation than natural selection), by the triumph of "more intellectual and moral" societies over the "lower and more degraded races." Other races were disappearing "from the inevitable effects of an unequal mental and physical struggle" with "superior" whites. This justification of Western imperialism did not sit easily with Wallace's earlier positive evaluations of indigenous peoples and cultures and his more critical view of colonial exploitation and dispossession. In South America, where Darwin encountered the "naked painted . . . hideous savage" whose insistent presence reverberates through his writings, Wallace, by contrast, met with "handsome, naked painted Indians," people of physical perfection living in social harmony. He remembered his "first meeting and living with man in a state of nature—with absolute uncontaminated savages" as a "most unexpected sensation of surprise and delight." It was this enduring memory of physical grace, sociability, and surprising intelligence and artistry, far beyond his expectations of savage encounter (primed, it should be noted, by his reading of Darwin's *Voyage of the Beagle*, among other travel narratives) that remained with Wallace to the end of his days. During his subsequent sojourn among the Dyaks of Borneo, Wallace famously wrote: "The more I see of uncivilized people, the better I think of human nature on the whole, and the essential differences between so-called civilized and savage man seem to disappear." His uneasiness with his 1864 Spencerian interpretation is evident in his response to Spencer's request for a follow-up article on "How to Civilize Savages" for the *Reader*: "White men in our colonies are too frequently the true savages."[86]

Wallace's discomfiture was heightened when in 1868 William Greg's controversial article, "On the Failure of 'Natural Selection' in the Case of Man," appeared in *Fraser's Magazine*. Greg drew extensively on Wallace's 1864 paper on human evolution to support his conclusion that a whole series of Victorian social reforms had compromised the law of natural selection by encouraging the survival of the unfit (a problem earlier identified by Francis Galton in his analysis of "Hereditary Talent and Character" in 1865). Wallace had argued that once racial physical characteristics had evolved, natural selection had acted on the human mind to promote the beneficial moral and social qualities of sympathy, cooperation, and altruism that characterized advanced societies. Greg upended Wallace's argument to claim it was those very qualities that were inhibiting the progress that untrammeled competition would ensure. The inferior (Greg, a Scot, had the "careless, squalid, unaspiring Irishman" in his sights) were protected from the consequences of their own improvidence and intellectual and physical inadequacies and were in the process of outbreeding the late-marrying "frugal, foreseeing, self-respecting, ambitious Scot." Greg recommended a form of eugenic control of reproduction that would weed out the undesirable: in an ideal society only those who passed rigorous competitive examination would be allowed to breed.[87]

As we have seen, Darwin, reading Greg, jotted his own sexual selection-ist remedy in the margin: "Live long & choose good wives." He was to draw directly on Greg's analysis (including the aspersions on the Irish) in the *Descent*. Wallace's response was to emphasize the uniqueness of man: man had the ability to transcend the brutish Hobbesian battle of all against all, and he might take the example of so-called primitives into account. Wallace concluded *The Malay Archipelago* (1869) by favorably comparing the state of justice, morality, and individual freedom in societies like that of the Dyaks with the social "barbarism" of Victorian England. Wallace's promotion of a humanitarian alternative to the Victorian ideal of competitive individualism led to correspondence with John Stuart Mill (whose *Subjection of Women* ap-peared in the same year) and an invitation to join Mill's Land Tenure Reform Association, sparking Wallace's lifelong commitment to the issue of land reform. Mill by this stage had lost his seat in Parliament. His efforts to se-cure an amendment to the 1867 Reform Bill (whose purpose was a modest expansion of the male electorate) to extend the same voting rights to women failed. But Mill's amendment received enormous publicity (though mostly mocking and adverse) and achieved an astonishing seventy-three votes in its support.[88] Wallace, not yet an outspoken advocate of women's rights, was moving steadily in that direction and well away from his earlier defense of the exclusion of women from the Anthropological Society.[89]

The Malay Archipelago (which Wallace dedicated to Darwin) went on sale early in 1869, and only a few weeks later Wallace went public in the *Quar-terly Review* with his belief that an "Overruling Intelligence" had guided the higher development of the human race. He had been handpicked for the *Quarterly* as a likely sympathetic reviewer of the tenth edition of *Principles of Geology* by the aging Lyell. Wallace obliged with this "*first time*" discussion of his own views on the "limitations to the power of natural selection." He prepared Darwin beforehand for his public defection: "I am afraid that Hux-ley & perhaps yourself will think [my views] weak & unphilosophical . . . but [they] are the expression of a deep conviction founded on evidence . . . which is to me absolutely unassailable."[90]

It is well known that Wallace explicitly rejected the view that the human moral sense and intelligence could be accounted for by natural selection. What is less known is that it was his positive evaluation of indigenous peoples that gave Wallace what he took as decisive evidence against that view and enabled him to rationalize supernatural guidance of human evolution. The intellect and bodily perfection of indigenes exceeded what was necessary for their survival, and this ran counter to the utilitarian principle that only those structures immediately useful to their possessors would be preserved and accumulated through natural selection. If the savage brain and body could not have been produced by natural selection, then it must have had supernatural help. Ideas of justice, abstract reasoning, and moral sensitivity were evident in both civilized society and that of the "lowest savages." The brain of the savage was virtually indistinguishable from "that of the average members of our learned societies" (surely a tilt at the Anthropologicals). Savages could, under appropriate conditions, be capable of the same outstanding intellectual and artistic achievements. Yet the mental requirements of savages were "very little above those of many animals." The human brain was overdesigned for the rudimentary requirements of savage life: the "utility principle" could have produced a brain only a little superior to that of an ape. Natural selection could not, therefore, account for the complexity of the savage brain (or, by implication, that of primitive man). Nor could it explain the refinement of the savage hand, which could be put to any of the advanced requirements of art, technology, and science; nor the erect posture of the savage, "his delicate and yet expressive features, the marvelous beauty and symmetry of his whole external form"; nor could his nakedness, his relative hairlessness, be considered of any utility to him, though it may have promoted the sense of personal modesty and so "profoundly affected our moral nature." Indeed, argued Wallace, the "supreme beauty" of the human form and face, though initially of no practical use, had probably given rise to those uniquely human aesthetic and emotional qualities that set humans apart from animals. If humans had retained the ugliness of an erect gorilla, these delicate sensibilities never could have evolved. These higher qualities and faculties (including human speech and music) could only be due to an "Overruling Intelligence" that had guided human development through a process of variation and selection similar to that exercised by breeders in the improvement of domesticates.[91]

With one sweep, Wallace removed human mental and moral progress from the vagaries of a merciless natural selection and put it within the reforming powers of an intelligent superhuman selector. And, at the very moment that Darwin was moving to finalize his exhaustive survey of sexual selection throughout the animal kingdom with particular reference to the role played by aesthetic selection in human racial divergence, Wallace consigned human beauty and the human aesthetic sense (not to mention the expression of the emotions) to the same supernatural agency. Even though Darwin was forewarned, the shock was considerable. Had he not "known to the contrary," Darwin would have "sworn" that this section of Wallace's review "had been inserted by some other hand." "You write like a metamorphosed (in retrograde direction) naturalist," he groaned, "& you the author of the best paper that ever appeared in Anthr[opological] Review! Eheu! Eheu! Eheu!"[92]

12.6 "But we shall never convince each other"

These larger differences between the two ran deep, and they are important. But they have distracted attention from another fundamental difference between the two naturalists that must be seen as more immediately pertinent to their dispute over sexual selection—and that is Darwin's saturation in and reliance on breeding lore and its practices; in particular, Darwin's identification of the selective practices of the breeder with the visual processes of mate selection. This in turn relates to their very different understandings of the selective process. Wallace, from his earliest paper on the origin of species of 1858, explicitly rejected the analogy with domestication and described selection as acting on varieties or races as well as individuals; whereas Darwin, with reference to the same analogy, always described selection as acting on individual organisms (except in special cases; for instance, sterile castes of social insects). Wallace had an environmentalist conception of natural selection: organisms were "brought into harmony with" their changing environments; the world of nature was "self-adjusting and capable of endless development." For Darwin, natural selection was an intensely competitive one-on-one process that not only adapted organisms to their changing conditions of existence, but that also (like the selective practices of the pigeon fancier) continued to act when these conditions were stable, driving divergence. Sexual selection, the competition for mates, is in this sense a "corollary" of this relentless competitive process, and it produces sexual divergence without any adaptive advantage.[93]

While Darwin, his travels and collecting experiences well behind him, was fraternizing with the fancy, breeding his own pigeons, and educating his fancier's "eye" for minute differences of individual form and color, Wallace—immersed in tropical forest—was honing his skills as a superb naturalist and an acute observer in the field, sensitive to the intricacies of animal and insect patterning and coloring viewed against their backgrounding foliage and forest-floor litter. His very success as a collector depended on this heightened perception of the visual engagements of living organisms with their environments. In his preface to The Malay Archipelago, Wallace enumerated the staggering total of his collection: "125,660 specimens of natural history" (310 mammals, 100 reptiles, 8,050 birds, 7,500 shells,

13,100 butterflies, 83,200 beetles and 13,400 other insects). While many were collected by paid assistants and indigenous hunters, Wallace's own part in their collection—particularly of those that could be profitably marketed (butterflies, beetles, and birds)—was critical. He "hunted specimens in the field for twelve years, six hours a day, spending weeks at a time in scores of different sites." This intensive and diverse collecting experience was brought to bear on Wallace's sophisticated interpretations of mimicry and protective camouflage: his collector's "eye" sought out and could appreciate the means by which even a seemingly gaudy animal or insect might disappear into its natural surroundings of shifting leaves, sticks, stones, and shadows. This emphasis is evident in his original 1858 paper, where Wallace drew attention to cryptic coloration ("the peculiar colours of many animals, especially insects, so closely resembling the soil or the leaves or trunks on which they habitually reside"), which he attributed to the action of natural selection: "Those races having colours best adapted to concealment from their enemies would inevitably survive the longest." We might say that where Wallace's focus was on the hard-to-see, on the subtle patterning and feathering that made a female bird sitting on her open nest blend into her environment ("by assimilating her to surrounding objects as the earth or the foliage"), Darwin's eye was caught and held by those more visible features of beak and tail length and shape, of color intensity, spots, and ocelli—those characteristics that made an individual stand out from its fellows, all detached from their environment and all viewed and valued as they would be by a fancier.[94]

This goes some way toward explaining the seeming paradox that Wallace— the collector of rare and beautiful butterflies and birds, including the stunning birds of paradise—should have so resisted the urge to fix on the bright and beautiful that dominated Darwin's conception of sexual selection.[95] Wallace was not immune to beauty. In *The Malay Archipelago* he recorded his extreme response to finally holding in hand the male of the elusive and extremely beautiful "bird-winged butterfly": a near-fainting experience so intensely emotional and physical that Wallace could liken it only to his previously experienced "apprehension of immediate death." Even so, he recorded the particular species of shrub that the butterfly frequented. Then there is the "wonderfully elegant image," the frontispiece in the same work, of the "Natives of Aru Shooting the Great Bird of Paradise," in which birds of ethereal beauty gather to display in the canopy above the bowmen who lie artfully camouflaged in the branches below. It is an image in which a Darwinian interpretation of male beauty conspicuously displayed for female admiration is brought into direct conflict with a Wallacean emphasis on its dangers and the need for protection from predation.

(It is also an image in which the indigenous hunters seem to share in the natural beauty of the birds. Wallace was full of admiration for the physical perfection of the Aru bowmen: "What are the finest Grecian statues to the living, moving, breathing men I saw daily around me?")[96] The dazzling birds' beauty notwithstanding, Wallace approached them with a practical collector's acumen, fully aware of their commercial and scientific value. The birds exhibited complex distribution patterns that featured in Wallace's biogeographical and evolutionary theorizing. He set out to collect every species of bird of paradise, both male and female specimens at different stages of maturity. But before all, the sale of their plumes bought Wallace five more years of traveling and collecting time.[97]

Darwin, in his youth, honed by his immersion in Humboldtian romanticism, could wax ecstatic at the beauty and grandeur of the Brazilian rain forest. By the 1860s, the closest he came to matching Wallace's intense physical response to inexplicable beauty was that he was made "sick" at the sight of the feather in a peacock's tail by his inability to explain it. By the end of the decade, both Wallace and Darwin were looking to explain the peacock's tail and the beauty of the butterfly in naturalistic, evolutionary terms. Both men had larger ideological concerns and commitments, but their dispute over sexual selection was carried out explicitly within the context of scientific naturalism and bolstered by such naturalistic arguments and instances as each could summon to his aid.[98] The major stages of their conflict preceded Wallace's public invocation of some higher power to explain the evolution of the human aesthetic sense. By October 1868, it was all over: both had agreed on their mutual inability to convince the other and to go their separate ways. Darwin's position was set firm and remained so until his death.

Darwin's long haul from his explicit identification of the manipulative pigeon breeder with the sexually selecting human male, on to its close coupling with an instinctive, efficacious appreciation of the beautiful by female animals and birds was complete. This was the crux of his construction of mate choice as the crucial component of sexual selection and the most original aspect of his theorizing. Male combat for possession of the females might be understood as an aspect of natural selection, but female choice could not be so interpreted. His explanation for nonfunctional male ornamentation or potentially harmful conspicuous coloration in animals dovetailed with Darwin's central explanatory fashion motif and with his long-held theories of inheritance and embryology. He might waver before the force of Wallace's cogent criticisms, but he was not about to cast this tightly interlaced complex of theory, practice, and analogy aside, nor even to modify it substantially. As often as Darwin teetered, he thudded back to his

Figure 12.3. "Natives of Aru Shooting the Great Bird of Paradise," the frontispiece to the second volume of Wallace's *Malay Archipelago* (1869). The beautifully plumaged males gather to display in the canopy above the artfully concealed bowmen. A Darwinian interpretation of male beauty conspicuously displayed for female admiration is brought into conflict with a Wallacean emphasis on its dangers and the need for protection from predation. Wallace 1869. State Library of New South Wales, Mitchell Collection.

grounding in pigeons, to his understanding of the fancier's practice of selecting to extremes, of pigeon development, and of sex-limited inheritance in
pouters and carriers: "I w[oul]d accept Wallace's view about [female] Birds
did I not remember pigeons."[99]

In January 1868, with *Variation* finally in print, Darwin recorded his
return to work on "Man & Sexual Selection." Almost immediately he told
Wallace: "I am fearfully puzzled how far to extend your protective views
with respect to females in various classes. The more I work the more important sexual selection apparently comes out."[100] He was making out a "strong
case" for sexual selection among insects through "fighting for the females, &
attracting them by music." His son Francis, now a Cambridge undergraduate
with a penchant for natural history, was collecting evidence on stridulation
and smell as attractants. Bates was convinced of the "gradual brightening
of the male beauty [in tropical butterflies] through the long ages of sexual
selection," and he could show Darwin cases "where the male varies in ornamental colours & where selection is in all probability *now* going on."[101]

But birds were the test case. Here Darwin was beginning to build an
impressive case for sexual selection, diverging more and more from Wallace. In March he reported on his "grand success this morning in tracing
the gradational steps by which Peacocks tail has been developed [through
female choice]. I quite feel as if I had seen a long line of its progenitors."
He could find no evidence of male birds ever selecting the most beautiful
females, and he could "see no improbability (but from analogy of domestic
animals a strong probability) that variations leading to beauty must *often*
have occurred in the males alone, & been transmitted to that sex alone. Thus
I sh[oul]d account in many cases for the greater beauty of the male over the
female, without the need of the protective principle."[102]

As Darwin's opposition to Wallace's protection thesis hardened, Wallace
became as convinced of its validity and increasingly critical of Darwin's intensifying reliance on female choice for the explanation of male ornamentation and color. While he agreed that sexual selection "may occasionally
produce brighter colours as well as additional ornaments & weapons in the
male," he extended his claim to argue that in "the great mass of cases in
which there is great differentiation of colour between the sexes, I believe it is
due *almost wholly* to the need for protection to the female." Darwin was not
to be moved. Even if he were to grant "(only for argument) that the life of the
male is of *very* little value," and if he were to grant further that the male did
not vary through female choice, then why, if both sexes inherited equally,
"has not the protective beauty of the female been transferred by inheritance
to the male? The beauty would be a gain to the male, as far as we can see, as

a protection; & I cannot believe that it would be repulsive to the female as she became beautiful.—But we shall never convince each other."[103]

Back came Wallace with a six-stage summary of his views on animal coloration: (1) color constantly varied and was generally inherited by both sexes; (2) color was protective in several ways—as camouflage, as mimicry, or as warning coloration; (3) color attracted the opposite sex; (4) it was therefore selected and accumulated; (5) the female was in greater need of protection and, more important, for the preservation of offspring, so the female alone often acquired protective coloration; (6) this occurred "either by subduing or checking the colour as acquired by the male, or by the accumulation of entirely distinct colours & markings." Darwin, in overt conciliatory mode, agreed "almost entirely"; nevertheless, he stubbornly maintained that he would "put sexual selection as an equal, or perhaps as even a more important agent in giving colour than natural selection for protection." Around the same time, he told Hooker (who had praised Wallace's recent paper on birds' nests) that while he agreed about Wallace's "wonderful cleverness," Wallace was "not cautious enough": "I find I must (& I always distrust myself when I differ from him) separate rather widely from him all about Bird's nests & protection; he is riding that hobby to death."[104]

To the death they went, Wallace going the harder against Darwin's toughening resistance. Alarm bells rang when in August Darwin wrote, "The further I get on, the more I differ from you about the females being dull-coloured for protection."[105]

Ultimately, for Darwin, it was the fixity of his long-held views on development and inheritance (views he considered to be experimentally and evidentially substantiated) that settled the matter.[106] Darwin had queried an array of correspondents—including Blyth, Tegetmeier, and Darwin's newest and most devoted advocate of female choice in birds, the customs official and part-time naturalist John Jenner Weir—as to whether they knew of any instances of vividly colored young birds that became duller when adult. None did. This, as far as Darwin was concerned, gave the lie to the protection thesis. If, as Darwin interpreted Wallace, conspicuous female birds were converted by natural selection into dull-colored ones through the need for protection, then the record of this modification should be preserved in their development: young female birds should be more conspicuously colored than mature dull ones. But, as Darwin had expected, they were not. There were no exceptions to his fundamental rule that, where adult males were more brightly colored than females, the young of both sexes resembled the duller females. It followed that the adult bright male alone had undergone modification, and, as Darwin had long assumed, that the dull female represented the earlier ancestral stage of the group: "The females of almost all the species in the same genus, or even family, resemble each other much more closely in colour than do the males; and this indicates that the males have undergone a greater amount of modification than the females."[107] This was consistent with his claim that sexually dimorphic characters were sex-limited from the beginning, and it reinforced his conviction against Wallace's notion of equal inheritance with the females being made dull for protection and inheriting this advantage.

Further to this, as his notes for the *Descent* show, by May 1868 Darwin had concluded, against Wallace, that it was not the form of the nest (open or concealed) that determined plumage, but plumage that determined the kind of nest: "Hence colour as transmitted to sex seems to have led to form of nest, & not the latter to that determination of colour of female sex.— *Very good.*" In the *Descent*, Darwin was to argue that originally both sexes were dull-colored and built open nests; female choice then made the male brightly colored; if this color was sex limited from the first, then the female remained dully colored and the nest remained open; if, however, the females inherited this bright color (equally or partially), then their nesting habits changed and they built domed or concealed nests to hide their conspicuous color.[108] This ingenious reversal of Wallace's argument produced the same results with respect to color and type of nest, but without the need to invoke natural selection for protection. It also conformed nicely to Darwin's general rules for the relation of adult to young plumage.

At this stage (early May 1868), contrary to the accepted account of their dispute, Darwin's notes show that he *did* consider the possibility that "Wallace w[oul]d say that [female] never had become beautiful but had been from first checked [by natural selection]."[109] But such a check, Darwin argued, could come "only at maturity . . . during nesting search" when the female was in need of protection, so some record of this still should be retained in development: "I believe that young w[oul]d sometimes show bright colours if changed, because in invariable instance, when [male] has become bright-coloured [young male] has retained previous dark-colours." Bright-colored males could never become dull, "as sex[ual] selection w[oul]d oppose change." He was inclined to think that natural selection came into play only "in checking differences which are dangerous being carried to an extreme." In any case, "we know" that with pigeons, the "wattles & colour have given with male alone"; that is, these characteristic sexual differences are limited to the males and inherited only by males. If the females were initially dull and kept dull through sex-limited inheritance, natural selection was superfluous, as Darwin went on to reason to himself:

Wallace's view requires sexual limitation just as much as mine—only more complex.—Beauty is checked in female, & the arrested beauty is transmitted to female alone—surely it is simpler to suppose that the initial variations followed the law of exact limitation.[110]

Darwin's arguments were well rehearsed, and he was certain enough of them to put them to the test by inviting Weir, Bates (who unfortunately could not accept Darwin's flattering summons, being otherwise engaged), Blyth, and Wallace to Down House for the weekend of September 12 and 13, 1868. It was a get-together of Darwin's leading informants on sexual selection, and its significance has been entirely overlooked by historians. Darwin left no identifiable record of their meeting (though some of his notes possibly relate to it[111]), but it is obvious what he was about with this carefully chosen gathering. All bar Wallace were strong advocates of female choice and might be counted on to give Darwin support from their various perspectives and expertise. "The thought makes me rather nervous," Darwin told Hooker, "but I shall enjoy it immensely if it does not kill me."[112]

Darwin was not slain by the experience (to the contrary, it made him uncommonly spritely), but it was the last gasp for Wallace's protection thesis. Wallace evidently did his best against the united opposition, causing Darwin "severe distress." A few days later, Darwin wrote to say that it had been a near thing, but his teeter-totter was over: "This morning I oscillated with joy towards you: this evening I have swung back to [my] old position, out of which I fear I shall never get." Nor did he, despite Wallace's carefully detailed fifteen-point "counsel on the other side," which concluded with this despairing cri de coeur:

Your view appears to me to be opposed to your own laws of Nat[ural] Select[ion] & to deny its power & wide range of action. Unless you deny that the general dull hues of female birds and insects are of *any use to them*, I do not see how you can deny that Nat. Select. must tend to increase such hues, and to eliminate brighter ones."[113]

Darwin did deny it: "I lay great stress on what I know takes place under domestication. I think we start with different fundamental notions on inheritance." Darwin found it "most difficult but not . . . impossible" to see how a "few red feathers appearing on the head of a male bird, & which *are at first transmitted to both sexes,* could come to be transmitted to males alone," if the females were made dull for protection; but, as a methodical fancier might set about the task, he had "no difficulty in making the whole head

red if the few red feathers in the males from the first tended to be sexually
transmitted." He could concede some points to Wallace, but he could not
persuade himself that "females *alone* have often been modified for protection." He could agree that open-nesting females were protected by their dull
colors, but he could only "wish to see reason to believe that each is specially
adapted for concealment to its environment . . . But I fear we shall never
quite understand each other."[114]

In this, Darwin was correct. To an extent, he and Wallace were talking
past each other. Wallace could not comprehend Darwin's view that the less
conspicuous female was simply the sex that did not have sexually dimorphic
characters added to its less developed form, so that natural selection for
protection was superfluous (not that Darwin ever put it quite this way in
his letters to Wallace, preferring to talk in terms of sex-limited inheritance).
Wallace had little appreciation of the embryological argument for evolution. He could not, for instance, understand Darwin's great enthusiasm for
Fritz Müller's *Facts and Arguments for Darwin*—which Darwin pressed upon
him—admitting to his "own dullness in matters of minute anatomy & embryology." Nor, he told Darwin, contra the domestication analogy, could he
believe that "cases of domesticated varieties will help us with facts, because
sexual characters have never been selected to anything like the enormous degree in which they occur in nature." Above all, Wallace found it impossible
to believe that the author of the *Origin of Species* could think that dull colors
in the female "have *no relation* to the environment"; and he was left spluttering by Darwin's new view that the nesting "*habits have altered* in consequence
of the danger of the gay colour."[115] Darwin, for his part (and somewhat
willfully), interpreted Wallace invariably to mean that females, originally
brightly colored and conspicuous, had been specially modified for protection through natural selection; whereas Wallace, though his expression was
at times ambiguous, meant that females generally were initially dull and
that natural selection acted to preserve this state by keeping the female dull,
rather than modifying her. Wallace belatedly became aware of this confusion and attempted to clarify his position, but by then Bates had already
brought the point to Darwin's attention without effect.

In January 1870, Bates, who had been reading the insect section of the
Descent manuscript at Darwin's behest, wrote to Darwin in an attempt to
bring his and Wallace's opinions "into harmony": "The necessity of females
being dull-coloured for protection is true, but they have not been *made* dull
from former brighter hues, but have simply been *kept* dull by natural selection eliminating all tendency to brightness. This will not disagree with
your clenching & true argument against Wallace that females of a genus are

truer in colour to the generic type than males."[116] Had Bates been present at the gathering at Down House in September 1868, events may have taken a different turn and the great debate been settled harmoniously; though, given Darwin's mind-set, this seems unlikely. Darwin, as we saw, earlier had considered the possibility that Wallace might say that the female "never had become beautiful but had been from first checked [by natural selection]," but had rejected it on the basis that it was superfluous to his own interpretation of females being kept dull through sex-limited inheritance.

Darwin's response to Bates has not been found; however, we do know that Darwin did not modify his manuscript in this respect. We also know his reaction to Wallace's subsequent complaint of misrepresentation. Wallace wrote to Darwin in January 1871, after reading an advance copy of the first volume of the *Descent*, which included the section on sexual selection among butterflies: "My view is, & I thought I had made it clear, that the female has (in most cases) been simply *prevented* from acquiring the gay tints of the male (even when there was a tendency for her to inherit it) because it was hurtful." Darwin made conciliatory noises, but he left unchanged the section Wallace complained of in the second edition of the *Descent*. His views were fixed, and he simply was not open to changing them. He acknowledged as much in his response to Wallace's critique of sexual selection in his review of the *Descent* (March 1871): "I will keep your objections to my views in my mind, but I fear that the latter are almost stereotyped in my mind."[117]

There is no "minor mystery" as to why Darwin leaned so far toward inheritance in opposing Wallace's "selection for protection" thesis, as Helena Cronin finds.[118] From his earliest evolutionary writings, Darwin had been concerned with issues of development and inheritance; over the years, he had integrated these into a tightly woven complex of theory, practice, and analogy that informed and sustained his theory of sexual selection; it was inevitable that he would privilege this over Wallace's single-minded focus on adaptation and natural selection.

Above all, Darwin was intent on preserving and keeping intact his cherished concept of female choice, which Wallace had placed under increasing threat. Two little-known letters from Wallace to Darwin (only recently published) show the extent of this threat. In February 1868, Wallace had begun to hint at a new theory that he believed would explain color better than sexual selection: "Colour in insects may be mainly produced by correlation with other parts of the organism. Thus, where the sexes of butterflies do not *materially differ*, the male is generally *smaller stronger*, a *quicker flier*, and more *highly coloured*." Some months later, on May 1, 1868, he refined on this explanation of male intensity of color in both insects and birds: "[It] may be due to greater vigour of male or be correlated with his sexual organs." In short, Wallace was beginning to formulate what became his later view—that there was a physiological correlation between color and the sexual vitality of the male.[119] Color was linked with strength, pugnacity, and other characters that would be of use to males in the general struggle for existence; it was connected with natural selection, not with sexual selection, and definitely not with female choice. It was at this very time that Darwin made his previously discussed series of notes that determined him to oppose Wallace's selection-for-protection thesis and committed him to sexual selection and female choice—the reasoning that preceded the September gathering of Wallace and Darwin's major informants on sexual selection at Down House.

Overriding all their differences was the critical issue of the divergence of the human races that Wallace also insisted on attributing to natural selection, whereas Darwin was driven by the need to explain racial divergence by sexual selection alone. He had not wavered from his 1864 declaration to Wallace that "sexual selection has been the most powerful means of changing the races of man." This was the raison d'être of his long-contemplated "essay on man" and the ostensible point of his exhaustive survey of sexual selection throughout the animal kingdom.[120]

As far as Darwin was concerned, the battle for beauty was over by October 1868, and he went on his way, writing the *Descent*; but for Wallace, it had only begun.

THIRTEEN

Writing the *Descent*: From Bird's-Eye View to Masterful Breeder

Sexual selection has been a tremendous job. Fate has ordained that almost every point on which we differ sh[oul]d. be crowded into this vol[ume].

—Darwin to Wallace, October 21, 1869

It w[oul]d not suffice to educate young girls to make ♀ = to ♂!

—Darwin's notes on Mill's *Subjection of Women*, July 18, 1869

Darwin recorded that he began writing *The Descent of Man* on February 4, 1868. He calculated that, allowing for illness and work on other projects, its writing took two full years. But Darwin had been writing the *Descent*, in one way and another, for almost a lifetime. By the end of 1869, he had almost completed a first draft and was "dull as a duck, both male & female." With Henrietta as copyeditor, he worked on its revision for much of 1870. It went into print early in 1871 in two hefty volumes priced at twenty-four shillings. It sold out fast, going into a second edition within just three weeks.[1] Darwin had anticipated "disgust" and "universal disapprobation, if not execration," and was pleasantly surprised by its comparatively mild reception. He welcomed this as "proof of the increasing liberality of England."[2]

The *Descent* arrived in an England well acculturated to evolution, its prosperous middle class grown complacent and tolerant of unorthodox opinion, when couched in such acceptable tropes. The reading public, inured to ape-men, well exposed to the notion of social progress through necessary and inevitable competition, took in with easy acquiescence the *Descent*'s familial narrative of racial improvement from ugly "barbarian" ancestry to morally and intellectually superior (middle-class) European manhood. Alongside this came reassurance of innate womanly modesty and domesticity and ethnic stereotyping, featuring degenerate "savages" and "squalid" Irish. A high-minded humanitarianism, looking to the survival of the "noblest part" of human nature—sympathy, altruism, and generosity—upheld all this, even as Darwin conceded that such evolutionary-based impulses enabled the survival of the unfit, endangering further progress. The extensive middle section on sexual selection—replete with allusions to "courtship," "love antics," "coy" choosy females, spousal fidelity, and "marriage arrangements," and featuring iridescent feathers, trilling song, and thrumming insect choruses—charmed and gently titillated an audience familiar with the courtship plots of Victorian fiction and the anthropomorphism of breeding and bird books.

What helped the whole go down was the increasing celebrity of its author, whose balding, beetle-browed image was fit matter for caricature and ready recognition as a "fine venerable old Ape." It was more the fame of the man than the novelty of his ideas on the animal ancestry of humanity that readers were buying. "This work," acknowledged Darwin, "contains hardly any original facts in regard to man." Within the last few years, as Darwin dwelled obsessively on "everlasting males & females, cocks & hens," a slew of works on the genealogy of humanity had appeared.[3] But Darwin was less interested in prehistoric humans, fossil primates, or newsworthy "missing links" than in racial origins and a courtship-by-courtship exposition of sexual selection from crustacea to humans.

With Darwin's growing fame had come his increasing detachment from those issues and areas of science that did not fit within his established framework. The living icon was slowly crystallizing into a living fossil, pursuing an agenda set more than a third of a century earlier. He went on assiduously collecting information—his worldwide correspondence was now enormous—largely to support long-held views, which, as he told Wallace, were "almost stereotyped." His later work would come to seem increasingly dated. His theory of pangenesis, by its time of publication, was already out of touch with new developments in cell theory. His views on the expression of the emotions harked back to his early notebooks, before he had arrived at the theory of natural selection. His primary target remained Sir Charles Bell (long dead and gone), chief exponent of the creationist view that humans are set apart from animals by possession of unique muscles, divinely designed for the communication of expression. Uninterested in any communicative function, Darwin focused on explaining the physical expression of the emotions through inherited habit, being comparable to and having evolved from animal responses and facial movements under the stimulus of primary emotions of fear, pain, and pleasure.

Darwin's "puzzling" failure to link expression with sexual signaling (or any other communicative function) is the more intriguing, given that he initially intended to include expression, alongside sexual selection, in the same work on "man," but portioned it off into a separate work as his material outgrew its space. Yet both were, for Darwin, necessary explanations for nonfunctional, nonadaptive structures and behaviors—explanations that he had held over a very long period, had cemented into place with his views on inheritance and development, and saw no reason to change.[4]

Additionally, by the late 1860s, natural selection was coming under intense scrutiny. Even Darwin's supporters disagreed about its extent and efficacy. Huxley, the skeptic, thought that Darwin's "hypothesis" could be vindicated only when the process had actually created new intersterile species. Lyell and Asa Gray had their reservations, with Gray looking to reconcile natural selection with theism. The Duke of Argyll's "creative evolutionism" was gathering adherents. His *Primeval Man* (1869) extended his critique to humans, arguing that humankind could not have made the step from savagery to civilization without supernatural help. In mid-1867, the Scottish engineer Fleeming Jenkin cut to the key issue with a "most telling" review "of the hostile kind." Jenkin was an associate of the physicist Sir William Thomson (later Lord Kelvin), who recalculated the earth's age, putting severe restrictions on the duration of the slow, gradual process of evolution by natural selection. Time was not on Darwin's side; there was not enough of it for the great changes he depicted to have occurred. He and his supporters responded by pointing to inconsistencies in calculations of the earth's age, thus prompting Jenkin's even more devastating criticism that a single variation, no matter how beneficial, could not survive in a large, freely breeding population. It would be blended out, or "swamped."[5]

Jenkin's swamping argument convinced Darwin that his emphasis on individual advantageous variations, such as the appearance of a bird with a longer beak, had to be modified: "I would now say that of all the birds annually born, some will have a beak a shade longer; & some a shade shorter, & under conditions or habits of life favouring longer beaks, all the individuals, with beaks a little longer would be more apt to survive." Only if a number of advantageous variations appeared simultaneously could the swamping effects of blending inheritance be overcome. Darwin, in revising *Variation*, had argued that his catchall pangenesis would permit the preservation of some favorable variations. They did not always disappear through blending, but might reappear through reversion or dormancy. He put progressively more emphasis on the effects of inherited habit, of use and disuse, for which pangenesis also could account. He tried ways to speed up evolution, revising the fifth edition of the *Origin* (1869) accordingly. He soon was to state publicly in the *Descent* that his previous work "probably attributed too much to the action of natural selection"; it "had not sufficiently considered the existence of many structures which appear to be . . . neither beneficial nor injurious."[6]

While Wallace was defending the utilitarian basis of all evolutionary change and extending the adaptive role of natural selection into sexual dimorphism, Darwin was retreating from natural selection, becoming more pluralist in his interpretations. Still, he held to his idée fixe: the origin of the human races through sexual selection and its necessary demonstration throughout the animal kingdom. As Darwin yielded ground on natural selection, he claimed more for sexual selection. In the midst of Darwin's writing, early in 1869, Wallace fired another shot across his bows: the coauthor of natural selection no longer believed that the higher human faculties, nor even the human hand, hairlessness, erect posture, beauty of countenance, or emotional expression, were explicable by natural selection, but required supernatural explanation—"i.e., miracles," Darwin stabbed contemptuously in the margin of the defector's paper. Wallace's apostasy gratified Lyell and gave advantage to Argyll and the opposition, while Wallace, as if to compensate, assiduously claimed back the nonhuman territory for natural selection that Darwin was intent on ceding to his principle of correlation, to acquired characters, and above all, to sexual selection.[7]

Throughout the writing of the *Descent*, the headbutting between the Darwinians and the Anthropologicals continued. Darwin felt the heat, complaining to Hooker in mid-1868 of the "last [number] of the Anthropological Review, in which I am incessantly sneered at." By this time, the Anthropologicals were in serious financial difficulties through their overly ambitious publishing activities and falling membership. Huxley seized his opportunity to put an end to this "scientific scandal." He accepted the presidency of the Ethnological Society on condition that its council support his efforts toward unification of the two societies.[8] One tactic he deployed was to initiate the exclusion of women from Ethnological Society meetings—a move contested by none other than that leading defender of Victorian prudery and propriety, Eliza Lynn Linton.

13.1 For (white) men only: The "new" evolutionary anthropology

Through her interest in human evolution and professed need for communication with "clever men" like Huxley, Linton, the ardent Darwinian, had become a regular "visitor" (as women attendants were known) to Ethnological Society meetings. On Huxley's notice of her imminent expulsion, she was forced to step outside her paid professional role of deriding and attacking the girl of the period, to plead her right to a better education and opportunities. "You know how few opportunities we women have for getting any serious or valuable talk with men," she wrote Huxley.

> We meet you in "Society" with crowds of friends about & in an atmosphere of finery & artificiality. Suppose I, or any woman—let her be as fascinating as possible—were to bombard you with scientific talk—would you not rather go off to the stupidest little girl who had not a thought above her pretty frock, than begin a talk on the Origin of Species?

If women were not to talk of science in society and were excluded from scientific societies, how were they to learn about science? In paraphrase of Huxley's own "Emancipation—Black and White," Linton sought to remind him of his liberal Darwinian obligations:

> What are the facts of woman's personal condition? We are thrown into an active hand to hand struggle for existence all the same as men—we of the middle classes have to earn our own bread—with very badly trained hands & brains . . . The battle of life is a very serious matter to some of us, and we are frequently hindered and heavily weighted . . . It is not fair to exclude us from the means of knowledge & of active thought, of extended views—such as we get from attending learned discussions—on the simple plea of our womanhood.[9]

She made her case according to her own precepts, with proper regard for the proprieties and none of the "shrieking" for which she castigated the new sisterhood. Nevertheless, Linton's powerful but womanly plea did not meet with the anticipated fair treatment. Her offer of voluntary female exclusion from "hazardous" discussions did not meet the real point at issue. If women controlled their own occasional exclusion, the Ethnological would still be a "ladies' Society." Huxley came up with the ingenious compromise of demarcating "Ordinary Meetings"—for "scientific" discussions to which "ladies will not be admitted"—from larger, popular "Special Meetings"—which "ladies" might attend "by special invitation."[10]

This inspired (and typically Huxleyan) solution reconstituted the Ethnological as a "gentleman's society" while paying lip service to the liberal principle of female admission. All were satisfied except Linton (and those she

represented), now relegated to the "popular element" and exiled from serious scientific discussions. But, as a leading public exponent of the separate
spheres ideology, she was hardly in a position to complain.

The exclusion of women served Huxley's purposes. It upgraded the professional status of the Ethnologicals and removed a major impediment to
their amalgamation with the Anthropologicals, who had become more ra-
bidly antifeminist in the face of John Stuart Mill's advocacy of women's
rights. The publication of *The Subjection of Women* early in 1869 signaled a
temporary suspension of their customary pursuit of racial issues by hard-
core fellows. They joined forces to make their "authoritative decision" on the
"claims by women to political power," seeking to exclude women from any
competitive role in the "ordinary struggles for existence."[11] The Anthropologicals' impressive co-optation of evolutionary arguments to ward off this
feminine threat demonstrates yet again the compatibility of Anthropological and Darwinian thought.

There was little difference between such arguments and Huxley's denial
to women of any "natural equality." He endorsed women's emancipation
on the understanding that they would not be able to overcome their biological limitations and compete effectively with men. Having demonstrated
his concurrence with James Hunt over female admission, Huxley pushed
hard for amalgamation of the two societies. With Hunt's sudden death in
mid-1869, the Anthropologicals lost their charismatic leader. Their new
president, John Beddoe, proved amenable to amalgamation; the remaining dissidents offered only token resistance to the forceful Huxley. At the
beginning of 1871, as the *Descent* went to press, the Ethnological and Anthropological Societies were united as the Anthropological Institute of Great
Britain and Ireland, with the Darwinians firmly in control of the science of
"man." Significantly, it was Huxley who came up with the name that "recognized the science but not the Society" of the defunct Anthropologicals. For
the next two decades, the Anthropological Institute was led by a succession
of Darwinians—John Lubbock, George Busk, Edward B. Tylor, and Francis
Galton, among others.

Lubbock's *Pre-historic Times* and Tylor's *Researches into the Early History of
Mankind* (both 1865) set the scene for the evolutionary "ascent" of humanity from a common stage of "primitive" ancestral savagery—comparable to
the state of existing indigenous people, such as the "Stone Age" Australian
Aborigines or the Fuegians—via the sequential stages represented by "lower"
races, to the civilized heights of Anglo-Saxon supremacy—an inevitable
progress toward Victorian values. The "new" evolutionary anthropology,
ratified by Darwin in the *Descent* and institutionalized by the Anthropo-

Figure 13.1. "The head," Darwin emphasized in the *Descent*, "is the chief seat of decoration" in both birds and "savage and civilized" humans. Polish fowl (from *Variation*, 1:229; University of Sydney Library, Rare Books and Special Collections) and modish woman's hat incorporating the plumage of a whole bird (*Harper's Bazaar*, February 19, 1876, 121; State Library of Victoria Collection).

logical Institute, integrated this value-laden interpretation with a Knoxian/ Darwinian insistence on the biological basis of distinctive racial moral and mental traits, the intractability of interracial competition and conflict, and the extinction of subject races at the further reaches of Victoria's empire. The once-divisive question of the specific distinctness of the human races was rendered redundant by an evolutionary process that made it a "matter of indifference," asserted Darwin, whether the "so-called" races were ranked as species, subspecies, or varieties: "When the principles of evolution are generally accepted," the dispute "of late years" between monogenists and polygenists "will die a silent and unobserved death."[12]

13.2 Confronting Wallace: Artful birds, noble apes, bestial savages, and good women

These issues and events framed the writing of the *Descent*. Of them all, it was his differences with Wallace that loomed largest for Darwin. With Agassiz vanquished, the coauthor of natural selection was emerging as Darwin's most "plausible" critic. Darwin was not unacquainted with criticism from his disciples: Huxley was a case in point, with his stand on natural selection as provisional until definitively proven. Nor had Huxley ever offered a scintilla of support for sexual selection. But Huxley was professional propagandist for the new creed of scientific naturalism, the moralizing "High Priest" of a "new evolutionary priesthood." Wallace, by contrast, challenged the best efforts of the Darwinians to accommodate this misfit: Was he insider or outsider, disciple or dissenter, outstanding naturalist and theoretician or the deluded defender of table-rapping frauds and charlatans? In the long run, they could only resolve the dilemma by splitting Wallace in two: there was Wallace the honored Darwinian, cofounder of natural selection, one of the anointed; and there was the "other" Wallace, who went in for such outré causes as spiritualism, phrenology, anti-vaccination, socialism, land nationalization, and feminism.[13]

Wallace's coincident enunciation of natural selection and his consistent deference had both validated Darwin's "discovery" of the concept and put Darwin in his debt. With his 1864 "man" paper, Wallace had seemed the most dedicated Darwinian of them all. His subsequent defection from the Darwinian line on human evolution shook Darwin like none other. Infanticide leaped to Darwin's mind: "I hope you have not murdered too completely your own & my child." He concurred with Bates that Wallace's "heterodox views" had done a "good deal of mischief to the cause"; yet Darwin eluded Bates's efforts to persuade him that he, Darwin, was the only one who might authoritatively "controvert" them in a specially commissioned article. At the same time, his clash with Wallace over sexual selection hardened Darwin's commitment to it. In any case, for Darwin, sexual selection was inextricable from human evolution. He resisted Bates's attempt to reconcile his and Wallace's views on sexual selection. The *Descent* was Darwin's answer to all of Wallace's heterodoxies: "Fate has ordained," he warned Wallace, "that almost every point on which we differ sh[oul]d be crowded into this vol[ume]."[14]

Their correspondence dwindled to a trickle, but Wallace kept tabs on progress and kept up the pressure: "I hear you have gone to press, & I look forward with fear & trembling to being crushed under a mountain of facts!" As late as November 1870, when Darwin, "half killed" by proofing his "confounded book," expressed his concern that it would "quite kill" him in Wallace's "good estimation," Wallace was sure they still agreed on "on nineteen points out of twenty, and on the twentieth I am not unconvinceable. But then I must be convinced by facts & arguments, not by high-handed ridicule."[15]

The *Descent* was structured by their differences. At its core, it turns on the most fundamental of these: their conflicting encounters with and interpretations of the "savage." Wallace's remembered "surprise and delight" at savage encounter stands against Darwin's fear and loathing, his deep and enduring antipathy to the thought of kinship with the "hideous" savage "other," a kinship that Darwin nevertheless is forced to acknowledge and rationalize through the exigencies of his argument for a common animal origin and for racial divergence through sexual selection: "Such were our ancestors." The savage is central to each of the aims of the *Descent* as Darwin set them out in his introduction: "to consider firstly, whether man, like every other species, is descended from some pre-existing form; secondly, the manner of his development; and thirdly, the value of the differences between the so-called races of man." The solutions to all these problems require sustained engagement with the savage. The savage is insistently present as stand-in for "primitive" ancestor, the bearer of low morals and lesser intelligence, as lowly link with bestial forebears, as foil for civilized man with his higher values and educated tastes, and as grab bag of physical and mental characteristics, practices, and beliefs that go to support the *Descent's* central and guiding thesis: that the human races are descended from a single primitive stock and that their characteristic differences are best explained by "one important agency, namely Sexual Selection."[16]

Even as Darwin worked to close the gap between people and animals—by invoking a barrage of anecdotes about the courage, affection, loyalty, sociability, intelligence, and, above all, the aesthetic sensibilities of animals—he insinuated the savage as less than the anthropomorphized animals that provided the key evidence for the animal origins of human moral, mental, and aesthetic capacities. The dogs that exhibited more self-consciousness, the birds that showed a more highly developed appreciation for beauty and music, those heroic, altruistic apes—all were grouped with civilized humanity while the savage was forced back into the wilderness, neither completely animal nor human, a Stone Age survival soon to become extinct because of a mysterious inability to survive and reproduce, "left without a place in the web of life."[17]

Darwin was outraged by Wallace's provocative claim that "natural selection could only have endowed the savage with a brain a little superior to an ape, whereas he actually possessed one but little inferior to that of the average members of our learned societies." He underscored this passage in his copy of Wallace's review and shouted an emphatic "No" in the margin, adding a cluster of dissenting exclamation marks. In the *Descent*, he enumerated the achievements of "man in the rudest state"—weapons, tools, traps, rafts, canoes, and the art of fire making, of identifying and rendering plant foods digestible and ridding them of toxins—and directly confronted Wallace: "These several inventions . . . are the direct result of the development of his powers of observation, memory, curiosity, imagination, and reason. I cannot, therefore, understand how it is that Mr. Wallace maintains, that 'natural selection could only have endowed the savage with a brain a little superior to that of an ape.'"[18]

Yet over and again in the *Descent*, the savage reasserted his demonic presence: guilty of "utter licentiousness" and "unnatural crimes," "utterly indifferent to the sufferings of strangers," delighting in torturing his enemies, taking a "horrid pleasure in cruelty to animals," showing sympathy and kindness only to his own yet witnessed dashing his child on the rocks for the most trivial of offenses—all attributable to his tribalism, to "insufficient powers of reasoning," and "weak powers of self-command." The gulf between the "lowest savages" and the "highest men of the highest races" was immense. And, although it was bridged by the "finest gradations" so that it was possible they "might pass and be developed into each other," the rise of the savage in the scale of civilization was not assured: "Many savages are in the same condition as when first discovered several centuries ago." His progress was not inevitable, he lacked the essentials for civilization, his future was problematic: "While observing the barbarous inhabitants of Tierra del Fuego, it struck me that the possession of some property, a fixed abode, and the union of many families under a chief, were the indispensable requisites for civilization."[19] Then came the final return of the degraded native, the ultimate rejection of these still vividly remembered people of Tierra del Fuego, as, in the penultimate paragraph of the *Descent*, Darwin reiterated his preference for an innocent ape as ancestor over kinship with such barbarous beings—the kinship of which, nevertheless, "there can hardly be a doubt."

As Desmond and Moore acknowledge, it is not easy to reconcile this consistent demonization of savage behavior with Darwin's "inviolate" antislavery and anticruelty ethic: it is an "incongruity . . . impossible to comprehend by twenty-first century standards."[20] Yet this incongruity is comprehensible as the outcome of the "double and contradictory place" of the savage in the *Descent*, as the savage is demonized and denied in order to make tolerable the bestial descent to which he bears witness.[21] This was Darwin's response to Wallace's elevation of savage morality and intelligence and Wallace's consequent repudiation of the role of natural selection in their evolution. It reverberated even more deeply through Darwin's assertion, against Wallace, of the role of sexual selection in the differentiation of the human races, where Darwin repetitively denied the savage "who knows no decency" as ancestor and promoted an apelike monogamous or polygamous progenitor in order to salvage sexual selection from the contemporary anthropological thesis of "primitive promiscuity" (see sec. 13.6).

If the savage was exiled from evolution in the *Descent*, the "good woman" of Darwin's lexicon fared little better. She was both there and not there. She was there as the choosy, sexually shy exemplar of animal female choice; she was there too as the chaste, choiceless object of the jealous, apelike primitive patriarch's selective practices in the formation of the human races; she inhabited the generalized "woman" who "differs from man in mental disposition, chiefly in her greater tenderness and less selfishness," a difference that "holds good even with savages"; but, as far as intelligence goes, she was equated with the "lower races" and relegated to a "past and lower state of civilization"; she was altogether missing from the struggle for existence and for mates that had sharpened male intelligence and left women forever lagging behind, dependent on men for subsistence and protection and for the inheritance of such intelligence as women possessed. Darwin's primary target here, of course, was not Wallace but Mill (see sec. 13.5). On Wallace's part there was, as discussed, a degree of overlap with Darwin's views on female sexuality, and Wallace was many years away from developing his mature version of rational choice by the educated woman of the future. Yet the seeds were sown in his commitment to the principle of utility and his growing resistance to the notion of female choice for mere aesthetic reasons as opposed to useful ones. And we may discern in Darwin's determined exclusion of females, animal or human, from any useful role in evolution, his continuing opposition to Wallace's forceful insistence on the centrality and utility of the female to the "continuance of the race" that had marked their earlier conflict over sexual selection.[22]

Where they clashed most overtly was at the nuts-and-bolts level: those sections of the *Descent* that deal with the theoretical and observational bases of sexual selection. Here Darwin was in his métier as he reprised the reasoning that had been tested and strengthened during their prolonged dispute and adduced a dazzling array of evidence in its support. It was "facts & arguments" Wallace asked for, and facts and arguments he got.

13.3 "Facts & arguments"

Darwin defined his primary concept and spelled out the embryological and hereditary processes that underpinned it in the introductory chapter to the central section of the *Descent*, "Principles of Sexual Selection." This largely theoretical chapter includes the previously discussed fundamentals of sexual selection: the greater variability, "greater passion," and sexually indiscriminate eagerness of the male for mating; the "comparatively passive" choice exerted by the "less eager" female; the differences between natural and sexual selection; the dominant form of sex-limited inheritance and the form of equal inheritance that many mammals exhibit. These went to support the basic premises of Darwin's theory of sexual selection: sexual selection could act only on sexually mature animals; from the "greater eagerness of the male," it had generally acted on males and not on females; the males had "thus become provided with weapons for fighting with their rivals, or with organs for discovering and securely holding the female, or for exciting and charming her." Non-useful or harmful variations (and sexual characters are often of this kind) acquired before maturity usually would be lost, while those appearing around maturity would be preserved (if sexually selected) and accumulated over successive generations.[23]

However, this chapter did include a new (and little-noticed) theoretical development. At the eleventh hour, Darwin introduced what amounted to a third form of sexual selection based on vigor and health. This shift was occasioned by his belated recognition of a crucial discrepancy in his theorizing that might compromise, if not negate, the effects of sexual selection: the ratio of the sexes.

Early in 1868 Darwin had suddenly become concerned about sex ratios. He had assumed that in sexually dimorphic species, males would outnumber females, assuring the essential competitive edge for both male combat and female choice. But he lacked information; and the difficulty, which had just occurred to him, was that if the sexes existed in equal numbers, sexual selection could not prevail. Even the worst endowed males would eventually find females (excepting among polygamous species) and leave as many offspring as the better-endowed ones. With help from Bates, he solicited information on sex ratios in a great variety of species, wild and domesticated. The results were not encouraging. As he explained in the *Descent*, "After investigating, as far as possible, the numerical proportions of the sexes, I do not believe that any great inequality in number commonly exists."[24]

His solution was one of timing: males were generally ready to breed before females, so they competed for a smaller supply of breeding females; the more vigorous and healthy females were ready to breed first. So the strongest, best-armed males would drive away weaker males and unite with earlier-breeding, vigorous females:

> Such vigorous pairs would surely rear a larger number of offspring than the retarded females, which would be compelled, supposing the sexes to be numerically equal, to unite with the conquered and less powerful males; and this is all that is wanted to add, in the course of successive generations, to the size, strength and courage of the males, or to improve their weapons.[25]

If female choice was added to the mix, it was "obviously probable" that the earliest breeding, most vigorous females would prefer the more vigorous and lively males—not just the best ornamented—and these would pair and reproduce; hence, over time, this would also augment the charms, strength, and fighting powers of the males. In the "converse and much rarer case of the males selecting particular females, it is plain that those which were the most vigorous and had conquered others, would have the freest choice," and it was "almost certain that they would select vigorous as well as attractive females." Such pairs would have an advantage in rearing offspring, and this would be enhanced in cases, such as with the higher mammals, where the male had the power to defend the female or to aid in providing for young. Where polygamy was practiced, as by many mammals (including some closest to man), it would have the same advantageous effect on sexual selection as a numerical preponderance of males: "If each male secures two or more females, many males will not be able to pair; and the latter assuredly will be the weaker or less attractive individuals."[26]

This late-introduced form of sexual selection had special implications for Darwin's theory of racial divergence (see sec. 13.6); it had larger and more immediate repercussions with respect to his overall thesis, which were never properly addressed by Darwin.

Darwin had always assumed an element of general vigor—of size, strength, and health—in the competition between males for possession of the females. This form of sexual selection not only explained the development of specialized weapons (horns, spurs, etc.) but also enhanced the action of natural selection by ensuring that the fittest males were reproduced. This was enunciated from the first edition of the *Origin*.[27] The continuity between natural and sexual selection did not, however, extend to the exercise of female choice, which depended on individual preference and the appreciation of beauty, without regard to male strength and vigor. It was this sense of nonadaptive—even counteradaptive—evolution that gave sexual selection its most distinctive element in Darwin's theorizing. His early education in the aesthetics of Burke and Erasmus Darwin had fostered an understanding of the perception of beauty as both sexually engendered and nonutilitarian. Yet by associating female choice with "vigour," with the selection of the "strongest and best-armed" males, Darwin weakened the distinction he otherwise insisted on between sexual and natural selection. Female choice, like male combat, was in danger of being subsumed into natural selection. This took Darwin a little too close for comfort toward Wallace's insistence on a necessary utilitarian, adaptive component to sexual dimorphism. Furthermore, the introduction of notions of vigor into what had always been a purely aesthetic choice destabilized the pivotal analogy that Darwin had been at such pains to establish and that is reiterated at key points throughout the *Descent*, between the pigeon fancier, who selects to extremes on the basis of appearance alone and according to his standard of taste; and those female animals that, "by having long selected the more attractive males, have added to their beauty."[28]

These contradictions, never acknowledged by Darwin, added a layer of complexity and uncertainty to his discussion of the "manner of action" of sexual selection in the *Descent*.[29] Nor did he consistently deploy this largely defensive third mode of sexual selection in his extended review of animal mating behaviors and selections.

His theoretical chapter was the springboard from which Darwin launched his comprehensive, purposeful excursion into mating behaviors and secondary sexual characters of the animal kingdom, leading up to "man" and racial divergence in the third and final section of the *Descent*. He did not confront Wallace directly until arriving at the butterflies and moths. While acknowledging the extensive role of protective and warning mimicry in these insects and the consequent difficulties of interpretation, Darwin asserted his opposition to Wallace's protective principle: "Mr. Wallace believes that the less brilliant colours of the female have been specially gained . . . for the sake of protection. On the contrary it seems to me more probable that the males alone, in the large majority of cases, have acquired their bright colours through sexual selection, the females having been but little modified." He followed this with a seven-page discussion of their respective positions on inheritance, reprising the main points of their earlier debate in arguing the "less probable" aspects of Wallace's view. Not content with this, Darwin again tackled Wallace in the section on birds with an "imaginary illustration" involving artificial pigeon selection, showing to his own satisfaction that sexual selection in conjunction with sex-limited inheritance from the first could explain sexual dimorphism without the need to invoke natural selection, as Wallace was insisting on. He next upended Wallace's much-prized correlation of the manner of nesting with coloration: it was not the form of the nest (open or concealed) that determined plumage, as Wallace claimed, but plumage (selected through female choice) that determined the kind of nest. He conceded the complexities of the relation of adult to young plumage in birds, enumerated and discussed his Blyth-inspired six "rules or classes" of resemblance, but held to his fundamental rule that the females and young were "left comparatively but little modified" and that "immature plumage approximately shews us the former or ancestral condition of the species."[30]

Darwin's quick trip through the mammals was accomplished with less reference to Wallace, though it held more significance for Darwin's case for human sexual and racial difference. He explained the fact that sexual dimorphism is far less pronounced in mammals than in birds in terms of the predominance of characters, acquired primarily by males, being, for reasons unknown, "transmitted equally, or almost equally to both sexes"; however, he argued, many sexual peculiarities of odor, voice, and call differences, hair development, and skin colors in mammals are attributable to female choice and to same-sex inheritance. Darwin made as much as he could of the specialized weapons—horns, tusks, and canines—and greater size and pugnacity of many male quadrupeds, acquired through the "law of battle." While conceding that "many quadrupeds have received their present tints as protection," he argued that "with a host of species, the colours are far too conspicuous and too singularly arranged to allow us to suppose that they serve this purpose." He dwelled on the different distribution of stripes and spots in the males and females of various antelopes in disputing Wallace's belief that the striped coat of the tiger "so assimilates with the vertical stems of the bamboo, as to assist greatly in concealing him from his approaching prey": "He who attributes the white and dark vertical stripes on the flanks of various antelopes to sexual selection, will probably extend the same view to the Royal Tiger and beautiful Zebra." Similarly, the "curious and elegant" hair crests and "extraordinary or beautiful" skin colors of many monkeys "have probably been gained through sexual selection, though transmitted equally, or almost equally, to both sexes."[31]

Sexual differences in monkeys warranted particular attention:

> The mental powers of the higher animals do not differ in kind, though so greatly in degree, from the corresponding powers of man, especially of the lower and barbarous races; and it would appear that even their taste for the beautiful is not widely different from that of the Quadrumana.

The facial cicatrices and painted decorations of some African tribes reminded Darwin of the facial furrows and arresting colored stripes of the male mandrill; while the African interest in posteriors, Darwin implied, was referable to the mandrill's even more brilliant "posterior end," which was "coloured for the sake of ornament." This, "no doubt . . . to us a most grotesque notion," was "not more strange than that the tails of many birds should have been especially decorated."[32]

There was "a striking parallelism between mammals and birds in all their secondary sexual characters," and Darwin stressed the telltale resemblance between the young and adult females in both classes of animals: "Considering this parallelism, there can be little doubt that the same cause [sexual

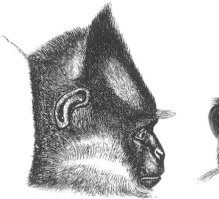

Fig. 71. Head of Semnopithecus comatus.

Fig. 72. Head of Cebus capucinus.

Fig. 73. Head of Ateles marginatus.

Fig. 74. Head of Cebus vellerosus.

Figure 13.2. Darwin argued that the "curious and elegant" hair crests and "extraordinary or beautiful" skin colors of many monkeys had been "gained through sexual selection, though transmitted equally, or almost equally, to both sexes." *Descent* 1871, 2:307. University of Sydney Library, Rare Books and Special Collections.

selection] has acted on mammals and birds." However, he had to admit one significant difference:

> With mammals we do not as yet possess any evidence that the males take pains to display their charms before the female; and the elaborate manner in which this is performed by male birds, is the strongest argument in favour of the belief that the females admire, and are excited by, the ornaments and colours displayed before them.[33]

This was the crux of Darwin's case against Wallace—that aspect of the *Descent* of which he was the proudest and on which he had worked the hardest.

13.4 From "just retribution" for a "wicked old she-bullfinch" to the "work of art" of the argus pheasant

The *Descent*'s four bird chapters constitute Darwin's clinching evidence for female choice, detailed over some two hundred pages. They are the culmination of months of work based on material accumulated over many years, supplemented by visits to the collections of skins, feathers, and stuffed specimens in the British Museum and to the living birds in the Zoological Gardens. Above all, they are testament to Darwin's transposition of the fancier's eye to the head and eyes of a discerning female bird. They are imbued with Darwin's conviction, against Wallace's renunciation, that aesthetic appreciation is continuous throughout nature, from birds to humans: the sense of beauty is not unique to humans, and what is beautiful to female birds is beautiful also to human females, who deck themselves with the plumes of birds in order to attract and entice men.

The scene was set early in 1868, when Wallace queried Darwin's efforts to trace the steps in the development of the peacock's tail: by contrast with natural selection, which worked on "constant *minute* variations," sexual selection seemed to require a "series of bold and abrupt variations. How can we imagine," he challenged Darwin, "that an inch in the tail of the peacock, or ¼ inch in that of the Bird of Paradise, would be noticed and preferred by the female?" In response, Darwin seamlessly swerved from the preferences of peahens to those of Victorian females and back again: "A girl sees a handsome man," he told Wallace, "& without observing whether his nose or his whiskers are the tenth of an inch longer or shorter than in some other man, admires his appearance & says she will marry him. So I suppose with the peahen; & the tail has been increased in length merely by on the whole presenting a more gorgeous appearance."[34]

In this exchange, both girls and peahens are exponents of the pigeon breeder's unconscious practice of selecting to extremes. But before long, as Darwin immersed himself in the finer points of plumage, assessing them with a fancier's flair for color, sheen, and design, he became convinced that birds admired details of plumage and not only overall appearance. Just as Eliza Lynn Linton's phenomenally successful journalistic creation, "The Girl of the Period," burst on the scene in all her manipulative persona and deviant finery, Darwin's peahens and hen pheasants began to take on all the attributes of a superior Sir John Sebright, able to turn mere spots of color into the "most refined beauty" of the peacock's tail or the artistic ocelli of the argus pheasant. Avian choice in the *Descent* is the equivalent of a skilled pigeon fancier crossed with the designing fast girl of Linton's manufacture, who lives only for dress, "to outvie her neighbours in the extravagance of fashion." *Punch* was ever ready to make the nexus clear with its own fixation on feminine frivolity, unfitness, and penchant for plumes (see fig. 8.5).[35]

Darwin had begun writing up his accumulated material on birds around mid-May 1868, stepping up his correspondence and reestablishing links with old informants. He appealed to the invaluable William Fox, ever "an acute observer of birds," for aid on "the females alluring the males—of victorious males getting wives—of attachment between individual birds—anything & everything." On visits to the British Museum, he made notes on plumage, with particular attention to the appearance and gradation of the ocelli (so harped on by Argyll) in a range of pheasants and peacocks. His informants ranged from specialist ornithologists, such as John Gould and Osbert Salvin, to breeders, zookeepers, gamekeepers, gardeners, and sportsmen. Most went out of their way to provide detailed information or specimens in support of his views. Salvin showed him several trogons in which the females and males "differed greatly in colour . . . yet these birds nest in holes. Very much opposed to Wallace." While many of Darwin's contacts, like William Tegetmeier, thought male beauty irrelevant to mating, assumed the sexual passivity of the female, and were doubtful of the efficacy of female choice, others like his long-term informant Abraham Bartlett—now superintendent of the Zoological Society's Gardens—thought otherwise. Bartlett, a man of "immense experience with birds of all kinds," Darwin stressed, "is fully convinced that the beauty of the male is admired by the female & serves to attract her."[36] Bartlett gave Darwin many details of plumage and display in male birds cited in the *Descent*. But Darwin's most devoted and compliant informant on birds had to be John Jenner Weir.

Weir is an outstanding instance of how effectively Darwin could use his fame, diffident charm, and sheer persistence to groom an informant into conformable behavior. Weir was enticed into time-consuming supportive observation and experiment that Darwin had neither the time nor inclination to undertake on his own behalf. At the beginning of their correspondence (initiated by Wallace early in 1868), Weir opposed female choice, citing his counterview of pugnacity among males and instances in moths in which color seemed to be the result of natural selection. A few months later, under Darwin's flattering tutelage, he had become convinced that "all vivid colours in birds are the result of sexual selection"; and he was there, a self-declared "disciple," ready to support Darwin against Wallace at the critical meeting in Down House in September 1868.[37]

Weir went to great lengths to collect corroborating information for Darwin, visiting bird catchers in Brick Lane and pigeon breeders in Spitalfields, even subjecting himself to a "Newgate cut" at the hands of a Tabernacle Row hairdresser, reputed a great authority on the "London Fancy," a breed of canary. "It was very kind, almost heroic in you," Darwin urged him on, "to sacrifice your hair and pay [threepence] in the cause of science." Weir supplied information on plumage changes and the timing of molting in various breeds, on the mating of starlings, kestrels, finches, and rooks, on inheritance of acquired characters, and reported on a trained bullfinch who piped a German waltz—a performance allegedly much appreciated by nearby caged linnets and canaries. Darwin took it all in, commenced an index of Weir's accumulating letters, and stepped up the "severe pumping process" for more such "curious information" relevant to sexual selection, most of which ended up in the *Descent*.[38]

Weir—along with his bird-illustrator brother, Harrison—brought support to Darwin's growing view that female birds admired actual "details of plumage." He reported on the courting presentation by male pigeons of their inflated iridescent neck feathers; on the "antics" of courting drakes, who bobbed their heads to display their brilliant green feathering to best advantage; on the wing actions and attitudes assumed by sexually excited goldfinches, bullfinches, and chaffinches, which turned to present the "beautiful parts" of their plumage to the females; and of the feather erections and colored wattle inflations displayed by mating pheasants and tragopans. An appreciative Darwin added these reports to his own observations and notes on details of individual feathers and their changing appearance, depending how they were held and presented during display. These included a series of experiments on feather color, markings, and sheen viewed at different angles to the sun and from different vantage points (see fig. 6.3). And always in mind were the comparisons to be made between the mating practices of humans and birds: "I am beginning to think that the pairing of birds must be as delicate and tedious an operation as the pairing of young gentlemen and ladies."[39]

It was Weir who gave Darwin his most "surprising" instance of female rivalry and choice among captive birds. Darwin by this stage had given up all hope of getting "direct evidence of the preference of the [domesticated] hens." Tegetmeier was still dragging his feet over Darwin's persistently advocated experiment of staining unmated male pigeons with magenta and observing the outcome. Darwin now turned to his new disciple and to the possibilities of experimentation with "wild" birds. At Darwin's urging, Weir darkened the breast of a young (supposedly) male bullfinch (which is normally red breasted and monogamous) and introduced it into his aviary. Events took an unexpected turn when this "dull-coloured and ugly" bird turned out to be a "ferocious" female who proceeded to give unequivocal evidence of female sexual preference. This was not quite what Darwin had intended, but as he told Weir, she "has done us a good turn in exhibiting her jealousy; of which I had no idea."[40]

In the *Descent*, Darwin edited out the initial mistake over her sex and gave an expurgated account of the bullfinch's "salacious" behavior:

> She immediately attacked another mated female so unmercifully that the latter had to be separated. The new female did all the courtship, and was at last successful, for she paired with the male; but after a time she met with a just retribution, for, ceasing to be pugnacious, Mr. Weir replaced the old female, and the male then deserted his new and returned to his old love.

In the original version of this avian morality tale, Weir reported that the rejected female had "become so excessively salacious, far worse than any bird I ever kept," soliciting every available male "to such an excessive degree, that all the feathers have been trampled off her back & she is now a most miserable spectacle." His account of this ornithological equivalent of the fallen woman of the streets was greeted with satisfaction by Darwin who was "glad to hear" of this fitting end for the "wicked old she-bullfinch."[41]

Darwin's very conventional views on sexual morality remained unchanged with the changing times. While Linton was denouncing the latest unwomanly escapades and extravagances of London's flirtatious fast girls, Darwin was much concerned with protecting the virtue of local maidservants from the advances of an amorous curate, spied "walking with girls at night." He did not believe there was "any evidence of actual immorality," Darwin wrote the absentee vicar John Brodie Innes, but the Down House maids had reported that "hardly anyone will go to Church now." Rumor ran rife: "Mr R" next was seen to go into the house of "some girl supposed to have a bad character." It was all hearsay, but a curate must be above suspicion if both the reputation of the church and the sexual innocence of housemaids were to be upheld.[42]

The sexually aggressive bullfinch that got her just desserts remained Darwin's outstanding instance of female rivalry and choice—an exception to the general rule of sexually passive but choosy female birds. It was not, however, an instance (as Wallace had suggested and Darwin had hoped for) of female aversion to male loss of beauty. Reluctantly, he conceded that captive birds were, for the most part, indifferent to color. Tegetmeier's belated experiment with the dyed pigeons, of which Darwin had such hopes, proved a failure: the magenta-daubed males excited neither interest nor aversion among the females.[43]

Nevertheless, Darwin was convinced that "it is impossible to doubt that the female appreciates the beauty thus carefully displaying before her." With the exception of man, birds were the "most aesthetical of animals," and their tastes were very like our own. In one new species of hummingbird shown him by Gould, the females had outer tail feathers tipped with white as was usually the case, while in the males the coloring was reversed: the outer feathers were black and the inner feathers were white tipped. This "curious case," Darwin noted, "shows that some change, without any addition of beauty, pleases the female, like change of fashion in dress with us." Fashion aside, in making his case for the prevalence of female choice in birds, Darwin relied heavily on stuffed specimens and individual feathers and on secondhand observations of the mating displays of living birds. Bartlett, for instance, showed him specimens of the male *Polyplectron* (peacock pheasant) "stuffed in position in which he has seen it court female: tail expanded & raised almost vertically & turned obliquely toward female, wing on near side expanded and lowered, wing on opposite side raised and expanded; in thus attitude every one of the innumerable ocelli are exposed in one expanse before the eyes of the female." Darwin related this display posture to those of the tragopan and the gold pheasant: "In all these cases & others the male raises his plumage to which ever side the female may be standing. Then give case of Argus [pheasant]. The evidence is [con]clusive."[44] Female choice could fly, it had wings, and it was a bird. But it had only one eye, and that eye was Darwin's.

Darwin's greatest triumph was less the famous case of the peacock's tail feathers than his reconstruction of how the exquisitely shaded "ball and socket" ocelli of the male argus pheasant (described by Darwin as "more like a work of art than of nature") had evolved from a "mere spot" through the accumulated sexual choices of female pheasants. He managed it all on the basis of a few wing feathers, the stuffed and mounted specimen in the British Museum (which "greatly disappointed" him, as it did not show the bird in the act of display), and Gould's sketch of the displaying bird.

No one, I presume, will attribute the shading, which has excited the admiration of many experienced artists, to chance—to the fortuitous concourse of atoms of colouring matter. That these ornaments should have been formed through the selection of many successive variations, not one of which was originally intended to produce the ball-and-socket effect, seems as incredible, as that one of Raphael's Madonnas should have been formed by a large succession of chance daubs of paint made by a long succession of young artists, not one of whom intended at first to draw the human figure. In order to

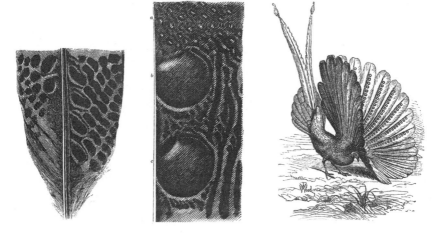

Figure 13.3. Depictions of argus pheasant feathers from *The Descent of Man,* showing the gradation in the development of the ocelli from (left) their "first trace" in the base of the secondary wing-feather to (center) "perfect ball-and-socket ocelli" (c) in the summit of a secondary wing-feather. The argus pheasant (right) was not shown in its display posture until the second edition of the *Descent. Descent* 1871, 2:144, 149; display posture from *Descent* 1874, 452. University of Sydney Library, Rare Books and Special Collections.

discover how the ocelli have been developed, we cannot look to a long line of progenitors, nor to variously closely-allied forms, for such do not now exist. But fortunately the several feathers on the wing suffice to give us a clue to the problem, and they prove to demonstration that a gradation is at least possible from a mere spot to a finished ball-and-socket ocellus.[45]

The major point of Darwin's detailed analysis, which extends over ten pages of the *Descent* and includes four finely drawn illustrative engravings, was that the full effect of the beautiful shading of the ocelli in its wing feathers was only properly revealed when the male was displaying with erect feathers, which were illuminated from above, just as they would appear to a discriminating female pheasant. Darwin, in considering how the spots would appear to the eyes of the female, subjected the plumage of the male pheasant to the same minute evaluation he had previously ascribed to pigeon fanciers, who study their birds "day after day to settle which to match together and which to reject."[46] Having thus transferred the eye and selections of the fancier via his own trained eye to the eyes and choices of the female argus pheasant, Darwin admitted that it would seem "utterly incredible" to many that "a female bird should be able to appreciate fine shading and exquisite patterns." It was "undoubtedly a marvelous fact" that the pheasant hen "should possess this almost human degree of taste, though perhaps she admires the general effect rather than each separate detail."[47]

The trouble was that Darwin's enthusiasm for his own powers of observation and reconstruction, together with his repeated invocation of the artist's eye over that of a lowly breeder, tended to undermine his contention that this "most refined beauty" had been achieved without intention through unconscious selection by a mere female bird. By the second edition of the *Descent*, he could even tie his bird's-eye view in with the artist's craft of producing optical illusions, and, moreover, verify it with the latest visual technology, the camera:

> In a photograph . . . of a specimen mounted as in the act of display, it may be seen that on the feathers which are held perpendicularly, the white marks on the ocelli, representing light reflected from a convex surface, are at the upper or further end, that is, are directed upwards; and the bird, whilst displaying himself on the ground would naturally be illuminated from above. But here comes the curious point, the outer feathers are held almost horizontally, and their ocelli ought likewise to appear as if illuminated from above, and consequently the white marks ought to be placed on the upper sides of the ocelli; and wonderful as is the fact that they are thus placed! Hence the ocelli on the several feathers, though occupying several very different positions with respect to the light, all appear as if illuminated from above, just as an artist would have shaded them.[48]

It was all too much. Wallace had earlier suggested to Darwin that the argus pheasants, male and female (which has a lesser tail and eyespots), were both "protected by their tails corresponding to dead leaves of the dry lofty forests in which they dwell," an argument repeated in his *Malay Archipelago*. It was the case of the argus pheasant, he later claimed, that "first shook" his belief in sexual selection. He refused to see the point of Darwin's attempted distinctions between conscious and unconscious artificial selection and their bearing on sexual selection. His criticisms of female choice were directed at it as a conscious process, as a critical act of aesthetic judgment and fine discrimination on the part of the female—as indeed many others beside Wallace understood it.[49]

The problem was largely of Darwin's own making. His language constantly betrayed his basic position that avian female choice was an instinctive appreciation of the beautiful, not to be confused with the higher aesthetic taste of the more cultivated members of society.[50] At the same time, readers were to understand the process as an application of the fancier's art and to empathize with the bird's aesthetic selections, which connected with the unconscious human choice of the racial beau ideal as exercised from the lowest savage to the aristocrat's privileged pick of the most beautiful of European women—all capped by the clinching claim that as "women everywhere deck themselves with these plumes, the beauty of such ornaments cannot be disputed." The argus pheasant hen, designer of the exquisite ball-and-socket ornament admired and replicated by human artists, was possessed of an "almost human degree of taste," a claim hardly counterbalanced by Darwin's hastily added disclaimer: perhaps she admired the "general effect" rather than the details. Darwin's account of the evolution of the ocelli necessarily was focused on the detail, and the devil was in the detail.[51]

The overall strategy of the *Descent* was to dehumanize the aesthetic basis of human mate choice, to give it a biological basis by extending it to animals—but animals were humanized in this exchange. Darwin's supreme instances of avian aesthetic faculty amounted to the extreme anthropomorphism of avian choice. Yet Darwin relied on the cumulative, persuasive effect of just such anthropomorphisms in making his case, not only for the continuity of aesthetic appreciation from animals to humans, but more fundamentally, for the animal origin of humanity, such as those pet dogs that were more capable of self-consciousness and forming abstract ideas than the average "savage" woman.[52]

Emma, who conscientiously read through the proofs of the *Descent* along with Henrietta, was more alive to the dangers of anthropomorphism than its author: "F. is putting Polly into his Man book," she wrote Henrietta (who was on a prolonged continental tour for the good of her always-questionable health), "but I doubt whether I shall let it stand." Polly was Henrietta's small terrier who, in Henrietta's absence, was becoming "as devoted to Darwin as Darwin was to her." A "fond Grandfather is not to be trusted," pronounced Emma. Nevertheless, Polly found her way into the *Descent*. She is there, for instance, as the little dog that fears the wind-blown parasol and triggers Darwin's account of the evolution of religious devotion from such animal fear of invisible agency, up through "savage" fear of invisible spirits held accountable for natural phenomena, on to belief in the existence of one or more gods. Polly is the epitome of doggy devotion and love for her master—a state of feeling, Darwin implied, not far removed from the reverence and gratitude, along with the fear, experienced by the devoted Christian.[53]

Henrietta, let off the tight leash of Down House invalidity and domestic routine, seemingly did not share her mother's concern that the sections on the comparable mental powers of humans and animals and the evolu-

tion of religious devotion were "again putting God further off." Caught up
in their sheer "interestingness," she made scant changes, all grammatical,
even flippantly urging her "Parson" Papa to expand his "sermon," given
that it concerned something as "unimportant as the mind of man." She
spent her mornings editing the *Descent* to Darwin's exacting requirements,
before donning shawl and bonnet (surely fashionably feathered) to stroll
the promenade of "wicked Monaco."

Already in her late twenties (on the shelf, by Victorian standards), Henrietta was alerted by Emma to the danger of turning into a "regular travelling old maid." She dutifully returned home to be courted by a worthy
though impecunious suitor, Richard Buckley Litchfield, a lawyer. Henrietta received a settlement of £5,000 on their marriage in August 1871. Her
younger brother George, the family mathematician, thriftily recommended
that his father make their yearly income "up to 400£ or 350£, as Litchfield
is not a grasping sort of man." These commercial essentials concluded, Henrietta daydreamed that the sober Litchfield might spontaneously appear at
Down House to sweep her into his arms, "to strike the match which is to
kindle me"—"what exquisite joy"—even wishing that she might make the
first loving advances. Nevertheless, Darwin's daughter was inhibited by her
well-inculcated understanding that those females, like their avian counterparts, who stepped outside the conventional feminine sexual coyness and
passivity (though it might be put to good account in the *Descent*), might
expect "just retribution." Henrietta aspired instead to become the "good
wife" her father advocated, to care (like "our dear old mother") more for
her husband's happiness than her own. Nor, though she contributed to the
Descent's "lucid, vigorous style . . . not to mention still more important aids
in the reasoning" (as Darwin assured her), did she lodge any objection to
Darwin's reasoning, in express opposition to John Stuart Mill, on the innate
and enduring intellectual inferiority of women. Those in the "high Radical
'woman' line" who were hopeful of winning the famous man's daughter to
the cause were wasting their time on one who thought the woman question
a "shibboleth."[54] All Henrietta's upbringing and the conventions of the day
saw to that.

13.5 "Thus man has ultimately become superior to woman"

Ironically, it was a prominent advocate of women's rights who provoked
Darwin's counterargument to Mill's *Subjection of Women*. This was the jour-
nalist Frances Power Cobbe, who had come to know the Darwins socially in
London at a time when Darwin was much engaged with his reconstructions

of peacock and pheasant tails. She was sympathetic to evolution, discussing it with Darwin—with whom, at that stage, she professed to be "enchanted" (disenchantment soon followed with their well-known conflict over vivisection). But Cobbe was also a committed theist who argued for the idea of a just and rational God whose moral law was evident to all through their intuition, not through revelation. It was her theism, her absolute belief in the moral autonomy of women, and her strong sense of their mental and moral difference from men that constituted the core of Cobbe's feminism.[55]

Cobbe and Darwin met again in the summer of 1869 on a hillside in Wales where both were holidaying, Cobbe with her long-term companion, the artist Mary Lloyd. Darwin, winkled out of his domestic routine, broadcast his "loathing" for the place, declaring himself "fearfully fatigued," barely able to crawl half a mile. On one of his crawls he stumbled upon Cobbe, who engaged him, across an impassible bramble patch, in a shouted discussion on Mill's views. Darwin had not as yet read Mill's *Subjection*; but, by Cobbe's account, he was aware of her recent review of it in the *Theological Review*. In any case, she brought him up to date across the blackberries.

Cobbe, lobbying hard in the press for the first Married Women's Property Act (which became law in 1870), had read *The Subjection of Women* with "intense satisfaction & exultation." Her review endorsed Mill's likening of marriage to domestic slavery for women and of the urgent need for just reform. The uxorious Darwin, who playfully styled himself Emma's "nigger," could hardly be expected to agree; but he did see eye to eye with Cobbe on some issues. The politically conservative Cobbe was critical of Mill's egalitarian liberalism, of his insistence that the supposed mental and moral differences between women and men were environmentally determined and not innate, and that, given equal education and opportunities, women would rival men in intellectual ability and vocation.[56]

While she supported the contemporary campaign for women's entry to university, Cobbe held to the more conventional view that women were morally superior but not the intellectual equals of men. She argued against Mill that *"inherited* qualities were more important than *education,"* a position with which, she was pleased to record, Darwin "intensely" agreed. Their agreement, however, was predicated on entirely different premises. For Cobbe these "intuitive" sexual differences were God-given, while for Darwin they demanded naturalistic explanation. Accordingly, he bellowed back across the blackberries that Mill "could learn some things" from science. Women's nature, like men's, was rooted in their biology: male superiority was the product of "the struggle for existence," while men acquired their special "vigor and courage" through battling for "the possession of women."[57]

Retreating to his lodgings, Darwin documented his more carefully considered position in a series of notes he made over several days while still in Wales, dated from July 16 to 18, 1869. In the wider world, the historic meeting at which women first spoke publicly on the suffrage took place in London on July 17, while Hitchin House—the forerunner to Girton College for women at Cambridge—was just about to open its doors with five students, after a prolonged campaign spearheaded by the indefatigable Emily Davies.[58]

Darwin's extended notes capture the essence of his preformulated views on the mental differences between men and women, the extent to which these were informed by his theories of inheritance and sexual selection and by his sense of his own peculiarly masculine "genius." They also document the pains Darwin took to get it right, to make unambiguously clear the evolutionary-based, inherited intellectual superiority of men and the futility of any attempts to equalize the sexes through equal education for women.

Their repetitive theme was that men who "when adult are strongest, biggest & most courageous *eager* lovers, & most energetic Hunters, get wives & bear & rear most progeny—They exercise their qualities & power . . . when adult & thus increase them. Hence [these qualities] tend to be transmitted to their male offspring." Females "do not exercise their qualities so much when adults," so these are not passed on to their female offspring. "Genius so [much] more depends on" the masculine qualities of "energy & perseverance & patient determination," that it "may be wholly due to the qualities which naturally follow from sexual selection." If boys and girls were "educated & trained whilst young," they would transmit these qualities "to both sexes." Women would have to "be improved when adult" to improve the overall intelligence of women compared with men.[59]

Back home, a copy of Mill's *Subjection* to hand, he refined this argument with particular reference to Mill's refutation of the anatomically based claim that men's larger brains were indicative of their greater intelligence. Mill had speculated that women's (relatively) smaller brains were quicker and more active, able to multitask, while men's larger brains worked more slowly but were capable of more effort and persistence in one line of thought without the fatigue experienced by women:

> And do we not find that the things in which men most excel women are those which require most plodding and long hammering at a single thought, while women do best what must be done rapidly?

Mill had put this claim forward as "entirely hypothetical"; any supposed "natural difference" in the mental qualities or capacities of the sexes was not open to investigation so long as the "psychological laws of the formation of character" had been so little studied and were habitually disregarded by those who were contemptuous of environmental or social explanations of racial and sexual differences.[60]

Mill's distinction, hypothetical or not, spoke as no more than the simple truth to one who had spent more than thirty years obsessively honing his seminal theory, plodding along and hammering away, producing his swelling stream of books and articles, steadily accruing fame and influence, while Emma busied herself with the mundane, diverse demands of domesticity. A year earlier, Darwin had congratulated his second son George on his "brilliant success" in the Cambridge mathematical honors examination. George's "indomitable energy & perseverance" had marked him out from infancy. Mill's "pure hypothesis" was for Darwin an incontrovertible given: "What I have said about difference of genius between men & women," he noted complacently, "agrees fairly much with what J. S. Mill . . . remarks." Mill, however, had no notion of the essential roles played by sexual and natural selection in promoting these crucial sexual differences, or of the laws of development and inheritance that underpinned them, or of the analogies with lower animals that sustained them. There Darwin, the observant naturalist, looking "from the outside thro' apes & savages," had the advantage of the philosopher. He could prove, against Mill, that "it w[oul]d not suffice to educate young girls to make ♀ = to ♂!"[61]

Pleased with this reiterated, scientifically based confirmation of his own genius and of the Darwin domestic and educational arrangements that kept wives and daughters at home and sent sons out for the best possible education to fit them up for the battle of life, Darwin jotted a few directions for writing up this section in the *Descent*:

> After stating mental difference of [male] & [female], say I am aware that some dispute whether there is any mental difference, but analogy comes strongly in support of the common belief—for with these animals in which the sexes differ in external characters (as understood in the case with mankind) the mental characters differ—compare for instance a bull & cow—a boar & sow—a stallion & mare & a cock & Hen.[62]

Which is precisely how the section on "Difference in the Mental Powers of the Two Sexes" was introduced in the *Descent*. Once there, Darwin leaped from the animal analogy of innate emotional differences between the sexes (all females, animal and "even . . . savages," were more nurturing, tender, and altruistic than competitive, ambitious, and selfish males) to the asser-

tion, against Mill, of the "higher [intellectual] eminence [of man], in whatever he takes up, than woman can attain—whether requiring deep thought,
reason, or imagination, or merely the use of the senses and hands." When
it came to male intellectual superiority, however, the essential animal analogy disappeared, and the claim was founded entirely on the social basis of
the lack of feminine eminence in the arts and sciences: "If two lists were
made of the most eminent men and women in poetry, painting, sculpture,
music . . . history, science, and philosophy . . . the two lists would not bear
comparison." Where Mill had devoted a whole chapter to interrogating and
critiquing such conventional reasoning on the grounds of the cultural and
legal restrictions imposed on women and their lack of a comparable education or opportunities, Darwin reaffirmed it in the one easy, polemical
affirmation of masculine preeminence.[63]

Mill's *Subjection of Women* was no mere polemic, but a work of political
philosophy that had its foundation in the basic libertarian values enunciated
in his earlier *On Liberty* (1859). His *Subjection* extended to women Mill's fundamental principle that "human beings are no longer born to their place in
life," as the old wisdom would have it; rather, "freedom of individual choice
is now known to be the only thing which procures the adoption of the best
processes, and throws each operation into the hands of those who are best
qualified for it." It is against this "general principle of social and economi-
cal science" to "fix beforehand, on some general presumption, that certain
individuals are not fit to do certain things . . . that to be born a girl instead of
a boy, any more than to be born black instead of white, or a commoner instead of a nobleman, shall determine the person's position through all life."
The indoctrination of women in the belief that they must submit to masculine authority, the lip service paid to the "empty compliment"—the "silly
panegyrics"—of women's superior moral goodness and virtue, the laws that
denied women economic and political independence—all conduced to the
"corruption" of both sexes:

> It would be a miracle if the object of being attractive to men had not become
> the polar star of feminine education and formation of character . . . [while men]
> represent to [women] meekness, submissiveness, and resignation of all indi-
> vidual will into the hands of a man, as the essential part of sexual attractiveness.

Women everywhere were campaigning for their liberal and just rights, which
had been too long denied them. "The time is now come for the morality of
justice," to entitle women, as well as men, to the "principle of the modern
movement in morals and politics," that of freedom of individual choice.[64]

Darwin had read Mill's *On Liberty* early in 1859: "very good." His son William described him as politically an "ardent Liberal" with a "great admiration" for Mill.[65] What is so striking is Darwin's calculated, utterly confident, biologically based trumping of Mill's extension of liberalism's principle of individual choice to women. His refusal to concede any but naturalistic explanations of human intelligence and morality was hardening into a biological determinism that rejected all social and cultural causation other than that which could be subsumed under natural laws of inheritance and thus become innate or fixed. By the time Darwin finalized the section on the mental and moral differences between the sexes in the *Descent*, further confirmation that genius was an inherited masculine prerogative was at hand and, moreover, with flattering reference to the Darwin sons. In 1869, Galton's *Hereditary Genius* (fittingly described as "a collective autobiography of the masculine Victorian elite") discussed the inheritance of superior mental ability through the male side of generations of eminent families. These included his Darwin relatives, who were singled out for their pedigree of inherited scientific achievement, traced from Grandfather Erasmus through to "the illustrious modern naturalist; author of the 'Theory of Natural Selection,'" and proven by the success of two of his sons, George and Leonard (who also had distinguished himself in university examinations). Darwin was puffed up with family pride:

> I must exhale myself, else something will go wrong with my inside. I do not think I ever in all my life read anything more interesting & original. . . . You have made a convert of an opponent in one sense, for I have always maintained that, excepting fools, men did not differ much in intellect, only in zeal & hard work; & I still think there is an *eminently* important difference.[66]

Whatever reservations Darwin had about Galton's arguments for inheritance as the decisive factor in intellectual ability, these were sufficiently diminished by the time the *Descent* went to press for him to assert in its pages that Mill's environmentalist interpretation of intellectual and moral differences was rendered "extremely improbable" by the "general theory of evolution." And Darwin could bring Galton's statistical "law of the deviation of averages" in support of his socially determined claim of male intellectual superiority: "If men are capable of eminence over women in many subjects, the average standard of mental power in man must be above that of woman."[67]

Building on his earlier notes, in the *Descent* Darwin turned Mill's argument back on itself. Was it not Mill, that "great authority," who had declared genius to be "patience"; and was not patience to be defined as "unflinching, undaunted perseverance"—just those qualities that Mill attributed to males and not to females, the very qualities that would be honed by sexual and natural selection, would be gained at maturity, and thereby transmitted more fully to the male than to the female offspring? "Thus man has ultimately become superior to woman." It was "fortunate," added Darwin fair-mindedly, that this law of inheritance did not always hold good, and that the "law of the equal transmission of characters to both sexes has commonly prevailed throughout the whole class of mammals"; otherwise "it is probable that man would have become as superior in mental endowment to woman, as the peacock is in ornamental plumage to the peahen."[68]

Having scientifically demolished Mill, Darwin could not forbear committing to print the policy conclusions he earlier had deduced in his notes, even though he conceded he was "here wandering beyond" his "proper bounds." Nevertheless, he proceeded to argue that even if women were granted the full liberal prerogatives of individual choice as Mill urged, were given the higher education and opportunities they were beginning to demand, they could never hope to equal men. The study of comparative embryology and the laws of inheritance that underpinned sexual difference and sexual selection made this impossible. True, if women were trained "when nearly adult" to "energy and perseverance," and had their "reason and imagination exercised to the highest point," they then would "probably transmit these qualities chiefly to [their] adult daughters." But, cautioned Darwin, in order for the general level of feminine intelligence to be raised, such highly educated women would need to produce more offspring over many generations than their less educated sisters. The implication was that this was unlikely. Meanwhile, although male combat was no longer in operation in civilized societies, male intelligence would be constantly honed through the severe competitive struggle that males necessarily underwent in order to maintain themselves and their families, and "this will tend to keep up or even increase their mental powers, and, as a consequence, the present inequality between the sexes."[69]

It was at this point that Darwin strategically called up Carl Vogt's anatomical argument to support his conclusion that the current sexual inequalities could only be enhanced rather than diminished by social progress:

An observation by Vogt bears on this subject: he says it is a "remarkable circumstance, that the difference between the sexes, as regards the cranial cavity, increases with the development of the race, so that the male European excels much more the female, than the negro the negress."[70]

Darwin earlier had invoked Vogt's assertion that the formation of the female skull is "intermediate between the child and the man" and backed it up with other citations gleaned largely from the pages of the *Anthropological Review*. Women's facial features, head shape, and general anatomy were more child-like, less variable, and closer to the "primeval" human form. The "children of the different races of man" more closely resembled one another than did the adults: "Some have even maintained that race-differences cannot be detected in the infantile skull."[71]

The embryological argument, given Haeckel's authoritative stamp, fig-ured prominently in Darwin's exposition of an animal ancestry for human-ity ("By considering the embryological structure of man . . . we can partly recall . . . the former condition of our early progenitors; and can approxi-mately place them in their proper position in the zoological series"); it un-derlay Darwin's conception of the evolution of human mental, social, and ethical faculties, which were recapitulated in individual ontogeny; it perme-ated his discussion of the mental differences between men and women; it predisposed Darwin to Vogt's view that sexual inequality was the hallmark of a more "developed" race; and it underpinned his explicit relegation of cer-tain "generally admitted" feminine traits ("the powers of intuition, of rapid perception, and perhaps of imitation") to those "characteristic of the lower races, and therefore of a past and lower state of civilization."[72]

With this, Darwin gave his overturning of Mill's egalitarian thesis the necessary racial dimension critical to the overall argument of the *Descent of Man*. Necessary because, let it be clearly understood, this was no mere prejudicial coupling of feminine with "lower" or more primitive intelligence on Darwin's part, but a necessary consequence of Darwin's theoretical views. Woman's intelligence was bound to her anatomy, which, in keeping with Darwin's fundamental developmental rule for sexually dimorphic species or races, was more childlike or juvenile than man's and therefore representative of the earlier ancestral stage of the group.[73]

And with this, Darwin put his imprimatur on Huxley's earlier assertion of the evolutionary impossibility of the liberal ideal of equality for women and blacks. Darwin was no salaried professional with an ax to grind; he had no professed desire to keep women and the "lower" races in their place. Nature, as interpreted by Darwin (and more and more of his fellow Darwinians), would do this anyway. It did not call into question Darwin's lived experi-ence of the displacement of their indigenous inferiors by intruding superior races, or of the strongly gendered roles of Victorian men and women, but rather, corroborated and naturalized existing social and racial inequalities and deeply entrenched Victorian values.

That Darwin never managed to transcend these conventions and take seriously Mill's critique of them should occasion no surprise. Very few Victorians managed to do so. Mill attributed his enlightened stand on gender issues almost entirely to his long-term companion and late wife, Harriet Taylor Mill. Although it was written after her death, *The Subjection of Women* was their "joint production": "All that is most striking and profound belongs to my wife." As he represented it, Mill was never nature's mouthpiece, but Harriet Taylor's. In the text of the *Subjection*, Mill pointedly argued that the "most favourable case which a man can generally have for studying the character of a woman" is that of his own wife:

> For the opportunities are greater, and the cases of complete sympathy not so unspeakably rare. And in fact, this is the source from which any knowledge worth having on the subject has, I believe, generally come. But most men have not had the opportunity of studying in this way more than a single case: accordingly one can, to an almost laughable degree, infer what a man's wife is like, from his opinions about women in general.[74]

It is a passage that might have been aimed directly at Darwin. But the man himself, buffered by his science and the unquestionable "nature" of his domestic relations, was beyond its reach. By the second edition of the *Descent*, Darwin was more certain than ever of Mill's unscientific misapprehension of the inherited bases of mental and moral differences, and said so in an uncharacteristically forthright footnote: "The ignoring of all transmitted mental qualities will, as it seems to me, be hereafter judged as a most serious blemish in the works of Mr. Mill."[75]

13.6 A pretty dilemma: Dealing with the anthropologists

Darwin's attempt to derive a biologically coherent narrative for male intellectual superiority and male choice in humans drove him into contradiction; it destabilized the very continuity between animals and humans that it was theorized to support; and it brought him into conflict with those same evolutionary anthropologists whose views might have been expected to be most congruent with his own. His trumping of Mill was a relatively simple matter compared with the complex maneuverings to which he had to resort in order to preserve his key thesis—of the primacy of beauty-based male choice as the determinant of human racial divergence—in the face of the gathering consensus among anthropologists that the primitive human condition was one of sexual promiscuity, in which women as well as men freely chose their sexual partners. This consensus was arrived at in reaction to those like Argyll who argued that present-day savages must have degenerated from the earlier higher state of morality and civilization originally bestowed on humanity by God. In response, the evolutionary anthropologists traced a triumphal narrative of civilized self-control over bestial sexual impulses and practices. This was based on the assumption that the "licentious" practices of contemporary savages, like the conduct of the Fuegians that had so shocked the young Darwin, were representative of those of the earliest humans.

John McLennan's *Primitive Marriage* (1865) and Lubbock's *The Origin of Civilization and the Primitive Condition of Man* (1870), with minor differences, tracked the evolution of sexual relations from a common origin in primitive promiscuity through a series of intermediary stages until the highest-evolved, civilized state of monogamous marriage was attained. The first stage was one of communal marriage with matrilineal descent; then came a stage of polyandry (the sharing of one wife by several husbands), brought about by a shortage of women through the practice of female infanticide during periods of population pressure; this was resolved by wife capture from outside the group (exogamy); then, with the accumulation of property, came polygamy, with the accompanying establishment of patrilineal descent; and finally, monogamy. Other contemporary marriage theorists like J. J. Bachofen and Friedrich Engels (who posited a primitive matriarchy) structured a similar narrative of increasing masculine control over female sexuality and reproduction as a defeat of female power. The evolutionary anthropologists, Lubbock in particular, tended to its more benign interpretation as a progression toward improved female prestige—one that, with increasing civilization, removed women from obligations of physical labor and sexuality, and opened up the possibility of marital and paternal love in the liberal Victorian ideal of companionate marriage.[76] It was an interpretation with which Darwin had a good deal of sympathy; but it was at odds with his theoretical need to demonstrate the continuity from animal to human courtship practices and to preserve his central thesis of animal female choice, which, in humans, was reversed to become the dominant and racially determinate male choice.

In Darwin's theorizing, the transition from female choice in animals to male choice in humans necessarily was pushed back to some very early stage of human evolution, before the emergence of the racial differences for which it was the determining cause. It was presumably during this racially indeterminate primeval period that man, being "more powerful in body and mind than woman," had "gained the power of selection." In the *Descent*, Darwin

AUSTRALIAN ABORIGINAL MARRIAGE CEREMONY.

Figure 13.4. An illustration, taken from contemporary "savage" practice, of the putative prehistoric stage of wife capture or "exogamy," from John Lubbock's *Origin of Civilization and the Primitive Condition of Man*. Lubbock 1870, 74. University of Sydney Library, Rare Books and Special Collections.

discriminated between contemporary "civilized" courtship practices, where "women have free or almost free choice, which is not the case with barbarous nations," and more primitive ("primeval") times, when male choice was the rule. As he acknowledged, the "only means" of assessing primeval practices was the new anthropological method of studying their commonality among "existing semi-civilized and savage nations"; hence Darwin's need to establish that in "all barbarous races ornaments, dress, and external appearance are highly valued; and that men judge of the beauty of their women by widely different standards."[77]

So, on the one hand, Darwin was heavily reliant on the decorative practices, aesthetic values, and wife choices of contemporary "barbarous races" in order to substantiate his explanation of the primeval divergence of the separate races through male aesthetic choice; on the other hand, he was faced with the need to dissociate the promiscuous sexuality assumed of contemporary savages from the sexual practices of primeval men and women, where, according to this same theory, women necessarily had little or no choice and the more powerful men were at least discriminating enough to select the more beautiful females according to their different standards of taste.

Darwin's extensive researches into mating behavior and sexuality across the animal kingdom, with few exceptions, all went to confirm his projection back onto nature of the Victorian conventions of sexuality and courtship. The language of the *Descent* reflects this. From the "marriage arrangements" of fish and reptiles, on through the "love antics" of birds to the "marriage unions" of mammals and the accounts of the romantic "love" and "devotion" of domestic bitches, so "enamored" of and faithful to particular dogs that they rejected all others, the female of the species was almost invariably sexually shy, timid, and discriminating. This casting of animal mating behavior in the acceptable terminology of courtship and marriage, where overt female sexuality was deviant and unnatural, smoothed the way for the continuity of animal to human sexuality, where the same conventions prevailed. Further to this, Darwin, in explicating male intellectual superiority, presumed a patriarchal model of primitive family life. Men had become more intelligent than women through a combination of natural selection and the law of battle, to which men were subject but women were not. This required that human females, even primitive ones, were exempt from the struggle for existence in important respects: locked out of the means of acquiring intelligence through the improving powers of natural selection, dependent on males for care, subsistence, and protection, and locked into a model of primitive family life that coincided with the Victorian ideal of middle-class patriarchy and the sexual division of labor. Yet Darwin's primitive patriarchal family was predicated on the same stereotypical sexuality that was fundamental to animal female choice. This was in tension with the supposed practices of early human progenitors where male choice was the norm, where males were still, by nature, indiscriminate sexual "seekers," ready to seize on any and every female, yet somehow, by virtue of their superior intelligence and physical strength, committed to a choice of mate (or mates) primarily on aesthetic grounds. Female choice in animals was an irrational, capricious, instinctive, purposeless fancy, driven by the perception of beauty in the prospective mate; primitive humans had "only feeble powers of reason"; yet male choice seemingly had followed on the development of a certain level of rationality in males.[78]

The contradictions piled up. One made much of by his more damaging critics (notably St. George Jackson Mivart) was that, as many male animals, like men, are larger and physically stronger than females (these differences in both instances, Darwin argued, being acquired through the "law of battle" by males for females), why should not their physical superiority have enabled such male animals to seize the power of choice from females? Yet there were only "exceptional" instances of male choice even among those animals closest to man, the quadrumana, whose sexual differences "were curiously the same" as those in humans, the males being "larger and much stronger" than the females. The rhesus monkey, in which the smaller female displays a distinctive patch of naked skin "around the tail, of a brilliant carmine red" (as Darwin circumspectly described this overtly sexual display), was the only such case deduced by Darwin.[79] If the issue was determined by the greater intelligence of men, how did Darwin's patriarchal family model, projected back onto a primitive ancestry, match Vogt's claim (otherwise relied on by Darwin) that primitive men and women shared essentially similar labors and had similar crania and similar intelligences and that the sexual division of labor was characteristic of more advanced races and highly evolved societies? Darwin himself argued that it was "not probable that the greater strength of man was primarily acquired through the inherited effects of his having worked harder than woman for his own subsistence and that of his family; for the women in all barbarous nations are compelled to work as least as hard as the men."[80]

These inconsistencies seemingly escaped Darwin, blindsided by his single-minded focus on substantiating his key theses of female choice in animals and male choice in humans, the driver of racial divergence. But he was all too aware of the major problem posed by those marriage theorists who were bent on deriving a late-arriving stage of patriarchal marriage from a common origin of primitive promiscuity or communal marriage, with intervening stages of infanticide, polyandry, and wife capture—all suppositions that undermined the applicability and efficacy of sexual selection and threatened to derail the whole point and argument of the *Descent of Man*.

To compound Darwin's dilemma, there was the above-discussed issue of sex ratios (sec. 13.3) and what it implied for his theory of racial divergence. In the case of humans, where Darwin assumed males and females to be about equal in number at birth and male choice to predominate, Darwin's need to uncouple the primitive human condition from the contemporary anthropological fixation on sexual promiscuity was all the more compelling.[81] Male choice could prevail only if these primitive men and women formed unions (at least temporarily) and succeeded in rearing a greater number of children to pass on the preferred qualities. Hence Darwin's desire to link this critical primeval stage of human development with a patriarchal monogamy (or, even better for his purposes, polygamy) in which the more powerful, vigorous males not only selected the prettiest wives but also defended and provided for their families and thus—as he had argued previously in the case of animals—enhanced their prospects of successfully rearing offspring. The effects of this would be negated if, as marriage theorists argued, there was an artificial shortage of women induced by female infanticide, or if communal marriage, polyandry, or exogamy were practiced.

These were the complexities with which Darwin juggled as he confronted the extent of the difficulties posed by the evolutionary anthropologists. He began to shape his answers in his copy of McLennan's *Primitive Marriage*:

> The scarcity of Women from infanticide of females leading to promiscuous intercourse & polyandry, w[oul]d make the selection of women very difficult . . . There must have been a time judging from lower animals, when men did not foresee [scarcity], when there was not infanticide & when sexes equal, & then sexual selection w[oul]d come in.[82]

But, oddly enough, it was Darwin's reading of Argyll's *Primeval Man*, in which Argyll argued against McLennan and Lubbock that it could not be assumed that the present-day degenerate practices and customs of savages were primeval, that offered Darwin his best strategy: "I must rest my conclusion on descent & not on types of savagedom.—Say animal nature—not necessarily like present Barbarians." In the back of his copy, Darwin, foreshadowing his conclusion to the *Descent*, wrote:

> If one of the lower animals c[oul]d reason & he heard that man was ashamed of being a co<descendant> with him he might laugh with scorn & ask what of [savage] practices.[83]

Following this, he dealt dismissively with Lubbock's extended argument for a primitive state of communal marriage: "Says so, but I am not convinced."[84]

In the *Descent*, Darwin conceded that the "licentiousness of many savages is no doubt astonishingly great," though perhaps their "existing intercourse" was not "absolutely promiscuous." Most, he claimed, and almost certainly the "chiefs," were polygamous. He ran through the evidence for existing—and therefore primitive—female infanticide, wife capture, polyandry, childhood betrothals, and the slavery of women, all "interfering causes" that would prevent or check the action of sexual selection.[85] In every instance, as he had determined, Darwin's tactic was to fall back on his animal analogy:

> Nevertheless from the analogy of the lower animals, more particularly of those which come nearest to man in the series, I cannot believe that this habit prevailed at an extremely remote period, when man had hardly attained to his present rank in the zoological scale.

Man was "certainly descended from some ape-like creature," and most existing monkeys and apes were either "strictly monogamous" or polygamous. Indeed, "from what we know of the jealousy of all male quadrupeds . . . promiscuous intercourse in a state of nature is extremely improbable." It followed that "if we look far enough back in the stream of time," when men were "governed more by their instincts and even less by their reason than are savages at the present day," they would have lived as monogamists (even if only temporarily or sequentially) or as polygamists; they would not have practiced infanticide or polyandry; there would be no infant betrothals; and women would not be valued as mere slaves.

> Their intercourse, judging from analogy, would not then have been promiscuous. They would, no doubt, have defended their females to the best of their power from enemies of all kinds, and they would probably have hunted for their subsistence, as well as that of their offspring. The most powerful and able males would have succeeded best in the struggle for life and in obtaining attractive females . . . Thus during these primordial times all the conditions for sexual selection would have been much more favourable than at a later period, when man had advanced in his intellectual powers, but had retrograded in his instincts.[86]

It was only by harking back to an idealized prelapsarian period inhabited by our "semi-human progenitors," when man was human but was not somehow fully human, was still guided by his "natural" instincts, and had not (by dint of developing reason and foresight) been "perverted" by unnatural savage practices like promiscuity, infanticide, and polyandry, that Darwin salvaged continuity between animal and human sexuality and preserved sexual selection—and male choice in particular—from the depredations of the anthropologists. But this came at the cost of destabilizing the moral progress that he elsewhere attributed to increasing rationality and to the acquirement of social instincts—of the parental and filial affections and obligations and of a moral sense or conscience—which even some animals exhibited.[87] And it led inexorably to the conclusion to the *Descent*, to that "heroic little monkey, who braved his dreaded enemy in order to save the life of his keeper," so much to be preferred as ancestor to the savage, who "practices infanticide without remorse, treats his wives like slaves," and "knows no decency."

Darwin's thesis of savage degeneration ended up strangely like that of his old enemy, the Duke of Argyll, a "new doctrine of the fall of man," as a reviewer of the *Descent* in the *Spectator* put it, a "far more wonderful vindication of Theism than Paley's Natural Theology, though we do not know, so reticent is his style, whether or not he so conceives it himself!"[88]

13.7 Keeping up appearances

None of these considerations deterred Darwin in the slightest from drawing copiously on contemporary practices of face and body decoration, wife choice, and widely different views on female beauty in demonstrating the primeval divergence of the separate races through male aesthetic selection. His continuing acquisition of this material led him into collaboration with some odd fellow travelers while writing the *Descent*.

One of the more intriguing was William Winwood Reade, African explorer, journalist and minor novelist, militant agnostic, and onetime card-carrying Anthropological. At the time he contacted Darwin in 1868, Reade's major claim to fame was his potboiler, *Savage Africa* (1864), which served up sensationalized accounts of those two perennials, gorillas and cannibalism; but Reade's more challenging opinions on the futility of the civilizing effects of Christian missions, on the necessity for polygamy and his qualified praise for Islam in Africa generated controversy. Earlier, Reade had been present to applaud Hunt's views on the "Negro's Place in Nature" at the infamous meeting of the new-fledged Anthropological Society in 1863. "The Negroes may be said to resemble schoolboys," Reade affirmed on that occasion; "it is impossible to make them work . . . You must flog them occasionally." His solution to slavery was to let slaves buy their way out of the institution: industrious ones would be liberated, with lazy ones remaining in slavery, "which is the proper place for them to be in."[89]

Yet Reade was also a monogenist and an evolutionist, shortly to identify himself as Darwin's "disciple"—yet another instance of the compatibility of Darwinian and Anthropological thought. Nor was Darwin too nice in his requirements to reject help from such a dubious source; when Reade was about to set out on his second expedition to West Africa, he offered to make any "special enquiries" that would "aid in elucidating or affording evidence for those grand problems" with which Darwin was engaged.[90] It sparked a correspondence concerning racial notions of beauty, marital preferences and practices, and their relation to sexual selection, skin color, and appearance, which did not end until Reade's early death, in 1875.

Although Darwin had "many facts" on racial notions of beauty (culled primarily from venerable works like Prichard's and Lawrence's and aging informants like Andrew Smith), he was "greedy for more." What he wanted from Reade was on-the-spot information on "whether negroes . . . admire a jetty black skin & woolly hair, & their own characteristic features," and *"more especially"* to find "how far in a quiet sort of way" African women had "any influence in leading particular men to woo them or purchase them from their parents." This last emphasis on the contemporary existence of "indirect" or passive female choice among Africans was to find support for Darwin's view that, in earlier and more primitive times, women, like birds, presumably exerted some degree of female choice (what Darwin did *not* want was evidence of free, overt female sexual choice that might verge on the theory of primitive promiscuity). While just about everything else that physically distinguished men from women—women's long tresses, sweeter voices, lack of facial and body hair, and greater beauty—had been acquired through male choice, Darwin assumed (as he had earlier told Wallace) that the masculine beard, that splendid emblem of Victorian virility, had been gained through female choice. Evidence for some limited form of female choice among "savages" was thus a point of "some importance" on which Darwin had collected some instances from his readings in the literature.[91]

With his first field report from Africa, his new disciple gave Darwin just what he wanted. In the *Descent*, Darwin quoted Reade to the effect that "the women, at least among the more intelligent Pagan tribes, have no difficulty in getting the husbands whom they may desire, although it is considered unwomanly to ask a man to marry them." On Reade's authority, African tribeswomen were also "quite capable of falling in love, and of forming tender, passionate, and faithful attachments." This meant, said Darwin, that even such subdued female preference, "steadily acting in any one direction, would ultimately affect the character of the tribe." Women would "choose not merely the handsomer men, according to their taste, but those who were at the same time best able to defend and support them." It all went to bolster Darwin's insistence on a primitive patriarchy, that women, "especially during the earlier periods of our long history," much like their animal counterparts, must have lived in relatively chaste relationships with more powerful, "well-endowed," protective males, who, for their part, naturally would select the "more attractive women."[92]

But he was not happy with Reade's considered opinion that "after a series of very cautious inquiries," he was "able to assert that the negro's idea of beauty is the same as ours, & not exactly opposite as so many have supposed." While it was generally the case that a black skin was more admired than lighter ones, Reade inclined more to the view that black skin, being associated with immunity to moist heat and coastal fever, was the result of natural selection rather than sexual selection—a view he continued to promote until his death. Reade was assuming less the role of disciple than rival theorist. He was, he told Darwin, with several years of African experience behind him, "beginning to understand this race." The "complex man of civilization is quite undecipherable: women and savages are a little easier & there is less variety among them."[93]

Darwin, who had never cared for field observers with opinions of their own, was taken aback. He pressed Reade for clarification, telling him (with no little understatement), "I have to touch on this point." The *Descent* was about to go to press, and he urgently needed more details on the African idea of beauty. He did not get these until Reade returned to London, having been forced to abort his search for the source of the Niger. "Respecting the beauty question," Reade wrote Darwin in November 1870,

> I would not venture to assert . . . that the Africans w[oul]d ever prefer the most beautiful European . . . to a good-looking negress. But . . . if Africans had to choose from a number of European women they would certainly select those whom we would select as the best looking.

Africans most admired faces "furthest removed from the prognathous type"; the most admired "points" were the same as in Europeans: small hands and feet, large eyes, a small nose, and a well-shaped nape of the neck. Reade also had an eye to the fashion analogy: the African "admiration of broad hips is carried to extravagance"; "with us" this same admired feature was achieved by "tight lacing," which served both to "compress the waist" and "broaden the shoulders and hips." Africans admired long hair and often wore false hair, "just as we do." The nose was the "only doubtful point" of difference, but a "*very* flat nose is not admired." As for the black skin, it was certainly admired: "but that is because the black skin is really the most beautiful." It was not admired through "force of habit, because black skins are the exception not the rule except among a few tribes." Reade, well aware of what Darwin wanted, was not to be moved from his belief ("& as the belief of those who have lived long among these people") that "their ideas on the subject of beauty are *on the whole* the same as ours." Reade had gone native, especially in finding black skins simply the best (his investigations into African sexuality were not as "cautious" as he represented them to be: "Kissing is unknown throughout West Africa").[94]

Darwin made the best of it. In the *Descent*, he opened his discussion of Reade's views with the assertion that he had "met with very few statements opposed to [his own] conclusion. Mr. Winwood Reade, however, who has had ample opportunities for observation . . " Darwin went on to give a fair summary of Reade's letter but ignored his contention that a black skin resulted from natural selection and omitted his view that it was admired because it was "really the most beautiful"; rather, he represented this as Reade's endorsement of the African admiration for a black skin. Darwin's own belief still stood: there was something "monstrous" in the African preference for "jet-black" skins.[95]

Reade's dissent, though he continued to maintain it, was manageable. Darwin had garnered plenty of supporting material over the years, and he had his outstanding instance of the Hottentot bottom, validated by Smith and Burton (though Burton also held to the universal ideal of beauty), the human corollary of pigeon selection for an admired but potentially debilitating female peculiarity. Darwin also had a good deal of carefully collated material on the decorative practices of "barbarous races," although this did not fit his notion that women "everywhere were conscious of their beauty" and were necessarily the more ornamented. Yet again this might be explained away by the "characteristic selfishness" and domination of savage men, who subverted the natural order by denying their women the "finest ornaments" and keeping them for themselves. His fundamental claim remained: that the variable dress and decoration of "semi-civilized and savage nations," as with the Hottentot bottom, bore out the "truth of the principle . . . that man admires and often tries to exaggerate whatever characters nature may have given him." This had been given substantial support early in 1868, when Darwin was contacted by another self-proclaimed "admirer," the Italian anthropologist and pathologist, Paolo Mantegazza.

Mantegazza had just finished reading Darwin's *Variation*, "a sublime monument to human intelligence." He dispatched a copy of his own recent record of his South American travels, *Rio de la Plata e Tenerife* (1867), helpfully marking those passages relevant to Darwin's "studies of races." Mantegazza's intellectual grounding lay in the traditional continental nexus of physiognomy, aesthetics and race. He later produced an "Aesthetic tree of the human race," which sprouted the least aesthetic Australians, Fuegians, and Hottentots on its lowermost, stunted, and unproductive twigs before ascending to the most beautiful Aryans on the loftiest, topmost branch. Darwin worked his way through Mantegazza's "full and excellent account" of the differential racial practices of hair eradication, nose, lip and ear piercings, tattooing, face and body painting, and head deformations. Inside the back cover, he wrote: "Seeing what a passion for [ornamentation] it is strange that races of man not more altered." In the *Descent* he stressed that Mantegazza "strongly insists" on the principle that "Man always exaggerates what he has." His Italian admirer brought timely reinforcement to Darwin's crucial alignment of the decorative choices of "savages" with those of fashionable

Figure 13.5. In the *Descent*, Darwin drew a parallel between the facial cicatrices, tattoos, or paint of "savages" and the male mandrill that "appears to have acquired his deeply-furrowed and gaudily-coloured face from having been thus rendered attractive to the female." (Left) "Head of a New Zealander," Hawkesworth 1773, 3: facing page 453; and (right) "Head of male Mandrill," *Descent* 1871, 2:292, University of Sydney Library, Rare Books and Special Collections.

European women and pigeon fanciers, all having the "same desire to carry every point to an extreme."[96]

One way and another, Darwin had more than enough supportive "facts" on contemporary savage decorative practices and aesthetic values, however conflicted his heavy reliance on these was with his insistence that the "extremely licentious" behavior of contemporary savages could have no correlation with primeval sexual practices. However, he ran into more contradiction when he turned from the remote past and from contemporary "barbarians" to the functioning of sexual selection among "civilized and semi-civilized nations." While claiming that civilized women have "free or almost free choice" of marriage partners, he focused almost exclusively on male choice. One promising investigation of male choice in action in contemporary British society led Darwin into association with another Anthropological stalwart, John Beddoe, the new president of the society.

Beddoe, a Bristol doctor who trained in Edinburgh, had begun his researches into British ethnicity under the aegis of Robert Knox. He formalized Knox's view that the "races" of Britain did not intermingle or successfully hybridize, but remained distinct, living on separately in a "kind of natural apartheid." Throughout his life, much like Knox before him, Beddoe traveled Britain, unobtrusively studying the physiognomies of the inhabitants and classifying them according to eye and hair color, charting the ethnicity of the population. His results were published in his *Contribution to Scottish Ethnology* (1853) and his major work, *The Races of Britain* (1885). In conjunction with these, Beddoe developed an "Index of Nigrescence," which purported to demonstrate the "tendency to melanosity" of the population, due to the intrusion of the "darker" Gaelic and Iberian races, which were allegedly outbreeding and swamping the "blond Teutons of England."[97]

Beddoe was another who integrated his views on racial separateness with Darwinian concepts. Darwin's interest initially was sparked by Beddoe's 1863 paper in the *Anthropological Review*. There, Beddoe had speculated that fair hair was becoming less prevalent in England due to the influx of darker-haired immigrants from Ireland, the Scottish Highlands, and Wales into the larger industrial towns where their "melanous temperament" was better able to endure the "anti-hygienic agencies" of the overpopulated cities; and "thus the law of natural selection operates against" the multiplication of the lighter-haired races. What caught Darwin's eye, though, was Beddoe's additional proposition that the "physical qualities of the race may be to some small extent moulded by the action of *conjugal* as well as of natural selection." His analysis of hair color and marital status of some seven hundred women patients in the Bristol Royal Infirmary showed that the percentage of married women with dark hair was significantly greater than fair-haired ones. The "mass of the population," he concluded, evidently did not sympathize with the more cultivated preference of artists and poets for fair hair; if this lowbrow trend in male choice continued, then over the next several generations, according to the laws of inheritance, the proportion of the dark-haired would considerably increase.[98]

Beddoe fell over himself in his eagerness to assist when Darwin contacted him in the summer of 1869. He supplied "several more years of observation on the point of conjugal selection" among "lower class" women, breaking his data down by age to exclude the possibility that hair simply might darken with age. Initially, Beddoe thought his earlier claim was confirmed and proceeded to break his data down further.

His new results, however, were "vexatiously subversive of [his] former inferences." Hoping to retrieve the situation with further extraction and analysis, Beddoe labored on until he finally conceded that his data, though suggestive, did not significantly support his thesis. Regretfully, Darwin agreed. "I must give up whole case," he penciled across the top of Beddoe's final report.[99]

Flawed though Beddoe's study was, it may have reinforced Darwin's underlying concern that sexual selection, in an advanced civilized society, worked to destabilize the effects of natural selection. Beddoe's study was not innocent; it was no simple, straightforward analysis of marital preference for a particular hair shade, but designed to support his thesis—with all that this entailed in racial and class terms—that "conjugal" selection of the darker haired favored the increasing "nigrescence" or "melanosity" of the population. His subjects were "lower-class" women among whom this trend was allegedly occurring, and Beddoe opposed the preference for dark hair to the highbrow aesthetic preference for fair hair; if such inferior male choice prevailed, it would result in a preponderance of the less desirable, lower-class, Irish and Welsh elements over their fair-haired betters and thus to social and intellectual degeneration.

It was an argument similar to William R. Greg's claim that, protected from the consequences of their improvidence by the social sympathies generated by natural selection, the inferior Irish were in the process of outbreeding the frugal, foreseeing middle classes. Beddoe's interpretation reinforced Darwin's imperative for middle-class survival: "Choose good wives." It was irresponsible sexual selection, along with natural selection, that was promoting the survival of the unfit (somewhat paradoxically, according to Beddoe, the inferior Celts were better fit to survive squalid, overcrowded, and unhygienic conditions than their class and racial superiors) and endangering social progress. The wrong choice of a wife had consequences for the future of society that Darwin had long been concerned with elucidating.[100]

Beddoe's failure to come good with significant evidence of ongoing male choice left Darwin with little on which to base his case in the *Descent* for the continuing action of sexual selection. There remained that old chestnut, the vaunted good looks of the aristocracy, originally brought to Darwin's attention by William Lawrence's *Lectures*, reinforced by his reading of Nott and Gliddon's *Types of Mankind*, and emphasized by Darwin in his very first letter to Wallace on the primacy of sexual selection in forming the human races. The egalitarian Wallace had put a dampener on this beauty myth: "Mere physical beauty . . . is quite as frequent in one class of society as the other." But Darwin was encouraged by Hooker who, like Huxley, had remarkably little to say in support of sexual selection in general, but evidently was privy to and shared Darwin's view that the lords were uncommonly "ken-speckle" through their continuous selection of the "most beautiful & charming women out of the lower ranks." Hooker, the elitist, also thought that "Blood, Blunt [cash], Brains, Beauty" (the "union of all must be irresistible, in every degree and condition of life, from Fuegia to London") would accumulate by natural selection and "must culminate in an Aristocracy—or there is no truth in Darwinism."[101]

Darwin was much amused; but in the *Descent*, true to its source in Lawrence's *Lectures*, he took the "Brains" out of Hooker's four *B*s that "regulate the development of an Aristocracy." Lawrence and others were convinced, he wrote, "as it appears to me with justice, that the members of our aristocracy," privileged by their wealth and standing, "have become handsomer, according to the European standard of beauty, than the middle classes . . . from having chosen during many generations from all classes the most beautiful women as their wives." Yet the inheritance and enhancement of aristocratic beauty was secured by primogeniture—that "scheme" for "destroying" natural selection, as Darwin earlier had told Wallace. Inheritance by the firstborn, without regard to physical or intellectual fitness, was a practice that ensured the perpetuation of unearned and thus socially and racially deleterious upper-class privilege. Middle-class men still had an eye for a pretty face and were superior to their animal and primordial forebears in valuing "mental charms and virtues" in women; but all too often they were seduced by "mere wealth and rank." Like the elder sons of the aristocracy, such ill-chosen wives had done nothing to earn their inherited wealth and status; their offspring could not improve the race. The marriage choices of civilized women, on the other hand (like those of their primitive sisters, who had some limited free choice and so looked for a powerful, protective partner), were "largely influenced by the social position and wealth of the men." Such successful middle-class men had proven heritable ability and intellectual powers gained through their own or their forefathers' struggles for existence.[102]

The door stood invitingly ajar to the possibility of racial improvement through female choice. Darwin firmly closed it—or, more precisely, did not even perceive that it had been opened. For Darwin, mate choice was correlated with the development of "mind." This explained why what was attractive to animals "generally coincides" with what was attractive to humans. But it also accounted for the superiority of the masculine mind and its potential for superior rational choice:

> He who admits the principle of sexual selection will be led to the remarkable conclusion that the cerebral system not only regulates most of the existing functions of the body, but has indirectly influenced the progressive development of various bodily structures and of certain mental qualities.[103]

A continuum of choice parallels this cerebral development. It runs from animal female choice—which is insistently represented as unconscious and intentionless, concerned with useless if beautiful (sometimes deleterious) ornamentation and the superficialities of appearance, through the aberrant choices made by "lower race" males with their penchant for female steato-pygia or "jet-black" skin, up to the superior choices of civilized European men, whose "cerebral systems" have been developed to the full; they alone are capable of conscious, rational, purposive choice.

The most explicit instance of contemporary choice by human females that Darwin discussed in the *Descent* is that famous (and much misunder-stood) one of the "pretty girl" at the fair, the lower-class country lass sur-rounded by "young rustics . . . courting and quarreling over [her], like birds at one of their places of assemblage." Like the birds, with which both she and her suitors are identified, we can only infer that she chooses through their strutting display and eagerness to impress, through their bumpkin behavior that allows her only passive, bumpkin choice. This is zoomor-phic, low-class, lowbrow, nonimproving "free choice," akin to that alleg-edly exercised by the inferior "melanous" (and melanous-preferring) men in Beddoe's study.[104]

The only kind of racially improving choice that Darwin can conceive of, and sees the necessity for, is that of better-educated, middle-class male choice, which operates within the constraints of the strongly gendered, pa-triarchal family structure that, by his account, both leads to and is the result of the evolutionary process. Hence his cautiously expressed hope in the *De-scent* that the same man who "scans with scrupulous care the character and pedigree of his horses, cattle and dogs before he matches them," but who "rarely, or never, takes any such care" when he comes to his own marriage, "might by selection do something not only for the bodily constitution and frame of his offspring, but for their intellectual and moral qualities." Only the right choice of a good wife, exercised by the "most able" men who have proven themselves through "open competition" and are "not prevented by laws or customs from succeeding best and rearing the largest number of offspring," could ensure racial progress.[105]

Darwin's venture into eugenic territory was no opportunistic gambit in-spired by Galton's[106] and Greg's recently published eugenic views, but, as I have argued, was integral to the genesis and structure of sexual selection (ch. 7). It was the culmination of Darwin's transference of the breeder's art of molding domesticates according to his taste, to humans who had thereby shaped themselves through a process of aesthetic self-selection. The pos-sibilities of selective human breeding were implicit in this transference and had been made overt many years before by Lawrence, who (along with the still-lingering shade of Alexander Walker) asserts his determining and continuing presence in Darwin's theorizing in this final section of the *Descent* in a burst of references and footnotes.[107]

Sexual selection had been a "tremendous job," and this was its climax. It remained but to take the *Descent* back to where it all began: to the shores of Tierra del Fuego, to the first sight of the savage—that shockingly unforgettable, hideous, wild, inhuman, inhumane, indecent, bestial being, whose brotherhood was undeniable yet utterly abhorrent, to set him down forever as ugly brother to the admirable ape.

FOURTEEN

The Post-*Descent* Years: Sexual Selection in Crisis, Female Choice at Large

Mr. Darwin's . . . explanation of almost all the ornaments and colours of birds and insects as having been produced by the perceptions and choice of the females has, I believe, staggered many evolutionists . . . It may perhaps be a relief to some of them, as it has been to myself, to find that the phenomena can be shown to depend on the general laws of development, and on the action of "natural selection," which theory will, I venture to think, be relieved from an abnormal excrescence, and gain additional vitality by the adoption of my view of the subject.

—Alfred Russel Wallace, *Macmillan's Magazine*, 1877

I saw a woman sleeping. In her sleep she dreamt Life stood before her, and held in each hand a gift—in the one Love, in the other Freedom. And she said to the woman, "Choose!"

And the woman waited long; and she said, "Freedom!"

—Olive Schreiner, "Life's Gifts," *Women's Penny Paper*, 1889

By the beginning of 1872, with *The Descent of Man* barely a year in print, Darwin faced the realization that "hardly any" naturalists agreed with him on sexual selection, "even to a moderate extent."[1]

For many, female choice simply attributed too great an intellect or aesthetic sense to female birds:

We must attribute to the hen Argus Pheasant the aesthetic powers of a Raphael in order to account for the decorations of her mate, or, more properly, we must assign to a succession of multitudes of generations of birds, a correctness of appreciation of the draughtsman's art, such as is a rare excellence among men.[2]

For others, the acquirement of such "refined and highly developed taste" by a mere bird was yet more evidence of divine intervention: "This instinctive selection of the beautiful leads to a theological inference a good way beyond that warranted by the selection of the useful."[3]

Even for sympathizers, the argus pheasant proved challenging. John Morley, leading utilitarian and associationist, favorably reviewed the *Descent* in the *Pall Mall Gazette*, yet criticized Darwin's phraseology on beauty: it was "loose scientifically and philosophically most misleading." When Darwin asked for clarification (it was "almost a life-time" since he had attended to the philosophy of aesthetics), Morley replied that the human sense of beauty connoted a "notional or rational element," while in lower creatures it must be "a simple sensation": that both the pheasant hen and the cultivated man admired ball-and-socket plumage was "an accidental occurrence." There was no evidence for the attribution of a "gratified sense of *beauty*" to a bird in which it was no more than an "inexplicable caprice"—not a taste at all, in the strict sense of the term. What was needed was a "neutral word" that would "not prejudge the philosophical question of the origin and composition of human preference in colour, sound, form &c."[4]

Morley's reassertion of associationist aesthetics came coupled with his claim that Darwin had "prolonged the operation of Natural Selection" into what was more properly "social selection." Like Mill, Morley argued that moral progress was achieved through the "medium of opinion" and "positive law"; it was transmitted by culture, not biology. Darwin, having directed the *Descent* against Mill and environmentalists as much as against traditional natural theologians, to bring both beauty and morality under the sway of natural law, was not about to be deflected by philosophical debate around this fundamental issue. He had held his views "for many years," he told Morley, and was "more and more satisfied" with them; he had "great confidence" that once the notion of sexual selection became more familiar it would gain acceptance.[5]

It was not simply the issue of animal beauty and its aesthetic appreciation that proved problematic. Few, if any, followed Darwin to the extent of endorsing his application of the beauty-making power of sexual selection to human racial diversity. Those who balked at the very idea of black beauty might reassert a universal aesthetic, common to black and white alike, a view attractive also to those who wished to restrict the sense of beauty to humans. Nor did a relativist conception of beauty mesh with an entrenched racial hierarchy. The *Saturday Review* delivered what Darwin declared a "capital" review, but the reviewer, while supportive of an animal origin for humanity and of natural and sexual selection, left no doubt of his objection to this aspect of the *Descent*'s thesis:

But by what exercise of taste are we to explain the colour of the negro? To Mr. Darwin himself it seemed at first sight a "monstrous supposition" that the jet blackness of the negro has been gained through sexual selection; and we confess that it seems to us monstrous still. To himself the view now appears supported by various analogies, and the negroes, he urges, are known to admire their own blackness. Pithecia Satanas [*sic*] . . . a perfect negro in miniature, may be an object of wondering admiration to his less dusky partner. Still our general impressions of the standard of taste in man . . . forbids our fancying the process of blackening consciously encouraged by the aboriginal savage.[6]

Darwin's prejudiced or overly enthusiastic language gave ready justification for dissent from his larger arguments. As he conceded to Morley, his expression had not been "sufficiently guarded."[7]

14.1 "The instability of a vicious feminine caprice"

It was not all Darwin's doing. The two leading critiques, by Wallace and by Huxley's erstwhile protégé, St. George Jackson Mivart, provided heavy ammunition for entrenched opponents. By comparison with Mivart's savage demolition job that sent Darwin into a tailspin of anger and chagrin, Wallace's review was generous to a fault, exciting a burst of praise from Erasmus Darwin: "The way [Wallace] carries on controversy is perfectly beautiful," he told Henrietta. Darwin, overtly appreciative of Wallace's "grand review," was privately less than enthusiastic, confiding in his daughter: "I see I have had no influence on [Wallace], & his Review has had hardly any on me."[8]

Wallace hailed sexual selection as "one of the most striking creations of Mr. Darwin's genius." He supported its male-male competition aspect and endorsed a degree of female choice, at least in birds: "Among birds is found the first direct proof that the female notices and admires increased brilliancy or beauty of colour, or any novel ornament; and, what is more important, that she exercises choice, rejecting one suitor and choosing another." However, among insects he found "no direct evidence but what tells against Mr. Darwin's view." And while Wallace praised the "acuteness and success" of Darwin's explanation of the "most marvelous developments of beauty in plumage by the constant selection of slight modifications," he yet could not believe that females could possess any common notion of beauty that would lead to such "beautifully definite colours and markings": the "action of an ever varying fancy for any slight change of colour" would "necessarily lead to a speckled or piebald and unstable result."[9] Similarly, human color and hairlessness would require the "very same tastes to persist in the majority of the race during a period of long and unknown duration." "Such identity of taste in successive generations" is against all analogy, "and this seems a fatal objection to the belief that any fixed and definite characters could have been produced in man by sexual selection alone."[10]

Wallace's agenda was, predictably, to assert the priority of natural selection: "To the agency of natural selection there is no such bar." He devoted the greater part of his review to an exposition of the protective and dominant role of natural selection in checking the action of sexual selection, arguing that Darwin "unnecessarily depreciates the efficacy of his own first principle when he places limited sexual transmission beyond the range of its power." But Wallace also went beyond an appeal to natural selection, hinting at his alternative interpretation: there were indications of "the existence of some laws of development capable of differentiating the sexes other than sexual selection." The "greater intensity of colour in the male" was "due perhaps to his smaller size and greater vigour." These hints were to become the basis of Wallace's later "vigor" theory for the explanation of brighter colors and conspicuous structures in males.[11]

Even favorable reviewers routinely called up "Mr. Wallace's" protection thesis, his criticisms of female choice and differences with Darwin over the evolution of mind and morality. Darwin responded mildly enough to Wallace and those who raised similar issues, but he was provoked beyond endurance when, a few months later, the turncoat Mivart picked up and ran with Wallace's criticisms in his damaging attack in the *Quarterly Review*.

Mivart, silver tongued and lawyer trained, was a former Darwinian who had become an expert on primates under Huxley's wing; but he was also a convert to Roman Catholicism and had been needling the brotherhood of the new secular faith since a falling-out with Huxley in 1869. His steady stream of articles on the "difficulties of the theory of natural selection," arguing for an element of design, of some higher guidance in the evolutionary process (which he conceived as a series of guided jumps), appeared in book form, *On the Genesis of Species*, just before the *Descent*. Darwin, thoroughly alienated by Mivart's two-faced professions of admiration while dealing out snide distortions and "unfair treatment" of his views, condemned the "accursed religious bigotry" he perceived behind Mivart's attacks. Mivart's long, anonymous, and hostile review of the *Descent*, which purported to lay bare the "entire and naked truth as to the logical consequences of Darwinism," was the last straw. Among his sins, Darwin was accused of undermining social stability. The *Descent*'s argument for the bestial origins of mind and morality was calculated to overturn the time-honored convictions of "the majority of cultivated minds" and to produce "injurious effects" on "our half-educated classes."[12]

Mivart went his hardest after the "new" concept, calling sexual selection the "cornerstone of Mr. Darwin's theory . . . and unless he has clearly established this point, the whole fabric falls to the ground." Male combat, pronounced Mivart, was undoubtedly a *vera causa*, it had real existence[13]— but it was properly a kind of natural selection. Female choice was the point at issue: "There is no fragment of evidence that [sexual characters] are in any way due to female caprice." Systematically, the primate expert reviewed Darwin's evidence, targeting its weaknesses and inconsistencies. Apes were "notorious" for their "vastly superior strength of body and length of fang," which would render resistance on the part of the female "difficult and perilous," even were we to adopt the "utterly gratuitous supposition, that at seasons of sexual excitement the female shows any disposition to coyness." As for the rhesus monkey, in which case the male selects, this "hypothetical reversion of a hypothetical process to meet an exceptional case" was "rash indeed." And what of the peafowl, of which the "first advances are always made by the female"? Then there was Darwin's instance of the old pied peacock preferred by the hens over the more beautiful japanned male: no one disputed the individual preferences of birds, but it did not follow that the pied plumage of the preferred male was "*the* charm which attracted the opposite sex." Even if this were the case, it would imply that

> the peahen's taste is so different from ours, that the peacock's plumage could never have been developed by it, or (if the taste of these peahens was different from that of most peahens) that such is the instability of a vicious feminine caprice, that no constancy of coloration could be produced by its selective action.

Males sometimes displayed to other males; peacocks even displayed to poultry or pigs. This undermined Darwin's insistence on the purpose of display:

> No doubt the plumage, song, &c., all play their parts . . . but to stimulate the sexual instinct, even supposing this to be the object, is one thing—to supply the occasion for the exercise of the power of choice is quite another.

As for Darwin's instance of the pretty girl at the fair, it assumed "the very point in dispute." An observer could attribute choice to the girl "only if he had reason to attribute to the rustics an intellectual and moral nature similar in kind to that which he possessed himself."[14]

The real point in dispute for Mivart, of course, was precisely Darwin's attribution of such a similarity to "rational beings and to brutes . . . Those who do not agree with him in this would require further tests than the presence of ornaments, and the performance of antics and gestures." The aesthetic sense was peculiar to humans, and "far from each race being bound in the trammels of its own features, all cultivated Europeans, whether Celts, Teutons, or [Slavonians], agree in admiring the Hellenic ideal as the highest type of human beauty." It "cannot be pretended that there is any evidence for sexual selection except in the class of Birds . . . and there is no *proof* that sexual selection acts, *even* amongst birds."[15]

With a lawyer's cunning, Mivart pitted Wallace against Darwin:

> Now as Mr. Wallace disposes of Mr. Darwin's views by his objections, so Mr. Darwin's remarks tend to refute Mr. Wallace's positions, and the result seems to point to the existence of some innate and internal law which determines at the same time both coloration and its transmission to either or to both sexes.[16]

Mivart previously had championed Wallace for his "noble self-abnegation" in ceding natural selection to Darwin and for addressing its limits. Now he rubbed salt in the wound by upholding Wallace's alternative explanation that "extra beauty of plumage, or of song, will accompany supereminent vigour of constitution and fullness of vitality." The colors and tufts of caterpillars could not be produced by sexual selection, but "by some internal spontaneous powers"; if these could produce such in caterpillars, "why not in perfect insects—or any other organism?" For Mivart, is was a "somewhat singular conclusion to deduce . . . that sexual selection is the one universal cause of sexual characters, when similar effects to those it is supposed to cause take place in its absence." Darwin's "power of reasoning seems to be in an inverse ratio to his power of observation. He now strangely exaggerates the action of 'sexual selection,' as previously he exaggerated the effects of the 'survival of the fittest.'"[17]

Mivart's incisive criticisms drew blood. Darwin had to admit the reviewer's "uncommon cleverness," and this exacerbated his rage. Instantly recognizing Mivart's "skill & style," he was cut to the quick, betrayed, "mortified": "He makes me the most arrogant, odious beast that ever lived," he fumed to Hooker. He took revenge by arranging for a highly critical review of Mivart's *Genesis of Species* to be reprinted, financing its publication and privately distributing copies. His pamphlet warfare had little effect. Darwin, despondent, became convinced that the "pendulum is swinging against our side." He rallied only when Huxley alerted him to his forthcoming savaging of Mivart's *Genesis* and his *Quarterly* article ("with some incidental touching up of Wallace") in an onslaught on "Mr. Darwin's Critics" for the *Contemporary Review.* Mivart, while "by no means a bad fellow," was "poisoned with . . . accursed Popery and fear for his soul." His "insolent attack" on Darwin obliged Huxley to "pin [him] out" like the insect he was. To the plaudits of the Darwinians, Huxley trumped Mivart's "presumptuous ignorance" with his own superior expertise in religious exegetics to assert that Catholicism and evolution were irreconcilable. As an aside to this, Huxley (with customary agility) deployed Wallace's own prior encomiums on savage life and its exigencies for enough intellect, artistry, and innovation to bat down Wallace's claim of overdesign of the savage brain for the simple requirements of savage existence.[18]

Darwin was delighted: "How you do smash Mivart's theology." But there was nothing better than Huxley's "argument v. Wallace on intellect in savages." Nevertheless, this left sexual selection still undefended. Huxley airily remarked that he had "kept pretty clear of that part of the matter," leaving it to Darwin to take up Mivart's "objections in detail."[19] Huxley, to oblige Darwin and counter Mivart, later contributed a section on the comparative anatomy of apes and humans to the second edition of the *Descent.* But at no stage did Huxley—famously cautious about endorsing natural selection— ever come out, publicly or privately, in favor of sexual selection.[20]

14.2 "My conviction of the power of sexual selection remains unshaken"

There was singularly little direct support for the concept among the Darwinians. They rallied round the *Descent* and around Darwin, but were curiously silent on sexual selection. Darwin made much of Ernst Haeckel's vaunted support, citing him as the "sole author" to have "discussed . . . the subject of sexual selection" and to have seen its "full importance." But, apart from an odd courteous assurance to Darwin of its importance and interest, Haeckel wrote remarkably little on sexual selection.[21]

That other self-proclaimed disciple, Paolo Mantegazza, who had given Darwin some of his best material on savage decorative practices, publicized "certain doubts concerning the capacity of sexual selection to produce all the differences between the sexes." Like Mivart, he contended that the more physically powerful male animal would dominate the female, precluding female choice. Mantegazza suggested that secondary sexual characteristics were due to the action of sperm that was absorbed by various tissues in the male, giving them heightened color and ornamentation. Darwin politely demurred: he had given "many clear proofs" of female preferences in domesticates, "and we can infer that they also happen in nature, facts which make me believe in the reality of sexual selection."[22]

Darwin's trickiest disciple had to be Winwood Reade, who, full of admiration for the *Descent*, had become a frequent correspondent. It was Reade who offered Darwin this anecdotal accolade, symbolic of the Victorian resonances of the *Descent*: a young lady of his acquaintance was shocked to hear Reade was a Darwinian—"Me a monkey indeed!"—but she had been "pretty much converted" after dining next to a naval man who "described the Fuegians to her—their habit of eating missionaries raw etc.—& her opinion seems to be now precisely what you express in your conclusion—that if we must own them for our ancestors, we can go a little lower down without feeling the drop."[23]

Reade, however, could not believe in sexual selection forming the "negro type." He continued to assert that black and white races shared the same ideal of beauty. Most alarmingly, he intended, as he informed Darwin, to devote his life to a "war on Christianity."[24] He was as good as his word. Reade's masterwork, *The Martyrdom of Man* (1872), was a paean to the progressive human struggle against religion and superstition and of the evolutionary rightness of imperialism, the subordination of women and the "inferior" races. Darwin's efforts to soften Reade's confrontational tone were ineffectual, and *The Martyrdom of Man* met with a thoroughly hostile reception in orthodox circles, condemned as both indecent and profane. It was not until well after his death that Reade's evolutionary epic became a best seller—a "substitute bible for secularists"—giving him an acknowledged place in secular evolutionary literature and future inspiration to the young H. G. Wells.[25]

In the meantime, Reade, undaunted, produced *The African Sketch-Book* (1873), a popular book on his travels, allegedly "written for women," which promoted Reade's view that the "jet-blackness of the negro" was to be attributed to "conditions of climate acting *indirectly* on the skin" through a process of natural selection, not to sexual selection, as Darwin claimed. No doubt, Reade appeasingly added, "comeliness of feature" was the result of sexual selection. Primeval men were probably dark skinned and polygamous, defending their harems "with tooth and claw"; initially, women, being slaves, would be selected for strength, but as a wife became an "article of luxury," women would be valued and selected for their beauty. "Female beauty, like female virtue, is a modern and gradual creation . . . as much an artificial production as the pouter breast of the pigeon, or the monstrous cabbage head. It is entirely the work of *women fanciers*, a creation of the male sex."[26]

It is not clear how this reworking of Darwin's arguments went down with Reade's intended audience of "gentle readers" who were exhorted to spare their blushes, abandon the "angel hypothesis," and take comfort from the revelation of the os coccyx, that little bone "modestly tucked in between the legs" ("The dreadful truth can no longer be concealed—*we are all of us naked under our clothes, and all of us tailed under our skins*"). But it can scarcely have added to Darwin's comfort.[27]

The evolutionary anthropologists remained a troublesome lot. Lubbock was "surprised" at Darwin's invocation of the analogy of the lower animals in opposition to "our views on Communal Marriage": "What monkey ever watched over the conduct of a daughter? or scrupled to carry off another's wife?" Communal marriage did not necessarily entail actual sexual promiscuity, but, said Lubbock, indicated the "retention by woman of all her rights over herself [including sexual choice], which she may exercise as she pleases; whereas marriage is the surrender of them to another." Andrew Smith was ready to pick a "little fight" with Darwin and a "great fight with Lubbock & McLennan over what constitutes marriage among primitive people." Darwin looked hopefully to E. B. Tylor to write the "sound estimation of the morals of Savages" that "Wallace, Lubbock &c &c" could not agree on. John McLennan, grateful for recognition from the famous man, initially did not quibble over the *Descent*'s treatment of his thesis of primitive promiscuity, but as the second edition loomed, he attempted, without success, to qualify Darwin's insistence on monogamy as innate and motivated by sexual jealousy evidenced by all male animals. Darwin, with female coyness and sexual selection under renewed threat, overrode McLennan's caution that monogamous exclusivity was relatively unusual not just among primitives, but "even in this country in late times." His revised discussion of the views of contemporary marriage theorists was more nuanced, but still culminated in Darwin's flat rejection of the theory that "absolutely promiscuous intercourse prevailed in times past." The animal analogy was downplayed, while sexual jealously became paramount: "The natural and widely prevalent feeling of jealousy and the desire of each male to possess a female for himself" ensured that monogamy and female chastity would prevail among primitive humans, at least on a temporary basis which "suffices for the work of sexual selection."[28]

One of the few to give unfailing support was Fritz Müller, who provided ongoing evidence for sexual selection—notably a case in which two forms of *Hesperiiadae* displayed their wings differently depending on which surface (the upper or lower) was colored (cited by Darwin in the second edition of the *Descent*).[29] Wallace, however, dealt dismissively with Müller's new evidence for butterfly sexual selection. Having had his say, he went quiet on the whole topic. Mivart had, in any case, taken Wallace further than he was prepared to go at this stage. In the face of Darwin's unabated hostility toward his most damaging critic, Wallace maintained a friendship with Mivart, who shared his interest in spiritualism.

In a period when natural selection was under devastating attack, primarily by Mivart, Wallace was emerging as its great champion. The priority was its defense. Darwin had to agree. He added a whole new chapter to counter "all Mr. Mivart's objections" to his primary mechanism to the sixth edition of the *Origin* (published in February 1872); however, he made only minor adjustments to the section on sexual selection. Its revision would have to wait for the second edition of the *Descent*, made all the more necessary by Mivart's "very sincere" best wishes for the New Year, accompanied by a brazen demand for a correction of the *Descent*'s "fundamental intellectual errors."[30]

Sick of controversy, "dead with fatigue," Darwin pushed on with the task of completing *The Expression of the Emotions in Man and Animals*. John Murray brought it out toward the end of 1872 to general acclaim. It made no calls on a contentious natural selection, resting its case entirely on inherited habit. It proved to be the most readable and successful of his publications, bringing Darwin a new readership captivated by blushes and amused by photographic images and line drawings of fractious babies, Polly the dog, and actors miming grief, pleasure, and disgust. "I don't think it is a book to affront anybody," said Emma complacently. Yet it too had implications for sexual selection: if humans were not unique in their ability to experience emotions, then the emotional experiences of love and the enjoyment of beauty were reducible to the animalistic urge to reproduce, stripping them of their traditional spiritual underpinnings. The close ties between the *Descent* and *Expression* were clear to those like John Ruskin, who joined Mivart and Cobbe in denouncing a "simious theory of morals" that would reduce beauty and the emotions it aroused to mere survival instincts.[31]

Darwin's series of evolutionary works was now complete. He was more courted and cultivated than ever. The famous and the ordinary converged on Down House to genuflect before celebrity. Letter writers begged for photographs and autographs. For the Victorian public, Darwin's was *the* face of evolution; Wallace at this stage rated scarcely a glance; Huxley, while assuming great authority in the spheres of professional science and technical education, counted for less in the appetite for celebrity. It was a celebrity that Darwin benignly accepted, even "helped to construct and codify." The honors flooded in; his legendary invalidism buffered him from unwanted intrusions; his routines were sacrosanct; and, with *Expression* behind him and selling well, he was free to potter around his greenhouse, where he fed,

A LOGICAL REFUTATION OF MR. DARWIN'S THEORY.

Figure 14.1. *The Descent of Man* in the Victorian parlor. The hirsute Jack, reading the *Descent*, teases his wife Mary about "Baby's descent from a hairy quadruped, with pointed ears and a tail." Mary refutes the notion with female logic: "Speak for yourself, Jack! *I'm* not descended from anything of the kind, I beg to say; and Baby takes after *me*. So there!" *Punch's* strongly gendered interpretation of responses to the *Descent* plays on Darwin's evolutionary defense of women's intellectual inferiority and essential femininity and domesticity. Many women, however, welcomed a naturalistic interpretation of sexual differences and a number of early feminists reinterpreted evolutionary theory and sexual selection in support of women's rights. *Punch*, April 1, 1871, 130. University of Sydney Library Collection.

tickled, and anointed a collection of insectivorous and climbing plants. "I have taken up old botanical work and have given up all theories," he told Wallace.[32]

The new edition of the *Descent* fell due toward the end of 1873, demanding more attention than Darwin felt inclined to give it. Astonishingly, it was Wallace to whom Darwin first turned for assistance. Wallace was financially overextended, struggling to survive on royalties and reviews. Darwin wanted to help, but perhaps he had more in mind the old dictum of the advisability of setting up the poacher as gamekeeper. His diffident offer to the needy Wallace was grasped with both hands: for seven shillings an hour (the higher rate was warranted by Darwin's atrocious handwriting) Wallace would do the necessary revision and editing. "Of course in such work I would not think of offering criticisms of matter," he assured Darwin. Within a few days, however, Darwin withdrew his offer. Family came first: "My wife thinks that my son George would be so much pleased at undertaking the work for me."[33] So it was, at Emma's insistence, that George, not Wallace, edited the revised *Descent of Man*.

The sickly George's legal career had foundered, and he was alternating between watering places, picking over his father's theorizing, producing the occasional journal article. One of these, "Development in Dress," argued that fashion followed the same evolutionary laws as organisms. Modes of dress yielded to better-adapted forms as habits changed: breeches and boots gave way to trousers as men gave up riding for railway travel, while hatbands and tailcoats persisted as vestiges of costume appropriate to former modes of existence. George took his cue from Tylor's anthropological "survivals," avoiding the delicate subject of the sexual significance of bonnets and bustles. His thesis evoked little paternal comment, other than a recommendation for the revision of one or two sentences.[34] George's more contentious production in the *Contemporary Review* (1873) followed from Galton's recently published eugenic proposal for a register of those "endowed above the average in mental and physical qualities," who would through "early intermarriage" diffuse these superior qualities through the nation. George, like his father, thought the promotion of such a superior "quasi-caste" unfeasible, and proposed instead a process of "unconscious," or negative, selection—like that carried out by savages who killed off their inferior dogs. This necessitated "beneficial restrictions to liberty of marriage," and George advocated legal changes making divorce easier on the grounds of hereditary disease (under which he categorized insanity, criminality and vice).[35] It was this article that was to lead to even greater fracas with Mivart.

George went to work on the *Descent*, interpolating Darwin's collected notes and corrections. On the side he pursued his interest in eugenics, investigating cousin marriage and congenital disease—issues of obvious concern to the inbred Darwin-Wedgwoods. With Henrietta engrossed in marriage and invalidism, Francis abandoned medicine and joined the family firm as Darwin's paid secretary, dealing with his enormous correspondence and aiding the greenhouse experiments. Darwin could not be extricated from these to oversee the proofing of the *Descent*, leaving the entire process to George. He had done with theory and had put what he now perceived as the "fiery ordeal" of controversial publication behind him. He would write no more books on evolution.[36]

The second edition of the *Descent* went on sale in November 1874. Murray packaged it as a cheaper edition, targeting those workingmen who reportedly clubbed together to buy the great man's works. In his scant preface, Darwin backed still further away from his earlier reliance on natural selection, calling more on the inherited effects of use and disuse and his principle of correlated growth; but he remained more committed than ever to sexual selection. He was undeterred by "the half-favourable criticisms on sexual selection," which struck him as very like those first made of natural selection: "My conviction of the power of sexual selection remains unshaken." Once naturalists had become more familiar with the idea, it would be, he believed, "much more largely accepted."

Aside from those discussed above, most of Darwin's revisions to the *Descent* concerned the updating of specific instances of sexual dimorphism. His discussion of Wallace's views remained largely unchanged, though he could not forbear a few hits at Wallace's inconsistencies.[37] He did add a lengthy paragraph to the section on birds in response to the criticism mounted by "several writers" (i.e., Wallace and Mivart) that "with animals and savages the taste of the female . . . would not remain constant for many generations." Taste was admittedly fluctuating, but it was not arbitrary. In humans it depended much on habit: "Even in our own dress, the general character lasts long, and the changes are to a certain extent graduated." Savages had admired for many generations the same artificially produced "deformities," and these presented "some analogy to the natural ornaments of various animals." Similarly, the taste of fanciers for particular breeds had remained constant over many generations and, while they "earnestly desire[d] slight changes," they rejected any "great or sudden change." Hence "there seems no improbability in animals admiring for a very long period the same general style or ornamentation or other attractions, and yet appreciating slight changes in colours, form, or sound."[38] It was a rationale drawn from Darwin's fundamental analogical association of fashion, savage decorative practices, artificial selection, and female choice. It was the best he could muster against Mivart's amplification of Wallace's criticism into the "instability of a vicious feminine caprice," a lurid phrase that conjured up just the kind of overcharged sexual imagery that Darwin was anxious to avoid.

Mindful of Morley's advice, Darwin toned down his expression in certain places. Reference has been made to his modification of the aesthetic powers of female birds, which were now depreciated to the level of savage aesthetics (though, as discussed, he compromised this by retaining contentious comparison with the art of Raphael and by introducing photographic evidence of the "wonderful" artistry of the placing and shading of the ball-and-socket ornaments).[39] In another critical instance, Darwin modified his introduction to his thought experiment of the pretty girl at the fair, so denigrated by Mivart. In the first edition of the *Descent*, Darwin had written: "With respect to female birds feeling a preference for particular males, we must bear in mind that we can judge of choice being exerted, only by placing ourselves in imagination in the same position." In the second edition, this became: ". . . we must bear in mind that we can judge of choice being exerted only by analogy."[40]

These changes were primarily cosmetic; while Darwin was willing to modulate his expression in response to criticism, he was not willing to make any substantive changes. He stood by his theory of sexual selection and its foundational theories of development and inheritance, along with those strategies adopted to protect them. The contradictions and ambiguities remained, leaving the *Descent* as vulnerable to criticism as before. More insidiously, they lay the *Descent* open to a variety of readings and evaluative positions, some of which threatened those very values that had informed its writing. It became increasingly obvious that Darwin could not control the diverse meanings and imputations attached to sexual selection, certain of them highly disturbing to the author of the "respectable and overtly patriarchal" *Descent*.[41]

14.3 A "delicate and difficult subject"

There was no getting round it: sex and sexuality, with all their complexities and connotations, lay at the heart of sexual selection. Well sensitized to the hazards any discussion entailed, Darwin trod warily from the first edition of the *Descent*, seeking to eliminate any suggestion of indelicacy or indecency. He and Murray then had scanned individual sentences with an eye to the proprieties, concerned with avoiding offending delicate sensibilities and compromising the *Descent*'s marketability—though, as Hooker remarked, its promisingly titillating content "no doubt promotes the sale," ladies being obliged to "order it on the sly."[42] Given its intense preoccupation with sex, most reviewers responded lightly enough to the bowdlerized version of the book—it was "as wonderful as a fairy tale" (*Field*), the stuff of romance and poetry (*Saturday Review*)—while some took to versifying as the appropriate response (*Blackwood's Magazine* and the *Gardeners' Chronicle*).[43] Even the *Times*, having linked the "disintegrating speculations" of the *Descent* with the immoral and atheistic "loose philosophy" of the Paris Communards, found "enchanting" Darwin's history of the "Loves of the Animals."[44] However, for the more fastidious, sexual selection was a "delicate and difficult subject" that reviewers scarcely knew how to deal with; its discussion was "not well-fitted for the pages of widely-read periodicals" (*Athenaeum, Literary World*).[45] Overtly hostile reviewers, like the geologist William Boyd Dawkins in the *Edinburgh Review*, played on the *Descent*'s alleged indecency—its incessant focus on sex was unseemly, not fit for the drawing room. Its thesis that morality was not divinely induced but the outcome of "a series of accidents" was akin to the corrupt philosophies of the pagan world. This was an allusion to the notorious Algernon Charles Swinburne's provocative poetic celebrations of sexuality (both homosexual and heterosexual) and a "mystical evolutionary pantheism." It was a charge reiterated by the egregious Mivart. In his critique of "Contemporary Evolution" in 1873, Mivart condemned those Darwinian theorists who articulated an "increasingly Pagan" naturalistic worldview, virtually identical with that conveyed by the "loathsome and revoltingly filthy verses of Swinburne." Adding to such associations with the worst excesses of pagan Rome were all those post-*Descent* lampoons and cartoons featuring hairy apes (many modeled on its author) that ranged from the merely amusing to the smutty and outright pornographic.[46]

The insinuations of impropriety, indecency, and sexual immorality attached to the *Descent* had their effect. Even as the second edition went to press, Darwin was thrown into turmoil by renewed assault by Mivart, this time targeting a Darwin son. A cursory reading of George's earlier essay "On Beneficial Restrictions to Liberty of Marriage" led Mivart to allege falsely that it endorsed prostitution as a check to overpopulation. With customary hyperbole, Mivart denounced the "hideous sexual criminality of Pagan days" to which the author and his school belonged, convicting George and his father of the "most profound moral corruption." Mivart later conceded that his potentially libelous allegation included an "incautious expression . . . which I much regret having used." It was too late for regrets. Darwin was beside himself, spluttering with rage, his son's good name impugned by Mivart's "infamous & explicit accusation." He unleashed his pent-up animosity and his attack dog, the equally outraged Huxley, on the squirming Mivart. Huxley, newly anointed "High Priest" of evolution, plotted "Mivart's excommunication from the church scientific." He cut the connection dead, and in a ferocious attack in the *Academy* turned the tables on Mivart: it was Mivart—not George, the spotless scion of the eminently reputable Darwin—who transgressed gentlemanly standards of taste and probity in his "blind animosity against all things Darwinian." Together, Huxley, Darwin, and Hooker saw to it that Mivart was socially excommunicated as well by blackballing him down the years from that bastion of gentlemanly science, the *Athenaeum*.[47] Mivart's ostracism was the price to be paid by those who insinuated that ultrarespectable Darwinians promoted vice and irregular sexual unions and practices. The whole sordid episode left the leading Darwinians more insistent than ever on rigid adherence to the Victorian moral code.

FUN.—November 16, 1872.

THAT TROUBLES OUR MONKEY AGAIN.

Female descendant of Marine Ascidian :—"REALLY, MR. DARWIN, SAY WHAT YOU LIKE ABOUT MAN; BUT I WISH YOU WOULD LEAVE MY EMOTIONS ALONE!"

Figure 14.2. Darwin takes the pulse of Victorian femininity. *Fun*, the main satirical rival of *Punch*, responds to the publication of Darwin's *Expression of the Emotions in Man and Animals* (1872). This widely circulated image of Darwin as an anthropoid ape with erect tail and hairy black hand on the dainty wrist of a "Female descendant of Marine Ascidian" (who is wearing an exaggerated version of the newly fashionable "crinolette," forerunner of the bustle) provokes amusement. But it also insinuates a sexualized interpretation and an allegation of indecency: Darwin may say what he likes about "man," but his discussions of woman's emotional responses, however circumspect, are indelicate (especially his allusions to the sexual implications of blushing: a "pretty girl" blushes under the intent gaze of a young man because she thinks of the "outer and visible parts" of her body and this alters their "capillary circulation"). *Fun*, 16 n.s. 1872, 203. © British Library Board/Robana/Art Resource, NY, AR9106460.

The other side to this coin was that, from its first edition, the *Descent* naturalized an implicit double standard: female sexuality obeyed the Victorian conventions—coy, passionless, and essentially maternal—while males were ruled by passion. Among primitive humanity, female virtue was inculcated by male jealousy and would "spread to unmarried females," leading eventually to the civilized "self-regarding virtues" of feminine "chastity" and "celibacy," although it did not, Darwin admitted, have the same salubrious effects on unmarried men, even in the "present day." In this, as ever, Darwin's concern lay less with the "implicit naturalization of nineteenth-century standards of marriage and sexual respectability" than with the defense of his theory of sexual selection.[48] Moreover, as discussed, his insistence on the immutability of male passion and female virtue conflicted with his argument for the predominance of male choice in humans and the superiority of intellect and reason in men over intuition in women. Nevertheless, Darwin's "paradoxical but ostensibly scientific" endorsement of the sexual double standard was seized on by an array of feminists, social purists, and eugenicists in the post-*Descent* period as a means of "reinvesting women with the agency of selection" on the grounds that only they were capable of making responsible sexual choices.[49]

Wallace's 1890 volte-face when he rested the future of human society on just such informed, essentially passionless female choice fits this genre of appropriations of female choice. It was indeed a "stunning reversal,"[50] because well before this, in mid-1876, Wallace had officially "given up" sexual selection in toto; as an added irritant, he did so in the context of a review of Mivart's new collection of essays, *Lessons from Nature* (1876).

14.4 "An abnormal excrescence"

Mivart's *Lessons*, which included his earlier review of the *Descent*, was replete with aspersions on Darwin's alleged dishonesty and deviousness. Wallace, while looking favorably on Mivart's criticisms of the animal origins of morality, strongly defended both natural selection and Darwin's "perfect literary honesty" against Mivart's "violent" and unfounded accusations: Darwin's "self abnegation in confessing himself wrong" was "rare," "admirable," and "truly moral." However, though Mivart's criticisms of natural selection were specious, he had "much more cogent arguments against the theory of sexual selection." Wallace had to concede their force: Darwin's views on sexual selection were "altogether erroneous"; there was "no evidence whatever that [female] choice is usually determined by small variations in the display." Females could not show the necessary unanimity and

constancy of choice. As for Mivart's criticisms of the relevance of sexual selection to human racial divergence, this had "generally been felt to be one
of the weakest parts of Mr. Darwin's book . . . If, however, the main theory as
applied to animals is unsound, its application to man will necessarily have
to be reconsidered."[51]

Darwin dealt with this treachery by thanking Wallace for his "generous
defence of me against Mr. Mivart," the only man who "has ever . . . treated
me basely," and by reminding Wallace of the "pain" inflicted by Mivart's
false accusation against his son of "encouraging profligacy." He and Huxley
had "cut" Mivart. It was clear where Wallace's Darwinian duty lay. In a postscript, Darwin made his position on sexual selection equally clear:

> I am very sorry that you have given up sexual selection. I am not at all shaken,
> and stick to my colours like a true Briton. When I think about the unadorned
> head of the Argus pheasant, I might exclaim, *Et tu, Brute!*

Darwin's shaft was a reference to Wallace's recently published *Geographical
Distribution of Animals* (1876), where Wallace had noted that, during display, the male argus pheasant's wing feathers concealed its head from observation and that, unlike other pheasants, its head was unadorned. This was,
Wallace had then remarked, a "remarkable confirmation of Mr. Darwin's
views, that gaily coloured plumes are developed in the male bird for the
purpose of attractive display."[52]

Wallace acknowledged the hit but did not take the hint. He maintained
friendship with Mivart, and he used his presidential address to the Biology
Section of the Glasgow meeting of the British Association for the Advancement of Science in September 1876 to detail a great array of instances in
butterflies, birds, and mammals in which colors and markings conferred
protection against predation, either through mimicry, through warning, or
by camouflage. He did not attack female choice directly, but his intention
was clear.[53]

This same meeting was the occasion of uproar over Wallace's sponsorship of a paper on thought transference in the Anthropology section. It
splashed over into the press when the young Huxley protégé, the naturalist
Edwin Ray Lankester, lambasted Wallace's "degradation" of the British Association in the *Times*, linking the "astounding credulity" of such a man of
science with Lankester's exposure of an American medium, Henry Slade, as
a blatant fraud. Slade, who enjoyed great success among British spiritualists, subsequently went to trial in the magistrates' court on Bow Street with
Wallace as key witness for the defense. Darwin—convinced, like Huxley,

that the phenomena evoked in the fashionable séances constituted "mere trickery" quietly bankrolled Lankester's prosecution as a "public benefit." Slade was convicted of fraud; his conviction was overturned on a technicality, and he headed home, his career as a leading communicator with the spirit world under a cloud. The trial did nothing to dampen Wallace's spiritualist ardor (or his continuing defense of Slade); to the contrary. But it was a "defining moment" in the relations between Wallace and the Darwinians. Wallace's application for the paid position of assistant secretary of the British Association was rejected, and he went on to help found the Society for Psychical Research. It was the end of his hope of establishing spiritualism as a *"new branch* of Anthropology." Huxley took the Chair of the Anthropology section of the British Association meeting of 1878 to reassert the primacy of scientific naturalism and rule spiritual phenomena out of bounds.[54]

Darwin's role in the affair marked his almost complete estrangement from Wallace. It came hard on the heels of his realization that the German Darwinian August Weismann—one of the very few to appear willing and of whom Darwin had retained high hopes—was by no means committed to sexual selection. Weismann's *Studies in the Theory of Descent* (1882, read by Darwin in the original German in 1875), strongly endorsed natural selection but gave only limited support for sexual selection: "Darwin ascribes too much power to sexual selection when he attributes the formation of secondary sexual characters to the sole action of this agency." Until experimentation had decided "how far the influence of sexual selection extends, we are justified in believing that the sexual dimorphism of butterflies is due in great part to the differences of physical constitution between the sexes." Weismann, like Mantegazza and Wallace, seemed to be hinting that the male condition or physiology somehow fostered sexual difference.[55]

Darwin, without substantive support for sexual selection among Darwinians, attempted some damage control of his own. It was at this juncture that he communicated his short paper to *Nature* on "Sexual Selection in Relation to Monkeys" (1876). It was a response to two of the major criticisms of sexual selection offered respectively by Mivart and Wallace. Mivart, as told, had claimed the inconceivability of female choice in monkeys and derided Darwin's obverse instance of the female rhesus, whose red bottom had evolved through male choice. Darwin (mindful of its potential for ridicule and insinuations of indecency) presented new evidence from Germany that males of certain monkeys habitually presented their rear ends in greeting and that this behavior was connected "with sexual feelings"; he reproduced (in German) an explicit observation of a female *Cynopithecus niger* (Celebes macaque) who showed the male a "strongly reddened posterior . . . at sight of [which] the male became visibly excited." Darwin also made another effort to counter Wallace's objection that "all the females within the same district must possess and exercise exactly the same taste." Female choice, Darwin reiterated, was like "unconscious selection by man"; over time, although its practitioners have not exercised deliberate selection of particular individual characteristics, domesticated animals common to a region will vary in ways different from those in another, just like humans:

> I presume that no supporter of the principle of sexual selection believes that the females select particular points of beauty in the males; they are merely excited or attracted in greater degree by one male than by another, and this seems often to depend, especially with birds, on brilliant colouring. Even man, excepting perhaps an artist, does not analyze the slight differences in the features of the woman whom he may admire, on which her beauty depends.[56]

This represented a retreat from Darwin's more extreme statements on the artistry and discrimination of female choice in birds (though he had made this point intermittently in his writings and in correspondence with Wallace). But it was not enough. Nor did Darwin's reemphasis on the "unconscious" aspect of female choice sink in. Not long after this, Wallace came out categorically against what he termed "voluntary sexual selection" in *Macmillan's Magazine* (1877) and, for the first time, offered a detailed account of his alternative vigor theory.

This time he pulled no punches: "I have long held this portion of Mr. Darwin's theory to be erroneous, and have argued that the primary cause of sexual diversity of colour was the need of protection, repressing in the female those bright colours which are normally produced in both sexes by general laws [of variation]." In the case of males, the "very frequent superiority of the male bird or insect in brightness or intensity of colour . . . now seems to me to be due to the greater vigour and activity and the higher vitality of the male." Male coloration was most intense during the breeding season, when vitality was at a maximum, and "would be further developed by the combats of the males for the possession of the females." Color was correlated with vigor and health, so such males would father and rear most offspring, and intensity of color would be increased by natural selection.[57]

Wallace explained the fact that only certain structures were highly developed or colored by arguing that increasing vigor would act "unequally on different portions of the integument" and often produce at the same time "abnormal developments" of hair, horns, scales, or feathers. This would "almost necessarily lead to variable distribution of colour, and thus to the production of new tints and markings." But in all cases, where increasing color became disadvantageous to the female, it would be checked by natural selection. The sexes or young of the same species also used color for recognition. As for "voluntary sexual selection, that is the actual choice by the females of the more brilliantly-coloured males, I believe very little if any effect is directly due." For all the "copious facts and opinions" collected by Darwin, there was a "total absence of any evidence that the females admire or even notice this display." The "plain inference" was that "males fight and struggle for the almost passive female, and that the most vigorous and energetic, the strongest-winged or the most persevering, wins her," and these were usually the most highly colored or ornamented.[58]

In the defining case of the argus pheasant, it was "absolutely incredible" that the beautifully shaded ocelli on the tips of the feathers "should have been attained through thousands and tens of thousands of female birds all preferring those males whose markings varied slightly in this one direction, this uniformity of choice continuing through thousands and tens of thousands of generations." Wallace pointed out that the pheasant's most richly varied markings and ocelli occurred on those parts of the plumage that have undergone the most modification and suggested that, during development, colors, spots, and bands may have developed unevenly and expanded "much as the spots and rings on a soap bubble increase with increasing tenuity." His vigor theory ensured that such developments were correlated with vitality and health, so that natural selection alone would enable their selection and accumulation over time: "The sexual selection of ornament for which there is little or no evidence becomes needless, because natural selection which is an admitted *vera causa* will itself produce all the results."[59]

This was, as Wallace represented it, the major advantage of his interpretation over Darwin's: it would be a "relief" to evolutionists to find that the phenomena could be shown to depend on "general laws of development" and on the action of "natural selection, which theory will," Wallace asserted, "be relieved from an abnormal excrescence, and gain additional vitality by the adoption of my view of the subject."[60]

Wallace's choice of the popular forum of *Macmillan's*, rather than a scientific journal like *Nature*, ensured that his rejection of sexual selection received wide publicity. His endorsement of Mivart's earlier dismissal of the validity of female choice challenged Darwin's commitment to the *vera causa* orthodoxy of Victorian science. And, by reducing male combat along with all color and ornamentation to natural selection, Wallace sought the theoretical high ground, not only of promoting a "real" efficacious cause, but also of extending and strengthening the "great principle of natural selection." In this he was, he affirmed complacently in his autobiography, "more Darwinian than Darwin himself."[61]

There was, of course, another dimension to Wallace's final rejection of sexual selection: his increasingly passionate support for spiritualism and its practitioners. Wallace concluded the second part of his 1877 paper, which analyzed the perception of color by animals and humans, by claiming that the "exquisite charm and pleasure" that humans derive from color "seem to rise above the level of a world developed on purely utilitarian principles."[62] His recent conflict with the Darwinians over the authenticity of psychical phenomena, their allegations of his gullibility, and Wallace's continuing efforts to rehabilitate the reputations of Slade and other mediums, had pushed him into closer alliance with theists, many of whom shared his serious interest in spiritualism. Many professional scientists (including the Darwinian Asa Gray) endorsed the notion that there was an integral spiritual or religious dimension to science. Like Mivart, they did not see Wallace's efforts to investigate the evidentiary claims of spiritualists or to articulate a theistic evolutionary biology as unscientific. Theistic science was a "powerful paradigm" and spiritualism a "potent force" in Victorian culture. Huxley and the scientific naturalists did not have it all their own way.[63] Wallace could share his belief that the sense of beauty was unique to humans and encompassed a spiritual dimension with a network of alliances that enabled him to maintain his independence of a strict evolutionary naturalism and reinforced his opposition to Darwin's insistence on a naturalistic aesthetics.

For their part, Darwin and his inner circle could not afford a definitive rupture with the recognized coauthor and great defender of natural selection as the true and exclusive cause of biological evolution; all the more so when Darwinism's defining mechanism was under sustained attack and insiders could not agree among themselves as to its efficacy and extent. More generally, "Darwinism" itself was under great pressure, its meanings proliferating beyond the control of its founders, taking on divisive religious, atheistic, idealist, materialist, capitalist, imperialist, and socialist interpretations, its elasticity stretched to bursting point. By the 1870s, the cold winds of change were beginning to blow about the ears of the British middle classes as the so-called "Great Depression" of 1873–96 undermined the foundations of mid-nineteenth-century liberalism. Attitudes hardened as economic, political, and social tensions intensified. This was the period when attempts to demarcate between scientific and nonscientific factors in the construction of scientific knowledge became the subject of bitter and divisive dispute. Huxley—that one-man think tank and political fixer, in the process of harnessing state power to create the modern laboratory—was hard pressed to sustain his strategy of an ideologically neutral, objective Darwinism, immune from the partisan political doctrines, religious affiliations, and social stances its various devotees would ascribe to it. What was at stake was the definition of Darwinism.[64]

Darwin read through Wallace's definitive rejection of sexual selection with resignation, offering occasional marginal protest: "I have said the most vigorous & most ornamented."[65] Well, yes, he had, but not consistently and only under pressure. The sting in the tail came with Wallace's denigration of his prized concept as an "abnormal excrescence" that was disfiguring and hindering the full theoretical expression of their joint production, natural selection. Their differences were unbridgeable, and, Darwin told Wallace, he was weary of conflict:

> You will not be surprised that I differ altogether from you about sexual colours. That the tail of the peacock and his elaborate display of it should be due merely to the vigour, activity, and vitality of the male is to me as utterly incredible as my views are to you. . . . I could say a great deal in opposition to you, but my arguments would have no weight in your eyes, and I do not intend to write for the public anything on this or any other difficult subject.

He made one last effort to put Wallace straight: "By the way, I doubt whether the term voluntary in relation to sexual selection ought to be employed: when a man is fascinated by a pretty girl it can hardly be called voluntary, and I suppose that female animals are charmed or excited in nearly the same manner by the gaudy males."[66]

Wallace, willfully obtuse to Darwin's recent reiteration in *Nature* of his critical analogy with "unconscious" artificial selection, replied that "perhaps 'conscious' would be a better word, to which I think you will not object . . . I lay no stress on the word 'voluntary.'" He offered consolation: Darwin had "brought such a mass of facts to support [his] view," and argued it so fully, that it was not necessary for him to do more. "Truth will prevail, as you as well as I wish it to do." With this, Darwin gave up. Further argument was futile:

> It is foolish in me to go on writing . . . thank Heaven, what little more I can do in science will be confined to observation on simple points. However much I may have blundered, I have done my best, and that is my greatest comfort.[67]

Wallace was not through with sexual selection. His article in *Macmillan's* was reprinted in *Popular Science Monthly*, in the *American Naturalist*, and in his *Tropical Nature and Other Essays* (1878). His influential *Darwinism* (1889),

which contained a whole chapter on "Colours and Ornaments Characteristic of Sex," was the culmination of his promotion of his vigor theory to
explain sexual dimorphism. Female choice, Wallace then pronounced, was
"now no longer tenable." The term "sexual selection" must be "restricted
to the direct results of male struggle and combat" and was "really a form of
natural selection." *Darwinism* reached a second edition in its first year, and
went through four reprints before a third edition in 1901, as much a testament to the tenacity of Wallace's campaign against sexual selection as to his
"eloquent defense of natural selection after three decades of criticism." So
closely did Wallace's views become identified with the rejection of sexual
selection that the zoologist George Romanes, who regarded himself as Darwin's true intellectual heir and was one of the few to give the concept any
credence, commented drily in 1892: "To consider the objections to sexual
selection . . . is virtually the same thing as saying that we may now consider
Mr. Wallace's views upon the subject."[68]

It is to Darwin's great credit that, despite his resentment of Wallace's
insistent campaign against the "abnormal excrescence" of sexual selection,
it was he who came to the rescue when Wallace was facing penury, with no
permanent job and only a trickle of income from books and articles. Darwin
persisted in the face of opposition from Hooker (who believed Wallace had
"lost caste" through his support for spiritualism and other dubious causes)
with a scheme for a government pension for Wallace. As a result of Darwin's
lobbying, in 1881 Wallace was awarded a civil service pension in recognition
of his "lifelong scientific labour," which freed him from constant economic
anxiety and enabled him to pursue his idiosyncratic blend of science, spiritualism, and sociopolitical activism.[69]

So well did Wallace spread his gospel that Darwin, in his final years, was
grateful for any public defense whatever of sexual selection, even from such
a questionable advocate as Grant Allen, "scientific popularizer, synthesizer,
and middle-man without peer" of the late Victorian period. The Canadian-
born Allen was by trade a journalist with scientific ambitions. His religion
was scientific naturalism, his gods Spencer and Darwin. He produced dozens of popular articles and a number of books on evolution, including two
specialist monographs on *Physiological Aesthetics* (1877) and *The Colour-
Sense: Its Origin and Development* (1879). These latter gave serious consideration to Darwin's sexual selection and support for the concept of female
choice, though they received scant recognition in scientific circles. Romanes
(also Canadian born) proffered condescending sponsorship. He sent a copy
of *Physiological Aesthetics* ("an entertaining little treatise") to Darwin, who
contributed to a fund for this latest recruit, who was ill and financially dis-

tressed.[70] Early in 1879 Darwin wrote to express his appreciation of Allen's championship of sexual selection and to dismiss Wallace's explanations for the same phenomena as "mere empty words." For many years, he confided in Allen, he had "quite doubted [Wallace's] scientific judgment."[71]

When, a few months later, Allen contested Wallace's explanation of human hairlessness in the *Fortnightly Review*, Darwin again wrote approvingly of Allen's counterexplanation, which incorporated an element of male choice in the evolution of this human characteristic: Allen's views were "very probable," but the best of it was his defense of sexual selection, which was "something wonderful" to hear.[72] Ironically, it was Allen who later was to contribute to the further fall of sexual selection from grace when he associated it with scandalous, socialistic free love, and it was this that provoked Wallace's retrieval of the concept in the service of a more respectably defined socialism and feminism.

What Darwin would have made of Wallace's extraordinary backflip we can never know—though we might make some informed guesses: Darwin had little empathy with feminism and none whatever with socialism.[73] Darwin died in 1882 and was interred in Westminster Abbey with full pomp and panoply of church and state—the ultimate Victorian accolade for this once-deemed-dangerous man. He died with a final tribute to his "good wife" on his lips and with a last defiant flourish of his conviction of the "truth" of sexual selection. This was read to a meeting of the Zoological Society on April 18, 1882, the day before he died. It was a preliminary notice to a manuscript "On the Modification of a Race of Syrian Street-Dogs by Means of Sexual Selection," relayed to him by a Dr. W. Van Dyck, a zoology lecturer in Beirut. Darwin had been "very anxious" that this testimony to the efficacy of sexual selection should be "published and preserved," choosing the Zoological Society over the "ephemeral" *Nature*, because "their journal goes to every scientific institution in the world, and the contents are abstracted in all year-books on Zoology."[74]

Van Dyck's paper claimed that over a twenty-year period, the "native bitches" that scavenged the streets of Beirut had mated preferentially with physically weaker but better proportioned "foreign" pointers, poodles and terriers, while introduced "pointers and other well-bred bitches" had shown a decided preference for feisty local street dogs, so that the present population of street dogs presented a mongrel, motley appearance, very different from the originals. In his introduction, Darwin cited verbatim as further "striking" evidence of such female preference, the Finnish naturalist Nils Adolf Erik Nordenskiöld's description of two Scotch collies on the voyage of the *Vega*, who "soon took the same superior standing" to the dogs of the indigenous people of the Chukchi peninsula "as the European claims for himself in relation to the savage." As their human counterparts were often alleged to do, the indigenous bitches gave sexual preference to the superior European race of dog, while indigenous dogs, recognizing their inferiority, failed to engage in the usual male combat for possession of the females: "The [male Scotch collie] was distinctly preferred by the female Chukch canine population, and that too without the fights to which such favour on the part of the fair commonly gives rise."[75]

For Darwin, this final set of doggy tales, with their all-too-familiar race, class, and gender resonances, validated his fundamental claim—which "many naturalists doubt or deny"—that females do exert a choice, that they often show preference "in the most decided manner" for particular males, while males usually accept any female. However, he conceded that it "would be more correct to speak of the females as being more excited or attracted in an especial degree by the appearance, voice, &c. of certain males, rather than of deliberately selecting them." Nevertheless, "having carefully weighed to the best of [his] ability the various arguments which have been advanced against the principle of sexual selection," Darwin remained "firmly convinced of its truth."[76]

14.5 Female choice at large

The accepted inspiration for Wallace's dramatic turnaround is his enthusiasm for the American socialist Edward Bellamy's promotion of the notion of free and informed female choice as an agency of human progress in his best-selling social utopian novel, *Looking Backward, 2000–1887* (1888). Bellamy's futurist novel sold hundreds of thousands of copies and captured an extraordinary following in America, inspiring a mass political movement based in cooperative, socialist principles. But it dealt only cursorily with female choice, being far more concerned with issues of nationalization and social cooperation.[77] Wallace's adoption of the idea had a longer genesis in his response to contemporary debates on the woman question and on "eugenics," the term coined by Galton in 1883. It also was contingent on the politics of late-Victorian science; specifically, the little-explored conflict between Romanes and Wallace over the ownership and definition of "Darwinism" and the place therein of sexual selection and the evolution of women. It was shaped more generally by the earlier history of radical attempts to offer a theory of human progress based on the Darwinian principle of female choice. There was history there—history that concerned not only Wallace, but also had embroiled Darwin and Huxley.

The great paradox was that, as its biological analogue went into crisis toward the close of the nineteenth century, under sustained attack by Wallace and without serious support among professional naturalists and biologists, sexual selection (and female choice in particular) was kept in play by an array of novelists, feminists, social purists, and more controversially, by assorted utopians, political radicals, and sex reformers.

George Eliot, Thomas Hardy, George Meredith, and a host of lesser writers of Victorian fiction were drawn irresistibly to the complexity and ambivalence of the associations of Darwinian sexual selection with love, beauty, sex, race, class, and breeding, finding in them a rich source of ideas, incidents, and plots. There was reciprocal influence here, in that the language of the *Descent* was inflected by the literary traditions and courtship plots of Victorian fiction. In turn, fiction writers reconfigured traditional tropes, drawing on their biological referents of male rivalry, strutting display, and feminine coyness and caprice, exploiting their ambiguities, playing with the possibilities of female choice, exploring potentialities that the two editions of the *Descent* struggled to suppress.[78]

To an extent, these fictional explorations of sexual choice overlapped with those of more didactic "improving" writers (particularly those concerned variously with issues of morality), with the rights of women, and with schemes for redressing the much-debated fear of racial degeneration. Here, as indicated, female choice was taken up by a number of feminists, often in association with the social purity movement, and so played a significant part in the post-*Descent* "biologization and feminization of morality" through their emphasis on the inherently "unhealthy tendency of men to promiscuity and vice, and the natural instincts of women to virtue." In the process, female virtue was recast as eugenic virtue: women's love choices were not only moral, but also the sign of rationally exercised and socially and biologically improving female will. This literature (which included the New Woman novels by best-selling authors such as Sarah Grand, whose courtship plots targeted male brutishness and exemplified the social/eugenic responsibilities of middle-class women) came to prominence in the last decade of the nineteenth century and is the focus of some searching historical analysis.[79] But, for the most part, these literary appropriations of sexual selection catered to and reinforced traditional sexual stereotypes—although feminism and social purity were not always coterminous, and many novelists had their own strategies and tropes for voicing "dissident desire."[80]

It was the appropriations of female choice by more radical "outgroups" and activists, intent on critiquing or overturning conventional morality (or represented as such by their detractors), that posed the more threatening challenge to Darwinian respectability and authority. Radicals and left-wing ideologues, drawn to Darwinism and its naturalistic premises, struggled (largely unsuccessfully) to reconcile the essential competition and conflict of natural selection with socialist, cooperative goals.[81] But as the issue of women's rights gathered pace and controversial eugenic schemes for the betterment of society multiplied, female choice presented fewer incompatibilities and promised more for radical and reform-oriented ends. Female choice was commensurate with a continuous tradition of sexual radicalism and feminist utopianism that ran from the time of the Owenite socialists to its reemergence among the more radically minded toward the end of the nineteenth century.[82] Sexual selection, already problematic, was made yet more problematic for professional Darwinians like Huxley, whose social and political goals ran counter to those of unruly radicals, disreputable sexual reformers, and dangerous New Women.

14.6 "The greatest of all possible evils to mankind"

In his last years, Darwin too had good intimation of the propensity of female choice to attract undesirable fellow travelers. Chief among these was that persistent and unwanted hanger-on to Darwinian coattails Edward Bibbins Aveling, whose version of radical female choice was proffered within a particularly obnoxious context of militant atheism, scandalous birth control, and free love.

Aveling has suffered such a bad press at the hands of vengeful Marxists as the philandering, money-filching has-been who allegedly caused and abetted the suicide of Marx's favorite daughter Eleanor, that it is difficult to recover the promising younger scientist, fellow of University College, and lecturer in comparative anatomy at London Hospital. He got into bad company when he joined the National Secular Society in 1879 and became a close associate of Annie Besant, the notorious leading lady of the Secularists, and their president, the radical atheist and republican Charles Bradlaugh, a man for whom Huxley professed his utter dislike, "personally, politically, and philosophically."[83]

Besant was a well-known radical—a feminist, freethinker, and neo-Malthusian. She became notorious when in 1877 she was prosecuted for "obscenity" at the Old Bailey with her codefendant Bradlaugh for openly distributing birth control literature—the reissue of a venerable work by the Massachusetts physician Charles Knowlton, denounced as a "dirty, filthy book" by the prosecution. Their intention was to emancipate women from the fear and bondage of childbearing, while, as neo-Malthusians, Besant and Bradlaugh saw birth control as the only rational "scientific" check to the otherwise inevitable overcrowding, hunger, and poverty of the lower classes. Before their trial, the accused subpoenaed a number of prominent theorists as witnesses in their defense that the "doctrine of the limitation of the family" was to be found in other works in general circulation and that many scientific texts offered explicit passages, open to the same construction of depraving vulnerable minds, as was being alleged against Knowlton's pamphlet. This severely tested the liberal principles of those whose writings were associated with neo-Malthusianism or, more generally, with naturalistic depictions of human reproductive anatomy and physiology, but who were aghast at the prospect of being forced into open defense of two militant atheists—one of them a woman living apart from her clergyman husband—in a sensational trial that was being splashed all over the press. Most notables ducked for cover, including an appalled Darwin.[84]

Darwin's letter to Bradlaugh cannot be traced, but the original draft in the Cambridge collection, with its many deletions and interlineations, recreates the urgency of his need to dissociate himself from any implied support for such a disreputable cause. It begins with the familiar litany of his ill health and the "great suffering" it would be for him to be a witness in court. Even were he able to appear, the defendants should know that his evolutionary views had long caused him to hold a "very decided" opinion against theirs. It was his conviction, stated in the *Descent*, that man's advance was dependent on his continuing "rapid multiplication" and consequent severe struggle for existence, and that this necessary struggle should not be impeded "by any means,"—meaning, Darwin now made explicit, "artificial means of preventing conception." But this evolutionary deduction did not represent the full horror of birth control:

> But besides the evils here alluded to I believe that any such practices would in time spread to unmarried women & would destroy chastity, on which the family bond depends; & the weakening of this bond would be the greatest of all possible evils to mankind; & [it] must be my duty to state in court. So that my judgment would be in the strongest opposition to yours.[85]

The trial went ahead without Darwin. The great scientific hero of radicals and freethinkers stood exposed as a typical middle-class conservative. The redoubtable Annie Besant, however, was equal to the challenge. She devoted much of her eloquent defense to contesting Darwin's arguments and arguing the indefensibility of his position on birth control and its implica-

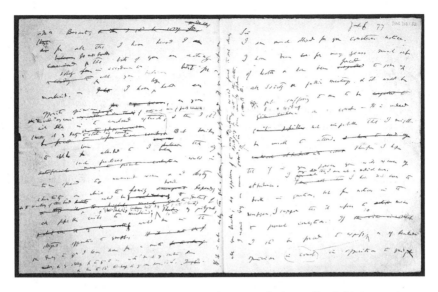

Figure 14.3. The original draft of Darwin's letter to Charles Bradlaugh, birth control advocate, written June 6, 1877, with its many deletions and interlineations, vividly recreates the urgency of Darwin's need to dissociate himself from any implied support for such a disreputable cause. Darwin Papers, Cambridge University Library, DAR 202:32. By permission of the Syndics of Cambridge University Library.

tions for female chastity: "That is Mr. Darwin's position, and putting aside for a moment the awful amount of human misery which he accepts as the necessary condition for progress, let us see if the position be defensible."[86]

Besant and Bradlaugh duly were found guilty of obscenity—their conviction later quashed on a technicality. The trial, credited with opening up public discussion of contraception, nevertheless brought great opprobrium down on the pair. Besant lost custody of her young daughter as a result. Shunned by the respectable Darwinians, she went on to become a leading reinterpreter of Darwinism for the secular and birth control causes. Her trial testimony became the basis of her sixpenny pamphlet, *The Law of Population: Its Bearing upon Human Conduct and Morals*, which was translated into several languages and sold hundreds of thousands of copies. Her relationship with Aveling enabled Besant to deepen her biological insights, and in return, her views on women and marriage and the benefits of birth control were incorporated into Aveling's prolific writings on Darwinism and his reinterpretation of the "revolutionary truths" of natural and sexual selection.[87]

With Aveling's help, Besant enrolled for a science degree at London University, which, against opposition, had just opened its doors to women. Together she and Aveling established, and for many years taught, the highly successful classes in science at the Hall of Science, the London headquarters of the National Secular Society—testament to their conviction of the radical-izing powers of science. These popular classes, designed to educate young women and men in the scientific method that purportedly underpinned the secularist platform and to prepare those interested for entry to London University, featured the Aveling-Besant version of Darwinism. In 1881, Aveling published these views in the collection entitled *The Student's Darwin*, which he proposed to dedicate to the great man himself, an honor that Darwin hastily repudiated.[88]

Before Aveling's emergence as a gazetted atheist, Darwin had encouraged his evolutionary writings. But by 1881, Darwin was chary of lending his name to a work to be published in a new series on science under the editor-ship of Besant and Bradlaugh, with the declared aim of spreading "heresy" among the "reading masses."[89] The scandalous Knowlton trial aside, in 1880 Bradlaugh had become the first declared atheist ever elected to Parliament on the Radical ticket and was embroiled in conflict over whether he would be permitted to affirm and, that rejected, over the refusal of the Commons to permit an atheist to take the requisite Bible oath. Besant and Aveling bat-tered on the doors of Parliament alongside Bradlaugh, denouncing Chris-tian bigotry and the denial of civil liberties to crowded audiences around the country.[90] This brought their attempts at the scientific reformation of society and the Hall of Science classes under parliamentary scrutiny. Besant's conviction for obscenity and the legal loss of her daughter were invoked to allege her unfitness as a teacher of the young and to make political capital out of connecting William Gladstone's ministry with the supposed authori-zation of public moneys for the dissemination of atheism, republicanism, Malthusianism, and worse. Aveling's recent evolutionary account of certain properties of the frog (in emulation of Huxley) was targeted as "condemna-tory of God." While the *Standard* wondered whether the frog was a Conser-vative or a Radical, the *Evening News* pointed out that "men like Professor Huxley, who are in receipt of large Government salaries, hold and teach the same doctrines on the evolution theory as Dr. Aveling"—a connection that may have gratified Aveling but hardly Huxley, who, among other things, was the director of the Science and Arts Department that had certified Besant's teaching qualification and eligibility for a government grant.[91]

All this was bad enough; but the central issue for the Darwinians was the correlation of atheism and birth control with free love. Darwin was scarcely alone in connecting the availability of contraception with female promiscu-ity. Birth control had a long association with free thought, but its indissoluble link with free love was forged by the explosive *Elements of Social Science* (1854), anonymously authored by George Drysdale (as it happened, brother-in-law of Darwin's hydrotherapist, the alleged adulterer Dr. Lane), cofounder with Bradlaugh of the Malthusian League in 1861. The antimarriage agenda of this so-called "Bible of the Brothel," its shocking advocacy of the "natural" expression of the sexual appetite for both sexes, along with "preventive" methods for avoiding pregnancy (and thereby the twin evils of overpopulation and poverty), exacerbated the deep-seated hostility of the religious and respectable to neo-Malthusianism. Bradlaugh repudiated free love, but could not escape the odium of his promotion of Drysdale's *Elements* in the *National Reformer*, the Secularist newspaper. It split the ranks of the Secularists and confirmed the worst prejudices of those who equated birth control with rampant immorality, godless cohabitation, and the general destruction of society. In England, atheism implied the doctrine of free love, and the Knowlton trial reinforced this widespread impression. Like it or not (and they did not), Bradlaugh and Besant were irrevocably tied to this promiscuous image.[92]

When Besant and Aveling began touting the Secularist variant of Darwinism, the reaction of the Darwinians was as much inspired by their fear of moral taint as of the expropriation of their creed to radical atheistic ends. It was for demarcating decent Darwinians from just such dangerous hangers-on that Huxley's neologism of 1867, the gentlemanly and neutral-sounding *agnostic*, came into its own. Agnosticism, Bradlaugh protested, was a "mere society form of Atheism." But there was a lot in a name. In 1879, Darwin, recoiling from the bellicose flag-waving atheism and neo-Malthusianism of the Bradlaughites, took shelter under Huxley's proffered standard: "I think that generally . . . Agnostic would be the more correct description of my state of mind."[93]

A reading of Aveling's *Student's Darwin* (which sat unmarked on Darwin's shelves alongside his famously uncut presentation copy of Marx's *Das Kapital*—later translated into English by Aveling) gives intimation of what Aveling and Besant were up to. In a radical departure from the *Descent*, Aveling urged the advantages of *rational* selection over natural selection, giving women a dominant and responsible evolutionary role via their intelligent exercise of female choice. As with animals, there was evidence among even the "very lowest tribes" that women exercised "a very considerable amount of choice." As women evolved an increasing complexity of brain structure and function, they would begin to admire qualities other than mere bodily size and strength, color and hairiness, and to choose between men on the basis of their mental and moral characters. Sexual selection for the advanced woman had therefore become "far more complex" and its results "far more momentous" for society:

To-day, woman has to consider beauty of face, beauty of form, social position, strength of mind, strength of moral character. She has to make her choice between men differing in infinite degrees in all particulars, and she has to make that choice conscious that her selection will influence not alone her life and his, but the lives which will spring from them.[94]

Aveling elaborated further on the eugenic potentialities of intelligent female selection in his *Darwinism and Small Families* of 1882. Besant's influence is evident in his argument that birth control was compatible with natural selection and racial improvement, but now Aveling contributed the vital added factor of an informed choice of mate. The important point that the Darwinian critics of neo-Malthusianism had missed was that it was not the *quantity* of the progeny that was crucial for social progress, but their *quality*. Women could play a powerful part in the "development of our race" through the "infinite care" they might exercise in their choice of partner.[95]

Besant's part in these theoretical constructions may be inferred from the pamphlet on women and marriage that she published around this time. It was a scathing denunciation of the Victorian marriage laws that made women the property and prisoners of their husbands and denied them their "rights" of control over their own bodies, personal liberty, property, and children. She put a Darwinian gloss on the history of marriage from its origins via the operation of "crude sexual selection" and looked to a future when women and men would freely choose their partners on the basis of intellectual as well as other attractions and be bound by "love instead of by law." While Besant made it clear that she was advocating "equal and monogamous" love, not free love, her plea for the recognition of unlegalized public unions between men and women laid her open to willful misinterpretation.[96]

It was hardly the sort of thing with which reputable Darwinians wished to be associated. Meanwhile, the Hall of Science classes were competing for just that freethinking working-class constituency that Huxley regarded as particularly his own. They had annual enrollments of some two hundred students (including significant numbers of women) and were maintaining above-average examination grades; many of their graduates were going on to teach courses up and down the country. As well, Aveling's prolific, ultracheap publications and pamphlets, with their inclusionary, attractive titles (such as *The People's Darwin; or, Darwin Made Easy* and *The Gospel of Evolution*) offered accessible introductions to evolutionary science to those with little education, who were hungry for new knowledge and impatient of old orthodoxies. The radical secular brand of Darwinism and sexual selection, connoting atheism and suspect morality, was reaching into territory unplumbed by professional Darwinians and middle-class popularizers.[97]

Huxley made his views on all this clear when Aveling tried to embroil him in the furor over Besant's continuing science studies when she (along with Bradlaugh's daughter Alice) was refused admission to botany classes at University College. The incident precipitated a student riot and was widely publicized by the Secularists. Aveling put pressure on Huxley—who had earlier lectured the honors-achieving Besant in botany—to sign a petition against her unwarranted exclusion. Huxley signed, reluctantly, in the interests of religious tolerance; but privately he indicated that the issue was not so much about Besant's religious beliefs as her reputation. He believed that the college council should be concerned "with the lives and character of those who are admitted." Huxley had a "strong feeling that freedom of thought should be carefully distinguished from laxity of morals." If Besant, as he had reason to think, was responsible for advocating "safeguards of sexual intercourse among unmarried people . . . we are out of the range of speculation and into that of practice—and I have no objection to her exclusion." "Freethinking," said Huxley censoriously, "does not mean Free Love."[98]

It was all grist to Huxley's conviction that "civilization depends very much on the maintenance of a decent standard of sexual morality." Like Darwin, although he recognized the pressing problem of overpopulation and its attendant "hordes of vice and pauperism" that would "destroy modern civilization," Huxley eschewed any consideration of contraceptive methods that might morally compromise reputable Darwinism and his credo of scientific naturalism.[99]

On the eve of the specially convened meeting of the college council to consider Besant's case, Huxley was elected president of the Royal Society, which gave him even more need to dissociate himself from such disreputables with their scandalous birth control, free love and subversive evolutionary interpretations. Huxley rose to the occasion (so to speak) by abstaining from voting. Aveling's motion lost by a large majority, and Besant's exclusion was ratified by the college. Her science degree was never completed.[100]

14.7 New Women, eugenics, and the socialist specter

Shortly after, Besant and Aveling went their separate ways, although both went to socialism—Besant with George Bernard Shaw, Aveling in the company of Eleanor Marx, a trailblazing New Woman with, according to a prudish Beatrice Potter (subsequently Webb), "peculiar views on love, etc. . . . and somewhat 'natural' relations with men!"[101]

If Huxley found the incendiary mix of free thought, free love, feminism, and Darwinism that radical Secularism stood for too much, insult was added to injury by its integration with socialism. While Besant, her enthusiasm for questionable causes and men undiminished, issued pamphlets proclaiming that she was a socialist "first and foremost because [she] was an evolutionist," Aveling united with Eleanor Marx to write the socialist version of radical sexual selection and its relation to the woman question. This was published in the *Westminster Review* (1886), later issued in pamphlet form. It was stimulated by the arguments of the German socialist and Darwinian, August Bebel, disciple of Friedrich Engels and inspiration of a militantly Marxist working-class women's movement.[102]

The 1885 English translation of Bebel's influential *Woman in the Past, Present, and Future* (first published in Germany in 1879, better known as the classic *Woman under Socialism* that ran through some fifty editions), created a great stir in left-wing circles and met with a "vituperative reception" in others. Bebel contributed the central socialist feminist argument that women could be genuinely independent of men and equal in rights only insofar as they achieved full economic independence. He integrated his socialism with a thoroughgoing scientific naturalism and evolutionism. In order to understand the present and future situations of women, "the same methods must be applied and the same laws examined as those by whose help modern science explains the origin of species and genera . . . We refer to the laws named Darwinian . . . based in the material conditions of existence." Bebel cited Mill against Darwin, Huxley, and Galton to argue that the average woman was inferior to the average man because "women are what men, their masters, have made them." What was required was the radical transformation of society if women were to realize their potential. If they were properly educated and if the competitive struggle for existence and for mates that they were currently obliged to endure was overturned, women's future would be assured:

> As all of these things depend on education and on the conditions of life, or in the plain language of natural science, are a question of "breeding," and as natural laws have already been applied in the case of domestic animals with startling results, there can be no doubt, that by the application of these same laws to the physical and mental development of mankind, even more unforeseen results will be attained.

Bebel made clear that he was not invoking a eugenic solution of social engineering by an authorized scientific priesthood like Galton and others;

rather, his idea was that in a socialist society their future evolution could be safely left to the "natural" sexual instincts and educated minds of ordinary women and men, and to this end Bebel urged women actively to pursue their sexual instincts and mate choices in the same way as men.[103]

The Marx-Aveling pamphlet of 1886 was an extension of Bebel's arguments by "two individual socialists . . . thinking and working together." Aveling (whose wife refused him a divorce) and Eleanor, now living openly together in "free marriage," deplored the contemporary double standard that permitted men ready sexual gratification but treated women who acted likewise as pariahs. They advocated sex education for boys and girls and argued that chastity was "unhealthy" and intellectually stunting for the women forced to endure it and for the society that refused them any relief. The solution lay first in the expropriation of all private property in land and all other means of production, followed by "equality for all without distinction of sex." Women and men would be free to make their own private contracts. Monogamy would probably be the outcome, but if not, women would be as free to choose and act as men, to fulfill their natural sex instinct and choose the man or men of their preference. In such a society there would be no need for divorce or prostitution. All would be carried out "fairly and openly," without the "hideous disguise, the constant lying, that makes the domestic life of all our English homes an organized hypocrisy."[104]

It was an assault on the Victorian double standard and sexual conventions, in which Aveling's earlier remedies of birth control and rational female choice were supplanted by a heady brew of Marxism, scientific naturalism, sexual selection, and free love. It contributed mightily to the popular perception that socialism advocated promiscuous free love and the dissolution of the marriage tie. Even those prepared openly to discuss the "sex instinct" with like-minded women and men, as in the newly formed "Men and Women's Club" (a loose enclave of radicals, socialists, and feminists, founded by the biologist Karl Pearson—later a leading eugenicist—in 1885), recoiled from Eleanor Marx's doctrinaire abandonment of bourgeois morality and "bourgeois feminism." Her membership was initially opposed, though some few New Women members came to accept free unions as the highest ideal of marriage, the legalized form of which was no more than prostitution or "parasitism" for women.[105]

One such was Olive Schreiner, author of the brilliant and unsettling *Story of an African Farm* (1883), intimate of Pearson and Henry Havelock Ellis (physician, evolutionist, and budding sexologist, also much interested in the psychological aspects of sexual selection[106]). Schreiner had already begun work on her better-known, much later published *Woman and Labour*

(1911), in which she integrated evolutionary and feminist principles with socialist economics in analyzing the sexual and social situation of women. Schreiner was determined to show on evolutionary grounds that many supposed gender differences were conventional, not innate, and to express the "great truth" that woman was "bound to lead the way, and man to follow" in the "higher development of sexual life." It was Schreiner (along with Mona Caird) who articulated the "man question" in relation to the omnipresent "woman question": the New Woman required a New Man if she was to realize her full potential and achieve the socialist ideal of the free and equal sexual union. The utopian ideal ignored the realities of the legitimacy and care of children and the personal consequences for women in a society that so inflated the value of feminine chastity and restricted access to contraception. Nevertheless, Schreiner held to her conviction that human sexuality, freed from some of the pressures for mere survival, was evolving toward the aesthetic and beautiful. Inevitably, New Women and New Men would be united in perfect mental and physical monogamous relationships in the coming sexual utopia:

> Always in our dreams we hear the turn of the key that shall close the door of the last brothel; the clink of the last coin that pays for the body and soul of a woman; the falling of the last wall that encloses artificially the activity of a woman and divides her from man; always we picture the love of the sexes, as, once a dull, slow, creeping worm; then a torpid, earthy chrysalis; at last the full-winged insect, glorious in the sunshine of the future.[107]

Schreiner's evolution-based arguments were largely directed against *The Evolution of Sex* (1889), authored by the Scottish biologist/sociologist Patrick Geddes (who had trained under Huxley) and his former pupil J. Arthur Thomson, which encompassed one of the more significant critiques of sexual selection.[108] Geddes and Thomson, while staunch evolutionists, pushed a line of their own, delimiting both Darwin's and Wallace's interpretations of sexual dimorphism: "According to Darwin, sexual selection, for love's sake, has accelerated the males into gay colours; according to Wallace, natural selection, for safety's sake, has retarded the females (birds or butterflies) and kept them inconspicuously plain." They advocated "a compromise":

> There is, in fact, no reason why both should not be admitted as minor factors; but the greater part of the explanation is to be found . . . in the physiological constitution of males and females themselves . . . The present position . . .

regards gay colouring as the expression of the predominantly katabolic or male sex, and quiet plainness as equally natural to the predominantly anabolic females.

Female choice came in for robust criticism. It carried with it "the postulate of a certain level of aesthetic taste and critical power in the female, and this not only very high and very scrupulous as to details, but remaining permanent as a standard of fashion from generation to generation." These were "large assumptions" and "scarcely verifiable in human experience." It was difficult to credit birds and butterflies with a degree of aesthetic development that few humans attained. Darwin's "essential supposition" was "too glaringly anthropomorphic."[109]

Geddes and Thomson have acquired historical notoriety for extending their fundamental physiological distinction between anabolic, quiescent, conserving females and active, passionate, variable, katabolic males to social theory, reinforcing gender stereotypes ("Man thinks more, woman feels more"), and pronouncing against egalitarian feminism on evolutionary grounds: "What was decided among the prehistoric Protozoa cannot be annulled by Act of Parliament."[110] This interpretation, however, requires qualification. *The Evolution of Sex* was based in a communitarian, anticapitalist ethos that rejected the Darwinian "internecine struggle" of individuals as deleterious to social progress and promoted the survival of the "truly fittest" through "love, sacrifice, and co-operation": "[The] ideal of evolution . . . is not struggle but love." This entailed a "new ethic of the sexes" involving "sympathetic co-operation" and mutual dependency. Biologically women were nurturers. Their ability to transform the world rested on their "civicism," which included not only love and mothering but also the activities women might usefully perform outside the home for the sake of the community.[111]

It was a position "at once radical and conservative," and it appealed to many socially conscious young women who attended Geddes's Edinburgh-based Summer Schools, participating in the exchange of ideas and associated community betterment schemes. Jane Addams carried back to America her Geddes-derived views on the ethical and philanthropic mission of women, inspiring a great deal of feminine involvement in a variety of benevolent and progressive causes deemed appropriate to women's sexual character.[112] Nevertheless, as its critics alleged, Geddes's stereotyping of women as nurturers discouraged their active political engagement and paid professional employment. More fundamentally, through its promotion of the notion of woman's biologically based "complementary genius," it contributed to the disengagement of nineteenth-century feminism from its Enlightenment, egalitarian roots. And, while it fostered a "highly moral, self-disciplined but intense pursuit of romantic love" as the proper basis for marriage and social advance, through its denial of an active sexuality to "anabolic" women and its rejection of the biological analogue of female choice, the role it played in late-century debates on these issues by more radically minded individuals was to provoke opposition, as with Schreiner and the American socialist feminists Eliza Burt Gamble (1894) and Charlotte Perkins Gilman (1898).[113]

It was the ubiquitous Grant Allen who really set the socialist cat among the Darwinian pigeons with his sensationalized free-choosing, free-loving New Woman. Allen, turning from penurious science to more lucrative journalism and fiction, had continued to proselytize sexual selection and female choice in his prolific writings.[114] Like Wallace to follow, the self-styled "Individualist Communist" was responding to the dominant eugenic schemes floated by Darwinians like Galton, Greg, and others as remedies for the racial degeneration sure to follow on the multiplication of the unfit and inferior, fostered by well-intentioned but ultimately suicidal liberal reforms. It was a reading of Darwinian social biology that Darwin, despite some protestations, had given the support of his immense reputation.[115]

Galton was one of the few certified Darwinians to uphold his famous cousin's notion of sexual selection after Darwin's death. Since his calculation of the dimensions of the Hottentot bottom, his passion for measurement had continued unabated. (Galton, the founder of biometrics, also invented the pocket "pricker," a device for unobtrusively ranking and recording the faces and forms of passing women for his proposed "beauty map" of Britain.) Although Galton viewed sexual selection as a necessary adjunct to racial progress, he saw little possibility of putting it to eugenic use. In 1883 he spelled out the necessary conditions for female choice, far more directly and uncompromisingly than Darwin had ever done and with little reference to the notion of morally superior "good woman" that had informed it. Women, alleged Galton, were characteristically "capricious and coy" and had "less straightforwardness" than men. This was particularly so "around the time of pairing, and there can be little doubt as to the origin of the peculiarity":

> The willy-nilly disposition of the female in matters of love is as apparent in the butterfly as in the man, and must have been continually favoured from the earliest stages of animal evolution down to the present time. It is the factor in the great theory of sexual selection that corresponds to the insistence and directness of the male. Coyness and caprice have in consequence become a heritage of the sex, together with a cohort of allied weaknesses and petty

deceits, that men have come to think venial and even amiable in women, but which they would not tolerate among themselves.[116]

Without female coyness, this vital evolutionary instinct, there would be no sexual choice and consequent degeneration of the "race"; however, in humans it had resulted in intellectually and morally deficient females who indulged their frivolity in mate choice as much as in ornament and dress. The urgent betterment of the race could not be left to ignorant feminine caprice while the less desirable types of humanity were outbreeding their betters. What was needed was the institution of some supervised system of breeding that would encourage and reward the early marriage and reproduction of superior types of men and women.[117]

Galton, Greg, and Karl Pearson were all advocates of such "positive" eugenics; others favored a form of "negative" eugenics by discouraging the reproduction of the "unfit" through education, moral suasion, or (in George Darwin's case) legal reform of divorce law, while more extreme proponents advocated the segregation or sterilization of the hereditarily unfit. Both Wallace and Allen agreed with the need for racial improvement, but rejected such coercive, legislative, or elitist solutions. In the tradition pioneered by Bebel and lately adopted by Bellamy, they advocated eugenics by social reform, and both turned to the emancipation of women and female choice as the effective agency for improving the human race.

Allen's version of reformist eugenics rested on the extension of this biological principle to the overturning of the convention of monogamous marriage in favor of "Free Temporary Unions" or "Free Relations" that would enable women to choose the "best fathers" for their children. In a series of articles, culminating in "The Girl of the Future" in 1890 (an obvious play on Linton's catchphrase "Girl of the Period"), Allen identified himself as an "enthusiast on the Woman Question" who wished to see woman "a great deal more emancipated than she herself at all yet desires." The currently advocated sterile, educational schemes that produced androgynous Girton graduates ("as flat as pancakes and as dry as broomsticks") must give way to a "new and daring" education that would not enfeeble women's sexuality and womanly instincts but develop their bodies and exercise their minds. "Women will then be as gods, having eaten of the tree of knowledge of good and evil." The scales would fall from their eyes, and they would see that marriage—which had its origins in capture and slavery and "crystallizes in its very form the brutal selfishness and jealousy of the stronger sex"—was but a form of prostitution. The truly emancipated, economically freed woman would "feel it incumbent upon" her to bring forth handsome, healthy children "for the State"; she would "use her maternity as a precious gift to be sparingly employed for public purposes," not squandered "irrevocably" on any one man: "The free woman will choose which lord she shall serve. And do you think her choice will be for the colonial broker?" A "practical system of eugenics, or nature's own plan of distinctive selection, will have automatically established itself," one that no "mere 'eugenic' system of marriage" based on "permanent and strictly monogamic union," as advocated by Galton and others, would ever achieve.[118]

It was a eugenic fantasy based on the assumption that women's instinctive powers of perception and sexual choice would do for society what peahens had done for the embellishment of the peacock's tail. And, while it seemed to give sexual passion and agency to an idealized free woman, Allen's notion of female choice was predicated on the same old assertion of feminine sexual passivity as the necessary basis for female discrimination. He agreed with Darwin and Galton that

> coyness and daintiness in the female is as necessary a part of sexual selection as strength or beauty in the male. The higher up we go in the scale of being, the more do we find such selectiveness prevailing. Physiologically speaking . . . it is in most cases the duty of the male to be aggressive and eager, the duty of the female to be coy and discriminating.[119]

Allen's hugely successful best seller, *The Woman Who Did* (1895), in which he fictionalized his sexually emancipated free woman, is riven by these contradictions. It is a morally ambiguous tale wherein his Girton-educated heroine is driven into penury and ultimately to suicide through her high-minded but parodic persistence in eschewing the hypocrisy of marriage and choosing to bear her child out of wedlock. It was a succès de scandale that structured public stereotypes of the sexually rebellious New Woman and reinforced the insistence of many feminists on the necessity of legal marriage and conventional family values.[120]

Wallace, the moralist, was among the many who pronounced Allen's audacious solution to the problem of racial deterioration "detestable." Wallace was almost as chary as Darwin and Huxley of any hint of moral laxity, although, in his essay of 1890 he could countenance the possibility, like Besant (whom he knew), that the true monogamous marriage he envisaged for the society of the future might be attained not by "rigid law" but rather by the "influence of public opinion." Nevertheless, he carefully avoided any mention of birth control (although he recognized the problems generated by population pressure), and he denounced any measures for social improvement, such as Allen's, that would favor the "increase of pure sensualism, the most degrading and most fatal of all the qualities that tend to the deterioration of races and the downfall of nations." No system could be "more fatal" to human happiness and the advancement of the race.[121] The association of free love with socialism and eugenics was giving both a bad name. Wallace's response was to banish the taint of free love from female choice and reinstate the "good woman" in rational socialist guise.

There was, however, another edge to Wallace's expropriation of female choice for socialist ends. This was his increasingly acrimonious conflict with Romanes, Darwinian heir apparent and expert on animal intelligence.[122]

Romanes, whom Darwin had taken under his wing in his last years, projected himself as the defender of "true Darwinism" against ultra-selectionists like Wallace and Weismann (who rejected the inheritance of acquired characters), for whom he coined the neologism "neo-Darwinians." Wallace was rendered extra-suspect through his denial of the fundamental Darwinian continuity of animal and human mentality. Romanes was, moreover, a political conservative, independently wealthy and offended by Wallace's public campaign for land nationalization and his burgeoning socialism. They initially clashed over Wallace's 1886 critique of Romanes's "physiological selection," a theory of the origin of infertile races. Romanes reciprocally attacked Wallace's explanation of sexual dimorphism in his review of Wallace's *Darwinism* in the *Contemporary Review* (1889). He admitted Wallace's "most powerful array of objections against the theory of sexual selection," but argued that the peacock's tail could only be useful in courtship and was "obviously incompatible with the supposition [of its] having been produced by natural selection." While not unequivocally endorsing female choice, Romanes opposed Wallace's claim that there was no evidence to support it by arguing that although little was known about the psychology of lower animals, "small details of mental organization are often wonderfully constant and uniform . . . even where it is impossible to suggest any utility as a cause." Their fundamental conflict was over Wallace's insistence that "natural selection must necessarily swallow up . . . sexual selection, as the fat kine swallowed up the lean in the dream of Pharaoh."[123]

Their running battle over the definition of Darwinism became particularly bitter in May 1890, just a few months before Wallace published his paper on "Human Selection." In an assault on "Darwin's Latest Critics" in the liberal periodical *Nineteenth Century*, Romanes publicly accused the "man of science" of becoming the "man of nonsense": "the Wallace of spiritualism and astrology, the Wallace of vaccination and the land question, the Wallace of incapacity and absurdity." There followed a heated exchange of letters in which Wallace accused Romanes of cowardice and duplicity: Wallace had viewed earlier correspondence in which Romanes had professed his belief that he was being contacted by otherworldly intelligences. There were "two Romanes as well as two Wallaces," but one Romanes kept the other secret while manipulating "ignorant prejudice" to attack Wallace, who had "the courage of his opinions." The rift between the two had not healed by the time of Romanes's early death in 1894; if anything, it had deepened and was marked by an intensification of support for sexual selection, as distinct from natural selection, by Romanes. Wallace refused reconciliation and nursed his resentment; he reprised the affair in his autobiography (1905) and published the hostile letters they then had exchanged.[124]

There is more: their very different interpretations of the "woman of the future," the catchphrase initiated by Romanes. As defender of the true faith, Romanes first had picked up the cudgels in defense of Darwinian sexual selection in 1887 in a widely read essay on the "Mental Differences between Men and Women," also published in the *Nineteenth Century*. He reaffirmed the major anti-Mill arguments of Darwin's *Descent* and brought them up to date with the most recent scientific "facts," which included an alleged five-ounce weight difference between the male and female brain. This constituted tangible evidence of women's intellectual inferiority and of the evolutionary processes that had contributed to the increasing divergence in male and female mental characteristics. Romanes attributed this primarily to sexual selection, which ensured that the female was attracted to the most "masculine" of her suitors, while the male preferred the most "feminine" female. As a result, women had a "deeply-rooted desire to please the opposite sex" and "little power of amassing knowledge." The "woman of the future" could not hope to recover the "ground lost in the psychological race by the woman of the past." Nevertheless, despite women's "natural inequality" and the impossibility of any educational process remedying the missing five ounces of the female brain, "whether we like it or not, the woman's movement is upon us." What men of science had to do was to "guide the flood into . . . the most beneficial channels." Romanes accordingly prescribed the kind of education that would not turn woman into an "unnatural copy of man" but would rather make her a companion, a complement, to her husband. Such education would not inspire women with any "unnatural, and therefore . . . impossible, rivalry with men in the struggles of practical life." He reprised the established Huxleyan line:

> The days are long past when any enlightened man ought seriously to suppose
> that in now reaching forth her hand to eat of the tree of knowledge woman is

preparing for the human race a second fall. . . . Then, I say, give her the apple, and see what comes of it.[125]

Wallace, his indignation with Romanes at its height, offered a diametrically opposed vision of feminine intelligence, educability, and social potential. For Wallace, contra Romanes, the improvement of the human race was utterly dependent on the "cultivated minds and pure instincts of the Women of the Future."[126] The overt target of Wallace's 1890 essay on "Human Selection" was Allen's free-loving "Girl of the Future"; its unacknowledged one was Romanes's "woman of the future," handicapped by her evolutionary past, unable to achieve intellectual and social parity with men. Wallace's essay was an assertion of his widely acknowledged role as the "greatest living champion" of evolutionary biology, of his accrued cognitive and cultural authority as keeper of the flame in opposition to the upstart Romanes, of his revived socialism and newfound feminism, and of his long-term spiritualist convictions. It was predicated on his endorsement of Weismann's recently formulated theory of the germ plasm and his rejection of the inheritance of acquired characters, which for Wallace meant that the old environmentalist solutions embraced by reformers and radicals could not bring about the desired social progress. This left only the possibility of "improvement by some form of selection." However, this could not come into play until a socialist society had replaced "our present phase of social development," which was "vicious and rotten at the core."[127]

Wallace confronted what he saw as the unequal distribution of wealth and advocated an end to the retrogressive struggle for existence through the creation of a socialist economy and state ownership of all land. The result would be, he argued, a "rational social organization adapted to secure the equal well-being of all." Only then, Wallace wrote,

> [will] the future of the race be ensured by those laws of human development that have led to the slow but continuous advance in the higher qualities of human nature . . . Then we shall find that a system of selection will come spontaneously into action which will tend to eliminate the lower and more degraded types of man, and thus continually raise the average standard of the race.[128]

A socialist system would free women from the obligation to marry for purely economic or social reasons, and as a result they would become more discriminating in their choice of husbands. Once relieved of "the struggle for mere existence," a woman would have no need to marry until she found the husband of her choice, if at all. In such a society, men would have no means of gratifying their sexual passion outside marriage, so almost every woman would receive offers, and thus a "powerful selective agency would rest with the female sex." The "bug-bear of overpopulation" would be eliminated through later marriage, higher education, and hence lessened fertility. The problem of excess females would be curtailed by lowered mortality of males through improved working conditions and the cessation of war, thus greatly improving the influence of women, who, being in the minority, would be "more sought after" and have "real choice in marriage." Women, along with men, would be educated to the highest levels of knowledge, attainment, and morality, and taught to look "with scorn and loathing" on the "idle," the "selfish," the "vicious," the "diseased," and the "weak in intellect." Female choice in marriage would then ensure the steady elimination of the morally and physically unfit and the overall improvement of the race. It was women, therefore, who, through their free choice in marriage, would bear the responsibility for future social progress.[129]

In effect, Wallace disconnected Darwinism from its Malthusian roots by insisting that social progress could occur only through the demise of the struggle for existence; but he brought human progress back under the sway of natural law by reasserting the process of sexual selection—specifically, female choice, which could only come into play after socialistic reform. Female choice had the further attraction for Wallace in that it was contingent on just those faculties and characteristics that he had long claimed owed their development to spiritual intelligences and that, against Darwin and Romanes, he had refused to allow in any degree to animals. His previous objections to the efficacy of sexual selection in the human realm were no longer valid. Only under socialism could female choice become a noncoercive, constant, cumulative, effective agency in evolution.[130] Furthermore, Wallace's woman of the future would work her social change in a very traditional feminine way, through monogamous marriage (whether legally sanctioned or not) and childbirth; by breeding a better race. Her dominant role in choice of partner was not related to any active sexuality on her part, but was accorded her because of her economically and socially enabled ability to reject those suitors who did not come up to her exacting standards and to exercise her superior moral and spiritual judgment in saying yes to the right one.

It was an interpretation of female choice that had no relevance to abstract notions of beauty, to the nonutilitarian aesthetics that imbued Darwin's sexual selection; and it had absolutely no bearing on Darwin's central concern, the divergence of the human races. While Wallace was writing "Human

Selection," he read and discussed the proofs of the *History of Human Marriage* with its author, the Finnish anthropologist Edward Westermarck, offering clarifying suggestions and "alterations." Westermarck's influential work was published in 1891, with a laudatory preface by Wallace. Adopting many of Wallace's arguments, he concluded: "Sexual selection of lower animals is entirely subordinate to the great law of the survival of the fittest." When it came to humans, Westermarck agreed with Darwin that "every race has indeed its own standard of beauty," but he inverted Darwin's argument to assert that the "different standards of beauty are due to racial differences." This meant that "physical beauty is therefore in every respect the outward manifestation of physical perfection, and the development of the instinct which prefers beauty to ugliness, healthiness to disease, is evidently within the power of natural selection."[131]

By the early 1890s, Wallace had covered all the bases. His continuing campaign against sexual selection as a viable evolutionary mechanism among animals and in the formation of the human races was capped by his appropriation and limitation of female choice to biological socialism. Wallace himself considered his principle of sexual selection under socialism "the most important contribution I have made to the science of sociology and the cause of human progress." It generated intense interest—particularly across the Atlantic, where his highly successful American tour (1886–87) had opened up a network of like-minded contacts and supporters and consolidated his international reputation. Wallace's essay was republished in *Popular Science Monthly* (1890) and reprinted in various other journals and collections. He followed it up two years later with an article for the Boston *Arena* that stressed the great importance for the future of society of giving women "real choice in marriage"; he was interviewed at length on the same topic for the English *Daily Chronicle*.[132] With Wallace's imprimatur, the notion of biologically improving free female choice was given enhanced validity in the close integration of late nineteenth-century feminism with eugenics: for the more didactic, the "question of eugenics" was "one with the woman question." In turn, as his feminism and socialism intensified, Wallace elevated the socialist women of the future beyond the mere equals of men into their "superiors." They would have full political and social rights and hold the balance of responsibility and power. They would be the "regenerators of the entire human race."[133]

14.8 The Woman of the Future meets Ethical Man

It was the aging, ailing Huxley who determined to rein in the Darwinian maverick. His famous "Evolution and Ethics," delivered in 1893 as the second of the lectures established by the wealthy Romanes at Oxford, was long understood as Huxley's celebrated stand against the laissez faire "oughts" of the evolutionary ethics of Spencer and the new-christened "Social Darwinists."[134] However, this time-honored interpretation has undergone considerable historical revision. In this revised interpretation, his Romanes Lecture was a "masterpiece of concealed debate" in which Huxley indirectly attacked Wallace and defended established social interests against the socialist menace by reinvoking those very tooth-and-claw processes he had earlier outlawed as the basis of an ethical code.[135] I want here to restore what these revisions overlook: the gender and sexual dimensions of the Romanes Lecture and its implications for sexual selection. For Huxley's "Evolution and Ethics" was as much concerned with contemporary sexual politics—with what the younger Huxley had denigrated as the "new woman worship"—as with politics at large.

Huxley was professionally opposed to the Spencerian laissez-faire that would deny the essential government backing for the scientific enterprise and technical education that he promoted, but at the same time, he was increasingly alarmed by the growth of socialism and the proliferation of various radicalisms that sought to appropriate or modify Darwinism for radical political ends. He was also old and depressive, marked by personal tragedy, the madness and death of his beloved artist daughter Marian (Mady).[136] His growing conservatism and the privations of the Great Depression of the eighties and nineties had put him out of step with the working-class audience he once claimed as his own. London riots and strikes (Besant, at her most radical, famously led the great Bryant and May match girls' strike of 1888), trade unionism, Gladstone's divisive Home Rule Bill, socialism, anarchism, land nationalization—all threatened the social stability that Huxley, the scientific bureaucrat, saw as essential to the scientific reformation of society. At stake was that scientific naturalism with Darwinism at its core that Huxley had spent the greater part of his professional life successfully promoting against an entrenched ecclesiastical opposition, and which he now sought to reclaim from its political hijackers of both right and left persuasions.

It was not just the socialist specter that haunted Huxley in the 1880s and '90s, but the equally dangerous New Woman in her more radical persona. She wanted a new ethical code, a new set of relations between men and women, an end to a whole array of inequities and moral and economic double standards. She threatened the old agreed lines of gender and sexuality. She evoked that "laxity of morals" Huxley had earlier censured in Besant and the radical secularists. Her more recent embodiments, with their evolutionary-sanctioned female choice, jeopardized "decent" Darwinism and all it stood for.[137] Wallace's version might not tout an offensive birth control and free love, but his socialist woman of the future was a double affront to Huxley, for whom it was inconceivable that women, socialist or not, could or should lead men (though they might be allowed into the classroom). Wallace was that worst kind of fanatical "philogynist" Huxley had inveighed against in 1865,

> who bid the man look upon the woman as the higher type of humanity; who ask us to regard the female intellect as the clear and the quicker, if not the stronger; who desire us to look up to the feminine moral sense as the purer and the nobler; and bid man abdicate his usurped sovereignty over Nature in favour of the female line.[138]

It was as if the woman "emancipated" by Huxley, endorsed by Darwin in the *Descent*, and ratified by Romanes had shrugged off the handicap of her evolutionary past, rejected the proffered liberal apple, and bolted off to a socialist future on the back of a biologically and morally suspect female choice.

Huxley was too seasoned a campaigner to confront Wallace directly, as Romanes had done. Besides, Romanes himself, now one of the Oxford establishment, was insistent that the lecture series funded in his name should avoid political or religious issues. Huxley's solution was to "peel ethics away from evolution." This allowed him to retain the competitive nature that he "needed [both] to undercut the socialist legions" and to "legitimate a meritocratic order," and at the same time, by setting ethical man in continuous struggle against ruthless amoral nature, to "neutralize nature's restraints to allow the technocratic transformation of society."[139] With the one hand, Huxley identified social progress with the "checking of the cosmic [evolutionary] process at every step"; with the other, he reintroduced the "Malthusian serpent" of overpopulation as the natural and inevitable check to the socialist "garden of Eden." In the face of the socialist threat, he urged the abandonment of Spencerian Social Darwinism and the adoption of more humane policies toward those who labored; but, Huxley warned, the result of the elimination of all social competition would be "unrestrained multiplication," which would force either artificial selection and tyranny, or scarcity and a return to struggle. Struggle indeed was necessary and ultimately beneficial, and Huxley deplored those "artificial arrangements" that impeded social struggle, because struggle ensured the dominance of that "exceptionally endowed minority" with whom Huxley identified.[140]

The "artificial arrangements" that most attracted Huxley's censure were those promoted by those who "contemplate the active or passive extirpation of the weak, the unfortunate, and the superfluous." Galton, Wallace, Allen, and the more extreme eugenicists were targeted indiscriminately as one, as promiscuous advocates of a "pigeon-fancier's polity" where, by some miraculous attainment of wisdom (a direct thrust at Wallace's spiritualist convictions), the pigeons can get to be "their own Sir John Sebright." This "collective despotism" by those would-be selectionists ("on whose matrimonial undertakings the principles of the stud have the chief influence") was unattainable "without a serious weakening, it may be the destruction, of the bonds which hold society together." For Huxley, and those who shared his prejudices and politics, eugenics by the 1890s had become "simply a synonym for sexual radicalism."[141]

Fortunately, asserted Huxley, the Malthusian inevitability of overpopulation would undermine the selective efforts of these self-appointed "saviours of society" who would disorder conventional sexual relationships, disrupt family bonds, and destroy society. It would ensure the competitive success of those it "must be obvious to everyone" should look after the interests of the family and society. These were, as Huxley defined them, those "endowed with the largest share of energy, of industry, of intellectual capacity, of tenacity of purpose, while they are not devoid of sympathetic humanity." In his Tennyson-inspired conclusion to "Evolution and Ethics," Huxley invoked the code of Victorian manliness, exhorting his meritocratic troops to individual moral and psychological effort against the conjoined seductions of socialism, sexual radicalism, and utopian eugenics: "We are grown men, and must play the man 'strong in will / To strive, to seek, to find, and not to yield.'"[142]

It was a reassertion of the superiority, inevitability, and naturalness of those timeworn masculine qualities that Huxley and his audience identified with leadership and social power, and it defined women out of any capacity for or share in that power. It was a resurgence of the liberal paternalist tactics that had served Huxley so well over the years. Biology could still ensure male supremacy in a competitive Darwinian universe.[143] It sought to turn the tables on Wallace by reinstating the Darwinian struggle for existence as essential to human progress, while denying biological and moral legitimacy to those applications of artificial selection and female choice that underlay Wallace's advocacy of woman's powerful and responsible role in a socialist society. It counterpoised ethical man to Wallace's morally superior, socialist woman of the future. It promoted an ethical code based firmly in traditional Victorian domestic ideology and conventional family pieties—one in which a patriarchal Huxley defended the social virtues of self-restraint, duty, and propriety against the incursions of those whose activities threatened to weaken the bonds of the family and therefore the fabric of society, to bring about what Darwin had so greatly feared—"the greatest of all possible evils to mankind."

Huxley's "egg-dance" of "Evolution and Ethics" was his last great effort in defense of the Darwinism he had done most to shape and codify—he died within two years—but it was too little and too late. It was essentially an "essay about limitation and failure . . . the failure of simple naturalism to light the way through the gathering social conflicts of the Victorian twilight," and Huxley, as he conceded, had only old "apples" to offer ("mostly of the crab sort").[144] By the last decade of the nineteenth century, Darwinism was in decline, natural and sexual selection were in crisis, and female choice was on the loose, dogged by its associations with radical New Women, scandalous free love, secularism, and socialism. The old certainties and values were in flux in a fin de siècle flaunting of preening men with flowers, knickerbocker-wearing women on bicycles, a sequence of sensational sex scandals, the emergence of a controversial new science of "sexology," gathering eugenic enthusiasms, and a renewed outburst of table rapping and spiritual encounters. Besant, weary of heartless Darwinism, atheism, and the sexual and social struggle, went "head over heels" into Theosophy, renounced neo-Malthusianism and earthly love, and opted for a kind of spiritual evolutionary progress via successive reincarnations. The old-guard Darwinians had gone or were about to go. Wallace died on the eve of the Great War (1913), still arguing indefatigably that the only selective agency capable of effecting permanent and progressive social and moral change was the principle of female choice under socialism.[145]

A century had passed since the daydreaming Melissa was offered the only "Female Choice" then thought possible, between the decorative but devious "Dissipation" and plain-clad, but beloved and useful, "Housewifery." A new vision of women's emancipation that prefigured a world able to grant women as well as men the liberal ideals of individual autonomy, choice, and meritocratic advancement had supplanted Melissa's; but still the choice for women lay between love and freedom.

Schreiner's allegory, "Life's Gifts," summed up the situation of the New Woman who had dreamed, "in eloquent Darwinian metaphors," of a sexual utopia in which both sexes would achieve their full evolutionary potential. "Life" offers the dreaming woman the gift of love or freedom. The woman ponders long and chooses "Freedom!"

And Life said, "Thou hast well chosen. If thou hadst said, 'Love,' I would have given thee what thou didst ask for, and I would have gone from thee, and returned to thee no more. Now, the day will come when I shall return. In that day I shall bear both gifts in one hand."

I heard the woman laugh in her sleep.[146]

Last Words

I remain firmly convinced of its truth.

—Darwin, 1882

Historians of science have fought hard to escape the constrictions of "great man" history by opening up conventional narratives of "discovery" and theory construction to competing players and alternative interpretations, giving them context and contingency, decentering the heroic author, and allowing play for other voices, practitioners, and readings in the construction of knowledge and its representations and appropriations; and rightly so. Sexual selection could never have been made without the ideas, practices, and labor, the intersecting networks of relations, of the many discussed in these pages. Its conflicted history bears witness to its context of struggles for authority over its definition and interpretation. Yet Darwin's centrality to its making is inescapable. It was Darwin who netted the multiple strands of sexual selection together, who identified patterns of connections between disparate areas and ideas, who correlated them with events, attitudes, and evidences from the biological and social worlds, imposing unity and continuity as best he could on a complexity of sources, notions, theories, and collated data. The trajectory of sexual selection was shaped by many thinkers and practitioners, by professional and institutional power plays and the larger issues of the day; but at every point, it bears the unmistakable impress of Darwin's individual experiences, cultural and social values, and intellectual commitments. It was Darwin, above all, who held fast to its "truth," who brought his great fame and his name to its promotion and defense against its many critics and detractors, whose self-declared "conviction of the power of sexual selection" remained "unshaken" unto death.

From the early stages of theory building, Darwin followed a particular guiding strand: his belief that the different human races had different inborn, heritable standards of beauty or aesthetic taste, the foundation of his theory of the role of aesthetic preference in the formation of the human races—a theory that Darwin extended to the whole animal kingdom and reconstituted through the practices of animal breeders. Crucial to this was Darwin's lived experience of savage encounter. The shock of the savage reverberates throughout the making of sexual selection: from Darwin's early adoption of an aesthetic theory of racial differences and origins that rationalized and made essential the moral and physical "ugliness" of the "savage" brother, through all the complexities and shifts of the connection of this aesthetic with other themes and intellectual concepts, its negotiation and reshaping under social, institutional, and peer pressures, onto the full-blown elaboration of sexual selection in the *Descent*.

The other component of sexual selection came from closer to hand. From the erotic evolutionary writings of his grandfather Erasmus, Darwin took the notion of sexual combat as a means of species benefit and as an explanation for the development of specialized male weapons. Darwin's early notebook theorizing on the sexual struggle merged into—likely primed him for—his earliest formulation of natural selection. Having disentangled the struggle for mates from the struggle for existence, Darwin then had to work through and develop the implications of his new theory of natural selection for what was to become the "law of battle" and to integrate both with his earlier views on adaptation, inheritance, and embryogenesis. This was achieved by the time of his sketch of 1842.

Female choice was not properly formulated until around 1856–58, when Darwin began to bind the composite strands of sexual selection together as the "secondary" principle of his theory of evolution. Yet fundamental to this principle was the essential Erasmus-derived understanding of beauty as naturalistic, as habituated and nonutilitarian, and as constituted by sexuality, emotionality, and the mating urge. This understanding was further shaped by the strongly gendered conventions of Darwin's society and by the readings in empiricist aesthetics and early ethnology (which had tight associations with physiognomy and aesthetics) that he adapted to the demands of his evolutionary theorizing.

The names line up like beads on a string: Reynolds, Burke, Alison, and Jeffrey, on whose various aesthetic writings Darwin selectively drew; and Stanhope Smith, Prichard, Lawrence, Blyth, and Walker, who all fostered Darwin's awareness of the potent analogy between artificial and aesthetic selection. Certain names stand out from the rest: Lawrence, whose pirated *Lectures* offered a uniquely materialist, monogenetic theory of racial origins, developed by analogy with domestication, that located mental and moral differences in anatomy and reinstated Stanhope Smith's original thesis of the role of aesthetic selection in the formation and stabilization of the human races; Blyth, who played the vital role of intermediary, reinforcing the notion of male combat and its role in reproductive success and integrating and bringing Lawrence's views to Darwin's notice at two critical stages of theory formation; and above all, Walker, whose *Intermarriage* placed explicitly before Darwin the reiterated comparison of wife choice with the selective procedures of animal improvers, and who made the masculine judgment of woman's beauty—a visually based, aesthetically informed process of "choice"—the agency of human progress or racial improvement. At the same time, Walker's explicit denial of sexual choice to human females reinforced Darwin's own conventional views on women's sexuality and essential passivity in any account he might render of human evolution or racial differentiation.

A key stage in the process of theory formation was Darwin's own immersion in the practices of animal breeders and the reconstitution of the notion of aesthetic preference by artificial selection. Faced with the looming problematic of useless or potentially harmful male beauty in birds that required explanation that could not be made consistent with his concept of natural selection or with his earlier views on inherited habit, Darwin shifted from an early inclination toward the dominant male role in any process of aesthetic selection to the realization that—in animals, at least—females must play the role of selector. He had to go against entrenched opinion in attributing not just a sense of beauty to animals, but further to insist that this aesthetic was primarily exercised through female sexual preference. He experienced real difficulties in eliciting confirmatory evidence from animal breeders whose lore and breeding practices ran contrary to the very notion of female choice. His own thoroughgoing identification of the process of aesthetic choice with the discriminating, artistic eye of the breeder was a stumbling block to the notion of analogous female choice in nature. Female choice, already a challenging concept for a Victorian, was made doubly so for Darwin, confronting the seeming contradiction of extending the masculine, manipulative art of breeding to the sexual preferences of female animals.

What enabled this transfer was Darwin's exposure to the perceived vagaries and excesses of the dress choices of Victorian women and his consistent recourse to the metaphor of fashion. His formulation of the threefold analogue of the selective practices of déclassé pigeon fanciers, of the frivolous dress choices of fashion-conscious women, and of the aesthetic choices of female birds—all of whom capriciously pushed their selections to nonfunctional extremes—was fundamental to Darwin's assertion of the efficacy of female choice among animals. Embryology was of particular importance in this.

Put simply, Darwin modified and extended Erasmus Darwin's embryological model for evolution. The general principle was substantiated in Darwin's eyes by his barnacle and pigeon researches and by his reinterpretations of more recent work in embryology, notably the creationist Agassiz's assertion of the parallel between the phases of embryonic development and the fossil sequence. Overriding Huxley's dissent, Darwin co-opted Agassiz's embryological authority to his own conviction that the sequence of embryological stages was the key to both classification and ancestry and that all were progressively divergent. Essential to this was Darwin's thesis that the separate sexes had evolved from ancestral hermaphrodites; that earlier embryonic stages represented ancient, ancestral adults, the progenitors of current species; and that species-altering variations generally occurred in mature organisms and were inherited at a correspondingly late stage of development—that they were "added" on to the end stages of development. His most compelling case for the terminal-stage addition of variations to embryonic development was that of sexually dimorphic species, particularly birds, where the resemblance between the young and females meant that the "masculine character" was "added" to species. In those cases where females and young were alike, where males were more ornamented than females, where such ornamentation had no connection with male combat and no utility in the struggle for existence, then, logically, it had to be the females who were doing the choosing and the "adding up" of mature masculine characteristics, a process repeated or "condensed" in embryonic development. Embryology, therefore, underpinned the common pattern underlying Darwin's three main evolutionary agencies—natural selection, divergence, and sexual selection: that of the breeder who favored the "most extreme forms" to make the divergent breeds of pigeon. Of these agencies, female choice was the most closely aligned with the breeder's selective practices. And it was Darwin's recently articulated analogy between pigeon breeding and the "extremes of fashion" that enabled this critical process of theoretical integration while Darwin was at work on his "big species book."

Its context was one of mounting racial and sexual tension in which a heightened sense of national identity reinforced the racial and gender superiority of the white middle-class male. As perceptions of racial and sexual difference intensified, embryology took on new political meaning. Gender and ethnicity met in the embryo in the pervasive equation of woman-as-child-as-primitive. This was interlinked with contemporary social, aesthetic, racial, and anthropological hierarchies and authenticated within the scientific racism that supplanted earlier environmentalist explanations of racial difference. Darwin drew on the writings of the new race determinists in excluding climate and habits of life as determining factors and reaffirming his earlier Lawrence-derived thesis of an aesthetic factor in the differentiation of gender, class, and race. The prominence accorded innate racial antagonism and antipathy by the radical anatomist and race theorist Robert Knox, endorsed by the polygenists Nott and Gliddon, concurred with Darwin's long-term assumption (for which he adduced support from pigeon breeders) of racial or varietal repugnance to interbreeding, which would prevent back-breeding and so permit the emergence of new species. Knox also emphasized the embryonic, or "generic," similarity of the divergent races and followed Walker in prioritizing external, visible form in the making of aesthetic judgments. His promotion of a racial hierarchy of beauty centered on woman's face and form meshed with Darwin's renewed conviction of the essential role of male aesthetic selection in the formation of race. Darwin's neologism "sexual selection" in the notes he made on Knox's *Races of Men* thus connoted the aesthetically determined choice of partners of the same race and the innate antipathy to interracial mating that would allow the long-term persistence of races insisted on by both the monogenist Knox and the polygenists. At this stage (1856) Darwin still allowed some role for natural selection, through resistance to disease. It was primarily the embryological argument that decided Darwin in late 1858, while writing the *Origin of Species*, that skin color in humans was, like the color of bird plumage, solely due to sexual selection. His axiom of plumage difference between bright-hued males and dull females and young rehabilitated Darwin's long-held view that women were more childlike than men and that this represented an earlier or more primitive stage of evolution, a stage from which the different races had diverged through aesthetically influenced sexual choice. The way was cleared for his *Origin* declaration on the determining role of sexual selection "of a particular kind" in the origin of the different human races.

The centrality of Darwin's interpretations of embryology and inheritance to his construction of sexual selection and their tight interdependency with his explanations of human sexual and racial divergence cannot be overemphasized. It was his unshakable faith in this developmentally centered understanding that drove Darwin's dispute with Wallace over sexual selection, that gave him the assurance to confront Mill's views on women's intelligence and capacities and, before all, gave Darwin the necessary confidence and self-belief to write the *Descent of Man*. This was the irreducible core, the raison d'être of the *Descent*.

Darwin's final assembly of the components of sexual selection and human sexual and racial divergence was shaped by his differences with Wallace, by the Huxley-led drive for Darwinian cognitive and social authority, and by institutional conflict over control of the politically sensitive science of "man"—personal and professional concerns and conflicts that refracted the larger dissonances of British imperialism and major industrial and social change.

The *Descent* was the completion of Darwin's long journey from his explicit identification of the manipulative pigeon breeder with the sexually selecting human male, on to its close coupling with an instinctive, efficacious appreciation of the beautiful by female animals and birds. Sexual selection, as presented in the *Descent*, dovetailed with Darwin's central explanatory fashion motif, with his theories of embryology and inheritance and his theory of pangenesis. It was instantiated with an enormous range of examples of sexual difference, ornamentation, courtship, and display, from crustaceans to humans, and by Darwin's detailed reconstructions of the evolution of exquisite ocelli and other beautiful patternings of feathers and fur.

It was a "tremendous job," as Darwin himself said, an enormous achievement. Yet he never managed to win its thoroughgoing acceptance, not even among his closest supporters. The reasons for this have emerged in the course of this account. They are implicit in its history. To begin with, sexual selection was, as Wallace intimated at the height of their conflict, "*altogether*" Darwin's "own subject." Darwin worked to keep it so, jealously guarding his exclusive property rights against incursions by Wallace or other interested parties. Those like Tegetmeier, Blyth, Bates, and Weir, who were recruited to the theoretical or evidential support of sexual selection, were in their various ways recipients of Darwin's patronage and not considered colleagues or theoretical equals. Tegetmeier's dissent from female choice was dismissed, even as he was pressed to its experimental demonstration; Weir was a pushover, flattered, and cajoled into an uncritical, hard-working disciple; what began as "Blyth's laws" were revised as Darwin's; the attempt by Bates to reconcile Darwin's and Wallace's views was rebuffed. Sexual selection had no other major stakeholders, no influential allies to fight for it against its many critics, to shore up those of its parts deemed shaky. It was very much Darwin's "owned" creation, an interlaced matrix of long-standing analogy, practice, and theory—certain aspects of which were out of touch with contemporary views or in dispute even as the *Descent* went to press. Its only voice was Darwin's. It was a celebrated voice, but, while he lived, Darwin struggled both to hold his creation together and to retain control of its component parts as it threatened to collapse under the weight of its own contradictions.

Chief among these was his naturalization of female choice in animals and his normalization of male choice among humans. This created unresolved tensions in his mature theory of sexual selection and threatened the very continuity between animals and humans that it was theorized to support. It has to be said that Darwin gave his critics plenty to work with. His paradoxical insistence on the immutability of male passion and female sexual coyness or virtue as the necessary preconditions both for female choice in animals and male choice in humans was at odds with his culturally inflected binaries of masculine rationality and feminine intuition, and it opened female choice to interpretations subversive of Darwin's own applications. It also brought him into conflict with the consensus on primitive promiscuity among contemporary anthropologists and marriage theorists and undercut Darwin's own narrative of human moral and mental progress. Darwin won virtually no support for his central explanation of human racial differentiation through aesthetically determined sexual choice. For Darwin, this was the crux of the issue, the fundamental argument of the *Descent*. Yet his own denigration of savage aesthetics and morality, his elevation of an ape ancestry over the savage forebear who "knows no decency," his assumption of the supremacy of the white European ideal of beauty, the language that played on his own and his readers' prejudices—all undermined his claims for the race-making power of aesthetic choice and contributed to its repudiation.

Adding to these problems, Darwin's transference of the discriminating artistic eye of a skilled breeder to female birds carried with it the attribution of a very high level of aesthetic taste to mere animals. It was indeed, as Darwin had foreseen in his 1864 letter to Wallace, an "awful stretcher" to which few could stretch. His focus on explicating the smallest details and subtleties of plumage and the high-flown language that invoked comparison with the art of Raphael made female choice vulnerable to criticism and negated his sporadic attempts to counteract this impression. His associated fashion metaphor was of immense help in theory building and in communicating his notion of female choice to his readers; but it too brought difficulties. How could fashion—notoriously fleeting, fickle, and frivolous—generate such sustained multigenerational selection as female choice required? It provoked disparaging and damaging critiques from those major detractors of sexual selection who refused to extend an aesthetic sense to animals.

Nevertheless, an animal aesthetics was vital to Darwin's defense of his theory of evolution against the established creationist view that the appreciation of beauty, like morality and intelligence, was God-given and unique to humans. Before all, beauty and its appreciation required naturalistic explanation that Darwin's other evolutionary agencies could not provide. For Darwin, beauty was, in the tradition of empiricist aesthetics that had shaped it, essentially nonutilitarian, and this helps to explain his resistance to linking it consistently with notions of adaptive "benefit," as Wallace sought to do. Sexual selection explained what natural selection could not: the evolution of beautiful, otherwise useless, even potentially harmful, structures and colors. It was this that distinguished the struggle for mates from the struggle for existence—the most distinctive aspect of his theorizing—and Darwin refused their integration or conflation. Yet, in the theoretical section of the *Descent*, Darwin himself contributed to this by his invocation of an element of fitness in aesthetic selection in order to overcome the problems posed by equal sex ratios in sexually dimorphic species. It was a move that, in ways unacknowledged by Darwin, complicated and compromised the distinctions he otherwise insisted on between sexual selection and natural selection. And it weakened his central explanatory metaphor of the breeder who selects to extremes without regard to utility.

Ultimately, sexual selection, so "intimately connected with the limits and efficacy of natural selection," was dependent on natural selection.[1] It was "secondary" to the primary principle, and stood or fell along with it. By the end of the century, this was by far the greatest challenge facing sexual selection.

With its illustrious figurehead gone, Darwinism went into long, slow decline, not to be retrieved until well into the next century. The last two decades of the nineteenth century constituted a particularly intense period of conflict, riven by theoretical, disciplinary, ideological, and political differences over the nature of the evolutionary process, which drove Darwinism, famously, into "eclipse." Natural selection came under hostile assault by assorted neo-Lamarckian evolutionists who rejected the ultra-selectionist stand of "neo-Darwinians" like Wallace and Weismann. Weismann's theory of the germ plasm (elaborated in the 1880s) insisted that there were two separate cell lines in organisms—the soma, or body cells, and the germ plasm that carried the hereditary material—and that changes in the soma could not influence or be transmitted to the germ line. Consequently, acquired characters could not be inherited. Lamarckism was impossible. Weismann put this to classic experiment by snipping off the tails of successive generations of mice that interbred without producing tailless offspring. The germ plasm theory divided those like Romanes, Spencer, and Haeckel, who allowed some role for the inheritance of acquired characters, from strict neo-Darwinians like E. Ray Lankester, Weismann, and Wallace, who proclaimed the "all sufficiency of natural selection." It generated a dogmatic, increasingly vocal neo-Lamarckian movement with a militant anti-selectionist agenda, culminating in the American school of "orthogenesis" led by the paleontologists Edward Drinker Cope and Alpheus Hyatt.[2]

In effect, the neo-Lamarckian onslaught united the neo-Darwinians in the defense of natural selection. While a few continued to research and promote sexual selection and female choice (notably the American biologists Elizabeth and George Peckham and—initially, at any rate—E. B. Poulton and C. Lloyd Morgan), for the majority, even if they were not persuaded by Wallace's vigor theory, sexual selection became "too contentious, too theoretical, and too easily attacked." By subsuming it within natural selection, it became easier to defend. Sexual selection might be reduced to male-male competition, and the controversial notion of an animal aesthetic sense—especially when exercised by females—might be forgone.[3]

For their part, the neo-Lamarckians, in attacking natural selection, targeted those nonadaptive structures that could not be explained on utilitarian principles, singling out gaudy colors and useless ornamentation, rejecting both female choice and Wallace's alternative interpretation, and arguing for the evolution of sexually dimorphic characters through a combination of inheritance of acquired characters and inner "laws of growth" or overgrowth.[4] Embryogenesis had long offered evolutionary explanation independently of natural and sexual selection and lent itself well to the inheritance of acquired characters through terminal addition to ontogeny. The late-century dominance of Haeckel's recapitulation theory (which formularized the claim that individual growth recapitulates the evolutionary history of the species), in eliciting hypothetical ancestries through ontogenetic analysis, strongly reinforced this neo-Lamarckian trend. It is arguable that Darwin's own emphasis on embryogenesis and terminal-stage addition in explicating sexual dimorphism had the perverse effect of rendering sexual selection redundant as explanation. Orthogenesis, the "more damaging alternative to Darwinism," was predicated on recapitulation theory and supposed that a pattern of development was imprinted on the individual organism so that it could only vary in a certain direction. It assumed nonadaptive parallel evolution, not Darwinian adaptive divergence. It offered an explanation of sexually dimorphic characters (such as the massive antlers of the Irish elk) as the result of "overdevelopment" or orthogenetic "trends," according to which the evolutionary growth of certain once-useful organs gained momentum that drove them in particular directions, regardless of adaptive advantage, even to extinction.[5]

By the turn of the century, sexual selection was all but dead. In the eyes of most evolutionary biologists—even those sympathetic to Darwinism—it was "practically discredited." In 1903, the Mendelian biologist Thomas Hunt Morgan, renowned for his experimental work in fruit fly genetics and sex-linked inheritance, subjected sexual selection to a "blistering critique," concluding: "The theory meets with fatal objections at every turn."[6]

These disputes over the existence and efficacy of sexual selection were never without ideological or political content and import. Sexual selection, inextricable from the politics of race, sexuality, and gender, had built into it the potential for providing answers to many of the perceived problems and challenges thrown up by the governance of empire, homegrown threats from labor unrest, socialism, Irish Home Rule, the suffragettes, and, last but not least, the proliferation of the "unfit." Darwin showed the way in the *Descent* with his promotion of a racial hierarchy, his demonization of the savage, his explicit justification of feminine domesticity and intellectual inferiority, and hints toward racial improvement through better informed choice of marriage partners. Female choice, as Darwin conceived and applied it, and as his fashion metaphor entailed, was delimited to the unconscious selection of the frivolous, fashionable, and unfit, even to the deleterious and dangerous. It was tailored to the ethos of the self-made man, who, through conscious, rational mate selection and competitive advance, would ensure future racial progress.[7] The trouble was that female choice's embedded conventional objectification of passive female sexuality, its implicit naturalization of the Victorian double standard, its close associations with the "good woman" of Victorian domestic ideology—cliché for woman's moral superiority and lesser sexual passion—invited its appropriation by those intent on investing women with eugenic agency on the understanding that only they were capable of making socially responsible, scientifically justifiable, morally pure sexual choices. Its more radical expropriators linked female choice with birth control, free love, and even worse, socialism, threatening Darwinian respectability and its growing political conservatism, inciting backlash from Huxley and further repudiation of sexual selection as a viable evolutionary mechanism.

It has been suggested that Darwinian sexual selection also offered naturalistic support for the "self-primping Dandy" of the aesthetic movement and its "semi-covert dissident sexual politics," an interpretation that also proffers explanation for the widespread rejection of sexual selection.[8] There are innuendoes in the literature that the strictly heteronormative courtship narratives of the *Descent* (with which Darwin, with his focus on sexual reproduction, was entirely preoccupied—there is no mention of homosexuality in Darwin's writings) might be so interpreted. Certainly, conservative critics of evolutionary science were keen to exploit any such associations, however obliquely they might be conveyed. And, for scientific naturalists like Huxley, concerned with maintaining a reputation for respectability and moral probity, the aesthetic movement—with its supposed overweening sensuality, deviance, and indecency—provoked strategic efforts to dissociate evolutionary science from its taint. Sexual selection may have helped to make sex sayable, but its name could be spoken only in circumspect and circumscribed ways.[9] The overt linkages of sexual selection with heterosexual free love, with birth control, eugenics, and socialism were warrant enough for censure without the need to invoke unspeakable homosexuality. More generally, it was its connotations of greater freedom, independence, and agency for women in the context of the rise of the women's movement and the emergence of the New Woman of the late nineteenth century that represented the greater challenge to the acceptance of sexual selection.

A striking feature of this period is the extent to which non-Darwinian theories of evolution were deployed in contesting women's higher education and the suffrage and reasserting female inferiority. As sexual selection went into abeyance, there was little inclination to draw on the anti-Mill arguments of the *Descent* in which Darwin had invested such confidence. Only Romanes tried to retrieve and extend the antifeminist potential of sexual selection, triggering Wallace's counterview. The role of Geddes and Thomson's alternative thesis of passive anabolic women and passionate katabolic men in reinforcing gender and sexual stereotypes has been discussed. But the most influential interpretation was promoted well before Darwin's death by the omnipresent Herbert Spencer. Despite Darwin's attempt to call him to account, Spencer shunted along his own intellectual railway tracks, finding plenty of biological justification for feminine intellectual and physical inferiority in his purpose-built integration of von Baerian individuation, the physiological division of labor, the conservation of energy, survival of the fittest, and the inheritance of acquired characters. Woman's "somewhat earlier arrest of individual evolution" enabled the conservation of her energies for reproduction, but at the expense of her physical and mental maturation. Both he and Darwin were committed to a fundamental embryological understanding of sexual difference, but, as Geddes and Thomson succinctly put it, in Spencer's view, "woman was underdeveloped man," whereas for Darwin, "man was evolved woman."[10]

Spencer's views were highly influential among American physicians and educationists, notably Edward H. Clarke (1873), who condemned those "mannish maidens" whose scholarship was achieved at the expense of their ovaries and whose marriages were sterile.[11] In England, they were adopted by Henry Maudsley, pivotal in the development of an evolutionary science of mind. Darwin had sought to recruit Maudsley's support for his argument for the evolution of female inferiority in the second edition of the *Descent*.[11] However, in the popular presentation of his views on "Sex in Mind and Education" in the *Fortnightly Review* (1874), Maudsley ignored this inducement, citing cumulative evidence offered by Clarke and others of the baneful effects of "excessive" education on the health and reproductive systems of young women. Women were "marked out by nature for very different offices in life from those of men." They had a clear biological and social duty to marry and produce healthy children; their own education should fit them for that role: "There is sex in mind and there should be sex in education."[13] Like Spencer's, Maudsley's interpretation made sexual selection and the theories of inheritance it embodied irrelevant to the divisive woman question.

Orthogenesis, the "most extreme form of anti-Darwinism," also lent itself to opinion on female inferiority and subordination. Hyatt's "degeneration" version theorized that, as the pattern of individual growth ended in decline and death, so too must each group of organisms degenerate (becoming more simplified) and eventually die out. To give women equal education and political rights would diminish the essential differences between the sexes and precipitate retrogression and degeneration. Hyatt's claim was predicated on Vogt's old thesis, taken up by Darwin, that increasing sexual differentiation was a sign of advancing civilization and evolutionary progress. If men and women were to adopt similar habits, this would result in effeminate men and "virified" women; the sexes would be driven back to their primitive hermaphroditic state, and racial degeneration would be accelerated. The only way to avoid such evolutionary retrogression was to impress on women the need to retain existing social arrangements that maintained essential sexual differences. Their demands for equality were biologically dangerous to the race.[14] Cope, the other major advocate of orthogenesis, who also had a great deal to say on women's unfitness for higher education and the vote, held to the less complicated thesis that women were less developed than men.[15]

At its simplest, the onslaught by non-Darwinian evolutionists on feminine intellectual and social parity is attributable to their rejection of Darwinian selectionist mechanisms and the perceived need to find other naturalistic explanations for masculine supremacy in a period when it was under strong challenge during the first phase, or "wave," of feminism; but it also implies resistance to the more radical implications of female choice. It wasn't merely, as has been claimed, that evolutionists resisted a biological concept that put "females in the driver's seat of evolutionary change,"[16] but rather because the notion of female choice lent itself far too readily to those who wished to claim a greater role for feminine will and judgment in the social and political arenas. The very term *female choice* (conferred on it by Wallace, not Darwin, it should be noted[17]), conjured up notions of feminine independence, freedom, and sexuality that threatened entrenched stereotypes and ran counter to Darwin's own applications and despite his maneuverings. The easier option was to disconnect sexual selection from its embryological substructure and revert to the fundamental woman-as-child-as-primitive equation that recapitulation theory also embodied and endorsed. The further attraction of Haeckel's formula that "ontogeny recapitulates phylogeny" was its well-documented usefulness in justifying racism and imperialism. Until its collapse in the 1920s, adherents of recapitulation theory, in Stephen Jay Gould's words, "collected reams of objective data all loudly proclaiming the same message: adult blacks, women, and lower-class whites are like upper-class male children."[18] Darwinian selection (both natural and sexual) was redundant to this dominant scientific validation of the supremacy of the white middle-class male.

It is telling that when, with the revival of interest in natural selection in the early part of the twentieth century, sexual selection again began to generate discussion, it was in close association with eugenics and women's rights. Both Ronald A. Fisher and Julian Huxley (grandson of Thomas Henry) played leading parts in the "evolutionary synthesis" of the 1930s and '40s, which integrated natural selection with Mendelian genetics. Their different interpretations of sexual selection were bound up with their divergent social and political agendas.

Fisher's thesis of "runaway" sexual selection (1915) was mooted during the First World War in connection with his eugenic commitments and concerns over the dysgenic implications of the war for the British middle class.[19] According to his "runaway" thesis (which Fisher set out fully in 1930 and is back in circulation today), females choose mates on the basis of "good points," or traits correlated with health and vigor. Over time, female preference for this trait sets up a feedback loop in which the female preference and the male trait evolve in tandem so as to "run away" from the control of natural selection. The initial benefit of the "point" is negated and the evolution of deleterious sexual ornaments and colors is explained. For Fisher, eugenics was dependent on the need to choose "good" mates and avoid the runaway effect. Postwar "racial repair" required the sexual preference of females for returning heroic soldiers who would marry and have large families to replenish the stock of courageous males. If women were to be distracted from such good eugenic choice by displays of wealth, the runaway effect would take over to generate increasing preference for ever-larger displays of wealth, leading to middle-class decay. Fisher therefore proposed a system of family allowances to encourage the production of greater numbers of high-quality, middle-class British offspring. Fisher's model of "sexual preference" (in which he was coached by Darwin's son Leonard, long-term mentor and president of the Eugenics Education Society) was also directed to maintaining conventional gender roles. In the ideal "eugenic marriage," men and women were chosen for their different kinds of "fitness"; that is, earning capacity in men and housekeeping in women. If women were selected for their earning ability, the runaway effect would ensure that they would soon lose the "finer female traits," the genetic ability to raise many children and manage the home, thus endangering social progress.[20]

Julian Huxley's early version of sexual selection, by contrast, initially was inspired by ideals of sexual equality and mutual choice in loving monogamous marriage. He advocated votes for women, "access to birth control, sexual satisfaction for women as well as men, and sexual pleasure for its own sake." His famous studies of the monogamous great crested grebe, which exhibits little sexual dimorphism and engages in elaborate courtship rituals, led to Huxley's concept of "mutual sexual selection," according to which characteristics are transferred from one sex to the other until the two sexes become similar. He took the line (first suggested by Mivart) that male display acts as a sexual stimulant that excites the female, making her emotionally and physically ready for mating. This meant that the problematic notion of female choice was delimited to an "intuitive, unreasoned, but none the less imperious [choice] . . . which reaches its highest stage of development in the intensely felt affinities of man and woman—in that condition known as 'falling in love.'" (1914).[21] However, by 1938 Huxley had retreated from this interpretation to claim that "it has now become clear that the hypothesis of female choice and of selection between rival males irrespective of general biological advantage is inapplicable to the great majority of display characters." Female choice was not the common phenomenon postulated by Darwin, but confined to a relatively few species practicing polygamy or with a high excess of males. It was maladaptive and dysfunctional, a "biological evil." Huxley repackaged sexual selection as "epigamic" selection, reducing it to species recognition, sexual excitation, and mating efficiency—in effect, to natural selection.[22]

Historians locate the primary reason for Huxley's minimization and marginalization of sexual selection in his privileging of monogamy as the role model for continuous, "harmonious," cooperative human progress. He was repelled by the mating behaviors of polygamous, pronouncedly dimorphic species like ducks that would mob and drown females in repeated mating attempts; it was "a painful and repulsive sight," indicative of "disharmony in the constitution of the species." Grebes, by contrast, had exemplary sex lives and domestic relations, based in cooperation, shared parenting, and mutual respect, celebrating their marital "joy" with beautiful and expressive "dances" that exhibited their emotional and mental similarity to humans.[23]

Both Fisher and Huxley, from their different perspectives, viewed female choice as potentially disabling and disruptive of progress, requiring control by the dominant principle: in Fisher's interpretation (which made little impact at the time), sooner or later runaway female choice must be curtailed by natural selection, while Huxley, more influentially, subsumed female choice within natural selection, giving it little room for maneuver. And there it remained, more or less, until its resurgence in the 1970s, in an explosion of interest that inflamed renewed controversy.

A complexity of reasons has been offered for this long hiatus: because of the lasting influence of Julian Huxley on evolutionary biology; because female choice invoked fears of anthropomorphism and had to be dissociated from the notion of aesthetic taste before it could gain acceptance; because of the dominance of laboratory-based physiology; because the evolutionary synthesis left out animal behavior or ethology; because sexual selection was too troublesome and distracted attention from the principal concern with reinstating natural selection as the main mechanism of evolutionary change; that it was subsumed under studies in population genetics and the mechanism of speciation; that its retrieval was the outcome of the resurgence of ethology in the form of sociobiology that came to prominence in the early 1970s; that it required new disciplinary interest in animal and human sexual behavior and an emphasis on individuation, or translation into modern terminology of "kin selection" and "selfish genes."[24]

The historian Erika Milam argues that the standard "eclipse narrative" is a "selective history," an artifact that served the interests of the new discipline of sociobiology as well as feminist purposes. Her analysis pays close attention to the ways in which sexual selection was conceptualized in the course of the twentieth century, its diverse research programs and methodologies and interrelated disciplinary struggles for prestige and funding. According to Milam, although most researchers contended that animals were not cognitively capable of rational choice-based aesthetic selection (an interpretation, it should be noted, that is at odds with Darwin's insistence on unconscious aesthetic selection), biologists' interest in female choice as a mechanism of speciation or animal diversity remained strong throughout the century. In the late 1960s and '70s, leading organismal biologists Lionel Trivers (1972), Edward O. Wilson (1975), and Richard Dawkins (1976) reinterpreted sexual selection in terms of rationalist individualist game theory. They promoted its study as a revolutionary way of explaining sex differences that avoided anthropomorphism and legitimated their overarching claims to explicating patterns of human social and sexual behavior through biological theory in the new sociobiology. The success of sociobiologists in defining sexual selection as an organismal field science structured an eclipse narrative that subsumed the earlier laboratory-based work of population geneticists and evolutionary biologists into natural selection until the "rebirth" of sexual selection in the 1970s. A further factor in the great upsurge of interest in sexual selection in this period was the highly visible involvement of a reinvigorated women's movement. The vehement reassertions by leading sociobiologists of the biological basis of indiscriminate male sexual aggression and choosy female sexual reticence and essential domesticity "seemed designed to quell claims of contemporary feminists that women deserved as much sexual and economic freedom as men." Second-wave feminists, reacting against a pervasive postwar "feminine mystique" that repressed women's sexuality and restricted their choices, strongly contested such claims, and they and other critics of "reductionist" sociobiology generated scientific controversy and enormous media and popular interest in sexual selection. By the 1980s, sociobiology had become the "dominant paradigm within which histories of female choice and sexual selection were written."[25]

Given the earlier history of sexual selection, one does not have to be in entire agreement with Milam's revisionist analysis to see the force of her thesis of the necessarily social and cultural components of its revival. Female choice did not come back forcefully onto the biological agenda until the sexual revolution of the 1960s, the availability of the contraceptive pill, a militant second-wave feminism, the abortion debates, and the studies of the female sexual response by Masters and Johnson (1966) signaled a new fascination with sex and sexuality and the beginnings of the erosion of conventional sexual and gender stereotypes.[26] Within this context, feminist critics and, perhaps more important, a growing number of women actively participating in traditionally male-dominated research fields were able to play a more significant role in resisting and reshaping conventional patriarchal evolutionary narratives. The conflict between sociobiologists and feminists became particularly acute in the 1980s, and gave rise to a number of countertheories of sexual selection and human evolution that gave females sexual agency and put them center stage. Since then debates and developments in the overlapping fields of evolutionary biology, anthropology, primatology, behavioral ecology, and developmental genetics have led to a broadening and reworking of mate choice and evolutionary themes that open up new interpretive possibilities and are more sensitive to culturally inflected stereotypes—racial, gendered, and sexual.[27]

This does not mean that these more nuanced reworkings escape disciplinary, methodological, or wider social and ideological pressures. As this and many other studies overwhelmingly demonstrate, all scientific concepts are context dependent and subject to constant ongoing negotiation and revision within and across their relevant research fields. While most biologists see the evolutionary role of male combat (or "intrasexual selection") as relatively unproblematic,[28] female choice (or "intersexual selection") remains in flux, conditional on the crisscrossing agendas of its advocates and disputants. There is no single agreed model of sexual selection, but a proliferation of models, many with conflicting assumptions. In some models, females choose mates for "good genes"; in others, such as in versions of Fisher's runaway effect, they do not. In some, sexual selection acts only on males; in others it acts on both males and females. But, in nearly all of these models, sexual selection is viewed as an aspect of natural selection. Where Darwin was convinced that animals, predominantly females, appreciate beauty as such, and exercise mating preferences on this aesthetic basis, contemporary theory presumes that they do not make an aesthetic judgment, but rather perceive "beauty," either directly or indirectly, as a signal of fitness or "good genes." Beauty is transposed into fitness, and sexual selection is thereby reduced to natural selection. Where Darwin insisted on a continuum of aesthetic appreciation from animals to humans, contemporary theory usually dissociates human aesthetic appreciation from any animal origins and delimits itself to examining how human aesthetic standards arose in the course of *human* evolution.[29]

With few exceptions, current theorists do not attribute human "racial" differences to the action of mate choice, as Darwin insisted, but to natural selection.[30] The standard view is that discernible ethnic differences in skin color, eye shape, and so on (and such differences are not considered as extreme or as distinctive as earlier theorists assumed) have evolved as adaptations to different environmental conditions as humans diverged from their common origin as *Homo sapiens* on the African savanna. Nevertheless, in speculative discussions of the evolution of human secondary sexual characteristics, many theorists still attribute human mate choice to human aesthetic standards or perceptions of female "beauty"; but unlike Darwin, they link the perception of beauty to reproductive fitness. Its evolutionary development in humans signals "youth, fertility and health." The aesthetic is again reduced to fitness criteria and so ultimately to natural selection. Moreover, most contemporary sociobiologists and evolutionary psychologists dispute Darwin's insistence on culturally or "racially" different standards of beauty and argue for a cross-cultural or "universal" standard of human beauty. Thus preferences concerning facial or bodily traits of the opposite sex (and predominantly the preferences of the male gaze on the female body), such as small and symmetrical facial features, the distribution of body fat (breasts and buttocks), or an "optimal" waist-to-hip ratio (the classic hourglass figure lives on in contemporary theory, even as it has lost its appeal in contemporary fashion) are argued to be reliable cues to the health and reproductive status of the individual female in all cultures surveyed.[31]

Perhaps most indicative of their distance from Darwin's interpretation of sexual selection is that its modern exponents make no use whatever of Darwin's fundamental analogy of artificial selection (nor indeed of his fashion metaphor), drawing their explanatory metaphors from contemporary political, marketing, or game theory. They invoke an "arms race" among competing males, or they view sexually selected variations as "innovations" and "diversifications"; they interpret courtship as a form of advertising, while assessing and analyzing "signals," "indicators," or "markers" of health and strength, or reproductive "investments," "costs," "incentives," and "resources," or "venture capital" and "profits"; and they would turn Darwin's principle of mate choice into "consumer preferences," mating "strategies," "bargains," "promises," "handicaps," "manipulations," and "trade-offs."[32] Since Darwin's time, the practices of horticultural and animal improvement, once so economically and socially significant, have lost their luster.[33] Modern interpreters of sexual selection require shinier tools for the job and more up-to-date, culturally meaningful metaphors. In the place of the breeder's eye invoked by Darwin, we are offered a "gene's-eye view," which supposedly divests the notion of "choice" of its troublesome anthropomorphic attributes.[34] Females don't get to choose; genes do it for them.

This last illustrates to perfection the profound shift in the metaphorical and theoretical underpinnings of what is left of Darwinian sexual selection. It is not simply that "selfish" genes and modern molecular biology rendered Darwin's basic theoretical tools of embryogenesis, pangenesis, and male-dominated inheritance obsolete. Embryogenesis, the linchpin of Darwin's system, found no place in the evolutionary synthesis of the 1930s and '40s— largely because it had become too closely associated with the much-reviled recapitulation theory and with non-Darwinian interpretations of the evolutionary process. It is only recently that developmental biology has been re-integrated into mainstream evolutionary biology (a process generally attributed to new insights generated by molecular developmental genetics). The new "evo/devo" models of evolution emphasize developmental dynamics in evolutionary explanation along with more traditional selective mechanisms. It is ironic, given the tight association of embryogenesis with sexual selection in Darwin's theorizing, that Joan Roughgarden, one of the more prominent theorists who identifies her interpretations with the "spirit of evo/devo," seeks to discredit both Darwin's and modern versions of sexual selection, substituting a complex alternative of "genial" genes, cooperative parenting, and "social selection" in which everyone gets to choose.[35]

One way and another, there is precious little left in modern evolutionary theory of what Darwin understood as sexual selection. Nevertheless, no matter how it is conceptualized, however it may be read, sexual selection still means Darwin.[36] His iconic status transcends his authorial death. In defiance of fashionable theory, this particular author has outlived his text in a variety of contradictory narratives, justifications, or critiques of sexual selection. Some of these interpretations simply recycle fragments of Victorian debates and read them into current controversies over sexual selection; some are directed against the "industry of Darwin apologia"[37] or, conversely, are concerned with rehabilitating or "re-enchant[ing]" Darwin and his secular science against the inroads of reductionists, the aspersions of creationists, and the imputations of feminists;[38] others are focused on the historical recovery of a significant moral and humanitarian dimension of Darwin's theorizing;[39] still others view sexual selection through the prism of Victorian fiction,[40] or from an unabashedly Whiggish perspective.[41]

The history of sexual selection is much more complex than these various readings suggest. To distill it into a simpler narrative is to leave out most of the surrounding clamor and color. This history has sought to recover the noise and to bring to light much hitherto hidden from view or misunderstood: the motivations and strategies, the borrowings and adaptations, the slow accumulation of details, the ongoing observations and experiment, the unremitting intellectual effort and commitment, and the assiduous networking; along with the cultural values and assumptions and the contradictions they engendered, the domestic, institutional, and social contingencies and contexts; and, above all, the fusion of ideology, metaphor, and theorizing that propelled the historical twists and turns of the plot.

The Darwin recovered here requires neither rehabilitation nor censure; there is no need to minimize, rationalize, or indeed, valorize those aspects of his thinking that do not mesh with present-day values, or might be seen to jeopardize Darwin's iconic standing. The history of science, as historians well know, is no place for icons, however dazzling their stature.[42] We may admit the humanitarian and compassionate qualities that led Darwin to struggle against the more overtly racist and sexist assumptions of his time and place, but it is crucial for us as historians to acknowledge the extent to which Darwin worked within as well as against these assumptions. And rather more is at stake here than determining whether, in the interests of historical accuracy, Darwin might be described as "great humanitarian," "racist," or "sexist." What is important is rather to determine how those same ideological assumptions informed Darwin's theorizing and (even more important) functioned within it. Ideology, as this history exemplifies, is a slippery concept. It not only works to obscure and oppress, but it also may creatively inform and even empower.[43]

Darwin, its central historical actor, was convinced of the "truth" of sexual selection. What Darwin thought to be vindicatingly true turns out to be culturally richer and far more historically rewarding than this eminent Victorian could ever have imagined.

And all the while, the peacocks strut, the bowerbirds build and rearrange their bowers, and the females, as ever, go about their business, inviting new conjecture and reinterpretation.[44]

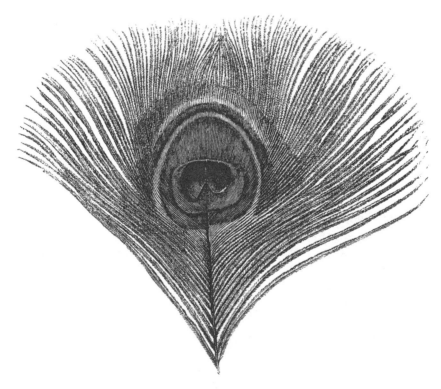

Peacock feather from *Descent* 1871, 2:137. University of Sydney Library, Rare Books and Special Collections.

NOTES

PROLOGUE

1. *Origin*, 136–38, 412.
2. *CCD*, 17:448.
3. In a representative sampling of the literature, Darwin is variously (and sometimes simultaneously) "lone genius" and "radical psychologist" (Miller 2001, 33, 45), misguided theorist (Cronin 1991), great humanitarian (Desmond and Moore 2009), early postmodernist (Kaye 2002, 91), even (by Victorian standards) some kind of feminist (Vandermassen et al. 2005). His theory of sexual selection was inspired by his search for a scientific explanation of animal ornamentation (Ghiselin 1972, 215–16; Ghiselin 1974; Cronin 1991) but masked by a pretense to account for human racial differences (Miller 2001, 36–37); his abhorrence of slavery and/or cruelty to animals (Desmond and Moore ibid.; Millstein 2012); the courtship plots of Victorian novelists (Yeazell 1991, 219ff; Yeazell 1989; Levine 2006); and/or fictive pretty women and a feeling for the organism (Levine, ibid.). His concept of female choice gave radical and un-Victorian agency to females (Levine, ibid.; Kaye 2002, 84–117); offered naturalistic support for the "self-primping Dandy" of the aesthetic movement and its semi-covert homoerotic politics (Kaye 2002, 99, 108–17); or perpetuated sexual and gender stereotypes (Roughgarden, 2011, 271–72).
4. *CCD*, 15:245.
5. Wallace 1877, 408.
6. Wallace 1864, clix, clx; *CCD*, 12:216–17, see also 182.
7. *CCD*, 6:515; *Origin*, 458, 227. On man as a "determining absence" from the *Origin of Species*, see Beer 1983, 58; for an excellent discussion of Darwin's intention to include "man" in his species book and his reasons for dropping it, see Desmond and Moore 2009, 281–93.
8. The phrase "go the whole orang" is Lyell's; *CCD*, 11:231. Lyell had greatly disappointed Darwin by adding a "creational law" to natural selection to explain the origin of human morality and intelligence in his much anticipated endorsement of evolution in his *Antiquity of Man*; C. Lyell 1863, 469, 495, 498.
9. *CCD*, 12:216–17.
10. Ibid., 12:220–21.
11. Ibid., 12:248–49.
12. Wallace 1869b, 391–92; Wallace 1890a; Fichman 2004; *CCD*, 18:17.

13. *CCD*, 16 (2): 763.
14. Bajema 1984, 150, 252–54; Bartley 1994a; C. Darwin 1882; Fichman 2004, 290. But see Milam 2010 for a dissenting interpretation.
15. *Descent* 1879, 4.
16. *Notebooks*, dating from 1836 to 1844.
17. Among the scant literature on its history, the interpretation offered by Helena Cronin in *The Ant and the Peacock* (1991) is skewed by her overt adoption of a Whiggish methodology and her total neglect of Darwin's application of sexual selection to the evolution of human racial and sexual differences. Carl Jay Bajema's *Evolution by Sexual Selection Theory Prior to 1900* (1984) offers a valuable collection of source material and brief commentary on the history of sexual selection, but has significant lacunae and is now badly out of step with more recent Darwin historiography. A limitation of both works is their failure to address the importance of aesthetics and animal breeding practices in the development of Darwin's views on sexual selection.
18. Moore and Desmond, introduction to *Descent* 1879 (2004); Desmond and Moore 2009. For critiques, see R. Richards 2009; Millstein 2012; for accounts centered on "pretty women," see Yeazell 1989; Yeazell 1991, 219–28; Kaye 2002, 84–117; and Levine 2006, 169–201.
19. *Autobiography*, 138–39.
20. *CCD*, 10:71.
21. *Origin*, 91; Secord 1981, 164; Sterrett 2002.
22. Feeley-Harnik 2004.
23. For an excellent discussion of studies of Darwin and visual culture, see Larson 2009; see also J. Smith 2006; Munro 2009.
24. R. Richards 2002, 518–26; R. Richards 2003. For a critique, see Ruse 2004.
25. Hobsbawm 1973, 318, 330; Eagleton 1990.
26. The phrase "visual ideology" is Norbert Finzsch's; Finzsch 2005.
27. Jane Munro has made a beginning in Munro 2009, 278–82. For discussions of Victorian fashion and its cultural resonances, see Halttunen 1982; Kunzle 1982; Steele 1985; Byrde 1992; Hollander 1993; Razek 1999; Kuchta 1996; Flint 2000; T. Logan 2001; Thomas 2004.
28. *CCD*, 2:445.
29. Nevertheless, it has been managed: Cronin 1991
30. *CCD*, 18:194, 255–56; Dawson 2007, 29–41.
31. It scarcely needs to be said that Darwin's circumlocutory discourse on sex is strictly heteronormative. The open discussion of homosexuality was not only deemed indecent and offensive, but Darwin's focus on reproduction meant that his discourse was centered exclusively on heterosexual courtship and sex.
32. Foucault 1980.
33. See R. Richards 1992; Nyhart 2009.
34. Browne 1995, 2002; Desmond and Moore 1991, 2009. On Darwin's correspondence as a "network of power," see James Moore 1985.
35. *Autobiography*, 137–38.
36. This can be likened to aspects of Foucault's archaeological method. In order to "uncover the regularity of a discursive practice," we must look for and analyze "a practice that is in operation in the same way in the work of their predecessors; a practice that takes account in their work not only of the most original affirmations (those that no one else dreamt of before them), but also of those that they borrowed, even copied, from their predecessors"; Foucault 2004, 161.

CHAPTER ONE

1. C. Darwin 1972, 197. E. Lucas Bridges (son of Thomas) disputes this claim, stating that no Fuegian would sell his child "in exchange for H.M.S. *Beagle* with all it had on board"; Bridges 1948, 30; Anne Chapman supports this, 2006, 84–85.
2. *Diary*, 119; FitzRoy 1839, 2:12; Anne Chapman 2006, preface.
3. *Descent* 1871, 1:34.
4. C. Darwin 1972, 485.
5. "The sight of a naked savage in his native land is an event which can never be forgotten"; *Autobiography*, 80.
6. *Descent* 1871, 2:404–5.
7. Duncan 1991, 24–25; see also Stocking 1987, 106–7.
8. C. Darwin 1972, 481.
9. *Diary*, 45; see also *CCD*, 1:312–13.
10. *Diary*, 79–80.
11. *Diary*, 43–46, 58; *Autobiography*, 73–74; Browne, 1995, 196–99.
12. C. Darwin 1972, 480–81; *CCD*, 3:241–42, 233, 235, 381; Desmond and Moore 2009, 192–98.
13. *CCD*, 9:163.
14. Moore and Desmond, introduction to *Descent* 1879, xiv. This claim is given extended treatment in Desmond and Moore 2009; for a critique, see R. Richards 2009; also Ruse 2015, 13n3.
15. C. Darwin 1972, 23.
16. *Diary*, 139.
17. Ibid., 100.
18. Duncan 1991, 20.
19. *Notebooks*, M 153; see also C 79 and C. Darwin 1972, 198.
20. Browne 1995, 248.
21. *Diary*, 124–25, 222–24.
22. Although this proved a trifle flexible: when the rather alarming warmth of their attachment was observed by English eyes, the very young Fuegia Basket was said to be "engaged to be married" to York Minster; Browne 1995, 238.
23. *Diary*, 143.
24. *Descent* 1871, 2:348.
25. *Diary*, 125, 226–27; FitzRoy 1839, 2:206–25, 323–27; Browne 1995, 252, 268–69.
26. *Diary*, 226–27; FitzRoy 1839, 2:324; *CCD*, 1:378, 80; FitzRoy makes the same qualification, good looking "for a Fuegian," with respect to "Jemmy's wife," FitzRoy 1839, 2:324fn.
27. *CCD*, 1:380; see also Beer 1996, 39, 64–68. For accounts of traditional Yahgan culture and life, see Bridges 1948; Anne Chapman 1982, 2006.
28. FitzRoy 1839, 2:175, 640–41.
29. Ibid., 1:415, 2:3, 186, 120–21, 192, 688.
30. Ibid., 2:120–21.
31. Ibid., 2:327.
32. There is some suggestion that FitzRoy's fundamentalist Christian beliefs were challenged by his geological discussions with Darwin, but that they revived in the aftermath of the voyage; FitzRoy 1839, 2:640–56; see Mellersh 1968, 128, 175–83.
33. C. Darwin 1972, 218; *Diary*, 227.
34. *Diary*, 223–24.
35. C. Darwin 1972, 219; see also *Notebooks*, E 47.

36. *Notebooks*, E 46–47.
37. In point of fact it was not the finches but the Galapagos mockingbirds, belatedly classified as separate species by the ornithologist John Gould, that led Darwin to open his first notebook on the transmutation of species in 1837; Sulloway 1982a and b.
38. Gruber 1974, 184; see also Beer 1996, 64–70.
39. *CCD*, 10:71.
40. *Notebooks*, C 154, 217, 223; see S. Herbert 1977, 207.
41. In Chiloe in 1834, Darwin had met with the surgeon of an English whaler who assured him that body lice from Sandwich Islanders could not survive on English sailors, dying within three or four days. At the time Darwin made a note that he later prudently deleted: "If these facts were verified their interest would be great.—Man springing from one stock according his *varieties* having different *species* of parasites. It leads one into . . . many reflections." Some ten years later, Darwin recounted the surgeon's claim to Henry Denny, then cataloging exotic Anoplura for the British Association, and provided him with specimens of lice collected during the *Beagle* voyage. Darwin told Denny, "I myself do not think our supposed knowledge of having come from one stock ought to enter into any scientific reasoning"; *CCD*, 3:38, 258. Darwin subsequently discussed the relevance of the different species of lice on the different human races in the *Descent*, pointing out that the evidence favored the classing of human races as distinct species; *Descent* 1871, 1:219–20. Darwin also made inquiries into the sterility of "negro-crosses"; *Notebooks*, B 34, 68–9, 86–87, 142, 189, 231; *CCD*, 3:38n1, 258. See also *CCD*, 13:359. Cf. Desmond and Moore 2009, 97–98, 115, 125–26, 193, 389.
42. *Diary*, 134, 223.
43. Duncan 1991, 17–18; Stocking 1987, 105–7; Stepan 1982, 29–46. On the role of imperial mapping and its ties with notions of race, colonization, and empire, see Burnett 2000.
44. Finzsch 2005.
45. *CCD*, 1:29; *Autobiography*, 51; *Descent* 1871, 1:232; Desmond and Moore 1991, 28. On "black London," see Gerzina 1995.
46. Lorimer 1978, 21–44. On the complex politics of the British abolition movement, see Blackburn 1988, chs. 2, 4, 8, 11; Davis 1975, 343–522.
47. Lorimer 1997; C. Herbert 1991.
48. For a history of ethnological show business, see Lindfors 1999; Qureshi 2011a.
49. Gould 1985; S. Gilman 1985, 83–88; Schiebinger 1993, 160–72; McClintock 1995, 42.
50. Magubane 2001; see also Qureshi 2004. On the dynamics of colonial economics and abolition during this period, see Blackburn 1988, ch. 8. We may discern similar equivocations in Darwin's description of the inhuman conditions of indigenous laborers in the deep copper mines of Chile and the appalling physical toll on "these beasts of burden" who "voluntarily" carried loads of two hundred pounds of ore up almost vertical notched poles from depths of more than six hundred feet; *Diary*, 329.
51. E. Richards 1994; Knox 1855
52. Browne 1995, 234; Anne Chapman 2006, 8–9.
53. Browne 1995, 234–25; Desmond and Moore 2009, 95.
54. Gould 1985; S. Gilman 1985, 83–88; Schiebinger 1993, 160–72. For a cogent critique, which disputes this interpretation and emphasizes the need to contextualize Baartman's story, see Magubane 2001; also Callanan 2006, 142–57; and Qureshi 2004.
55. Knox 1850, 23–24, 179. On the "Celtic Calibans," see McClintock 1995, 52–53.

56. Stocking 1987, 16–19, 34–37; Lorimer 1978, 1–44. On the Reform Bill crisis and the class and property lines drawn by the Act of 1832, see E. Thompson 1975, 888–909.
57. Prichard 1835, 175. On the cult of domesticity as central to British imperial identity, see McClintock 1995.
58. See Finzsch 2005; McClintock 1995, ch. 1; C. Herbert 1991, 60–68, 72–73.
59. S. Gilman 1985; Stocking 1987, 106; Robert J. C. Young 2006; McClintock 1995; Schiebinger 1993.
60. *CCD*, 1:303–4, 306n5; Darwin 1972, 204; *Variorum*, 111; *Variation*, 2:214–15. Darwin records with horror that one Fuegian informant fell about laughing at the thought of the old women roasting over their own family fires. This may well have been a joke that played on the credulity of the ignorant British; see Bridges 1948, 33–34; Anne Chapman 2006, 46–49.
61. *Notebooks*, C, 79; Finzsch 2005, 106; see also Radick 2002; Radick 2007, 29–31, 35–39; and Radick 2008 on Darwin's view that primitive peoples have correspondingly primitive languages.
62. See Brantlinger 2003, chs. 1 and 2; Butcher 1994; Finzsch 2005.
63. *Diary*, 402, 408.
64. C. Darwin 1972, 486; see also Barta 2005; Butcher 1994.
65. Sturt cited in Finzsch 2005, 97, 113. Sturt, with the benefit of more direct experience, later modified his view of the lowly status of Aboriginals; Butcher 1994, 376. On "abjection" as characterizing those whom industrial imperialism rejects but cannot do without—slaves, prostitutes, the colonized, the insane, the unemployed—see McClintock 1995, 71–72.
66. Robert J. C. Young 2006, 138 and passim; Schiebinger 1993, chs. 4 and 5. For an analysis of gender, race, and colonialism, see McClintock 1995. Distinctions between races were also gendered; see Sinha 1995.
67. T. Mitchell 1838, 1:304.
68. Strzelecki 1845, 343; Butcher 1994, 376.
69. *Marginalia*, 1:790–91.
70. *Notebooks*, D 99.
71. *Diary*, 75; *CCD*, 1:192–94, 208, 211, 213–14, 219–20; Browne 1995, 199–201.
72. *CCD*, 1:276–79; *Diary*, 115.
73. *Diary*, 257, 293, 349–50.
74. *Diary*, 202, 115; Browne 1995, 262–63.
75. *CCD*, 1:416, 460.
76. *Diary*, 79.
77. Ibid., 104–5.
78. *CCD*, 1:472; Browne 1995, 310–13; Browne 1994; Desmond and Moore 2009, 99–101.
79. *Diary*, 169–80.
80. Ibid., 190–98.
81. C. Darwin 1972, 486; *Diary*, 366–87; Browne 1994.
82. *Diary*, 398–99, 411–12.
83. Darwin's interchangeable use of the terms *barbarian* and *savage* is interesting. For an analysis of their etymology and use, see Kuper 2005, ch. 2.
84. In the second edition of his *Journal*, Darwin edited out his vision of colonial productivity as justification for genocide, retaining his horror at the slaughter and settler acquiescence in the extermination of the Indians: "Who would believe in this age

that such atrocities could be committed in a Christian civilized country?" C. Darwin 1972, 97; see Barta 2005.

85. For an excellent critique of the concept of genocide and the appropriateness of its application to the British destruction of indigenous Tasmanians, see Docker 2015; also Curthoys and Docker 2008.

86. *CCD*, 1:482–85.

87. *CCD*, 1:490.

88. *Notebooks*, OUN 24, 8.

89. *Notebooks*, N 26–28, written October 27, 1838; see also M 31–32.

90. Darwin was responding to the *Discourses* of Sir Joshua Reynolds, where Reynolds used the instance of the "Ethiope" to argue that the idea of beauty was formed by habit; see sec. 4.4.

91. *Notebooks*, N 26–28; M 31–2; OUN 8, 11, 14, 20–21, 22–24.

92. See also Tipton 1999.

93. Duncan 1991, 22.

94. *Notebooks*, B 232, C 76–79; *CCD*, 2:80–81; Duncan 1991, 25; see also Schmitt 2009, 32–56.

95. *CCD*, 8:20.

96. Ibid., 8:400. Darwin did not directly contact Bridges, but via the South American Missionary Society; he did not receive Bridges's response until 1867; see Radick 2010b. Bridges was the foster son of the leader of the mission whose Fuegian branch was massacred at Woolya in Tierra del Fuego in 1859. His only contact with Fuegians was with those who were brought to the Keppel Island mission to be Christianized, including a reluctant Jemmy Button and his family. Bridges became fluent in the language of Jemmy's tribe, but did not actually visit Tierra del Fuego until 1863. He took orders in England in 1868 and returned with his family to the finally successful Fuegian mission at Ushuaia on the coast of the Beagle Channel in 1871; see Bridges 1948.

97. *Descent* 1871, 2:351.

98. *CCD*, 18:193. Various contradictory accounts of the massacre and Jemmy's role in it have been offered. Jemmy died in an epidemic that raged through Tierra del Fuego in 1863; see Nichols 2003, 267–76; Mellersh 1968, 241–55; Bridges 1948, 40–46.

99. C. Darwin 1972, 218fn; *CCD*, 5:164. This was not, however, Fuegia's last recorded appearance. Thomas Bridges met her in 1873, when Fuegia was in her mid-fifties. She told him that York, by whom she had two grown children, had been killed in retaliation for the murder of a man. Ten years later, Bridges met Fuegia again, ill and miserable, and futilely proffered biblical comfort to a dying woman who had forgotten her early religious training; Bridges 1948, 83–84.

100. *CCD*, 18:182, 192–93, 195–96. For Darwin's note on the "improvable" but "not improving" Fuegians, see *Notebooks*, C 79. Darwin also contributed to a fund established by Sulivan in 1879 for the "adoption" of Jemmy's orphaned grandson, Jemmy FitzRoy Button; DAR 177, *Calendar* 12258. On Darwin's alleged change of mind on the Fuegians, see Radick 2010b.

101. Arthur Mellersh to Darwin, 1872, DAR 171.

102. Anne Chapman 2006, 40–96. It was thanks largely to Darwin that in the new evolutionary anthropology that took off in the 1870s, the Fuegians displaced the ubiquitous Hottentots or Australian Aborigines at the bottom of the infamous scale of the evolution of the human races; see Duncan 1991, 14–15, 37–41, 43n26. E. Lucas Bridges (son of Thomas), who was born in Tierra del Fuego and grew up among

the Yámana, was scathingly dismissive of Darwin's account: "These were the people whom Charles Darwin had labelled, if not the missing link, then the next thing to it"; Bridges 1948, 136. Of course, just as with the classic case of the Tasmanians, "extinction" depends on the ideologically and politically laden notion of a "pure-blooded" descendant; as with the Tasmanians, there are extant claimants to Yámanan genetic descent; Anne Chapman 2006, 121–26.

103. C. Darwin 1972, 418–19; *Notebooks*, E 63–65, *CCD*, 2:19–20; Duncan 1991, 31–32. In contrast to Darwin, John Lort Stokes, the mate of the *Beagle* during Darwin's voyage and its commander during its later Australian survey voyage of 1837–43, confronted this widespread view that an "all-powerful law" confirmed by "history" necessitated the "depopulation of the countries we colonize." There was no "mysterious dispensation of Providence" that could rationalize the extinction of indigenous people; the "moral responsibility on the part of the whites" must be acknowledged; see Douglas 2008, 71.

104. *Descent* 1871, 1:201.

105. Ibid., 1:237–40.

106. Ibid., 1:239.

107. To the contrary: in his private correspondence with Lyell (with whom Darwin kept up a running disagreement about Lyell's equivocations over slavery), Darwin endorsed Lyell's "simile of man now keeping down any new man which might be developed" with the blunt, if ironic, assertion: "White man is 'improving off the face of the earth' even races nearly his equals." *CCD*, 8:373; see also 7:345–46.

108. *Descent* 1871, 1:101, 167–70.

109. Ibid., 1: 94; see Barta 2005. Darwin discussed the "slave-making instinct" among ants in detail in the *Origin*. Desmond and Moore argue that Darwin did naturalize the ants' slave-making instincts in the sense that he argued, "What applies to one animal will apply throughout all time to all animals"; Desmond and Moore 2009, 303–4. Even here, Darwin's description is inflected by notions of British moral superiority: the British *Formica sanguinea* are "tyrannical" but vigorous and more independent of their slaves than their decadent continental relatives, *F. rufescens*, who are "abjectly dependent" on their slaves and cannot survive without them; *Origin*, 243–447.

110. Mellersh 1968, 207–35; Amigoni 2007, 87; C. Darwin 1972, 409–10.

111. *Descent* 1871, 1:167.

112. *CCD*, 2:444–45.

CHAPTER TWO

1. Greg 1868; Darwin's annotated copy of Greg's paper is in DRC. See also Greene 1977, 9–14.

2. *CCD*, 2:444–45.

3. *Descent* 1871, 2:326–29; E. Richards 1983.

4. *ED*, 2:45.

5. See McNeil 1987, 9–15, 19, and passim.

6. Seward 1804. The best analysis of Erasmus Darwin's life and work in social context is that of McNeil 1987; for biographical detail, see King-Hele 1977.

7. E. Darwin 1797, 127; King-Hele 1977, 210, 234–37; Wedgwood and Wedgwood 1980, 55; Uglow 2002, 180.

8. Both poems were issued anonymously, and although he was widely known as their author, Darwin did not officially confirm his authorship until 1794; Browne 1989, 595n5.

9. R. Edgeworth 1820, 2:266.

10. Seward 1811, 6:91–100.

11. Ibid., 2:311–12, 3:154–56, 6:91–100; Seward 1804, 125–132.

12. Seward 1811, 5:55, 6:82.

13. Tomalin 1974, 119.

14. Seward 1811, 3:47, 75–76, 159–60, 202–8.

15. Ibid., 3:117.

16. Ibid., 5:73–74.

17. Ibid., 2:387; 6:141–45; Seward 1804, 86–87, 217.

18. Seward 1811, 6:84.

19. Apart from her supposed romance with Erasmus Darwin, Seward had a number of intense emotional relationships with members of both sexes. She caused scandal in Lichfield through her close friendship with a choral singer, who was married but separated from his wife; Uglow 2002, 192.

20. Weeks 1981, 19–33; Poovey 1984, 3–30; Davidoff and Hall 1987, 401–3; C. Hall 1992, 82–92.

21. Quotations taken from the uncut version of Darwin's *Life* of Erasmus; King-Hele 2003, 5, 36, 75–78; DAR 210.20: 48–49.

22. DAR 210.20:48–49. Seward referred to the Misses Parker in her biography of Darwin, but, perhaps as a result of the "softening" of the text imposed on her, primly described "these ingenious and good young women" as Erasmus's "relations"; Seward 1804, 390.

23. DAR 210.20:44; King-Hele 2003, passim. It was Henrietta who, in an excess of editorial zeal, deleted Darwin's reference to Krause's revision of his essay, thus provoking the great row with Butler; see ch. 3, note 73.

24. King-Hele 2003, 33.

25. Browne 1989, 614–18.

26. Healey 2002, 29.

27. Beddoes, while not formally a member of the Lunar Society, had strong links with it. Beddoes's radical political views were a "matter of public notoriety" and he was forced to resign his chair as chemical reader at Oxford; Stock 1811, 68, 131, 139, 144, 145.

28. Day became author of an educational best seller for children, *Sandford and Merton* (1783–89), which extolled a sort of primitive socialism (to each according to work done)—without, however, upsetting the established order; see Rowland 1996, 207–48.

29. In *Émile*, nature acts as a guide for Émile's future state as a free man, but nature becomes the reason for Sophie's subjection. Rousseau relied on a biological justification of these differences, arguing that nature had made men and women different but perfect, each in their own way. They are intended to complement each other. These natural differences between the sexes are represented and amplified in the form of moral differences that society ignores or erases only at its peril. See Rousseau (1762) 1981, 5, 357–63; Lacqueur 1990, 198–200; Bloch and Bloch 1980, 30, 35–36; Schwartz 1984; Landes 1988, 66–89. Rousseau's educational system aroused the interest of radicals throughout Europe, including other Lunar Society intellectuals; Schofield 1963, 54–55; Uglow 2002, 309–22.

30. Maria Edgeworth, a successful woman writer in her own right, recorded Day's argument that the "extent of a woman's erudition should consist in her knowing her simple letters, without their mischievous combinations"; R. Edgeworth 1820, 2:341–42.

31. Day's "superior abilities, lofty sentiments, and singularity of manners" disarmed any feminine criticism of his "breeding up a young girl in his house, without any female to take care of her"; R. Edgeworth 1820, 1:236, 239, 226. Day apprenticed the rejected Lucretia (her surname is not known) to a milliner in London and provided her with a marriage dowry; Seward 1804, 37–38; Rowland 1996, 17–21; Uglow 2002, 185–88.

32. And thus distressing Maria Edgeworth, to whom Sabrina made the complaint; see Rowland 1996, 350. Day's antislavery poem, *The Dying Negro*, was famous among abolitionists.

33. Seward 1804, 50, 37–41; R. Edgeworth 1820, 1:339.

34. R. Edgeworth 1820, 1:346; Rowland 1996, 61–62, 277–79, 356–57.

35. M. Edgeworth and Edgeworth 1798, 1:168, 526. It is significant that of those women who left accounts of the Lunar Society gents, only Maria Edgeworth escapes the censure of historians. Maria was almost insufferably loyal in the face of the paternal eccentricities and domestic tyranny of Richard Lovell Edgeworth, who ignored her as a child to concentrate on the upbringing of his firstborn son after the ideas of Rousseau; R. Edgeworth 1820; Butler 1972.

36. E. Darwin 1797, 10; C. Darwin 1887, 115. It is doubtful that the real-life pupils of the Misses Parker were actually exposed to Darwin's ambitious program. The school, which cannot have conflicted too much with socially accepted standards, was run successfully by the Misses Parker until the late 1820s. The prospectus for their school promises only the standard girls' school curriculum: "Embroidery and Needle-work of all kinds both useful and ornamental, reading with propriety, grammar, a taste for english [sic] classics, an outline of history both ancient and modern, with geography and the use of the globes," E. Darwin 1797, 128; King-Hele 1977, 234.

37. Austen (1816) 1966, 52.

38. Healey 2002, 46; Desmond and Moore 2009, 60–61. On the radical Unitarians and woman's role, see Gleadle 1995.

39. *Autobiography*, 28–29, 39; *Notebooks*, M 149; Healey 2002, 24–30, 42, 46, 57–60, 109–10, 104.

40. *ED*, 1:1–14; Healey 2002, 52–53, 31–35, 44–45; Browne 1995, 111; *Autobiography*, 54–56.

41. *ED*, 1:14.

42. *CCD*, 1:518–19, 524, 2:80–81, 86, 115; *Autobiography*, 95–96; Desmond and Moore 1991, 216–17.

43. On her marriage to Darwin, Josiah Wedgwood settled £5,000 on Emma (arranged as a trust fund, since married women could not hold investments), plus an allowance of £400 a year; Robert Darwin contributed £10,000 in stock invested in canals, land, and railways, which gave an income of around £600 per year; *CCD*, 2:119; Browne 1995, 392–93.

44. *CCD*, 2:319, 313; Randal Keynes 2001, 161–98.

45. *CCD*, 3:xiv–xv, 43–45.

46. *CCD*, 2:127–28, 132, 166; *ED*, 1:272; 2:13, 24.

47. *Notebooks*, C 165–66; *CCD*, 2:171–73, 122–23; see also *CCD*, 9:155–56 (written June 1861); Browne 1995, 395–99.

48. Darwin's "Recollections," written in 1876, were published posthumously as his *Autobiography*; see C. Darwin (1876) 2010, 397–98; *ED*, 2:56; *Autobiography*, 96–97.

49. *ED*, 1:61–62.

50. Healey 2002, 70–72, 295; *ED*, 2:211.

51. Healey 2002, 112–14, 134; Desmond and Moore 2009, 60–64, 117, 261; C. Hall 1992, 156.
52. Healey 2002, 02, 06, 101, 113–24, 237.
53. *ED*, 2:40–41.
54. Healey 2002, 223–34; DAR 119:21b; *CCD*, 4:477; *ED*, 2:48.
55. *CCD*, 13:309–10, 315; *ED*, 2:40–41, 48, 172; Browne 1995, 534.
56. *CCD*, 12:212, 319, 387; 11:582, 556, 564, 695. After Darwin's death, Emma became an ardent opponent of home rule for Ireland as the issue threatened to tear the Liberals apart, delighting in "baiting Parnell," gloating over the "rage of the Irish," and celebrating her son Leonard's election win as a Liberal Unionist in 1892; Healey 2002, 336–37; DAR 219.9:289; *CCD*, 10:275, 331.
57. *CCD*, 11:776–81, 557, 623, 624, 643, 689–91; 12:184–85; Healey 2002, 265–68.
58. DAR 219.9:135. On Snow Wedgwood and Darwin, see Joy Harvey 2009, 200.
59. DAR 144; DAR 202; F. Darwin and Seward 1903, 2:49; Browne 2002, 443–44. On the contemporary use of a sponge soaked in vinegar or douching with vinegar or alum as contraceptives, see McLaren 1992, 185–86; Mason 1995, 57–64.
60. *CCD*, 5:197; 2:312; 4:302–3, 311, 335; Healey 2002, 195–97; Randal Keynes 2001, 146–47. Emma may have received chloroform during the birth of her seventh child, Francis, in 1848, but this is not certain; *CCD*, 4:140, 160; 6:318.
61. Charles Waring died of scarlet fever at the age of two. See Randal Keynes 2001, 224–26; *CCD*, 5:100. On Darwin's concern that his children had inherited his ill health, see *CCD*, 7:158, 164–65.
62. *CCD*, 13:482.
63. *CCD*, 4:147; see also 2:293–94. For an analysis of Darwin's symptoms and the various diagnoses proposed, see Colp 1977; Janet Browne gives possibly the best account of Darwin's illness and its consequences in Browne 2002, ch. 7.
64. DAR 219.9:95
65. Emma was highly indignant when the barrister and civil servant Sir Thomas Henry Farrer initially opposed the marriage of his daughter Ida to their son Horace, condemning Horace as "too dammed hypochondriac"; Healey 2002, 307.
66. On the relation of "manliness" to the ideal of domesticity, see Tosh 1999; on "muscular Christianity," see the essays in D. Hall 1994; see also P. Anderson 1995, ch. 3. On Victorian notions of sexuality, see Mason 1994, 1995.
67. See Tosh 1994, 181, 183; Tosh 1999, ch. 2.
68. Tosh 1999, 68, 59. On Bank's notoriety as a Pacific rake and "Botanic Macaroni," see Fara 2003, 7–16.
69. Anon. 1860, 294.
70. Davidoff and Hall 1987, esp. chs. 6 and 8; C. Hall 1992, ch. 10. Paul White, in his study of Thomas Henry Huxley, emphasizes the significant role played by Henrietta Huxley in fostering Huxley's own identity as a "man of science," the ways in which domestic values permeated that identity, and Henrietta's contributions to Huxley's professional work; P. White 2003.
71. *Descent*, 2:328–29; *Autobiography*, 78, 139–45; see also Darwin's answers to Galton's questionnaire of 1873 on his intellectual powers, F. Darwin 1888, 2:357.
72. Mill 1970, 181; Rossi 1970, 58–59. On the understanding of masculinity as a manifestation of social power, see Tosh 1994.
73. Tosh 1994, 185; Tosh 1999.
74. Davidoff and Hall 1987, chs. 1 and 2; C. Hall 1992, 266.
75. Tosh 1999, 54, 60–61, 77.

76. Browne 2002, 231; F. Darwin 1888, 1:136, 137, 159.
77. Raverat 1960, 99.
78. *ED*, 2:178; Joy Harvey 2009, 199, 209–10; Browne 1995, 533–34.
79. Raverat 1960, ch.7; DAR 219.11:16; DAR 219.9:111; see also DAR 219.8:21.
80. Raverat 1960, 121, 177; *CCD*, 5:142n2; Browne 2002, 227 44; Randal Keynes 2001, 216–17.
81. Raverat 1960, 162, 169.
82. See Healey 2002, 291–92; Browne 2002, 145; Randal Keynes 2001, 255–57.
83. DAR 219.9:72; *CCD*, 18:380; Browne 2002, 341–42; Joy Harvey 2009, 200, 206.
84. *Autobiography*, 11–13; DAR 112.A:19–22; James Moore 1989, 199–209; Healey 2002, 215–17, 280–81.
85. DAR 219.9:74; *ED*, 2:196; DAR 185:58; DAR 219.9:77. For some indication of Henrietta's conventional views on the differences between men and women, see DAR 219.10:22 (written in 1867).
86. *ED*, 2:204–5.
87. *Descent* 1871, 2:326.
88. On the "triumph of the Ladies at Cambridge," see Darwin to George Darwin 1881, *ED*, 2:245; on Huxley and the exclusion of women from the Ethnological Society, see E. Richards 1989a.
89. Darwin to Caroline Kennard 1882, DAR 185:29; Kennard to Darwin 1881 and 1882, DAR 185:28, 31; see also Charlotte Pape to Darwin 1875, DAR 174:27; Joy Harvey 2009.
90. *CCD*, 2:169–70.

CHAPTER THREE
1. DAR 226.2:71.
2. *Variorum*, 60–61; *Recollections*, 371.
3. *Recollections*, 371; Seward 1804; Browne 1995, 83–88. For an excellent analysis of Grant's politics and related evolutionary views, see Desmond 1984; on the relationship between Grant and the young Darwin, see Sloan 1985.
4. My interpretation contradicts that of two leading interpreters of Charles Darwin's concept of sexual selection, Michael Ghiselin and Carl Bajema, who have argued that Erasmus Darwin's "theories about sexual combat were not the source of his grandson's theory of sexual selection"; Bajema 1984, 8; Ghiselin 1976, 127. But Ernst Krause, who wrote the essay on Erasmus Darwin's evolutionary biology, published with introductory essay by Charles Darwin in 1879, was in no doubt about the younger Darwin's debt to his grandparent: "With the most perfect certainty we [here] have the principles of a theory of sexual selection laid down." Krause 1879, 132–33, 190, 207–8.
5. E. Darwin 1794, 1:502–3, 480. See also McNeil 1987, 100–102; R. Porter 1989, 45–58.
6. E. Darwin 1800, 33; E. Darwin 1803, additional note 1; R. Porter 1989, 59–60.
7. Cited in J. Logan 1972, 16. On the late eighteenth-century craze for botany and its study as a means of educating the young and female in normative sexual relations, see Shteir 1996, 21–25; Fara 2003, 40–46.
8. Browne 1989, 614–18; see also Schiebinger 1993, 30–33.
9. E. Darwin 1789–91, canto 4, lines 401, 406; Browne 1989; McNeil 1987, 86–122; R. Porter 1989.
10. A good deal has been made of Erasmus Darwin's influence on the younger generation of romantic poets, notably Keats and Shelley (King-Hele 1977, 303–6), but little

attention has been given to his relation to the diffusion of romantic concepts in British and German biology and medicine; see E. Richards 1976, 40–44, 197–202. A number of leading German romantics, including Friedrich Schelling and Goethe, read and drew inspiration from *Zoonomia*; see R. Richards 1992, 29, 37; R. Richards 2002, 300–301.

11. A. Cunningham and Jardine 1990; Desmond 1989; E. Richards 1989b, 1990; R. Richards 2002.

12. E. Richards 1990.

13. *Zoonomia* was published in three parts, the first volume or part appearing in 1794, the second and third parts in 1796, along with the revised or second edition of part 1. A third and final revised edition was issued in four volumes in 1801.

14. This was the physiological doctrine that fetal growth amounted to no more than the enlargement, unfolding, and hardening of already preformed parts. Preformationists, who included most influential naturalists of the eighteenth century, believed that each individual egg (the "ovist" school) or spermatozoon (the "animalculists") contained all subsequent generations in miniature. The theory was compatible with contemporary belief in the immutability of species. All species were preordained by God, being fixed for all time in the first members of every created species.

15. E. Darwin 1794, 502, 505.

16. E. Darwin 1794, preface; E. Darwin 1801, 2:318; E. Darwin 1803, canto 1, lines 247–50.

17. Hobbes 1651 cited in Bajema 1984, 12–15; see also 3–4; Laqueur 1990, 156.

18. Rousseau (1761) 1971, 81–82, quoted in Bajema 1984, 36–62.

19. Rousseau (1761) 1971, 79, 85 in Bajema 1984, 36–62, also 6–7; Laqueur 1990, 198–99.

20. Bloch and Bloch 1980, 28, 31; see Schwartz 1984 for an analysis of the contradictions in Rousseau's sexual politics.

21. E. Darwin 1794, 503.

22. E. Darwin 1803, 67–69.

23. Darwin's "sentimental love" bears some relation to Rousseau's "moral love," but is much closer to Edmund Burke's "sentiment of beauty," a debt that Darwin acknowledged; E. Darwin 1803, 91; see text.

24. E. Darwin 1803, 113.

25. Ibid., 123–24; 129–34.

26. McNeil 1987, 185–86; E. Thompson 1975, 70–73.

27. Winch 1987, 24–32; Robert M. Young 1985, 24–26.

28. Winch 1987, 16–35.

29. Robert M. Young 1985, 23–55; Nicholson 1990, 403; see also the introduction to Dawson and Lightman 2014.

30. Malthus 1976, xxii, 134–36; Winch 1987, 32–35.

31. Laqueur 1990, 201; Nicholson 1990, 412–13. On the paradoxes in Malthus's idea of desire, see also C. Herbert 1991, 106–15.

32. It was apparently Godwin who persuaded Malthus to adopt the notion of moral restraint that Godwin himself had suggested; Mason 1995, 262–68.

33. McNeil 1987, 110–13.

34. E. Darwin 1800, 271; E. Darwin 1803, 159, 165–66.

35. Hobsbawm 1973, 73; Royle and Walvin 1982, 43; McNeil 1987, 59–85.

36. There is of course more than a little sexuality to such imagery; see, for instance, E. Darwin 1789–91, canto 2, lines 187–92. King-Hele 1981, 53, 193–94, 196–200; McNeil 1987; Uglow 2002.

37. Hobsbawm 1973, 33–35; King-Hele 1981, 225–26.
38. Uglow 2002, 258–60, 410–14; Browne 1995, 196; McNeil 1987, 27–28, 60–85; King-Hele 1981, 225, 235; E. Thompson 1975, 194–95, 199, 216–17.
39. The three-day riots, ostensibly sparked by a Lunar dinner to commemorate the fall of the Bastille, were instigated by the Birmingham establishment, grown increasingly hostile against the domination of the economic, intellectual, and corporate life of the city by a wealthy, entrepreneurial, and largely Dissenting middle class. The urban poor who comprised the "mob" possibly had their own grievances against wealthy employers such as Boulton and Watt, who had to fortify their Soho foundry against the rioters. See McNeil 1987, 80–82; Uglow 2002, ch. 37.
40. McNeil 1987, 80–85; E. Thompson 1975, 216–17, 79–85, 101–2, 194–95.
41. McNeil 1987, 122–24.
42. King-Hele 1981, 279–82; Butler 1984, 1–17.
43. Anti-Jacobin 1799, 1:7; Garfinkle 1955, 384–86.
44. Anti-Jacobin 1799, 2:164. On the counterrevolutionary hostility toward mingling eroticism with radical republicanism, see Priestman 1999, 60–75.
45. Anti-Jacobin 1799, 2:162–74, 200–205, 274–80; Seward 1811, 5:113–15, 136–41.
46. Polwhele had earlier written a complimentary "Idyllium" on the Botanic Garden, which Darwin included in later editions of his poem; King-Hele 1981, 223–4.
47. Polwhele 1798, 7–9n. For Wollstonecraft's views on sexuality, see Nicholson 1990; Poovey 1984, 69–81; Jane Moore 1999, 12, 93, 99–101; B. Taylor 1992.
48. "As she was given up to her 'heart's lusts,' and let 'to follow her own imaginations,' that the fallacy of her doctrines and the effects of an irreligious conduct, might be manifested to the world; and as she died a death that strongly marked the distinction of the sexes, by pointing out the destiny of women, and the diseases to which they are liable"; Polwhele 1798, 28–30n. For perceptive views on the reception of Wollstonecraft's writings, see Janes 1978; Poston 1988, 221–356; Kelly 1993, chs. 1 and 2.
49. Poovey 1984, 1989.
50. There is a very large literature on the "moral panic" and the "making of" Victorian sexuality, including those studies that have sought to destabilize both historiographic constructs. For excellent overviews, see C. Hall 1992; Mason 1994, 1995.
51. C. Hall 1992, ch. 3; V. Jones 1990, 131–39.
52. E. Darwin 1801, 2:261.
53. E. Darwin 1794, 1:489.
54. E. Darwin 1801, 2:313; see also E. Darwin 1800, 126–29; Ghiselin 1976, 129.
55. E. Darwin 1801, 2:263; also E. Darwin 1794, 1:519.
56. E. Darwin 1801, 2:269–70; R. Porter 1989, 53. See also E. Darwin 1794, 1:523–24.
57. E. Darwin 1794, 1:144–45; see J. Logan 1972, 57–64.
58. E. Darwin 1794, 1:144, 146.
59. See ch. 4.
60. Burke's politicized aesthetics and his associated defense of the ancien régime inspired the opprobrium of the two leading contemporary radical defenders of the rights of man and woman, Tom Paine and Mary Wollstonecraft; Whale 2000, chs. 2 and 3. Erasmus Darwin's response was clearly more complex. Although his debt to Burke's aesthetics is evident in Zoonomia, he does not mention Burke (who at that stage was denouncing Priestley as the leader of a plot to subvert the English social order), but attributes the more liberal-minded Hogarth; however, by the time of The Temple of Nature of 1803, Erasmus does cite Burke's "sentimental love" as the source of his aesthetics; "Additional Notes," 91. See also J. Logan 1972, 57–64; Klonk 1996, 13–18.

61. E. Darwin 1794, 1:163–83.
62. Ibid., 1:141, 144.
63. On the iconic status of the female breast in Western cultures and its representation in eighteenth-century natural history and anthropology, see Schiebinger 1993, 40–74.
64. Adam and Eve make a cursory appearance in *The Temple of Nature*, E. Darwin 1803, 55–56. There are, aside from his condemnations of slavery (which are not structured in terms of race), surprisingly few references to human racial variety in either *Zoonomia* or *The Temple of Nature*, and Erasmus makes no attempt to account for the origin of the different races, which he refers to as "nations," other than the suggestion that it is the development of "REASON" that divides "man from brute, and man from man"; ibid., 117–18. The major distinction he drew was the typical Enlightenment one between "civilized" and "savage" man.
65. See Bajema 1984, 4–5, 24–25.
66. Addison 1970, 179–80; Addison's essays were first published in the *Spectator* in 1712; Guyer 2005, 21–24; Burke 1823, 50–51. On the relation of Erasmus Darwin's views to those of Addison, see J. Logan 1972, 50–52.
67. E. Darwin 1803, 113; E. Darwin 1794, 1:152, 509–10.
68. On the vogue of Lavaterian physiognomy in England, see R. Porter 1985; on its particular appeal to radicals, see Juengel 1996, 2001.
69. See Hartley 2005, 32–43; Norton 1995, 176–209; Gray 2004, 1–55.
70. Significantly, Erasmus did not cite Lavater in this connection, but imported Burke's aesthetic physiognomy into his discussion of expression and its relation to mental states; E. Darwin 1794, 1:146–47; Klonk 1996, 13–15.
71. E. Darwin 1794, 1:147–53; E. Darwin 1797, 17.
72. For a persuasive argument to this effect, see Desmond 1989, 398–414.
73. For accounts of Darwin's conflict with Butler, see King-Hele 2003, 149, 152, 156, 157; *Autobiography*, 134–35, 167–219; Browne 2002, 472–75.
74. Just when and how often he read them is not easy to say. According to his records of books read, Darwin reread and annotated both works in February 1842; DAR 119:12a. But the notebooks demonstrate a much earlier and detailed post-*Beagle* reading than this. See also *CCD*, 2:19 [dated May 19 to June 16, 1837], where Darwin, at work on the publication of his *Beagle* diary, asks his sister Caroline to check whether a particular passage on acquired instincts in crows is in *Zoonomia* or the notes to *The Botanic Garden*. Darwin, pre-*Beagle*, was also well versed with *The Botanic Garden* and *The Temple of Nature*. In a letter to his sister Susan written from Sydney in 1836, Charles refers to Erasmus Darwin's verses on the Wedgwood medallion of Sydney Harbor in *The Botanic Garden*; *CCD*, 1:482; in his *Beagle* diary he refers to the primeval forests of Brazil and Tierra del Fuego as "temples filled with the varied productions of the God of Nature"; *Diary*, 444. Darwin's annotated copies of Erasmus Darwin's books are in the Cambridge collection; for his annotations, see *Marginalia*, 1:183–88.
75. *Notebooks*, B 1.
76. Hodge 1985; R. Richards 1992.
77. This is the opinion also of Robert Richards; R. Richards 1992, 92–98.
78. *Notebooks*, D 57, 76 and "proved facts relating to generation" D 167–79, 174–175, 152e–162, 168–173e.
79. *Marginalia*, 1:183–88.
80. *Notebooks*, C 61.
81. Ghiselin, in the process of denying that Charles Darwin's concept of male combat was derived from his grandfather, makes the point that Charles's pre-Malthusian

formulation of the adaptive or progressive role of male combat "could be translated into natural, but not sexual, selection, but only if the population model were understood"; Ghiselin 1972, 218. Richard Burkhardt points to Darwin's prior formulation of the struggle for mates in the period leading up to the Malthusian formulation; R. Burkhardt 1985, 341–42. Gruber also makes the point that "in its nascent [pre-Malthusian] form, the idea of natural selection is suffused with the special notion of sexual selection"; Gruber 1974, 167; note also Kohn and Hodge 1985, 190–91, 197.

82. *Marginalia*, 1·185. There is, however, the possibility that this notation was made much later, when Charles Darwin was working on his "Life" of Erasmus. It is not possible to determine the dates of his various annotations.

83. This point is well made by Hodge 2003, 60–61; see also Radick 2003, 149.

84. "The passion of the doe to the victorious stag, who rubs the skin off horns to fight, is analogous to the love of women (as . . . seen in savages) to brave men." September 13, 1838; *Notebooks*, D 99.

85. *Notebooks*, D 103.

86. September 17, 1838; *Notebooks*, D 113–14; see also his earlier entry C 178, written mid-May to June 1838.

87. *Notebooks*, M 149, dated September 23, 1838. According to Carl Bajema, this is where Darwin first "explicitly mentioned both forms [i.e., male combat and female choice] of what he was later to define as sexual selection"; Bajema 1984, 99. Promising as it is, this "mention" is hardly an explicit formulation of male combat, lacking any sense of its reproductive advantage, while the notion of female or aesthetic choice has to be read into it—and then only by default from the failure of the male to attract the female.

88. *Notebooks*, D 134–35.

89. For a succinct account of these stages in theory construction, see Hodge 2003.

90. It is to be noted that the very next entry after Darwin's famous Malthusian insight was on the sexual interest of male monkeys in other female animals and especially in their closest relatives, women: "These facts may, be turned to ridicule, or may be thought disgusting, but to philosophic naturalist pregnant with interest"; *Notebooks*, D 136–39.

91. In the so-called "Pencil Sketch," when Darwin wrote his first tentative summary of his selectionist views in 1842; see *Foundations*, 10.

92. This point was made by Roy Porter, though he did not link it with Charles Darwin's ideas on beauty; R. Porter 1989, 51.

93. The phrase comes from Faas 2002, 136.

94. *Descent* 1879, 247.

95. J. Logan 1972, 69. On the conventional relation of beauty to utility, see Guyer 2002.

96. For instance, see *Notebooks*, E 58, where Darwin, summarizing the three principles of natural selection, wrote "(1) Grandchildren like grandfathers."

97. See Hodge 2010; Endersby 2003, 78–82; Ghiselin 1976, 127, 129.

98. The phrase, a good one, is taken from Endersby 2003, 69.

99. Hodge 2003, 45–46, 64–67; 2009b. On the family *Weltanschauung* of the two Darwins, see also Gruber 1974, ch. 3.

CHAPTER FOUR

1. *CCD*, 1:67, 393; F. Darwin 1888, 1:145–46.

2. *Autobiography*, 61; F. Darwin 1888, 1:103, 146; see Browne 1995, 105.

3. *CCD*, 1:150.

4. Ibid., 1:396–97. In her beautifully nuanced analysis of the imagery of this letter, Gillian Beer lays bare the "primary nakedness" of Darwin's three images: the "inscrutable," "natural," male body of the Fuegian, the untouchable, "encultured," erotic female body of Titian's *Venus and Cupid with a Lute-Player*, and the "known" body of the gutted small animal that Darwin touches and invades in his practice as a zoologist. She relates these to his future evolutionary theorizing in which "sex and sexual congress is central," but largely ignores their racial dimensions; Beer 1996, 14–30.

5. CCD, 13:35.

6. *Notebooks*, B 33–34, 71, 68–69, 93, 189 (written about December 1837); B 161 (around November 1837); see also C 71 (around April 1838); see also Gruber 1974, 187.

7. CCD, 2:444–45; Gruber 1974, 188.

8. E.g., *Notebooks*, D 137–39, M 85.

9. R. Richards 1987, 83–98; R. Richards 2003.

10. *Notebooks*, B 4.

11. See Greene 1961, 222–29; Wood 1996; Bindman 2002, 58–78 and passim.

12. CCD, 1:345; *Diary*, 42; see Sloan 2003; R. Richards 2002, 518–26; R. Richards 2003. Not all Darwin's romanticism was Humboldtian in origin; there were homegrown influences at work here too, notably that of Erasmus Darwin: e.g., *Diary*, 444. Kohn makes the point that Darwin's "enthusiastic reading [of Humboldt] is only the most familiar part of a much richer chain of connections between Darwin's cultural and educational exposure and European romantic natural history"; Kohn 1996, 17.

13. See annotations by Darwin in CUL Humboldt 1819, 4, 466–522; *Marginalia*, 1:417, 418; *Descent* 1871, 2:351, 352; *Expression*, 481; Sloan 2003, 34.

14. Stepan 2001, 25, 35–43, 67; Leask 2003, 28–29; Lindquist 2004; Rupke 1999, 335; Rupke 2005.

15. *Diary*, 75; CCD, 1:397.

16. *Diary*, 424.

17. This is most evident in Darwin's well-known metaphor of the "entangled bank" in the conclusion to the *Origin*, the metaphor David Kohn sees as encapsulating the "destructive creative balance of natural selection"; Kohn 1996. Kohn sees Miltonic influences at work in Darwin's imagery of Tierra del Fuego (Robert Richards also invokes Milton in explaining the "tangled bank" metaphor, though without reference to Tierra del Fuego; R. Richards 2002, 537–39), while Leask, recognizing its non-Humboldtian aspects, yet depicts Darwin's descriptions of Tierra del Fuego as a "nightmarish inversion" of the more favored Humboldtian tropics; Leask 2003, 23. For Darwin on the scenery of Tierra del Fuego, see *Diary*, 124–27, 219–20.

18. Hobsbawm 1973, 318, 330; Eagleton 1990.

19. Humboldt apart, Darwin's sole gesture toward German aesthetics was to read Gotthold Ephraim Lessing's famous *Laocoon*, which had appeared in a recent English translation of 1836, because "[brother] Erasmus thinks I should lik[e] it"; *Notebooks*, C 266.

20. Dillon 2004, 498; Hobsbawm 1973, 307–33; Bindman 2002, 17–21; Eagleton 1990, 3–4, 7–9, 31–66.

21. The *Discourses* comprised a body of lectures delivered in the Academy from its establishment in 1769 until Reynolds's retirement in 1790; they were translated into Italian, French, and German during his lifetime and reissued in countless English editions after their author's death; Wark 1975, v, xvi, xxiii; R. Jones 1998, 117–52, 207; Barrell 1986b; Barrell 1986a, 63–72.

22. Hume, "Of the Standard of Taste" (1742), quoted in Gurstein 2000, 207.

23. Sir Joshua Reynolds 1798, 1:212–13, 219–20; *Notebooks*, OUN 8, 24.

24. Sir Joshua Reynolds 1798, 1:219–20. On taste and the "conversable world," see Gurstein 2000.
25. *Notebooks*, N 26–29.
26. Sir Joshua Reynolds 1798, 1:199–200, 231–32, 235–37. See Leypoldt 1999, 339; for a more political interpretation, see Uphaus 1978, 62.
27. Sir Joshua Reynolds 1798, 2:240–41. This is the edition given as read by Darwin in the bibliography to the edited *Notebooks*, and, from the content of his notebook references to Reynolds, Darwin clearly read this *Idler* essay along with the *Discourses*, though this is not recognized by the editors of the *Notebooks*.
28. *Notebooks*, M 31–32.
29. See Dabydeen 1985, 41–51; Meijer 1999, 160–66; S. Gilman 1982, 19–34.
30. Burke 1823, 211–12; S. Gilman 1982, 19–34; Armstrong 1996, 219–20.
31. "We know how filthy the Hottentots are, and how much there is which they esteem as delicate and holy which only excites in us disgust and horror: a squashed nose, flabby breasts hanging down to the navel, the whole body anointed with a varnish of goat's fat . . . : conceive this to be the object of an ardent, reverential, tender love: let us imagine these details expressed in the noble language of earnest admiration, and abstain from laughing"; Lessing ca. 1874, 186–87; S. Gilman 1982, 25–29. Reynolds, like Lessing, is drawing on Chesterfield's tale published in the English weekly the *Connoisseur*, which offered a skeptical recategorization of standards of beauty, opposing the Hottentot beauty Knonmquaiha to the painted and bewigged Mynheer van Snickersnee; see S. Gilman 1982; Hudson 2008, 28–30.
32. *Diary*, 125, 225.
33. *Descent* 1871, 2:353.
34. *Descent* 1871, 1:64.
35. *Notebooks*, OUN 11; Sir Joshua Reynolds 1798, 2:119–20; Barrell 1986b.
36. See Leypoldt 1999, 342–45; R. Jones 1998, 88–89; Bindman 2002, 67–70.
37. Sir Joshua Reynolds 1798, 1: dedication. The notorious footnote appended by Hume to the 1754 edition of his essay makes just this point (though in language that Reynolds nowhere adopts): "I am apt to suspect the Negroes to be naturally inferior to the Whites. There scarcely ever was a civilized nation of that complexion, nor even any individual, eminent either in action or speculation. No ingenious manufactures among them, no arts, no sciences . . . Such a uniform and constant difference could not happen, in so many countries and ages, if nature had not made an original distinction between these breeds of men." Quoted in Bindman 2002, 69–70.
38. *Descent* 1871, 1:169.
39. DAR 119:3a; DAR 91:10–11; *Notebooks*, M 31–32; OUN 10–11; N 26–31. Darwin's *Beagle* diary contains an entry on the Tahitians that suggests an earlier direct Reynolds influence: "In my opinion, [the Tahitians] are the finest men I have ever beheld . . . It has been remarked that but little habit makes a darker tint of skin more pleasing & natural to the eye of an European than his own color.—To see a white man bathing alongside a Tahitian, was like comparing a plant bleached by the gardeners art, to the same growing in the open fields"; *Diary*, 366. Only the Tahitians lived up to Darwin's expectations of the noble savage.
40. *Notebooks*, M 31–32 [<< >> indicates Darwin's later insertion]; see also C 255–26, M 58 and ZEd 19, 20, on the differences in birdsong in America and England.
41. E. Darwin 1794, 1:155. Around this same time, Charles Darwin was extending the notion of sexual competition to song, and more tentatively, to plumage; *Notebooks*, D 113–14. See ch. 3.

42. *Notebooks*, M 85 (August 1838). Darwin had also been reading Martineau's observations on national character, which argued that the moral differences among the different races or nations were the "result of the particular circumstances amidst which the society exists"; *Notebooks*, M 75–77; see also M 142.
43. *Notebooks*, OUN 10–11b, written before September 6, 1838.
44. *Notebooks*, N 19.
45. See *Notebooks*, N 18, 107–9; *Descent* 1871, 2:337, 330–7; 1:64.
46. *Notebooks*, N 26–28, written October 27, 1838; see also M 31–32.
47. Sir Joshua Reynolds 1798, 1:238–39. This was a caveat that gave Reynolds cause for self-congratulation in the wake of the French Revolution: nowhere in his *Discourses* had he "lent [his] assistance to foster newly-*hatched unfledged* opinions"; see Uphaus 1978, 62.
48. R. Richards 1987, 2003.
49. See also *Notebooks*, C 221: "Male glow-worm knowing female good case of instinct." By the second issue of the *Descent*, Darwin was able to explain the puzzle of the superior luminosity of the female glow worm not through sexual selection, but through mimicry; *Descent* 1879, 319.
50. *Notebooks*, D 76, written September 9, 1838.
51. *Notebooks*, N 64 (March 16, 1839).
52. *Notebooks*, N 59 (January 6, 1839).
53. On the erotic symbolism of Titian's various depictions of Venus, see Faas 2002, 90–92. Its eroticized musicality may explain the jump from the evocation of the Titian *Venus* to the desire to hear some good music in Darwin's letter to Whitley. Music is the analogue of the visual experience; see Beer 1996, 21–30.
54. *CCD*, 2:155, 147–48.
55. Cited in Faas 2002, 132–33; see also Whale 2000, 26–27; Eagleton 1990, 54; Eagleton 1989.
56. Darwin's 1864 recollection to Kingsley of the "naked painted, shivering hideous savage" whose morality is less than that of the good-hearted ape, is an almost irresistible reminder of Burke's well-known denunciation of the savagery of the regicidal sansculottes and those revolutionary philosophes who would "rudely" tear off "all the decent drapery of life" in the new era of "light and reason": "All the super-added ideas, furnished from the wardrobe of a moral imagination, which the heart owns and the understanding ratifies as necessary to cover the defects of our naked, shivering nature, and to raise it to dignity in our own estimation, are to be exploded as a ridiculous, absurd, and antiquated fashion." I am indebted to Ofer Gal for this observation; Burke 1864, 515–16; *CCD*, 10:71; see Eagleton 1989. In a radical revision of Burke's aesthetics and his "politics of pain," Luke Gibbons argues that far from being an uncritical apologist for the existing social order, as this "emblematic figure of conservative thought" is conventionally depicted, Burke's Catholic Irish background gave him an abiding distrust of British colonialism and its excesses of repression and vengeful reprisal against the rebellions of the colonized. We must, argues Gibbons, reassess Burke's aesthetics and its politics against the background of the "complicity of liberalism with the despotism of colonialism"; Gibbons 2003, xii, 12, and passim. Whale also argues the need for reassessment of the politics of Burke's aesthetics, seeing it as not simply serving the dominant, but offering as many opportunities as disadvantages to the political imagination; Whale 2000, 19–41.
57. Burke 1823, 47, 125.
58. Ibid., 166.

59. Ibid., 50–51.

60. Ibid., 157–71; Armstrong 1996, 215–17.

61. Most of these references relate to Burke's notion of the "sympathetic sublime" which Burke directly opposed to Adam Smith's theory of sympathy with its detached and judgmental "impartial spectator." Darwin, mulling over the problem of the evolution of altruistic behavior through inherited habit, a problem which he only much later resolved, was attracted to Burke's view that the emotional rapture of the sublime is associated with the proximity of danger or pain, which nevertheless excites ambivalent, agreeable sensations. *Notebooks*, M 51, is the earliest reference to Burke, made around July 1838; for Darwin's annotations in his copy of the *Inquiry*, 1823, CUL (Rare Books), see *Marginalia*, 1:102–3. On Darwin and Burke's "sympathetic sublime," see Larson 2013.

62. Burke, 1832, CUL; *Marginalia*, 1:102; *Notebooks*, B 161 (around November 1837); see also C 71 (around April 1838).

63. Ironically, it was Burke who offered the figure of the grandfather as the "reconciling image," enfeebled by age into a "feminine partiality," that might soften the filial fear and rebellious resentment aroused by the terrible authority of the towering masculine sublime; Burke 1823, 160; Eagleton 1990, 55. Surely not coincidentally, Charles Darwin offers a notebook evocation of "my F." in a rage, rubbing his hands, stamping his feet, and grinding his teeth to heighten his fury, an entry that reduces this display of paternal terrorism to physiognomic proportions as indicated by Grandfather Erasmus and Burke; *Notebooks*, M 149.

64. *Notebooks*, M 71; see also M 38–39 for Darwin's notes on Erasmus Darwin's connection between breasts and the line of beauty; M 108–9, dated August 26 [1838].

65. *Notebooks*, M 82, 84.

66. *Notebooks*, N 41, dated November 27 [1838].

67. *Notebooks*, N 45, 51–52, 56–59 (note the specific reference to Darwin's notes in "Burke's essay on the sublime & Beautiful"); M 146, 150, 153; Burke 1823, 219.

68. *Notebooks*, M 51–53; N 68; OUN 39.

69. *Notebooks*, N 6–10. Darwin also scored the margin of his copy of Burke's *Inquiry* where Burke discussed the practices of Campanella; Burke 1823, 191–92; *Marginalia*, 1:103. On Burke's 'psycho-physiology," see Eagleton 1990, 57; Eagleton 1989; Klonk 1996, 13–14; Gibbons 2003, 116–18.

70. *Notebooks*, M 107; E. Darwin 1794, 1:147–53. On Burke's "politics of pain," see Gibbons 2003.

71. *Notebooks*, M 128.

72. *Notebooks*, M 138; see also D 136–39.

73. *Notebooks*, M 153; see also C 79 and M 142.

74. *Notebooks*, N 87; see also N 63.

75. *Notebooks*, N 64.

76. *Notebooks*, N 107, 109. Note Darwin's remarks on insects.

77. Darwin's annotated copy of Bell is in the Cambridge Library; CUL Bell (1806) 1844, 83, 121, 141; *Marginalia*, 1:47–49.

78. *Expression*, 355. On the blushing of the Fuegians, see *Notebooks*, N 15, *Expression*, 318.

79. R. Richards 2003, 110; see also R. Burkhardt 1985, 360; Hartley 2005, 146, ch. 5.

80. *Expression*, 353, 361.

81. Even before then, in his *Beagle* diary, Darwin had indicated his awareness of the aesthetic debate about the relation of beauty with utility, noting "how little" in general "utility explains the delight received from any fine prospect"; *Diary*, 366.

82. On Erasmus Darwin's rejection of the inclusion of fitness or utility in his aesthetics, see J. Logan 1972, 69. On the eighteenth-century debate over the relation between beauty and utility, see Guyer 2002.
83. Burke 1823, 149–52.
84. Ibid., 173, 146. Burke's overtly feminized, sexualized, and Europeanized definition of beauty implies an ugliness to non-European, dark-skinned people. But if so, in Burke's thinking, as we have seen, this ugliness was associated with sublimity. That Darwin took account of the Burkean claim that blackness is inherently both ugly and fearful, is evidenced in his notebook entry on the terror engendered in his small nephew by the approach of his sister who had covered her head with a veil, even though he knew it to be his sister: "Is this part of same feeling which make us think anything ugly—a beau-ideal feeling?"; *Notebooks*, N 121.
85. *Notebooks*, OUN 14. There was a contemporary literary substream of "agreeable ugliness" devoted to sermonizing the superior virtues of feminine morality, modesty and docility over looks; where it was the plain but good, sweet maid who charmed the hero away from the vain, dissipated, socially ambitious beauty; R. Jones 1998, 170–88.
86. *Notebooks*, C 109–12; D 147, 160.
87. *Notebooks*, E 146–47; written around June 1839.
88. Addison 1970, 179–80 (see sec. 3.7). Other than Burke's reference to Addison in denying a sense of beauty to animals (Burke 1823, 50–51), I have not been able to link Charles Darwin's reading in aesthetics to Addison.
89. For instance, *Notebooks*, E 160.
90. *Descent* 1871, 2:92.
91. *Variorum*, 371.
92. *Notebooks*, OUN 14; Jeffrey 1811.
93. *CCD*, 2:80–81.
94. Hipple 1957, 158–81. For an analysis of Alison's aesthetics, see Dickie 1996, 55–84; on Jeffrey and the *Edinburgh Review*, see Rendall 2004; on Jeffrey's "laissez faire" aesthetics, see Kaiser 1996, 104.
95. Jeffrey 1811, 18.
96. Ibid., 17, 10, 12, 16.
97. The connections between Alison's associationism and the aesthetics of Erasmus Darwin are explored in J. Logan 1972, 66–69.
98. *Notebooks*, OUN 8.
99. According to the editors of the published *Notebooks*, Darwin's entry possibly dates from before September 6, 1838, but according to Darwin's reading notebooks, he read Jeffrey's review toward the end of 1840 (though this possibly refers to a second reading); DAR 119:9b. I have found Darwin's reading notebooks a very unreliable guide to the actual dates of his readings as indicated by the contents of the transmutation notebooks.
100. Jeffrey 1811, 10, 12, 18, 44; see Camfield 1988, 335; Klonk 1996. 23–24.
101. Quoted in Rendall 2004, 366.
102. R. Jones 1998, 80–81, 114.
103. Faas 2002, 136–37.
104. Darwin's notes on Stewart's views on the sublime and taste are in *Notebooks*, OUN 18–21.
105. See sec. 14.2; Budge 2003, v–xxv; Klonk 1996, 22–26.
106. Camfield 1988; Dillon 2004.
107. *Descent* 1871, 1:65–69; R. Richards 2003.

108. It is tempting, but far too facile, to say that while the Fuegians may not have known the devil, Darwin saw the devil in Tierra del Fuego and he was a Fuegian. *Descent* 1871, 1:67; *Notebooks*, C 244, M 135–6.
109. *Descent* 1871, 2:350.
110. Tipton 1999, 127 28.
111. *Autobiography*, 138–39.

CHAPTER FIVE

1. Lawrence 1822; *Marginalia*, 1:485–86; Desmond and Moore 1991, 253.
2. Wells 1971; Desmond 1989, 117–21; Jacyna 1983, 312–13.
3. See Wells 1971.
4. On the alleged critical role of Smith's formulation in Darwin's theorizing, see Alter 2007, 246–49; for claims of the significance of Prichard, see Moore and Desmond, introduction to *Descent* 1879, xxix–xxx.
5. The work of David Bindman is exemplary; Bindman 2002. See also Robert J. C. Young 2006, 94–98 and passim; Meijer 1999; Callanan 2006; Hudson 2008.
6. Bindman 2002, 11–21; Callanan 2006, 5–6; Eze 1997, 98; Stepan 1982, xii.
7. Bindman 2002, 90, 81–92; Leypoldt 1999, 342–43.
8. Norton 1995, 177; Hartley 2005, 42; Bindman 2002, 92–4.
9. Gray 2004, 5, 1–55; Bindman 2002, 94–99; Hartley 2005, 32–43.
10. See Cooter 1984; Gray 2004, 35–53, 57–69; Norton 1995, 181; Bindman 2002, 100, 104–5, 108.
11. Lavater ca. 1800, 2:23–26; Gray 2004, 103–5.
12. Lavater 1789, 3:85, 85–106; Bindman 2002, 96.
13. Lavater ca. 1800, 1:116; Lavater 1789, 2:223–26; Gray 2004, 86, 104–7; Bindman 2002, 102, 108–13.
14. *Autobiography*, 72; Juengel 1996; Juengel 2001.
15. Bell (1806) 1844, 52–53, 144–45; Hartley 2005, 36, 44–79; J. Smith 2006, 202–4.
16. Darwin read Lavater in both the 1804 English edition and the ten-volume French edition of 1820 (a copy of which he had owned pre-*Beagle*) toward the end of 1838; *Notebooks*, C 270, D 164, 165, M 145, N 6, 9, 10. Certain of his annotations relate explicitly to "Sexual Selection." These are contained in the 1820 French edition of Lavater, and must date from a later period when the concept had been fully formulated by Darwin (see ch. 10); Lavater 1820, 4:17, 120; see *Marginalia*, 1:484–85.
17. Camper 1794, 1:9, 33–36; Bindman 2002, 202–4.
18. In a major reassessment of Camper's anthropology, Miriam Meijer argues that the intention of his facial angle was "to demonstrate the equal worth of facial features among the human races"; see Meijer 1999, 154–60, and passim.
19. Darwin's copy of Lavater includes a volume on Camper, including this well-known diagram: 1820, 4: plate 179 (opposite p.119). Both Meijer and Bindman stress that for Camper, aesthetics was not separate from anthropology—his lecture on the facial angle was intended for both artists and anatomists; Bindman 2002, 206; Meijer 1999, 5 and passim; see also Grindle 1997.
20. Bindman 2002, 208–9; Meijer 1999, 118–21, 172.
21. In Darwin's copy of Lavater 1820, this was reproduced in vol. 8, plates 527–28 (opposite pp. 12–13). See also Gray 2004, 110–111; Hartley 2005, 34–35; Bindman 2002, 210–11; Meijer 1999, 121.
22. Bindman 2002, 200–201; Greene 1961, 223; Blumenbach 1865, 102, 186; *Marginalia*, 1:61.

23. Blumenbach 1865, 265, 223, 269; Schiebinger 1993, 129–31.
24. Bindman 2002, 201, 209; Meijer 1999, 168–171.
25. Schiebinger 1993, 131–32; *Descent* 1871, 2:356–57.
26. Morton instrumentalized the measurement of the facial angle with his "goniometer," an instrument improved on by Paul Broca, one of the founders of French anthropology, who standardised its use in his "craniometry," the scientific measurement of skull shape and cranial capacity; see Meijer 1999, 171–77; Haller 1995, 9–34; Gould 1981, 82–88.
27. C. White 1799, 2:42; Bindman 2002, 214.
28. C. White 1799, 134–35; Bindman 2002, 214–18.
29. Schiebinger 1993, 64. Darwin read White's *Regular Gradation* in 1843, dismissing it as a "foolish book with some odd facts"; DAR 119:14a; *CCD*, 13:359–60.
30. Lawrence 1822, 290–91; Schiebinger 1993, 133, 154–55; Meijer 1999, 140–43; Bindman 2002, 196–200; Hudson 2008.
31. C. Lyell 1830–33, 2:2, 60–61.
32. *CCD*, 3:38n1, 258; *Descent* 1871, 1, 219–20; Desmond and Moore 2009, 92–94, 97–98; Desmond 1989, 288–91.
33. *CCD*, 1:560. For a full listing of works on board the *Beagle*, see *CCD*, 1:558–65. Darwin had his own annotated copy of Forster's *Observations*, owned pre-*Beagle*. It does not appear to have been the one on the *Beagle*, which possibly belonged to FitzRoy; *Marginalia*, 1:240.
34. Forster had himself translated into English and published Bougainville's *Voyage* in 1772; this was the version on the *Beagle*; *CCD*, 1:558.
35. Forster (1778) 1996, xliv, 154, 174.
36. Ibid., lxxviii, 180, 192–93, 202–8. Perhaps through his radical connections, Forster accorded women an unusual degree of agency in social change, viewing the women of Tahiti as having achieved "greater equality" and therefore a greater civilizing influence on Tahitian society. However, this was not without the usual contradictions; see Guest 1996.
37. Forster (1778) 1996, 171, 180; *Diary*, 169. Note also the conclusion to the *Origin of Species*, where Darwin deployed Forster's measure of Fuegian intellectual deprivation and cultural inferiority as the foil against which he might counter pose the intellectual promise of an advanced evolutionary outlook; *Origin*, 456–58; cf., Forster (1778) 1996, 193.
38. See Forster (1778) 1996, 164–65, 173–74; Bindman 2002, 127.
39. C. White 1799, 2:137.
40. Winthrop Jordan, introduction to S. Smith (1810) 1965, v–vii; Paul Wood, introduction to S. Smith (1788) 1995. On Smith's Common Sense agenda and its pervasive influence in early American politics and ideology, see also Livingstone 1999, 107–9; Livingstone 2005.
41. Burrow 1966, 10–23; Stocking 1987, 16–30; Wood, in S. Smith (1788) 1995, vi–xvi; Wood 1996.
42. Kames 1774, 1:168–69, 183, 216; Wood 1996, 206–7.
43. Kames 1774, 1:12–13; *Notebooks*, C 257.
44. Kames 1774, 1:37–38.
45. Winthrop Jordan, introduction to S. Smith (1810) 1965, xxviii, xxxiii–xl; 151–86. On the ideological role of Smith's ethnology, see Livingstone 1999, 108–9; Livingstone 2005, 324–28.
46. S. Smith (1810) 1965, 97–98, 109, 149. All quotations are from this reprint of the second 1810 edition of the *Essay*.
47. Ibid., 74–78, 160–63.

48. Ibid., 33–40, 71–72, 105. On Smith's "magnificent and pathetic" proposition, see Winthrop Jordan's introduction in ibid., xlii, xlv, xlvi.
49. Ibid., 97n, 69.
50. Ibid., 109, 98.
51. Ibid., 114, 126, 109, 98.
52. Ibid., 115.
53. Ibid., 117.
54. Ibid., 118, 119.
55. DAR 119:7a, 13a.
56. S. Smith (1788) 1995; S. Smith (1810) 1965, xvii. Smith's essay was generally known in Britain as an outstanding representation of the argument that human variations were caused by environment, especially by climate. At the popular level, the article on "Complexion" in Rees's *Cyclopaedia* drew heavily on Smith's account of the environmental causes of skin color, but not on his idea of aesthetic selection; Rees 1819, ix, "Complexion"; see Juengel 2001.
57. Stocking 1971; Stocking 1987, 242–44; Bravo 1996.
58. Biographical detail on Prichard taken from Stocking's "Introductory Essay," Prichard (1813) 1973; see also Stocking 1987, 48–53; Augstein 1999.
59. Curiously, Stocking, in his otherwise comprehensive discussion of Prichard's sources, makes no mention of Smith. Augstein mentions Smith, but does not give him any influential role in Prichard's theorizing; Stocking in Prichard (1813) 1973; Augstein 1999, 124n30.
60. Stocking in Prichard (1813) 1973, xxvii–xx, xliv–xlix; Augstein 1999, 108–11.
61. Prichard (1813) 1973, 13–14, 84–85; Augstein 1999, chs. 3 and 4.
62. Prichard (1813) 1973, xlix–1, 227, 233–39. Others before Prichard had argued for an original blackness of primitive man, including Camper and John Hunter; see Augstein 1999, 131–33.
63. Stocking in Prichard (1813) 1973, lxv–lxx; Augstein 1999, 112.
64. Prichard (1813) 1973, 38, 41, 41n.
65. Ibid., 231–32, 235; Augstein 1999, 110–11.
66. Prichard (1813) 1973, 71–76.
67. Ibid., 41–42. Prichard established the axiom of innate or instinctive repugnance against interspecific breeding in the wild, arguing by analogy that since there was no "invincible" repugnance between men and women of different races and since human "mixed breeds" were invariably prolific, hence "the several tribes of man are but varieties of the same species." From the time of the notebooks, Darwin adopted this definition in preference to the standard one of hybrid sterility; notably Notebooks, C 161e; see Augstein 1999, 75–76, 235; Prichard 1826, 1:95–98, 126–28; Prichard 1836–47, 1:147, 150; Prichard 1843, 11–26.
68. Prichard (1813) 1973, 44–45.
69. In this second edition, Prichard suggested a differentiation of sex roles in the hereditary process: the father contributed "hereditary conformation" or stability, while congenital variations arose via the mother, primarily through the influence of changing climatic conditions during pregnancy, Prichard 1826, 2:536–55; see also Stocking in Prichard (1813) 1973, lxix; Augstein 1999, 113–14.
70. Stocking in Prichard (1813) 1973, lxxviii–lxxx; Augstein dissents in some particulars from this interpretation; 1999, 111–22.
71. Browne 1995, 421; Augstein 1999, 145–46; Desmond and Moore 2009, 150–51; DAR 119:6a. Prichard's "On the Extinction of Human Races" (1840) reflects the dual

benevolent and scientific mission of the Aborigines Protection Society in which Prichard was active; see Bravo 1996. What impressed Darwin was Prichard's "profound consideration" of the "method by which races of men have been exterminated . . . owing to immigration of other races"; see *Notebooks*, TAN 81.

72. Moore and Desmond; introduction to *Descent* 1879, xxix–xxx; this position is less pronounced in Desmond and Moore 2009, 282–84.

73. Darwin owned and annotated the five volumes of the 3rd ed. of Prichard's *Researches* (Prichard 1836–47) and the first two volumes of the posthumous 4th ed. (Prichard 1851). See *Marginalia*, 1:680–86. He abstracted vols. 1 and 2 of the 3rd ed., possibly in 1839 or '40: DAR 71:139–42; 119:6a; 120; 71:4. He read vols. 3, 4, and 5 of the 3rd ed. in 1847: DAR 119:20a. In 1856, Darwin also read and abstracted Prichard's *Natural History of Man*, 3rd ed. 1848; DAR 71:143–45; 128:18.

74. In a note in the back of his copy of the 4th ed. of 1851, Darwin reminded himself, "If I ever consider Man look over other & earlier Edition." This note in Darwin's hand on the back of SB2 in Darwin's copy of Prichard's *Researches*, vol. 1, 4th ed. 1851, CUL, has been overlooked by the compilers of the *Marginalia*: see *Marginalia*, 1:680–86. Only the third and fourth editions are cited in the *Descent* 1871, 1:115, 119, 146; 2:340, 345, 349, 352.

75. Wells 1971, 359.

76. Lawrence had previously translated and edited Blumenbach's *Short History of Comparative Anatomy* (London: Longman, 1807).

77. Cooter 1984, 28–35, 44, 45, 312n50.

78. Lawrence 1822, 473, 445.

79. Ibid., 387, 441.

80. Ibid., 390; note Darwin's speculations on excluding and isolating "peculiar" people, *Notebooks*, B119–20.

81. Lawrence 1822, 473.

82. Ibid., 418–25.

83. Ibid., 432. Lawrence, in relating skull shape to cerebral organization and mental faculties, was to some extent influenced by Franz Joseph Gall's phrenology or craniology which became a popular movement in early nineteenth century England and was taken up by the more reformist members of the medical profession; see Cooter 1984.

84. Lawrence was a scathing critic of White's "false" interpretation of Camper's facial angle that "condemns the poor African to the degrading situation of a connecting link between the superior races of mankind and the orang-utan"; the "moral and political consequences, to which it would lead, are shocking and detestable"; Lawrence 1822, 107–8. However, it must be said that Lawrence's expression at times lends itself to the very racism he denies.

85. Lawrence 1822, 434, 426–29. Note Darwin's similar comparison: "Must grant that the conscience varies in different races.—no more wonderful than dogs should have different instincts"; *Notebooks*, M 75.

86. Lawrence 1822, 441.

87. Ibid., 393. Lawrence evidently worked from the 1788 Edinburgh edition of Smith's *Essay*; cf. S. Smith (1788) 1995, 112–13.

88. Lawrence 1822, 393, 397; *Marginalia*, 1:485–86. John Waller, in his survey of hereditarian and eugenic thought in the pre-Galton period, emphasizes the historical importance of Lawrence's two page discussion of regulated breeding: "[It] foreshadowed the profound importance that the professionalization of science, medicine and government would have for the credibility of eugenical thought"; Waller 2001, 469.

89. Lawrence 1822, 398.
90. Desmond 1989.
91. DAR 119:iv; *Marginalia*, 1:485.
92. *CCD*, 2:141–2.
93. *Notebooks*, C 204. In this same entry, Darwin also noted Lawrence's references to the elongated nymphae of "Hottentot women" and noted that in "some monkeys clitoris wonderfully produced," reminding himself to "make abstract on this subject from Lawrence, Blumenbach & Prichard." See CUL Lawrence 1822, 366–68. Note also *Notebooks*, C 178, where Darwin, speculating on the hairiness of men and on whether cock birds attract females by beauty as well as song, refers to the "analogy of man" in this connection. For the dating of these entries, see *Notebooks*, 237.
94. Lawrence 1822, 451–52 (CUL).
95. *Notebooks*, M 32 (around July/August 1838), N 26 (October 26, 1838).
96. *Notebooks*, M 85 (August 1838)
97. *Notebooks*, C 166; M 19, 57. I am not arguing that Darwin became a materialist through reading Lawrence, but rather that he was receptive to such argumentation. Similarly, Darwin's extensive readings in aesthetics around this time had to do with his thinking on instinct and his attempt to explain instinct as the effects of inherited habit, see ch. 4.
98. *Notebooks*, C 163, 166; see also M 73, 74.
99. *Notebooks*, M 57. See Gruber 1974, 202–5; Desmond 1989, 412–14; cf. R. Richards 1987, 152–54.
100. Lawrence 1822, 393 (CUL).
101. Darwin (1876) 2008, 423; *Autobiography*, 140. The sequence of events that led Darwin back to Lawrence's *Lectures* around 1856 is discussed in chs. 10 and 11.
102. *CCD*, 11:217.
103. *Descent* 1871, 2:357, 318, 338, 349.
104. That said, it is quite probable that Darwin, at this early stage, was even more interested in Lawrence's speculations on the porcupine family and his views on the inheritance of congenital, as opposed to acquired, variations. These intermeshed with Darwin's accumulation of readings and notes on variation, inheritance, monstrosity and embryology, and their relation to adaptive species change. Darwin read and absorbed Lawrence's *Lectures* during 1838, the most intense and creative phase of his theory construction, when he arrived at his theory of natural selection. Wells in 1971 drew attention to the similarity of Lawrence's and Darwin's early views on variation and inheritance, but was misled into thinking that Darwin's copy of the *Lectures* was not annotated and that Darwin first read the *Lectures* in 1847; Wells 1971, 337.
105. Desmond 1989, 120.
106. Prichard 1836–47, xii, 1.
107. *CCD*, 5:401.
108. *CCD*, 12:99; Brandon-Jones 1997; Sheets-Pyenson 1981; *CCD*, 6:38–39, 43.
109. *CCD*, 4:139; *Notebooks*, D 29e, 95e; *CCD*, 5:317n25.
110. Brandon-Jones 1997; *CCD*, 8:232n7, 233.
111. Blyth 1835.
112. Blyth seems to have been familiar with the original version of Prichard's *Researches* as well as its subsequent editions, see his letter to Darwin, January 23, 1856, *CCD*, 6:26–27. Although he did not cite it, Blyth's essay was also obviously indebted to Sebright's 1809 paper on the "Art of Breeding" (see ch. 6); and, no less obviously to

the *Zoonomia* of Erasmus Darwin (also uncited), from which Blyth quarried certain ideas on protective coloration, male combat and the struggle for existence.

113. Blyth 1835, 42, 47.
114. Ibid., 47–48.
115. Ibid., 48–49.
116. Ibid., 46.
117. Blyth 1836, 397.
118. *Notebooks*, C 70, 71, see also C 82, 84, 85; Sheets-Pyenson dates early readings from July 1837, and rereadings in 1840 and as late as 1856; Sheets-Pyenson 1981, 238; DAR 119:10a.
119. But it is going too far to claim, as did Loren Eiseley, that Darwin "passed by way of the stepping stone" of Blyth to his concepts of natural and sexual selection. As Eiseley himself shows, Blyth's conception of the struggle for existence and sexual selection was very different from that Darwin came to hold. Eiseley's further claim that Darwin later suppressed or carefully covered the tracks of his debt to Blyth is not borne out by the evidence of the *Notebooks* or by Darwin's many later attributions to Blyth in his published works; Eiseley 1979; see Sheets-Pyenson 1981, 240–43.
120. Blyth 1835, 46, DRC. One of the notes in the back of the volume refers to Darwin's proposed ch. 6 on sexual selection, a definite reference to his "big book"; see also Sheets-Pyenson 1981, 238.
121. In his notebook, immediately after his earliest reference to this same paper, Darwin wrote thoughtfully: "Can be said animals no notion of beauty, when does prefer most powerful buck"; *Notebooks*, C 71.
122. *Notebooks*, C 204 (early June 1838). The earlier entries, C 84–85 on albinism, are also references to Blyth's 1835 paper, and the previous note (C 83–84) refers to the heritability of polydactyly which Lawrence, but not Blyth, discussed. Wells also argues the importance of Blyth as an "intermediate link between the views of Lawrence and those of Wallace and Darwin"; see Wells 1971, 344–48.
123. *Notebooks*, C 198–99; Blyth 1837.
124. *CCD*, 5:445.
125. *Notebooks*, E 71.

CHAPTER SIX
1. *CCD*, 3:79.
2. Corsi 2005, 75–76.
3. *CCD*, 2:446–49, 448; *Notebooks*, E 136
4. "Questions for Mr. Wynne," written sometime in the first half of 1838, *CCD*, 2:71.
5. *Notebooks*, E 71, D 147; *Origin*, 4. See Kohn in *Notebooks*, 396; Hodge 2003, 60–64. Bert Theunissen argues that Darwin's comprehension of breeding practices was moulded by his understanding of the work of natural selection in nature, rather than the other way round, as is often stated; Theunissen 2012.
6. *Origin*, 91; Secord 1981, 164; Sterrett 2002.
7. *Origin*, 90. See Darwin's copy of Youatt 1837, CUL; *Marginalia*, 1:890.
8. *Origin*, 94.
9. *CCD*, 7:364; see also *Variation*, 1:133, 177–79; 2:217–22.
10. *Descent* 1871, 2:258–59; Ghiselin 1972, 218; see also Ruse 1979, 179.
11. *Descent* 1879, 246.
12. *Descent* 1871, 2:370; see E. Richards 1983, 78.
13. See sec. 11.8.

14. On Darwin's concerns with inbreeding in his own family, see James Moore 2001.
15. Ritvo 1987, 45–81; Secord 1985.
16. *CCD*, 6:236; Secord 1985, 525–27; Hodge 2009.
17. *CCD*, 2:182, 187–89, 190–92, 446.
18. Secord 1981, 1985; Desmond and Moore 2009, 253–54.
19. Secord 1985; Secord 2000, 426–29.
20. *CCD*, 7:404; *Notebooks*, C 133–34; *Origin*, 90; Secord 1985.
21. *Variation*, 2:3; Secord 1985, 534–35, 539; see also Sandow 1938.
22. Ritvo 1987, 52–63.
23. Sebright 1809, 7; see also N. Russell 1986, 19.
24. Sebright 1831, 6–8.
25. Sebright 1836; *Notebooks*, C 134, N 63.
26. Desmond 1987, 1989; P. White 2007.
27. *Notebooks*, D 13, 16.
28. *Notebooks*, E 3–6; Hodge 1985, 226; Hodge 2003, 51, 61.
29. *CCD*, 7:452–53; 8:252–53; Hodge 2003, 65–67; Hodge 2009; Feeley-Harnik 2004, 353; Desmond and Moore 1991, 324.
30. *CCD*, 5:509; Feeley-Harnik 2004, 331, 332, 334, 353; Brent quoted in Secord 1981, 173.
31. *CCD*, 4:63–64, 248–49; Sebright 1809, 28–29; *Notebooks*, D 43–44, 88; *Foundations*, 92–93; *Variation*, 1:302–3; Secord 1985, 537; P. White 2007.
32. Darwin's copy of Sebright 1809, DRC, see cover and 26–27; *Notebooks*, B 71, 189; C 30; Hodge 2003, 50–51. For Darwin's experiments on crossing, see Secord 1981, 177–78. On the Victorian obsession with hybridity and its implications for theories of race, see Robert J. C. Young 2006; cf. Theunissen 2012, which argues that crossing was an accepted practice among breeders.
33. Sebright 1809, 9–11; Ritvo 1987, 62.
34. Darwin's copy of Sebright 1809, DRC, 14–15; *Notebooks*, C 133, n5.
35. *Notebooks*, C 133n3; see also D 179.
36. *Notebooks*, D 163.
37. *Descent* 1871, 1:273.
38. Sebright 1809, 15; DRC.
39. *Descent* 1871, 1:259–60.
40. Ibid., 1:272.
41. *CCD*, 2:188.
42. *CCD*, 16, 1:446–47, 452, 459; *Descent* 1871, 1:405–10.
43. N. Russell 1986, 93–121.
44. *Notebooks*, D 171.
45. *Variation*, 1:397–405; *Notebooks*, D 168, 169, 176; R. Burkhardt 1979; Endersby 2003, 78–82, 87.
46. *Notebooks*, N 10; *CCD*, 18:78; R. Burkhardt 1979, 7–8.
47. *Descent* 1871, 2:113–24.
48. *CCD*, 8:108; *Descent* 1871, 2:117–22; 270–73; C. Darwin 1882, 368.
49. *CCD*, 9:93, 148–53.
50. *CCD*, 9:149; *Descent* 1871, 2:117–18; Hodge 2003, 51–53.
51. *CCD*, 5:509; 6:236; *Origin*, 90, 91.
52. Feeley-Harnik 2004; *Variation*, 1:138.
53. *CCD*, 7:428–29; see also Secord 1981, 169. Boldface taken from *CCD*, 7:429 presumably indicates that Darwin double-underlined the word in emphasis.
54. *CCD*, 7:428.

55. *Origin*, 93, 97, 90.
56. Secord 1981, 185; see also Feeley-Harnik 2004, 327. Another reason for Darwin's distinction between the two modes of selection was that he wished to relegate large variations (saltations or mutations) to methodical selection and the production of "monstrosities," leaving the minute variations he favored as the raw material of natural selection to the realm of unconscious selection; see ch. 9.
57. *Variation*, 1:215; see also *Origin*, 112; *NS* 227–28. On Darwin's selective interpretation of breeding practices and preference for those of pigeon breeders in forging his analogy between artificial and natural selection, see Theunissen 2012.
58. *Variation*, 2:226; Eaton 1852, 33.
59. Eaton, *Almond Tumbler* 1851, as cited by Secord 1981, 169–70.
60. *CCD*, 5:492; 7:429; CUL Eaton 1852, vi; see *Marginalia*, 1:217.
61. *CCD*, 8:255; see also 253–55, 258–59, 260–61; Robert M. Young 1985, 79–125; Secord 1981, 185; Tipton 1999, 130n.
62. See *Variation*, 2:248–49, 430–32, 193, 211.
63. Compare *Variation*, 1:216 with *Variation*, 2:197–89; see also *Origin*, 155. In illustration of the confusion surrounding Darwin's distinction of methodical from unconscious selection, Sterrett sees Darwin's use of Eaton's description of divergent long and short-beak pigeon fanciers in the *Origin* as an instance of *methodical* selection causing divergence, see Sterrett 2002, 158; see also Tipton 1999. My interpretation is very different from that offered by Alter 2007, who insists on the consistency of Darwin's thinking and applications of unconscious selection.
64. *Variation*, 1:215–16.
65. Ibid., 2:208.
66. Ibid., 2:239–40; see also 248.
67. *Descent* 1871, 2:369.
68. Ibid., 2:345–46.
69. *Origin*, 78, 101; *Notebooks*, D 107; E. Richards 1994, 407–11.
70. *Descent* 1871, 2:381–82.
71. *Descent* 1871, 2:354. On Victorian notions of beauty, see Cowling 1989.
72. *Descent* 1871, 2:370.
73. *CCD*, 8:140.
74. *CCD*, 16 (2) :762.
75. Wallace 1889, 285–86; Cronin 1991, 187–88.
76. *Variation*, 2:74; DAR 84.2:6, 9.
77. See ch. 12; Kottler's account of the dispute over sexual selection is the fullest and best, Kottler 1980; see also Cronin 1991, 147–49.
78. *Descent* 1871, 2:93; 1:273.
79. Wallace 1858, in *CCD*, 7:513–14; see Hoquet and Levandowsky 2015, 22. On Darwin's "spurious" rationale for female choice, see Cronin 1991, 115.
80. Jonathan Hodge argues that Wallace was only briefly wary of the domestication analogy, and had ceased to be so by around 1860 after reading the *Origin of Species*. He points out that Wallace deployed the analogy in a strict Darwinian sense in his *Darwinism* of 1889 (personal communication). See Wallace 1889, ch. 5.
81. *Descent* 1871, 2:140–41.

CHAPTER SEVEN
1. Dickens n.d., 785–86, 790–91, 948–49; see Langland 1992, 298–99.
2. Kincaid 1971, 167–69; McKnight 2006.

3. Ritvo 1987.

4. CUL Dixon 1851, 2, 55–56. For Darwin's annotations, see *Marginalia*, 1:199–201. I am indebted to Jim Secord for bringing this reference to my attention.

5. Dixon 1851, 144–45; *CCD*, 4:293, 231, 169. Darwin's esteem for Dixon faltered only when he became an "excommunicated wretch" after Dixon's credentials came under a cloud and he began to publish under a pseudonym; *CCD*, 5:294–95, 391.

6. *CCD*, 2:80; *Notebooks*, M 107, C 79, M 138–40.

7. Cited in Secord 2000, 443; Ritvo 1987, 37.

8. *Notebooks*, D 139.

9. C. Darwin 1876, 18–19; *Descent* 1871, 2:293, 313.

10. *Calendar* 12230, 12220. On Ruskin's critique of Darwin's aesthetics, see J. Smith 2006, 125–36; Prodger 2009.

11. Dawson 2007, 58–74. On earlier depictions of the relations between humans and apes, see Schiebinger 1993, 75–114.

12. Huxley, in fact, denied that he ever said this, see Desmond 1994, 278–81, 417n18.

13. *Descent* 1871, 1:62 (Darwin is here citing Büchner); see also 1:67, 68; *Variation*, 2:220; Ritvo 1987, 21.

14. G. Romanes 1896, 441–42; Ritvo 1987, 91, 107.

15. Ritvo 1987, 3–4, 180–86; Ritvo 1988; Nead 1990, 120–22.

16. Dickens n.d., 65, 78, 103, 116, 151, 731–32, 865, 942.

17. *Descent* 1871, 1:68, 40, 78.

18. Ritvo 1987, 20–23.

19. *CCD*, 2:444.

20. Ritvo 1987, 6; C. Hall 1992, ch. 10.

21. *Calendar*, 10546; see E. Richards 1997, 130–33; Elston 1987.

22. *Autobiography*, 138–39; Beer 1983, 62; McKnight 2006.

23. For Darwin's remarks on the need to marry only "good women," see *Notebooks*, OUN 25n1. *Intermarriage* was reviewed in Anon. 1839; see also DAR 119:5v; Alexander Walker 1838. Darwin's extensively annotated copy of *Intermarriage* is in CUL; see *Marginalia*, 1:832–35. William Erasmus Darwin was born on December 27, 1839.

24. Until the comparatively recent analyses of Walker's writings on women by the art historian Robyn Cooper (1992, 1993), Walker was better known to literary historians for his English translation of Benjamin Constant's *Adolphe* in 1816, or to medical historians for his controversial claims over the discovery of the functions of the roots of the spinal nerves (see Alexander Walker [1839] 1973). Courtney (1975) makes the only effort to reintegrate Walker's literary and medical activities. As yet, although art and fashion historians as well as historians of heredity and physiognomy have begun to take notice of those of Walker's writings that fall within their ambit (see besides R. Cooper, Montague 1994, 98–100; Hartley 2005, 115–21; Hartley 2001, 24–27; Waller 2001, 464), there is no proper historical assessment of the whole man and his writings.

25. Alexander Walker 1836, vii, 9.

26. Alexander Walker 1838, xvii, xx, 150ff.

27. Courtney 1975, 149; see also Cranefield, introduction to Alexander Walker [1839] 1973, v.

28. On the Grub Street hacks and the explosion of scientific and medical journalism and publications in the second quarter of the nineteenth century, see Desmond 1989; Secord 2000.

29. See Alexander Walker 1838, ix; Alexander Walker [1839] 1973, x–xi. The first two English editions were brought out by John Churchill, the highly reputable medical

publisher who was primarily responsible for the promotion and marketing and hence the great popular success of *Vestiges*; Secord 2000, 111–25.

30. Alexander Walker 1838, Dedication and frontispiece. On Carlisle's conservative anatomy and his role in thwarting the democratization of the College, see Desmond 1989, 111–12, 419.

31. Foucault 1980, 124.

32. Waller 2001; 2003; Hilts 1982.

33. E.g., Combe 1840; see Hilts 1982.

34. Alexander Walker 1836, 148, vii–viii.

35. Alexander Walker 1834, 7. Walker also produced *The New Lavater* in 1839; see the bibliography in Alexander Walker [1839] 1973, vii–xi. Walker had earlier assisted with the English text of Spurzheim and Gall's *Physiognomical System* of 1815; see Courtney 1975, 139. For an excellent discussion of Walker's relation to the physiognomic tradition, see Hartley 2001.

36. Alexander Walker 1838, 149. Walker was almost certainly personally acquainted with Lawrence, who in the early 1820s was closely involved with Thomas Wakley in running the *Lancet*, but who had by the 1830s, as a result of career threats and professional inducements, retreated from his earlier radicalism; see Desmond 1989, 15, 257–58.

37. Alexander Walker 1836, 3, 12–13.

38. Ibid., 148–61. Most women do not represent such "pure" species of beauty, being composites of the three kinds. But invariably, one "species" will predominate.

39. Ibid., 224.

40. Ibid., 179–80, 188, 190, 192, 194, 200, 263–64, 366–68; see R. Cooper 1993, 43.

41. Alexander Walker 1838, 19, 76–77, 92–94.

42. Alexander Walker 1839, 169–71.

43. Robyn Cooper discusses a number of reviews; R. Cooper 1992, 359; 1993, 46; see also Alexander Walker 1838, appendix.

44. On the politics and styles of the various London medical journals, see Desmond 1989. On Walker's relations with the *Lancet*, see Courtney 1975.

45. Mason 1994, 190–91; Gates 1998, 184–85.

46. Smith 1996; Secord 2000, 439–40.

47. Walker had also published an earlier anatomy text and atlas "adapted to the use of professional students . . . and artists" (1813); cited by Courtney 1975, 148n15.

48. Quoted in Montague 1994, 100. On Goodsir, see Desmond 1989, 422; Rehbock 1983, 91–98.

49. Spencer (1854) 1966, 2:387–94.

50. Cited in Paxton 1991, 31.

51. Spencer 1859, 395; see also Spencer 1855, 600–601; Hartley 2005, 110–12. On the relationship between Evans (George Eliot) and Spencer and on Spencer's gendered evolutionism, see Paxton 1991.

52. Hartley 2005, 112; 2001, 31.

53. S. Gilman 1995, 92.

54. See Poovey 1989, ch. 1; Davidoff and Hall 1987, esp. chs. 1–3; C. Hall 1992, esp. ch. 3.

55. Carlile 1838, 8; W. Thompson [and Wheeler] (1825) 1983; Fryer 1965, 43–86. On the feminism of the Owenite Socialists, see B. Taylor 1983; on that of the radical Unitarians, see Gleadle 1995.

56. Royle 1971; Royle and Walvin 1982; B. Taylor 1983; Gleadle 1995.

57. Alexander Walker 1836, 251; 1838, 381.
58. Ibid., 250; Alexander Walker 1839, 67.
59. Alexander Walker 1839, 38, 41, 43; Alexander Walker 1838, 406, 422–25, 431.
60. Alexander Walker 1839, 129, 147.
61. Ibid., v.
62. See R. Cooper 1992, 362–4.
63. Tyrell 2000, 195; C. Hall 1992, ch. 7.
64. B. Cooper and Murphy 2000, 4, 17, 18; Peterson 1990; *CCD*, 1:524; Easley 1999, 80–81, 83; Martineau (1877) 1983, 1:400–401.
65. Martineau (1838) 1989, 184–85.
66. *Notebooks*, M 75–76; DAR 119:2a.
67. For Darwin's notes on Aimé-Martin, see *Notebooks*, OUN 8. Darwin read the 1837 French edition. An English translation became available in 1842, but by then its arguments had been made known to a wide transatlantic readership through Sarah Lewis's derivative and extraordinarily popular *Woman's Mission* of 1839. On its influence on the "woman question," see Richardson 2003, 45; Alison Chapman 1999.
68. Quotations taken from the English edition of (1834) 1842, 24, 31.
69. Aimé-Martin (1834) 1842, 38, 49, 64, 66, 72, 157, 322.
70. Ibid., 30, 45–47.
71. *Notebooks*, OUN, 8.
72. *Descent* 1871, 2:326.
73. *Notebooks*, C 220 (written June 1838); note Darwin's reference to miscegenation, on the need to avoid crossbreeding of classes in accordance with the practice of breeders of keeping breeds separate. For radical Lamarckian opinion on the necessity for the education of women, see Desmond 1989, 60.
74. *Notebooks*, OUN 25n1.
75. Alexander Walker 1838, 435, 176.
76. Ibid., 188.
77. Ibid., 147.
78. Anon 1839, 385.
79. See Walker 1838, 138–44, 149, 357–360. Darwin's extensively annotated copy of Walker's *Intermarriage* is in the Cambridge University Library; see also *Marginalia*, 1:832–36.
80. CUL Alexander Walker 1838, 112, 115–6, 118, 120, 124–5.
81. Ibid., 416, 51–58, 120–25.
82. Ibid., 215, 224, 271.
83. Ibid., 199, 232, 362, 369. Walker was, like Lawrence, a monogenist. Darwin noted especially Walker's views "on advantages of crossed races of Man" in the back of his copy of *Intermarriage*; see also Desmond and Moore 2009, 146–47.
84. CUL Alexander Walker 1838, 150ff, 196, 204, 285.
85. Ibid., "Dedication," 158–59, 368, 415.
86. Ibid., 156.
87. Ibid.; see *Marginalia*, 1:833.
88. *Notebooks*, C 275; E 113, 169; QE4; *CCD*, 2:185; 6:210–11; *Marginalia*, 1:682; see also Bartley 1992, 318–27; Cowan 1972, 401–2. For Darwin's notes on Orton, see Orton 1855, DRC; see also Lewes 1856.
89. Cf. Carl Bajema, who alone among historians accords Walker's *Intermarriage* any significance in the formation of Darwin's evolutionary theorizing. Bajema suggests that Walker's consistent use of the term "selection," rather than the more customary

breeder's term of "picking," may have alerted Darwin to the "value of selection as a metaphor in communicating his ideas" on natural selection; Bajema 1984, 100. However, the term *selection* was also used by Blyth, Sebright, and other breeders, such as Youatt.

90. R. Cooper 1993, 51.
91. *CCD*, 18:194, 255–56; Dawson 2007, 36–41.
92. *Descent* 1871, 2:356–57.
93. Durant 1985, 301.
94. Ibid., 299.
95. E. Richards 1983; Paul and Day 2008.
96. *Descent* 1871, 2:326–27.
97. *Descent* 1871, 2: 354, 381–82.
98. Ibid., 2:92–93, 141–51; Levine 2006, 195–97; see also Kaye 2002, 104–5. In contradistinction to my interpretation, Levine argues that with female choice, Darwin "introduces aesthetic taste and female intention as driving forces in the evolutionary process"; Levine 2006, 199.
99. *CCD*, 19:208; see also Kaye 2002, 103–4.
100. *Descent* 1879, 246.
101. *Descent* 1871, 2:371–72, 39.
102. *Descent* 1871, 2:352.

CHAPTER EIGHT

1. *Evenings at Home* had run through thirteen authorized editions by 1823, Levy 2006, 129; other collections for children "borrowed" from it; e.g., "The Female Choice" was also included in Lindley Murray's popular, often reprinted *Introduction to the English Reader*; see Murray 1831, 37–39.
2. Biographical detail on Barbauld from McCarthy 2008; see also Robbins 1993, 139, 147–48; Watts 1998; George 2005.
3. Levy 2006, 123, 127; Bradshaw 2005.
4. Barbauld and Aikin 1805, 3:161–62.
5. Ibid.:163–64.
6. Ibid.:165.
7. Carlyle (1833–34) 1987, 30–31. *Sartor Resartus* was first published in serialized form in *Fraser's Magazine for Town and Country* (1833–34).
8. Ibid., 207; Carter 2003, 10–14. On the enormous impact of *Sartor Resartus* on the reading public, see Secord 2014, 205–35.
9. Kuchta 1996, 65.
10. Steele 1985, 51–62; Hollander 1993, 127–34; Byrde 1992, 23–84; John Harvey 1995, 23–26.
11. Davidoff and Hall 1987, 410–15; Byrde 1992, 88–109.
12. Cited in Steele 1985, 107.
13. On the masculine creed of "inconspicuous consumption," see Kuchta 1996; on vests, see Razek 1999, 8–9; Davidoff and Hall 1987, 412.
14. John Harvey 1995, 197.
15. See Kunzle 1982, 136–48; Razek 1999, 11.
16. Victorian girls played with adult dolls, rather like modern-day "Barbies," with nipped in waists and wide hips—"baby" dolls only appeared toward the end of the century, see Razek 1999, 10. On dressing up by Henrietta and Annie Darwin, see Healey 2002, 197, 214.

17. *ED*, 1:9–10, 276; Healey 2002, 154–5, 197.
18. *CCD*, 1:276–9.
19. *CCD*, 5:541.
20. *CCD*, 2:296.
21. Raverat 1960, 266; Halttunen 1982, 80–89; Steele 1985, 127.
22. Langland 1992, 296; Byrde 1992, 117–44, 148–59.
23. Davidoff 1973, 93; see also Steele 1985, 132–36.
24. Langland 1992, 294.
25. Raverat 1960, 152; *CCD*, 6:394–95. William received an allowance of £100 per annum at Cambridge; *CCD*, 7:146.
26. Langland 1992; T. Logan 2001, ch. 1; Briggs 1989, 211–59; Richardson 2003, 37.
27. *CCD*, 2:155, 147–48; see Bryant 2009.
28. Raverat 1960, 197; Cohen 2006, ix–62; T. Logan 2001, chs. 2 and 3.
29. *CCD*, 3:248. Darwin had little interest in the valuable Wedgwood medallions and slate reliefs that were used to mold the pottery designs that he and Emma had inherited; he sold the latter off for £150 in 1858 to purchase a new billiard table; *CCD*, 7:153.
30. Cohen 2006, 89–104; T. Logan 2001, 89–104.
31. T. Logan 2001, xiii, 105–202; see also Tosh 1999.
32. Steele 1985, 113; P. Anderson 1995, 8–11, 21–22; T. Logan 2001, esp. ch. 3; Alison Smith 1996.
33. Ruskin cited in T. Logan 2001, 223. On the instability of Victorian pictorial and literary representation of domestic and adulterous women, see also Thomas 2004, ch. 2; Nead 1990, 48–86; Reynolds and Humble 1993.
34. Cited in Thomas 2004, 10–11; see also Flint 2000, 214–16.
35. T. Logan 2001, 86–88.
36. Linton 1868; Broomfield 2001.
37. Moruzi 2009.
38. P. Anderson 1995, 11 and passim; Hollander 1993, 69. On Victorian sexuality and its expression, see Mason 1994.
39. Raverat 1960, 256, 260.
40. Hollander 1993, 360–61; Razek 1999, 12–13; Kunzle 1982; Steele 1985, 90–95, 161–91.
41. Hollander 1993, xi, xv, xvi, 91, 448, 450.
42. T. Logan 2001, 205, 100–104.
43. Raverat 1960, 256.
44. Anon. 1847.
45. Halttunen 1982, 82–91; Steele 1985, 121–42.
46. Flint 2000, 216–17.
47. See Razek 1999, 16; Ogden 2005, 14–16.
48. Altick 1997, 17; Noakes 2004.
49. "Crinolineomania," *Punch*, December 27, 1856, 253.
50. J. Thomas 2004, 78.
51. Ibid., 79–87.
52. Ibid., 94, 91, 77.
53. Ibid., 102–3.
54. Raverat 1960, 108–10. Other accounts suggest that men and women alike were well accustomed to the sight of nude bathers of both sexes, ostensibly confined to separate beaches, but well within viewing range. Seaside voyeurs were not confined to those males who trained their spy glasses on the women's bathing machines, but, according to the *Saturday Review*, "strings of respectable women" positioned themselves near

the men's bathing areas and "placidly and unmovably," and with no apparent notion of impropriety, unblushingly gazed at "hundreds of males in the costumes of Adam," see P. Anderson 1995, 15–16.

55. J. Smith 2006, 30–31.

56. T. Richards 1991, 16. On the Darwin visit to the Great Exhibition, see *CCD*, 5:53.

57. J. Thomas 2004, ch. 2; Flint 2000; P. Anderson 1991; T. Logan 2001, 137–40; Cohen 2006, 44–62.

58. J. Thomas 2004, 74–75; Alison Smith 1996.

59. Desmond 1997, 129; see also 160–61, 175–77, 294n. The painting "By the Tideless, Dolorous, Midland Sea," was dismissed by the *Spectator* as an "objectless and futile" attempt by a woman artist to offer what was little more than an ordinary nude study; Alison Smith 1996, 195. On the campaign for women's access to life classes, see ibid., 37–44; Alison Smith 2005, 160–61. On the career and other difficulties encountered by Marian Huxley Collier, see Sanders 2013.

60. On "visual acquisition" and Victorian "ways of seeing," see Flint 2000, esp. 217, 311; and Briggs 1989, ch. 3; on Worth, Briggs 1989, 277.

61. T. Logan 2001, 103.

62. Linton 1868, 340. On Linton's earlier comparisons of women with birds, see Broomfield 2001, 273; on the sexual and social freedoms available to young English girls, see Mason 1995, 116–24.

63. *Variation*, 1:215, 2:208, 215–16, 226; *Descent* 1871, 2:369.

64. *Descent* 1871, 2:352.

65. "Mr. Darwin on Domestication," *Saturday Review*, March 14 and 28, 1868, 358–59, 423–44; Linton 1868.

66. Against the unnatural "Girl of the Period," Lynn Linton upheld the "natural" ideal of the inherently modest, domestically oriented girl who "when she married, would be her husband's friend and companion, but never his rival; one who would consider his interests as identical with her own . . . who would make his house his true home and place of rest." The bold designing woman who disdained her designated role of household angel would not be chosen as a wife: such women, "bold in bearing" and "masculine in mind" and in futile and "wild revolt against [their essential feminine] nature" and biologically determined role in society, would be denied the chance of reproduction and participation in the evolutionary process. Linton 1868, 339, 340; see E. Richards 1989a, 1997.

67. DAR 226.1:148, 149, 150. On the phenomenal impact of Linton's essay, see Broomfield 2001, 279 and passim; Moruzi 2009.

68. *Descent* 1871, 2:39, 371–72; see also 1:63.

69. Ibid., 2:38.

70. Ibid., 2:230, 74.

71. Ibid., 2:230–31; see also 1:64–65.

72. T. Logan 2001, 144–45, 160, 164–65, 172–73; see also Cohen 2006, 128–30; Munro 2009, 260–61.

73. Fagan 2008, 67–68.

74. Gates 1998, 114–22; Ehrlich et al. 1988; Forbes and Jermier 2002; Birdsall 2002.

75. *Descent* 1871, 2; 71.

76. *Punch*, April 23 1870; see also *Punch*, December 21, 1867, 256; Munro 2009, 280–82.

77. It appeared next to an article "Most Natural Selection," advocating the evolutionary advantages of human-ape marriages, *Punch* 1871, 127; see also Ogden 2005, 135–36; Bernstein 2007, 69–70.

78. *Descent* 1871, 2:141.
79. Ibid., 2:381.
80. See also Ogden 2005, 136. The role of *Punch* in shaping scientific discourses is discussed by Noakes 2004.
81. *Descent* 1871, 2:120–21.
82. Ibid., 2:135–41.
83. *CCD*, 16 (1):369; *Descent* 1871, 2:120. At Darwin's urging, Weir undertook certain experiments devised by Darwin to illustrate the existence of female choice (see sec. 13.4). On Gould's domesticated depictions of bird life and Darwin's deployment of them, see J. Smith 2006, 92–125.
84. *Descent* 1871, 2:108.
85. J. Smith 2006, 120–22.
86. *Descent* 1871, 2:69–70. My interpretation departs significantly from that of Jonathan Smith, who argues that this particular instance of male bird behavior disrupts Victorian notions of courtship and sexuality, and together with other disruptive depictions in the *Descent*, accounts for the dismissal of sexual selection by Victorian readers. Smith, while he acknowledges that Gould also quoted what he describes as the "ridiculous and menacing" courtship behavior of the male (which somewhat blunts his point about Darwin's quotation of the same extract), neglects to include the final part of the quotation referring to the "gentle" behavior of the female; J. Smith 2006, 122, 116.
87. *Descent* 1871, 2:112–13; see also 1:63.
88. Ibid., 2:118.
89. *Descent* 1871, 2:122, 124; cf. Levine 2006, 192–95.
90. *Descent* 1871, 2:343.
91. *Descent* 1871, 2:338. Darwin did not refer to Carlyle by name but as "an English philosopher."
92. Carlyle (1833–34) 1987, 18; Carter 2003, 1–4; see also Secord 2014, ch. 7.
93. *Descent* 1871, 1:64; *CCD*, 4:462.
94. Carter 2003, 26, 61, 153.
95. *Descent* 1871, 2:342, 343–54.
96. Ibid., 2:129–30, 296.
97. Ibid., 2:343, 344.
98. Ibid., 2:351–52.
99. Ibid., 2:372–4.
100. Ibid., 2:356.
101. Ibid., 2:371–72.
102. Ibid., 2, 370; see E. Richards 1983, 78.
103. Briggs 1989, 206–8, 284–88.
104. Herzog 1998, 338–39; see also Vickery 2001, 117.
105. Linton 1868, 339.
106. Wallace 1889, 286–87.

CHAPTER NINE
1. Anon. 1839, 372.
2. Bajema 1984, 99–100.
3. Ruse 1979, 179.
4. *Foundations*, 10; Aiken 1982, 5–6.
5. *CCD*, 4:65, 67, 84–85; Desmond and Moore 1991, 314–15, 323–324.
6. *Foundations*, 92–93.

7. This is all the more intriguing, as his terminology, "use of a choice male" by agriculturalists, so readily lends itself to analogous selection by females of a choice male in the case of birds, and so to female choice. It supports my contention of Darwin's initial reluctance to depart from the cultural convention of male sexual choice that had been reinforced by the overt statements of such masculine sexual prerogatives in his readings in aesthetic theory, ethnology and the literature on breeding. Interestingly, the extract from Darwin's theory that was published in conjunction with Wallace's paper in the Linnean Society Journal in 1858, substituted "mate" for "male"; see Bajema 1984, 108.

8. *Origin*, 88–89.

9. *Origin*, 89.

10. *Origin*, 199.

11. *CCD*, 8:350, 356; 7:432; *Autobiography*, 125.

12. E. Richards 1976, 358–85; Ospovat 1981; Hodge 1985; R. Richards 1992, ch. 5; Nyhart 2009.

13. Many prominent historians (notably E. S. Russell 1916; De Beer 1962; Gould 1977; Ospovat 1981, 152, 162) fought hard to dissociate Darwin from the embryological views of Fritz Müller and Ernst Haeckel, particularly Haeckel, whose name is most closely associated with the pernicious theory of recapitulation or "biogenetic law." It was Haeckel who popularized the view that the embryo, as it develops, "climbs up its family tree." He coined the notorious formula: "Ontogeny recapitulates phylogeny." And it was Haeckel who led the search for *adult* ancestral portraits in developing embryos. The search for ancestors came to dominate embryology until Haeckel's recapitulation was discredited in the early twentieth century. His theory then was reformulated to state that ontogeny recapitulates the *embryonic* forms of certain ancestral types. The conventional view became that Darwin presciently had rejected recapitulation theory for the more acceptable views of the earlier nineteenth century embryologist Karl Ernst von Baer, and therefore had advocated something akin to modern embryological theory. However, a revisionist historical interpretation, spearheaded by Robert Richards, is now in formation; R. Richards 1992, 111–15, 169–80; see also E. Richards 1976, 358–93; Nyhart 2009.

14. E. Richards 1990; Appel 1987; terminology and distinctions drawn are from the classic study E. S. Russell 1916.

15. Desmond 1982, 115–21; Desmond 1984; Desmond 1989, 58–62, 66–67, 89–90, 300–302, 395–97, 398–405.

16. C. Lyell 1830–33, 2:62–64; E. Richards 1976, 207; Hodge 1982; Hodge 2003..

17. Ospovat 1976; Desmond 1982; Desmond 1989, 322–27, 333–51, 376–77; E. Richards 1987; R. Richards 1992, 102.

18. Ospovat 1976; E. Richards 1987.

19. E. Richards 1990, 135–37.

20. Agassiz 1849, 26–27; Agassiz 1850; Lurie 1960, 20–63; E. Richards 1990, 137–38,

21. Secord 2000, 332–35 and passim; Wallace 1905, 1:254, 355, 362. On *Vestiges* and the universal gestation of nature, see Hodge 1972.

22. [Chambers] (1844) 1969, 201–2, 226–27; Ospovat 1976; E. Richards 1987; Secord 1989.

23. Sedgwick 1845; Secord 2000, 231–47; E. Richards 1990.

24. [Chambers] (1844) 1969, 199–201, 214–16, 306–7, 310.

25. Secord 2000.

26. R. Richards 1992, 133–34.

27. Ibid., 116–23, 134, 137.

28. Desmond 1994.
29. L. Huxley 1900, 1:162–63, 166; T. Huxley 1855a; Desmond 1982, 37–51, 84–101; Desmond 1994, 223–24; Lyons 1999, ch. 2.
30. T. Huxley 1854b; Secord 2000, 498–504.
31. T. Huxley 1855b, 242–47; Carpenter 1854, 134, 95 117; Richards R. 1992, 148–49.
32. CCD, 5:213–14, 351; T. Huxley 1855a, 1855b; Darwin's notes on Huxley's review of Vestiges are in DAR 205.6:56–57.
33. Desmond and Moore 2009, 242 and passim; Gould 1981, 42–50.
34. Foundations, 221.
35. Baer 1853; see E. Richards 1976, 364–65, R. Richards 1992, 124–25.
36. Foundations, 225–26, 242.
37. Notebooks, B 161; see E. Richards 1994, 407–11.
38. This was reinforced by Darwin's adoption of "Yarrell's law" of heredity, which, as we saw in Ch. 6, meant that in any cross, the parent whose characters were older or more "fixed" in the breed would have a greater influence on the offspring than the parent whose characters were those of a younger breed. As Darwin put it: "an animal is able to transmit only those peculiarities, to its offspring, which have been gained slowly"; E. Richards 1994, 408; Hodge 2003.
39. Notebooks, E 83–84.
40. Ibid., E 151–52; D 107; NS, 319; Origin, 16, 44.
41. Origin, 443.
42. Notebooks, E 6e; Hodge 1985, 226; E. Richards 1987.
43. See Notebooks, D 113e–14e, D 161–62, also D 95e-96e; Hunter 1837, 46–49; Hodge 1985, 224.
44. Notebooks, C 178.
45. Ibid., D 76. See also D 57, 147e, 161, 162, and "proved facts relating to generation" D167–79; 174–175, written before September 8, 1838. Darwin's theory of separate sexes is developed in ch. 3 of Natural Selection—but not included in the Origin.
46. Ibid., D 147e
47. NS, 42–46, 362–63; CCD, 4:388–409; 5:118.
48. His discovery of up to twelve males associated with one hermaphroditic female also went against the grain of his later rejection of the contemporary anthropological thesis of an early stage of polyandry among primitive humans. Hooker, at that time collecting plants in the Himalayas (1849), actually extended the analogy from barnacle polyandry to the marital practices of the "uncouth" Bhothea, the Himalayan tribe who provided his coolies: "The supplemental males of Barnacles are really wonderful, though the supplemental males in the Bhothea families (a wife may have ten husbands by Law) have rather distracted my attentions of late from cirripedes"; CCD, 4:128, 140, 159, 169, 180, 204. Roderick Buchanan has analyzed the ways in which Darwin's Victorian assumptions about gender and sexual roles confused his interpretation of some critical barnacle observations; Buchanan 2016.
49. Origin, 137. Note that Darwin's formulation of the relation of embryogenesis and inheritance to sexual selection leaves some room for maneuver. The modifications are inherited "either by the males alone, or by the males and females." Darwin was well aware that there were many exceptions to his "well-known laws" on bird plumage, and his concession to females inheriting some sexually selected modifications was an attempt to account for these.
50. Autobiography 120–21; Milne Edwards 1844; DAR 72:117–19v; CCD, 5:197–98; R. Richards 1992, 134–36; cf. Ospovat 1981, 159–65.

51. *CCD*, 5:201; DAR 205.5:149, 201–2; see also DAR 205.9:250, where Darwin equates "complexity" and "perfection" with progress; Kohn 2009, 105; Hodge 2005, 118; *NS* 233; *Origin*, 115–16.
52. *NS*, 249–50, 272–3, 252, 235, 335, 165, 308; Kohn 2009, 93, 96, 103–7.
53. Kohn 2009, 93–95; see also Tammone 1995, 131.
54. *NS*, 227–28.
55. Divergence is, Darwin himself agreed, "closely analogous" to artificial selection; and, as I have argued, mate choice is *constituted by*—that is, it is patterned directly on—artificial selection. Kohn makes the point that Darwin "made a similar move when he developed sexual selection in *The Descent of Man*. There, females, typically, are the external selectors . . . There, too, the selection is creature-on-creature." Both forms of selection "extended the explanatory domain of natural selection by structurally mirroring artificial selection"; Kohn 2009, 94; *NS*, 273. Robert Richards also stresses that Darwin's divergence was patterned on the fancier's practice of selecting to extremes, and that this had a "special initiating cause" in Darwin's pigeon breeding experiences, which were concurrent with the forging of his principle of divergence. I would concur with this, but Richards interprets this as analogically conferring intentionality and intelligence on natural selection and divergence, assuming that Darwin interpreted the fancier's selecting to extremes as a rational, goal directed process; R. Richards 2012. However, as I have been at pains to emphasize, Darwin insistently stressed that selection to extremes was an "unconscious," irrational, capricious process, analogous to the extremes of fashion, and thus to be distinguished from "conscious" goal-directed or "methodical" selection.
56. *CCD*, 7:102; Kohn 2009, 89, 92–93.
57. *CCD*, 7:340.
58. *Origin*, 425, 449–50; Nyhart 2009.
59. *Origin*, 338, 435, 449, 450.
60. R. Richards 1992, 152–58. So fixated on Agassiz's interpretation was Darwin in the *Origin*, that he even misattributed von Baer's story of a missing label that had made it impossible to determine whether the unlabeled embryo was that of a "mammal, bird or reptile" to Agassiz. This was not corrected until the third edition of the *Origin* in 1861; Oppenheimer 1967, 252–53; E. Richards 1976, 377–78, 410–11.
61. This point is well made by Nyhart 2009, 207; see also Hodge on divergence and progress, 2005, 118–19; Kohn 2009, 105.
62. *NS*, 73, 308.
63. Nyhart 2009, 213–14; see also E. Richards 1976, 385–97.
64. *Descent* 1871, 2:187; Ghiselin 1972, 227–29.
65. *Descent* 1871, 1:286, 296–97.
66. Ibid., 1:286–87.
67. *CCD*, 2:169–70; 4:410–33, 419; *Notebooks*, M 129.
68. C. Darwin 1877, 288.
69. *CCD*, 4:422; C. Darwin 1877, 287–88.
70. *Descent* 1871, 2:326–27.

CHAPTER TEN

1. *NS*, 28, 213, 317–18; *CCD*, 6:523.
2. The missing pages are DAR 85:B25r, B34r. The nearest Darwin gets to the concept of female choice (but without actually articulating the concept) in the extant folios of his "big species" manuscript is the following excerpt from the draft of ch. 8, "Difficulties on the Theory of Natural Selection," which Darwin completed on September 29,

1857. Note Darwin's reference of sexual selection to those animals which may exert "will" or "choice": "In the case . . . of those animals, which possess will & choice, we must not forget 'sexual selection,' which may modify parts of little general importance, namely such as favour the struggle between male & male, or such as serve to charm the females"; *NS*, 376; see also 270.

3. Possibly by September 1857, as the above note suggests.
4. C. Hall 1992, 209, 255–75.
5. Mill 1850, 31, 29; Carlyle 1849, 671, 675; C. Hall 1992, 273–75; Robert J. C. Young 2006, 127–28. For Darwin's views on Carlyle, see *Autobiography*, 407.
6. Carlyle 1849, 672–73; Robert J. C. Young 2006, 5–6.
7. C. Hall 1992, 265–68. On racial theory as a form of cultural self-definition, see Lorimer 1978; Stocking 1987; Stepan 1982; Robert J. C. Young 2006; Callanan 2006.
8. C. Hall 1992, 279–81; Russett 1989, ch. 2.
9. Pieterse 1995, 61, 45–57; Lorimer 1978, 82–83; Desmond and Moore 2009, 245.
10. Knox 1850, v; E. Richards 1989b; Robert J. C. Young 2006.
11. Healey 2002, 223–34; Hobsbawm 1975, 47.
12. Qureshi 2011b, 149 and passim; Wallace 1905, 1:321–23.
13. Dickens 1853, 337–39; *Pictorial Times*, 1847, cited in Secord 2000, 358, 441; Skotnes 2001; Kirby 1965, 269; A. Bates 2010, 122.
14. Cited in Robert J. C. Young 2006, 120–21; Story and Tillotson 1995, 8:459; Callanan 2006, ch. 3.
15. Skotnes 2001, 301; Bank 1996, 401–2.
16. Andrew Smith 1831, 121, 124, 125, 339, 342, 119; Knox 1850, 182, 158; Kirby 1965, 222–23, 144, 266, 116; Bank 1996, 393–94; Hudson 2004; Desmond and Moore 2009, 104–5; Dawson 2007, 37–38.
17. Robert J. C. Young 2006, 120; Douglas 2008, 53.
18. Hobsbawm 1975, 268; Lorimer 1978, 1997.
19. Callanan 2006, 14, 59 and passim; Robert J. C. Young 2006, 92–93; C. Hall 1992, 206–89.
20. Browne 1995, 525; *CCD*, 6:58; 5:482, 294, 250–251.
21. *CCD*, 5:337, 294, 288–89.
22. Darwin also had pickled or skeletonized the young of bulldogs and greyhounds, ducklings, geese, chickens, rabbits and turkeys—all to be measured and compared. His notes included detailed measurements of diverse embryos, young and adults, even foals and dams of draught and racehorses. DAR 205.6:9–10, 40–44; *CCD*, 5:386; R. Richards 1992, 143–44.
23. *CCD*, 5:391, 492, 508, 510–11, 521, 523; 6:290.
24. *Origin*, 424–25; Bartley 1992.
25. *Origin* 82–83; *NS* 443; Desmond and Moore 2009, 255–56.
26. Desmond and Moore 1991, 430; *CCD*, 7:404–5; 6:236.
27. *Variation*, 1:151–53; Bartley 1992, 317–18.
28. L. Wilson 1970, 52–55, 87.
29. K. M Lyell 1881, 2:128.
30. Desmond and Moore 2009, 234–35, ch. 9; *CCD*, 5:492–93.
31. *CCD*, 5:492–93. For Darwin's abstracts of Bachmann's pamphlets, see DAR74:157–58, 159–60; Bachmann 1850, 135, 307, 215, 209; Desmond and Moore 2009, 244, 262–64. By contrast with the analysis of Desmond and Moore, both Stanton and Stephens are agreed that polygenism was not of crucial importance in justifying slavery: Stanton 1960; Stephens 2000, 266–67; see also Livingstone 2011.

32. Gould 1981, 50–72.
33. S. Morton 1847; Desmond and Moore 2009, 201–4, 243, 226; Nott and Gliddon 1854, 397–98; Lurie 1954, 229–33; Robert J. C. Young 2006, 1–28; Kenny 2007, 372–74.
34. Desmond and Moore 2009, 163, 161, 168–71, 199; Nott and Gliddon 1854, 56; Haller 1995, 74–79; Robert J. C. Young 2006, 122–24.
35. Agassiz in Nott and Gliddon 1854, lxxvi, lviii–lxxvi.
36. CCD, 3:242; Lurie 1954, 239n60 and passim; Nott and Gliddon 1854, xiv; see Desmond and Moore 2009, 263 and passim; Menand 2001–02.
37. CCD, 5:186–87, 167, 118; see also 6:176, 178; Lurie 1960; Lurie, introduction to Agassiz (1857) 1962; Desmond and Moore 2009, ch. 9; D. Porter 1993.
38. As transcribed by Desmond and Moore 2009, 265; Darwin's annotated copy of Nott and Gliddon 1864 is in CUL, see Marginalia, 1:603–6.
39. Marginalia, 1:603–4.
40. As he later explained to Lyell (1859) in discussing his views on "man," it had taken Darwin "so many years" to overcome the habit of attributing "too great importance" to climate; CCD, 7:346, 336–37, 339, 353.
41. Nott and Gliddon 1854, 54–58; Marginalia, 1:604.
42. Nott and Gliddon 1854, 182–83, 416, 312, 269–70, 458–59; a "portrait" of the "Hottentot Venus" is on 431.
43. Nott and Gliddon 1854, 415, 411–12, 69; see also 316–17.
44. In a passage unmarked by Darwin, Nott went on to claim that according to Agassiz, the brain of an adult Negro never developed beyond that of a Caucasian boy, and that the Negro brain bore a "marked resemblance to the brain of the orang-outan"; Nott and Gliddon 1854, 415, 414; Darwin's annotations, Marginalia, 1:606.
45. Nott and Gliddon 1854, 405–8.
46. CCD, 6:69; 5:309, 352, 394, 432, 457–58. Darwin may have been earlier in contact with Blyth and this correspondence lost; see CCD, 5:315n1.
47. Darwin's marginalia in J. Moore, Columbarius 1735, 85 (see Marginalia, 1:219); CCD, 5:398–99; 6:57, 62–63; Desmond and Moore 2009, 256–57, 261–62.
48. CCD, 5:444, 448, 449.
49. Ibid. 5:445. Darwin evidently received Blyth's notes in December 1855; ibid.:454n.
50. See also Blyth's follow-up letter, in which he discussed male combat and the aversion of herds or races to intermixing or interbreeding; ibid.:473–77; DAR 128:4, 165, 16.
51. As discussed in sec. 5.7, one of Darwin's two explicit references to "sexual selection" in his notes on this paper alludes to his proposed ch. 6 on Sexual Selection, which dates it to sometime in 1856 when Darwin was at work on his "big book" on species. The other may have been made earlier, in late 1855, before Darwin had begun writing and was still collating his old notes; DRC, Blyth 1835; see also Sheets-Pyenson 1981, 238; CCD, 5:215. This supposition is strengthened by the fact that Darwin again discussed with him one of Blyth's old papers in the Magazine of Natural History (most likely this one) while he was at work on sexual selection in March 1867; CCD, 1 5:119.
52. CCD, 5:475.
53. Powell 1855, 395, 399–400; read by Darwin in January 1856; DAR 128:14. Darwin's notes on Powell's Essays are in DAR 71:43–50.
54. Biographical detail on Knox from Rae 1964; Lonsdale 1870; A. Bates 2010; see also Bank 1996, 393. On William Edwards and Knox, see Robert J. C. Young 2006, 14–17, 120–21; see also Kenny 2007.

55. Desmond and Moore 2009, 39–40. By my account, Knox's mature views on race were provoked by his translation and editing of Quetelet's *Sur l'homme* in 1842; E. Richards 1989b.
56. Richards 1989b.
57. Stepan 1982, 51, and chs. 2 and 3; Robert J. C. Young 2006, 12–16; E. Richards 1989b; Kenny 2007.
58. Lonsdale 1870, 294–95, 383; BL Peel Papers 40601:50–51; E. Richards 1989b, 392n. See also Knox 1863, 138, where Knox referred to Walker as "a human anatomist second to none."
59. Knox 1850, 1, 2, 245.
60. On this basis Knox could explain the inability of the Celtic Irish to endure Saxon government and Saxon laws. He claimed to have predicted the 1848 revolution as an irresistible European-wide racial convulsion as the various tyrannized races struggled to throw off their alien rulers and reconstruct their own government and laws in accordance with their innate racial predilections; Knox 1850, 22, 15, 21, 2–3; see also Biddiss 1976.
61. Knox 1850, 456, 243–44, 88–89, 65–68, 149.
62. Ibid., 456–67, 302, 313–14, 147, 43–44, 223–24, 230, 243–4; Knox 1862, 576.
63. For instance, Callanan 2006, 52–54; but see A. Bates 2010, 124–125, 127–129.
64. Knox 1850, 23–24; Knox 1862, 552; A. Bates 2010, 126–7.
65. Quoted in A. Bates 2010, 127, 126–29; Knox 1850, 39–45, 46, 49, 53–54, 57, 157–59, 194, 196, 374.
66. This confusion is also attributable to the promotion of aspects of Knox's views after his death by the polygenist-oriented Anthropological Society of London; E. Richards 1989b.
67. Darwin's earlier notes on Knox's *Great Artists and Great Anatomists* (1852) dismissed Knox's scattered embryologically based hints as "opposed to my view" and as so much transcendental "Rubbish" (though he noted the work as a reference on "History of development," and readily appropriated Knox's leading instances of the cuticular fold and "tiger arm" to his own embryologically based view of human evolution). By "Rubbish" Darwin was referring explicitly to Knox's explanation of rudimentary structures, such as the "tiger arm" or supra-condyloid foramen that sometimes occurs in adult humans but is a regular structure in felines, as dependent on a "great plan or scheme of Nature . . . to which all organic beings are moulded"; see DAR 71:2, 60; DAR 128:4; *Descent* 1:23, 28. On Knox's tiger arm and its place in his theorizing, see E. Richards 1994.
68. Knox 1850, 100–101. In this sense, Knox *was* opposed to Darwin's views, and he remained so after the publication of the *Origin*; Lonsdale 1870, 386; Knox 1862, 570, 589, 594; Richards 1989b, 410.
69. Knox 1850, 28–29, 100–101, 35, 431–46; Knox 1852a, 60–63; see Richards 1989b.
70. Knox 1850, 215, 217, 444, 448, 477; Richards 1989b, 398–99.
71. Cited in Douglas 2008, 53, 63–64.
72. DAR 71:62.
73. Knox 1850, 410, 411, 228–29, 431–33, 443–44, 421–22, 415, 419; see also Knox 1852b, 164–66, 99–100; A. Bates 2010, 139, 146–51, 155. Knox jettisoned the associationist aesthetics of Alison and Jeffrey and poured Burkean scorn on the "Saxon" utilitarian, the "clock-regulated" mind's notion of beauty: "Why may not a wheelbarrow lay claim to the title of beautiful? or a pigsty? or a pair of Jack-boots? . . . But what say men of taste . . . who love *form* for the sake of the form itself? *Woman* presents the

perfection of that *form*, and, therefore, alone constitutes 'the perfect.' It is not youth, nor intellect, nor moral worth, nor associations of any sort, which constitute the beautiful and the perfect; nor is life required, nor complexion, nor motion; it is *form* alone which is the essential"; Knox 1850, 410–14.

74. Knox 1850, 403, 400, 410, 38, 414, 227–28; Callanan 2006, 70–74, 44–46; B. Cooper and Murphy 2000, 31–32; A. Bates 2010, 146–51; Neher 2011.

75. Knox 1850, 227–28; Alexander Walker 1838, 232, 369. Walker was, like Knox, a monogenist.

76. Knox 1850, 50, 197–99.

77. DAR 71:62; Knox 1850, 197–99, 201, 152; Robert J. C. Young 2008, 90–91.

78. See Darwin's references to Knox's *Races of Men* in his annotations of *Types of Mankind* (146, 148) in *Marginalia*, 1:604–5; note also Darwin's discussion of this in *Descent* 1:217n.

79. DAR 71:62–63.

80. Ibid.:64–65; see also Darwin's discussion of these issues and references to Knox and Smith in *Descent* 1871, 1:241–46.

81. Van Amringe 1848, 38–39, 654–57.

82. The races were ranked on the basis of their different "sexual relations," by which Van Amringe meant the "history and condition of women" which was specific to each race. In the "dark races," women are "slaves and articles of merchandize"; while in "all the white races," with the exception of the Circassians and Georgians, "they stand on the same platform with man." It is only in civilized societies, where men and women have highly developed aesthetic standards and the "freedom of selection or rejection" that social and moral improvement might result. Mr Walker, albeit largely through his analysis of animal breeding, has established laws of heredity that undermine any notion that intermarriage between the human races is of lasting benefit; Van Amringe 1848, 41–44, 639, 656–58, 676, 700, 703, 708–11.

83. Hotze in Gobineau 1856, 381n. On Gobineau's views, Hotze's translation, and Hotze's influence in London anthropology, see Robert J. C. Young 2006, 99–116, 125–26, 133–34; Desmond and Moore 2009, 235, 332–43.

84. Lewes did not cite Knox, but his *Races of Men* is an obvious source; Lewes 1856, 161–62; *CCD*, 6:210–11; Bartley 1992, 318–23.

85. On the reception of Knox's racial views, see Richards 1989b, 406–10; A. Bates 2010, 126–29; Robert J. C. Young 2008, 71–87, 94–95.

86. Douglas 2008, 60–63; Augstein 1999, 75–76, 235; Prichard 1826, 1: 95–98, 126–28; Prichard 1836–47, 1:147, 150; Prichard 1843, 11–26; Robert J. C. Young 2006, 6–19, 79–80; Robert J. C. Young 2008, 82.

87. Latham 1851, 97 (CUL); *Marginalia*, 1:483; Darwin read Latham in August 1856, DAR 128:20. *Notebooks*, B 189, 33–34, 68–69, 71; C 161e; E 63; *Foundations*, 95. The leading point Darwin drew from one of his earlier readings of Prichard's *Researches* was that "some birds—owls for instance are like men cosmopolites & do not vary like man because partly they interbreed, not having like man hostilities"; DAR 71:139–142. On Latham's views on racial purity, see Augstein 1999, 234–35.

88. The dovecote pigeon was, in Darwin's opinion, "the ancestor of all the breeds"; *NS*, 257–58, 441; *CCD*, 6:21, 34–35, 45. See also Darwin's marginalia in his copy of Riedel 1824, 158–59: "same coloured pigeons pair more readily (ch. 6)," *Marginalia*, 1:708–10.

89. *CCD*, 6:62–64.

CHAPTER ELEVEN

1. *CCD*, 6:109, 106–7, 265, 135, 89.

2. Darwin's marginalia in Westwood 1839–40, read at the end of 1854 (DAR 128:10); *Marginalia*, 1:861–64.

3. Darwin's marginalia in Rudolphi 1812, 184; *Marginalia*, 1:716, 718; *CCD*, 4:493.

4. Knox argued that the majority of women were forced to prostitution by unemployment and want, and urged that the only remedy for poverty and the reform of working conditions was not piecemeal philanthropy or legislation, but "combination," i.e., unionism, a radical solution that went against the grain of the great crusade against the regulation of prostitution, led by Josephine Butler; Knox also believed that most women had a tendency to "licentiousness," especially the French; Knox 1857, 142, 197, 49–50, 56. On Butler and her crusade, see Walkowitz 1980.

5. Mason 1994, 49–63.

6. Darwin was hoping for evidence against jellyfish inbreeding in line with his views on the necessity for an occasional cross; *CCD*, 6:161, 456, 178. On Robert Taylor as Darwin's "Devil's chaplain," see Desmond and Moore 1991, 70–73, 84–85; on Carlile's views on female choice, see Carlile 1838, 8.

7. *CCD*, 6:135, 217, 238, 438–39.

8. *CCD*, 6:372. On Huxley's marriage, see Desmond 1994, 206–15.

9. *CCD*, 6:369, 385–86, 394–95, 439, 476, 345, 343–44.

10. *CCD*, 6:130, 70–71, 74–75; DAR 85:A99. For Darwin on the aristocracy, see *CCD*, 3:228–29; 10:48.

11. Lawrence 1822, 393 (CUL); *Marginalia*, 1:485–6. This chronology is reinforced by the relation of Darwin's other annotations in the *Lectures* to the content and context of his other readings around this time. There is an explicit reference to "sexual selection," suggesting that this rereading postdates his reading of Knox's *Races of Men*.

12. Lawrence 1822, 394 (CUL); Darwin's note in Prichard 1851, 1 (CUL); *Marginalia*, 1:485–86, 684.

13. Darwin's annotations in Lawrence 1822 (CUL), 366–68; *Marginalia*, 1:486. On Lawrence's "high-mindedness," see Hudson 2004, 323–24. Lawrence suggested that the fat buttocks of Hottentot women were "most analogous" with those of fat-bottomed sheep, a comparison that drew on his domestication analogy; Lawrence 1822, 368–69.

14. Kevles 1985, 7–8. On cultural and aesthetic images of the Hottentot Venus, see Qureshi 2004 and Hudson 2008.

15. DAR 85:A91; Galton 1853, 87–88; Darwin's unmarked copy (read in August 1853) is in CUL, see *Marginalia*, 1:248; DAR 128:5.

16. Lawrence 1822 (CUL), 276–77n, 317–19, 272, 274, 276, 337, 356, 357, 368.

17. Prichard 1851, (CUL) 1: rear notes, 4:407, 519, 534, 535; *Marginalia*, 1:683–86; Wells 1971, 359; Augstein 1999, 228–30; Desmond and Moore 2009, 282–84. Desmond and Moore date Darwin's 1857 rereading of Prichard antecedent to his reading of Lawrence, largely on the basis of Darwin's annotation of his earlier edition of Prichard's *Researches*, where in March 1857 he noted that he had "gone through the later edition [of the Researches]"; ibid, 408n39; *Marginalia*, 1:681. As with Lawrence's *Lectures*, Darwin read through Prichard's *Researches* (both editions) several times, and these rereadings are difficult to date. Sometime between readings, Darwin changed his mind about his proposed note on man in his "big book": "If I ever consider Man look over other and earlier edition"; Prichard 1851 (CUL) 1: verso of rear note, not transcribed in *Marginalia*, 1.

18. Prichard 1851 (CUL), 5 note in back, see *Marginalia*, 1:686.

19. *CCD*, 6:112, 109, 106, 179, 199, 201, 259, 266–67; Desmond and Moore 1991, 397.
20. E. Richards 1989a, 258. On the decline of Chartism and the rise of the new liberalism see M. Taylor 1995; on Huxley's role, see Desmond 1994.
21. *CCD*, 6:180, 216.
22. *CCD*, 6:235–37 (Darwin's emphasis); Desmond and Moore (2009) interpret this as Darwin's expression of his support for Garrisonian "Disunion," 267–68, 276–79.
23. *CCD*, 6:299–300; R. Richards 1992, 150–55; cf. Desmond and Moore 2009, 286.
24. *CCD*, 6:259, 260; 5:339; Carpenter 1854, 107; *Marginalia*, 1:155; Desmond 1982, 37–51; Desmond 1994, 229–30; Owen 1855, 645–48 (*Marginalia*, 1:653); Owen 1849 (*Marginalia*, 1:655).
25. *CCD*, 6:191–92; Lucas 1847–50, 2:848, 576, 137. For Darwin's extensive annotations, see *Marginalia*, 1:519, 522, 513–23. On Darwin's use of Lucas's views on inheritance, see Noguera-Solano and Ruiz-Gutierrez 2009.
26. Notably the influential naturalist, Charles Hamilton Smith, who drew on the Meckel-Serres law to support his view of the separate emergence in time of the different human "species": the "successive cerebral appearances of the foetus" suggested that the "woolly-haired" first appeared, followed by the "intermediate Malay and American," then the "Mongolic," and finally the "Caucasian"; C. Smith 1852 (CUL), 126–28; *Marginalia*, 1:766; Desmond and Moore 2009, 204–9; Kenny 2007, 375–76, 379.
27. CUL Lawrence 1822, 452; *Marginalia*, 1:486; *Descent* 1871, 2:318; Knox 1850, 289.
28. *CCD*, 6:241–42; 7:201–2, 322–23; 8:28–29. On nineteenth-century theories of race, climate, and disease in the Indian context, see Harrison 1996.
29. *CCD*, 6:184; L. Wilson 1970, 96; T. Huxley 1854a, 249, 250–51, 253; Desmond 1994, 205; Desmond and Moore 2009, 270–75.
30. "Shows what a good man will write when involved in controversy"; Darwin's annotation on Bachman's pamphlets, DAR 74:157–58, 159–60.
31. *CCD*, 6:249, 256; 4:444; CUL Eaton 1852; *Marginalia*, 1:217; DAR 128:8; Yarrell 1843, 1; annotation in *Marginalia*, 1:883.
32. *Notebooks*, D 147e.
33. *CCD*, 6:335.
34. *CCD*, 6:214, 253, 375.
35. Tegetmeier's *Poultry Book* was issued in 11 parts from May 1856, but Darwin does not appear to have received them all; he read the whole in June 1857; *CCD*, 6:416; Tegetmeier 1856–57, CUL, part 8; Darwin's annotations in *Marginalia*, 1:798–803.
36. *CCD*, 6:420–21, 424–47, 427–28; *NS*, 275–79, 303–4. On Darwin's visits to Moor Park, see Browne 2002, 63–73.
37. *CCD*, 6:456, 460, 461, 463; Desmond 1994, 234–36; Desmond and Moore 2009, 307–8; *NS*, 379. On Huxley's early views on beauty, see Kottler 1980, 205.
38. Wallace 1855; *CCD*, 6:315, 325, 340, 413, 423, 431–32, 445–50; D. Porter 1993.
39. *CCD*, 6:387–88; 290; 5:519–22; Desmond and Moore 1991, 456.
40. Wallace's query on man is missing from the portion of his letter retained by Darwin; *CCD*, 6:457. It is inferred from Darwin's response; *CCD*, 6:514–15.
41. *CCD*, 6:457; Camerini 1993; James Moore 1997; Wallace 1905, 1:288, 361–62; G. Jones 2002; Fichman 2004.
42. *CCD*, 6:514–15.
43. *CCD*, 7:89, 412–13; 6:445; Desmond 1994, 241–42, 246; Desmond and Moore 1991, 457, 464.
44. CUL Agassiz 1857, 60, 121; Darwin's annotations in *Marginalia*, 1:9–11

45. Agassiz 1857, 112, 115, 172, 225; *Marginalia*, 1:9–11. On Agassiz's "superficiality & wretched reasoning," see *CCD*, 6:456; on the clash between Huxley and Owen over the ape and human brain, see Desmond and Moore 1991, 464–66; Desmond 1994, 239–41.

46. *CCD*, 7:262.

47. *CCD*, 7:62; L. Wilson 1970, 247. On Lyell's racial concerns, see Desmond and Moore 2009, 270–71.

48. The missing manuscript pages are on the verso of two of a series of calculations on sex ratios in a variety of domesticated animals; one of them is numbered as "7 (g)"; the missing pages from the big book were numbered from 7b to 7k. Both have been crossed through, indicating that they were used in the preparation of the section on sexual selection in the *Origin*; see verso of DAR 85:B25, B34.

49. DAR 85:A95; *Descent* 1871, 2:346–47.

50. DAR 85:A38, 93; 128:23; *Descent* 1871, 2:346–47; *CCD*, 6:445.

51. *CCD*, 7:80, 83, 116, 148–49, 84, 503, 63; Browne 2002, 71–72.

52. *CCD*, 7:89, 107–8, 115–25, 521; Desmond 1994, 244–45; Desmond and Moore 1991, 467–70; Browne 2002, 33–42; Darwin's abstract and Wallace's essay are reprinted in *CCD*, 7:507–20.

53. *CCD*, 7:165, Browne 2002, 43–50.

54. This was reproduced from Darwin's 1844 "Essay" as evidence of Darwin's priority over Wallace; see Bajema 1984, 108; *CCD*, 7:507.

55. *Origin*, 89, 199.

56. Darwin's abstract of Yarrell 1843 is in DAR 71:166–79, extracts on ff. 167, 167r; Darwin recorded reading this work on November 12, 1858; DAR 128:22.

57. DAR 71:176, 176r.

58. Ibid., 178.

59. Ibid., 179, 179r.

60. Ibid., 179r.

61. See Darwin's extended discussion of the various theories on skin color in the *Descent*, where he gives priority to sexual selection and argues that "the immunity of the negro is in any degree correlated with the color of his skin is a mere conjecture"; *Descent* 1871, 1:240–50, 244; although in his 1864 letter to Wallace, Darwin did concede some correlation of complexion with constitution and referred to a questionnaire on this aspect that he circulated to all army surgeons in tropical climates, Darwin still assumed sexual selection to be the "most powerful means of changing the races of man"; *CCD*, 12:216–17.

62. DAR 71: 192–214; citations at ff.194, 196, 199, 202, 202r. Darwin's marginalia in Audubon 1831–39; *Marginalia*, 1:21–23. Emma's undated notes may have been made in the late 1860s when Darwin was at work on the *Descent* (they are discussed in the *Descent* 1871, 2:116–17), though Darwin recorded that he reread Audubon's ornithology toward the end of 1858, and his own annotations were presumably made around this time; DAR 128:22..

63. DAR 71:194.

64. *Origin*, 88–90.

65. I am indebted to D. Graham Burnett for his discussion of "collapsing analogy" in Darwin's *Origin*, though he does not extend his analysis to sexual selection as I have done; Burnett 2009.

66. Desmond and Moore 2009, 313. Cf. Stepan, who argues that sexual selection "tended to strengthen rather than overthrow racial ideas"; Stepan 1982, 59.

67. *CCD*, 9:80; 8:28.
68. Ibid., 7:427.
69. Nead 1990, 5–7, 52–56; Phillips 2004, 54–64. On differing constructions of women's sexuality, see Mason 1995, 195–205.
70. Darwin's letter to Fox is all the more remarkable for its having been written just after his receipt of Wallace's letter and at the height of his distress over the illnesses of his children; *CCD*, 7:116. On Lane and the Robinson trial, see Colp 1981; Janice Allan 2011; Summerscale 2012; on the Victorian "anti-sensual mentality," see Mason 1995, 7 and passim.
71. *CCD*, 7:249, 331; Browne 2002, 71–72.
72. Nead 1990, 52–56, 5–7; Holmes 1995; Phillips 2004, 54–64.
73. Mason 1995, 195, 191–93; Phillips 2004, 60–64; Acton (1857) 1865, 112–13, 95; Greg 1850, 456–57; C. Hall 1992, 189, 197 and passim; Crozier 2000.
74. For details, see Kate Summerscale's wonderfully contextualized analysis of the Robinson case, 2012, citations at 47–50 and passim.
75. Ibid., 181–84.
76. Ibid., 190–200, 203–9; see also Janice Allan 2011; *CCD*, 7:247.
77. Phillips 2004, 81–95; Walkowitz 1980; Caine 1992, 150–95.
78. *CCD*, 7:240–41, 166–67, 107. On Huxley and the "young Guard" Darwinians, see Desmond 1994.

CHAPTER TWELVE

1. *CCD*, 15:141; see also 109.
2. Desmond 1994, 310–63. On the membership and role of the X club, see Barton 1998; Barton 2002; Desmond 2001.
3. *CCD*, 15:xvi–xx, 9, 74, 141.
4. C. Lyell 1863; *CCD*, 11:148, 173–74, 223; Browne 2002, 217–23; T. Huxley 1863, 69–70, 125ff; Desmond 1994, 292–97, 312–15; Wallace 1864; *CCD*, 12:173, 216–17, 220–22, 248–49.
5. E. Richards 1989b, 410–28; Stocking 1971. On the connections of the ASL with the Confederacy, see Robert J. C. Young 2006, 133–40; Desmond and Moore 2009, ch. 12.
6. *CCD*, 10:104; 11:582, 556, 564; 12:212, 319, 387; Desmond and Moore 2009, 324–31; Browne 2002, 214–17. On British responses to the complexities of the Civil War, see Foreman 2011.
7. Wallace to Huxley, February 26, 1864, Huxley Papers, 28.91, ICL; *CCD*, 13:256, 278; Hunt 1863 and 1864; Stocking 1971; E. Richards 1989a, 1989b; Robert J. C. Young 2006, 135; Desmond and Moore 2009, 232–35; Vetter 2010.
8. E. Richards 1989a, 1989b.
9. Wallace 1864, clxv–vi, clxix–x, and discussion, clxxviii–clxxxi; Hunt 1867, 113; see Benton 2009, 29–32.
10. *CCD*, 12:67n23, 93n18, 98, 100n8; Darwin's annotated copies of the *Anthropological Review* are in DRC; on Darwin's correspondence with members of the ASL, see sec. 13.7.
11. *CCD*, 13:278; 14:130; Semmel 1963.
12. William's recollections in DAR 112.2; Desmond and Moore 1991, 540–51; Desmond and Moore 2009, 348–57; Desmond 1994, 351–54. On the Eyre affair and its repercussions in England, see Semmel 1963; Lorimer 1978, 131–200; C. Hall 2002.
13. E. Richards 1989b; quotations from Hunt at 425–27.

14. Wallace further offended by endorsing the Anthropological takeover of the *Reader*, the generalist magazine in which the Darwinians (including Darwin) had invested much effort and hard cash, intending it as a vehicle for Darwinism; *CCD*, 13:256, 262–63, 278; Desmond and Moore 2009, 345–7, 416n9. Jeremy Vetter argues that by 1866 Wallace had become disenchanted with the ASL, but could not find common ground with the liberal, free-marketeering Ethnologicals; Vetter 2010.
15. *CCD*, 13:278; Wallace 1866a; E. Richards 1989a, 423; E. Richards 1989b.
16. Although he did not accept the polygenist evidence for this, Huxley was "*A priori* . . . disposed [on Darwinian grounds] to expect a certain amount of infertility between some of the extreme modifications of mankind; and still more between the offsprings of their intermixture"; T. Huxley (1865) 1873, 134, 155, 158–59, 163.
17. T. Huxley (1865) 1968, 67.
18. Ibid., 73–74; E. Richards 1989a. On Huxley's stand on abolition, see Desmond and Moore 2009, 335.
19. *CCD*, 14:435–36; 15:68–69; Desmond 1997, 65–69, 246–47; Shteir 1996; on the scientific and medical opposition to the claims by nineteenth century feminists and their response, see Alaya 1977; Russett 1989; Caine 1992; E. Richards 1989a, 1997; Gates and Schteir 1997; Gates 1998; Kohlstedt and Jorgensen 1999.
20. Vogt 1864, 81; Hunt, "Editor's Preface" in Vogt, xii–xiii.
21. Martineau (1838) 1989, 179, 184–85; see sec. 7.3.
22. *CCD*, 13:231–230, 309–310.
23. *CCD*, 15:96, 220; *Marginalia*, 1:824; *Descent* 1871, 1:1. On Darwin's esteem for Vogt and his work, see Amrein and Nickelsen 2008.
24. Hunt 1866, 322, 325–6; E. Richards 1989b, 425.
25. *CCD*, 15:93.
26. Darwin's marginalia in DRC Wallace 1864, clxv.
27. *CCD*, 8:xxiii, 292, 591; 10:xix, 292, 331; Ghiselin 1972, 134–47; J. Smith 2006, 140–41.
28. *CCD*, 8:20, 86, 132, 140; 9:93, 148–53. On Darwin's correspondence with Bridges, see Radick 2010b.
29. *CCD*, 9:74–75.
30. *CCD*, 9:xvi, 286–87; Bowler 1992a, 29–30; Browne 2002, 223–26; Caro, Merilaita, and Stevens 2008, 149.
31. *CCD*, 9:68, 75, 80–81, 360; see also 14:54, 55n6.
32. *CCD*, 9:363–64; 10:59–61, 540; H. Bates 1863; Browne 2002, 224.
33. *CCD*, 13:xx–xxi, 22–23, 34–36; Darwin's notes on Argyll's essays in *Good Words* are in DAR 47:20.
34. *CCD*, 13:311–13, 316; 14:42–43, 54, 57–58, 135–37, 144, 379–80, 397.
35. *Variorum*, 367, 369, 370, 371.
36. Müller (1863) 1869, ch. 10, esp. 97, 108–10.
37. *Variorum*, 702; *CCD*, 8:29; E. Richards 1976, 385–91; R. Richards 1992, 158–61; R. Richards 2008, 152–56. Huxley, after "nailing his colours to the mast" in dissent from the embryological section of the *Origin*, had gone remarkably silent on the issue; *CCD*, 7:432.
38. *Variorum*, 703–4.
39. Müller (1863) 1869, 20–24; see *Descent* 1871, 1:297, 328–29.
40. *CCD*, 13:212–213, 238; 15:187.
41. Ibid., 15:xvii, 144, 145n6; 16 (2): 664, 665n 3; *Descent* 1871, 1:205–6.
42. *CCD*, 15:210, 218–19, 271; R. Richards 2008, 148–56; E. Richards 1990, 140–41. Haeckel visited Down House toward the end of 1866; Desmond 1994, 354–60, 365.

43. *CCD*, 13:xix–xx, 130–31, 196–97, 202–3; 15:504, 431; *Variation*, 2:360, 385–86, 397–99; Hodge 1985, 227–37; Hodge 2010; Browne 2002, 285–86.
44. *Variation*, 2:74.
45. Ibid., 2:71–75; *Descent* 1871, 1:286–87.
46. *Variation*, 2:360, 385–86, 397–99; DAR 71:178; Endersby 2003.
47. *Variorum*, 371–72.
48. Kottler 1980, 1985.
49. *CCD*, 15:83, 88, 96–97; DAR 84.1:179–80; DAR 84.2:3–4; *Descent* 1871, 2:187.
50. *CCD*, 15:99.
51. Ibid., 15:105–6.
52. Wallace 1865b, 10, 22; Wallace 1905, 2:3–7; Kottler 1980, 206–8.
53. *CCD*, 15:99, 105–6, 109.
54. Ibid., 15:92–93, 128, 188–89, 190–91, 337; 16 (1): 156, 277–78.
55. Argyll 1867, 137, 202–3, 231, 233–48, 245.
56. *CCD*, 15:295; 16 (1): 516–17, 527. On Kingsley's significant role in the popularization of evolution and the accommodation of natural theology with evolution, see Lightman 2007, ch. 2; Hale 2012; on Kingsley, "muscular Christianity" and the "sexual self," see Rosen 1994.
57. *CCD*, 15:297–98.
58. Ibid., 15:236–38.
59. Ibid., 15:238.
60. Ibid., 16 (1): 406; Kottler 1980, 208.
61. *CCD*, 15:240–41.
62. Ibid., 15: 245–46.
63. Ibid., 15:250–51, 240; Kottler 1980, 209–10.
64. *CCD*, 15:308, 493; DAR 82:B5–6; see also *CCD*, 16 (1): 459–60, 472–73, 491–94, 499.
65. *CCD*, 15:240.
66. Ibid., 15:120; Wallace 1905, 1:421–22.
67. *CCD*, 8:400; 15:25–26n3, 179–80.
68. Burton 1864, 237–38, 245 (DRC); *Descent* 1871, 2:345–367, 345n.
69. Of the thirty-six replies Darwin received, only one was written by a "native": Christian Gaika, or Ngqika, a missionary-educated Xhosa member of the colonial police force, who to Darwin's amazement could actually read and write. Darwin interpreted this as evidence of the "progress of civilization," a view discouraged by his Anglo assistant naturalist in South Africa, John Weale, who had distributed Darwin's questionnaire; *CCD*, 15:360; 16 (1): xxiv–xxv; Shanafelt 2003.
70. Wallace 1867b, 2, annotated copy in DRC; Wallace 1867c, annotated copy in DRC; Kottler 1980, 210–11; Gould 1980b, 43–51.
71. Wallace 1867a.
72. *CCD*, 15:421, 394–95, 405–6; Wallace 1867c, annotated copy in DRC.
73. *CCD*, 10:241; Joy Harvey 1997, 77; see also chs. 4 and 5 for a discussion of the exchanges between Royer and Darwin.
74. Royer cited in Joy Harvey 1997, 66; *CCD*, 13:465–66n5.
75. Darwin possibly had heard of Royer's irregular union with the married Marcel Duprat, prominent republican and freethinker, and of the birth of their child; see Joy Harvey 1997, 76–79.
76. *CCD*, 17:489, 492.
77. *CCD*, 15:329–30.

78. For a revisionist analysis of Darwin's relations with Vogt, see Amrein and Nickelsen 2008.

79. *CCD*, 15:210, 223, 217–19. On the "terrible 'Darwinismus'" and the problems it posed for Darwin and Huxley in the 1860s, see James Moore 1991.

80. Wallace 1905, 1:433; Wallace 1866b; *CCD*, 15:272; Fichman 2004, 142–43; James Moore 2008; Marchant 1916, 2:187–88.

81. Wallace 1905, 1:409–12; A. Owen 1990; Fichman 2004, 142, 256, and passim; G. Jones 2002; Claes 2008; James Moore 2008.

82. *CCD*, 16 (1): 446–47, 452, 459; see ch. 6.

83. Wallace 1867b, 36–37; *Descent* 1871, 1:405–10; Wallace 1890a.

84. Quotation from Fisher 1930, cited in Kottler 1980, 225.

85. Aspects of this kind of thinking may be found in writings Wallace produced before 1858. Fichman argues that from the time of his earliest evolutionary speculations in the 1840s, Wallace had been sympathetic to the idea that other than strictly mechanistic forces were operative in the history of human evolution, that natural selection was always only a "partial solution"; Fichman 2004, 194–95. This is also the view of Charles Smith (2008); but see Benton 2008.

86. *CCD*, 14:227–29; Wallace 1865a, 672a; Wallace 1853, 280–82; Wallace 1864, clxix–clxx; Wallace 1905, 1:288, 342–43; Durant 1979; C. Smith 2008, 398–400; Schmitt 2009, 57–90; Benton 2009; Vetter 2010.

87. Greg 1868, 358, 359, 361, 362; Galton 1865.

88. Darwin's annotated copy of Greg 1868 in DRC; Wallace 1869a, 2:459–63; Wallace 1905, 2:235–40; Paul 2008; Paul and Day 2008, 226.

89. Vetter 2010.

90. *CCD*, 17:155; Wallace 1869b; Desmond and Moore 2009, 367.

91. Wallace 1869b, 391–94. Wallace reiterated much the same arguments in his follow-up, better-known article "The Limits of Natural Selection as Applied to Man" (Wallace 1870); see Fichman 2004, 192–95; Schmitt 2009, 61–62; Vetter 2010, 35–37; Benton 2009. Darwin's annotated copy of Wallace's 1869 paper, with its showers of dissenting exclamation marks and final dismissive annotation "i.e. miracles," is in DRC, DAR 133:14.

92. *CCD*, 17:206; 18:17, 90.

93. Wallace 1867a, 479–80; see also Bowler 1992b, 43–45; Kottler 1985; Ghiselin 1972, 215; Fagan 2008, 85–86; Gayon 2010, 142–43; Ruse 2015; Hoquet and Levandowsky, 2015; but see Bulmer 2005 for an alternative interpretation of Wallace's views. Hodge also argues against the conventional view of Wallace's rejection of the domestication analogy (personal communication), see ch. 6, n80.

94. Wallace 1858, 61; Wallace 1868, 82–83; Wallace 1969a, 1:xiv. Melinda Fagan sees the different collecting practices of Wallace and Darwin as fundamental to their theoretical differences; see Fagan 2007; Fagan 2008, 73–77; see also Caro, Merilaita, and Stevens 2008, 125–26; Wyhe 2013, ch. 5.

95. Wallace himself favored phrenological explanation: "The shape of my head shows that I have *form* and *individuality* but moderately developed, while *locality, ideality, colour,* and *comparison* are decidedly stronger"; Wallace 1905, 1:25–26.

96. Stepan 2001, 72–73; Wallace 1869a, 2:50–51, 254; but see Wyhe 2013, ch. 8, for a different reading of Wallace on Aru.

97. Stepan 2001, 72–73; Fagan 2008, 73–74, 79.

98. This point has been well made by Durant 1979, 45.

99. DAR 84.2:6, dated May 4, 1868.

100. *CCD*, 16 (1): 161; (2): 974.

101. Ibid., 16 (1); 246, 261, 266

102. Ibid., 16 (1): 278, 406; Darwin's notes on the gradations of the peacock's tail made in the British Museum from March 17 to 19, 1868, are in DAR 84.2:60–67.

103. *CCD*, 16 (1): 448, 452.

104. Ibid., 16 (1): 460, 472, 528; Kottler 1980, 212–16.

105. *CCD*, 16 (2): 689.

106. Kottler 1985, 426; cf. Cronin 1991, 150, 155. Cronin entirely ignores the critical role of embryology in Darwin's thinking.

107. *Descent* 1871, 1:296–97, 285–87, 410; DAR 84.2:172, 173.

108. DAR 84.2:8; see also DAR 84.2:7, 214; *Descent* 1871, 2:171.

109. DAR 84.2:8 verso; cf. Kottler 1980, 204; Cronin 1991, 153.

110. *Descent* 1871; DAR 84.2:8 verso, 2, 9, 15, 163, 216.

111. Notably DAR 84.2:215, 216.

112. *CCD*, 16 (2): 731, 732, 739, 714, 731, 732, 975; Wallace 1905, 2:1.

113. *CCD*, 16 (2):746, 754.

114. Ibid., 16 (2): 762–63.

115. Ibid., 16 (2): 784–86; 17:153, 155; Kottler 1980, 217–21.

116. *CCD*, 18:12–13.

117. Ibid., 19:46, 50–51, 185; Wallace 1871; Kottler 1980, 221–22.

118. Cronin 1991, 150, 155; see Bartley 1994a, 29.

119. These letters were not available in previous collections of Darwin or Wallace correspondence and were published for the first time in 2008; *CCD*, 16 (1): 171, 459; see also Bartley 1994a, 32; Kottler 1980, 222–23.

120. *CCD*, 12:216–17.

CHAPTER THIRTEEN

1. *CCD*, 17:476; 18:254; 19:785.

2. *CCD*, 18:104, 192, 323; F. Darwin 1888, 3:133; *CCD*, 19:205; Browne 2002, 351–53.

3. Darwin listed them in the *Descent* 1871, 1:4n1; the most notable was Haeckel's *Natürliche Schopfungsgeschichte* (1868, second edition 1870, published in English as *The History of Creation*, 1876), a work much admired by Darwin; *CCD*, 17:476; 19:169.

4. See Radick 2010a for a discussion of the various interpretations of Darwin's failure to link expression with natural selection. He concludes, following Desmond and Moore (2009), that it stemmed from Darwin's concern to defend human unity; but see R. Burkhardt 1985 and R. Richards 2003.

5. Jenkin 1867, 289, 291–94; *CCD*, 15:xxiv, 298–99; Browne 2002, 282–83.

6. *CCD*, 17:339; *Descent* 1871, 1:152–53, 2:125n32.

7. *CCD*, 17:155, 157, 211; Wallace 1869b; Wallace 1905, 1:427–28; Darwin's copy of Wallace's 1869 paper, with its showers of dissenting exclamation marks and final dismissive annotation, is in DRC, DAR 133:14.

8. *CCD*, 16 (2): 645; E. Richards 1989b.

9. ICL Huxley Papers 21.223–26; E. Richards 1989a, 273–76; E. Richards 1997, 122–28.

10. E. Richards 1989a, 275.

11. Pike 1869, xlvii, liii, lix; J. McGrigor Allan 1869, ccxii, ccxiii; Harris 1869.

12. *Descent* 1871, 1:228, 235, 180–84; Stocking 1971, 383–86; E. Richards 1989b, 432–35.

13. Fichman 2004; *CCD*, 16 (1): 266; 18:124, 125. On Huxley as "High Priest," see Desmond 1997, xi and passim.
14. *CCD*, 16 (2): 763; 1:266; 17:157, 175, 206, 448; 18:12–13, 124, 125, 204, 206, 304.
15. *CCD*, 18:204–5, 303.
16. *Descent* 1871, 1:2–3, 249.
17. The phrase is Schmitt's, in his excellent analysis of the role of the savage in Darwin's work; Schmitt 2009, 32–56, quotation at 55.
18. *Descent* 1871, 1:137–38; Darwin's annotated copy of Wallace 1969b in DRC, DAR 133:14.
19. *Descent* 1871, 1:35, 94, 96–98, 166–67.
20. Desmond and Moore 2009, 318, 370.
21. Schmitt 2009, 51; see also Duncan 1991.
22. Wallace 1867b, 36–37.
23. *Descent* 1871, 1:272–79, 285–91, 297–99. The impact of this theoretical section of the *Descent* was dampened by a rather panicked postscript to vol. 1 in which Darwin explained that he had overstated the "singular coincidence in the late period of life at which the necessary variations have arisen . . . and the late period at which sexual selection acts." It now seemed clear that "variations arising early in life have often been accumulated through sexual selection, being then commonly transmitted to both sexes." It appears that this addendum arose from Darwin's realization, while proofing the second volume, of his extensive recourse to this latter form of equal inheritance to explain the similarity of male and female forms among mammals; *Descent* 1871, 2:ix, 161, 237.
24. *Descent* 1871, 1:261.
25. Ibid., 1:261.
26. *CCD*, 16 (1): xxii, 93–94, 129; *Descent* 1871, 1:262–63, 266, 271, 300–315; Cronin 1991, 114–15.
27. *Origin*, 136–38, 412.
28. *Descent* 1871, 2:259.
29. See Gayon 2010, 139–40.
30. *Descent* 1871, 1:339, 403–410; 2:155–59, 166–171, 185, 187–223, 238, 303.
31. Ibid., 2:298, 299, 302, 307–8, 313; Wallace 1867b, 5.
32. *Descent* 1871, 2:296.
33. Ibid., 2:296–97.
34. *CCD*, 16 (1): 278, 27, 283, 291–92.
35. *Punch*, April 23, 1870.
36. *CCD*, 16 (1): 174; DAR 84.2:198, 208; *CCD*, 18:410n8; 19:10–11; *Descent* 1871, 1:238.
37. *CCD*, 16 (1) 197–98; 236–37, 408, 472; DAR 82.1:109–12; DAR 84.1:139–40.
38. *CCD*, 16 (1): 329–30, 561–62, 568, 372–73, 402, 369, 412.
39. Ibid., 16 (1): 258, 265, 273–4, 295, 408, 412, 413; Darwin's observations on plumage in relation to the angle of the sun are in DAR 84.2:28–30.
40. *CCD*, 16 (1): 156, 197–98, 237, 247, 277–78, 369, 413, 445, 471, 477, 511.
41. *Descent* 1871, 2:121; *CCD*, 16 (1): 589–90, 591.
42. *CCD*, 16 (2): 888, 898–99, 901; 17:330.
43. *Descent* 1871, 2:118, 63; *CCD*, 16 (1): 369.
44. DAR 84.2:190 verso, 193 verso, 205–6, 224; *Descent* 1871, 2:89–91; Gould's case of the hummingbird is discussed in *Descent* 1871, 1:151. On the "taste for the beautiful" in bowerbirds, see *Descent* 1871, 2:112–13.

45. *Descent* 1871, 2:141–42.
46. *Variation*, 1:216; *Descent* 1871, 2:91–93, 141–51. See also DAR 84.1:167, 174; DAR 84.2:67–69, 116, 117.
47. *Descent* 1871, 2:92–93; see Munro 2009, 266–68.
48. *Descent* 1889, 2:441; see also Ghiselin 1972, 222–23.
49. *CCD*, 16 (2): 784; Wallace 1869a, 1:51–52; Wallace 1891, 374; Cronin 1991, 165–67.
50. In the conclusion to the *Descent*, Darwin explained the consistent preferences of generations of argus pheasant females by claiming that this continuity was achieved through the "aesthetic faculty of the females having been advanced through exercise or habit in the same manner as our own taste is gradually improved"; *Descent* 1871, 2:401.
51. *Descent* 1871, 1:63; 2:401, 92–93.
52. Durant 1985.
53. *Descent* 1871, 1:62, 65–69; Browne 2002, 349.
54. DAR 219.9:73, 74, 77; 219.11:16; *CCD*, 18:25, 55; 19:199, 550, 804–6; Browne 2002, 347–49, 356–58.
55. Cobbe 1894, 1:124; Caine 1992, 103–49.
56. Cobbe 1869; S. Mitchell 2004, 191.
57. Cobbe 1869. For Cobbe's account of her discussion with Darwin, see Cobbe 1894, 2:123–25; see also E. Richards 1997, 128–30.
58. S. Mitchell 2004, 191; Caine 1992, 77; Phillips 2003, 105. Darwin's notes on Mill and the mental and emotional differences between the sexes are in DAR 85:A10–12; these notes indicate he previously had not read Mill's *Subjection*, but subsequently obtained a copy, interpolating the necessary references to Mill's text in pencil at a later date; the references to Mill's *Subjection* are on the verso of DAR 85:A10.
59. DAR 85: A12.
60. Mill 1869, 119–24.
61. DAR 85:A10 verso, dated July 18, 1869; 85:A11; *CCD*, 16:33, 38. Darwin's claim that he was no introspective "great philosopher" but a "degraded wretch" who looked "from the outside thro' apes & savages at the moral sense of mankind," is taken from his exchange of views with Cobbe on the "moral sense," see *CCD*, 18:81–82, 84–85.
62. DAR 85:A10.
63. *Descent* 1871, 2:326–27. Darwin's wording here bears a striking similarity to that of George Harris, Barrister at Law and President of the Anthropological Society of Manchester, published in the October 1869 issue of the *Anthropological Review*. Harris also identified "an essential and extensive difference and inequality between persons of different sexes . . . mental and moral as well as material, arising from a difference in material structure, in texture and temperament as well as organic, which no similarity of education can ever remove, no identity of circumstances can ever annihilate"; Harris 1869, cxii, cxciv–cxcv; cf., Mill 1869, ch. 3.
64. Mill 1869, 31–32, 141–42, 213; Rossi, introduction to Mill 1970, 58–60; Paul and Day 2008, 226–27.
65. DAR 128:25; *CCD*, 4:496; ED 2:169.
66. *CCD*, 17:530–31; Galton (1869) 1950, 202–3; Browne 2002, 290.
67. *Descent* 1871, 1:71n, 2:327.
68. Ibid., 2:328–29.
69. Ibid.
70. Darwin cited further evidence from measurements of "Negro" and German skulls, but scrupulously added Vogt's qualification that more observations were requisite before it could be accepted as generally true; ibid., 2:329–30, n24.

71. Ibid., 2:316–18.
72. Ibid., 2:326–27, 389.
73. To claim that there were other possibilities open to Darwin is to miss the point of the centrality of his interpretations of embryology and inheritance to his construction of sexual selection. For Darwin this *was* the point. It stood at the heart of his dispute with Wallace, and he was committed utterly to its defense. And, of even greater import for Darwin, it was indissolubly linked in his theorizing with the predominance of beauty-based male choice that was responsible for racial divergence. Darwin's enduring resistance to any revision of this interpretation is evidenced by his response to S. Tolver Preston's claim in 1880 that Mill's "magnificent theoretic analysis" was upheld by Darwin's law of equal inheritance for mammals of both sexes: in spite of millennia of oppression, women's intelligence was not greatly different from men's; if women were properly educated and their brains developed, men would benefit also through the law of equal transmission of characters to both sexes; women's education was thus vital to the future progress of both sexes. Darwin's brief letter of dissent invoked Vogt's old claim of progressive sexual divergence: the "sexes have equal brains amongst savages," hence Preston's claim for progressive equalization of the sexes and social progress through female education and equal inheritance was without foundation. See Preston 1880; Darwin's copy is in DRC; *Calendar* 12775; 12776.
74. Mill 1869, 43–44; Rossi in Mill 1970, 56–57.
75. *Descent* 1879, 121n5.
76. Jann 1994, 290–91; Stocking 1987, 145–170; McLennan 1865; Lubbock 1870.
77. *Descent* 1871, 2:371, 356, 355.
78. Jann 1994, 295–96; Russett 1989, 82–84; *Descent* 1871, 2:270–72, 367.
79. *Descent* 1871, 2:293, 318, 371; [Mivart] 1871a (see sec. 14.1); Jann 1994, 295.
80. *Descent* 1871, 2:325–26.
81. Ibid., 1:300–3. Curiously, Darwin made no reference, published or unpublished, to the highly politicized contemporary debate about the problem of redundant women, attendant on the discovery in the 1850s that females greatly outnumbered males in the British population. In 1862, Greg suggested that "surplus" women, who would otherwise be a burden or cause social upheaval, should be shipped off to the colonies to find husbands, a suggestion vigorously countered by Cobbe, who argued that single women could lead full and satisfying lives; Greg 1862; see Caine 1992, 114.
82. Darwin's copy of McLennan 1865 in CUL, annotation at 176; *Marginalia*, 1:560–61.
83. Darwin's copy of Argyll 1869 in CUL, marginalia at 136 and inside back cover; *Marginalia*, 1:16–17.
84. Darwin's copy of Lubbock 1870 in CUL, marginalia at 67; see *Marginalia*, 1:511–12. See also Darwin's notes on L. Morgan 1868, DRC.
85. However, Darwin did suggest that where polyandry prevailed, the women therefore would select the handsomest men, and if women were valued merely as slaves or beasts of burden, "men would still prefer the handsomest slaves according to their standard of beauty"; *Descent* 1871, 2:358, 363, 366.
86. *Descent* 1871, 2:361–62, 367–68.
87. This argument is well made by Jann 1994, 293–94. On the "perversion" of the natural instincts in savages, see *Descent* 1871, 1:134–35.
88. Anon., "Artistic Feeling of the Lower Animals," *Spectator*, March 18, 1871, 355; Durant 1985, 298–99; Jann 1994.
89. Hunt 1864, xviii–xix; Reade 1864. Ian Hesketh shows that Reade had fallen out with Hunt by 1868; Hesketh 2015; see also Driver 1999, 90–116.

90. *CCD*, 16 (1): 513–14; 17:533; 18:94.
91. Ibid., 16 (1): 279–80, 529; *Descent* 1871, 2:372, 382–83.
92. *CCD*, 17:22–23; *Descent* 1871, 2:374–75. Darwin also welcomed Reade's next communication on the custom of the "Jollofs [Wolof or Ouolof] a tribe . . . of uniformly fine appearance," of "picking out [their] worse looking slaves and to sell them." This practice of "rouging" out the ugly and less desirable, although it was "not done [consciously] to improve the race," was, Reade pointed out, "very suggestive" of the "negro's knowledge of the power of selection." Darwin reproduced it as such in the *Descent*; *CCD*, 17:533; 18:93, 150; *Descent* 1871, 2:357–58.
93. *CCD*, 18:94, 150.
94. Ibid., 18:284; see also 241, 288; and DAR 85:109–12; 176:65.
95. *Descent* 1871, 2:350–51, 381–82.
96. *CCD*, 16 (1): 281–82, 2:922; Mantegazza 1867. Darwin's heavily annotated copy of Mantegazza's *Rio de la Plata* is in CUL; see *Marginalia*, 1:563–65; McClintock 1995, 38; *Descent* 1871, 2:338n36, 341, 351–52, 351n63. On Mantegazza, see Moruno 2010; Sigusch 2008.
97. Robert J. C. Young 2006, 17, 72, 76, 80–81, 123; Young 2008, 131–39, quotation at 137.
98. Beddoe 1863; Darwin annotated Beddoe's paper in his copy of volume 1 (2) of the *Anthropological Review*, and made a memorandum on the back cover: "On hair colour different in married & single women"; DRC.
99. *CCD*, 17:382, 365–68; DAR 85:A13–28; see also DAR 85:B1–17.
100. *Descent* 1871, 2:403; Darwin's annotated copy of Greg's 1868 paper is in DPC. See also Greene 1977, 9–14.
101. *CCD*, 10:48, 127; 12:216–17, 221; 14:385, 393. On Hooker's identity as a gentleman first and a professional scientist second, see Endersby 2014.
102. *Descent* 1871, 1:169–70; 2:122, 338, 356–57, 402–3; *CCD*, 12:217.
103. *Descent* 1871, 2:401–2.
104. *Descent* 1871, 2:122–23; cf. Levine 2006, 192–95.
105. *Descent* 1871, 2:124, 402–3.
106. In his 1865 article, Galton linked the beauty-making power of aristocratic wife selection with his proposed competitive selection of those young ladies possessing "grace, beauty, health, good temper, accomplished housewifery . . . in addition to noble qualities of heart and brain" to be mated with men of "talent, character and bodily vigour" to produce a "highly-bred human race," fit for a future Utopia; Galton 1865, 165, 319.
107. *Descent* 1871, 2:318, 338, 349, 357. Note the strong resemblance between Darwin's passage on the formation of different standards of beauty in dispersed tribes and Lawrence's on the same: ibid., 2:370; Lawrence 1821, 391. There is only one reference to Walker's *Intermarriage* in the *Descent*: 1871, 1:117.

CHAPTER FOURTEEN
1. F. Darwin and Seward 1903, 2:95, 97; F. Darwin 1888 3:155.
2. Anon, 1871. *British Quarterly Review*, 53:565–69.
3. Anon, *Spectator*, March 11, 1871, 355.
4. [Morley] 1871; *CCD*, 19:208–9, 238–40.
5. *CCD*, 19:286–87, 301–2; see also 245. Darwin took enough notice of Morley's arguments to pass his letter around family members for comment, and he had William check Mill's *Utilitarianism* to see that Darwin had correctly represented his position on the moral sense in the *Descent*; *CCD*, 19:244, 246–47.

6. Anon., *Saturday Review*, March 4 and 11, 1971, 276–77, 315–16, at 316, in DAR 226.2:102–4. See also *Observer*, March 19, 1871, in DAR 226.2:101; *CCD*, 19:146.

7. *CCD*, 19:208.

8. *CCD*, 19:185, 199, 200; Marchant 1916, 1:127.

9. Wallace 1871, 177, 179, 181, 182.

10. Ibid., 179, 180.

11. Ibid., 178, 181, 182; Bartley 1994a, 39–41; Kottler 1980, 222.

12. [Mivart] 1871a, 47, 52, 85, 87, 90; [Mivart] 1871b; *CCD*, 19.xxiii–xxiv, 570; Desmond and Moore 1991, 568–71, 582–83; James Moore 1991, 395–402; Browne 2002, 329–31; Dawson 2007, 42–43.

13. On the significance for Victorian scientists, including Darwin, of the *vera causa* or "true cause" principle, see Hodge 1977, 238–44.

14. [Mivart] 1871a, 53, 55, 57–59, 62.

15. Ibid., 60, 62, 63.

16. Ibid., 60.

17. Ibid., 61, 87. For Mivart's previous praise of Wallace, see Mivart 1871b, 10, 278–81; *CCD*, 19:51.

18. *CCD*, 19:484–85, 574, 586, 605–6, 610; T. Huxley 1871; Browne 2002, 353–55; Desmond 1997, 25–26.

19. *CCD*, 19:602, 605–6.

20. I am indebted to Adrian Desmond for this information. Huxley's only known reference was facetious: on receipt of the *Descent*, he told Darwin he would try to "pick out from 'Sexual Selection' some practical hint for the improvement of gutter-babies & bring in a resolution thereupon at the Schoolboard" (Huxley was a member of the London School Board); *CCD*, 19:81; *Descent* 1879, 230–40.

21. *Descent* 1871, 1:5. In fact, Darwin had been disconcerted to find "nothing about sexual selection" in Haeckel's *Generelle Morphologie*, However, Haeckel did argue there for the role of female choice in improving racial intelligence, while male choice would enhance female beauty; Haeckel 1866, 2:244–47; *Marginalia*, 1:358. For Haeckel's remarks to Darwin on sexual selection, see *CCD*, 16 (2): 923; 19:753; see also Desmond and Moore 2009, 418n34; Bajema 1984, 259; R. Richards 2008, 158.

22. *CCD*, 19:437, 594, 776–77; Mantegazza 1871, 317. Mantegazza, undaunted, went on to advocate a "sexual hygiene" aimed at the eugenic improvement of the race through the elimination of the "bad & ugly," extending the concept of sexual selection into a beauty-based "ethnography of love" which accorded heightened sexual pleasure and choice to women yet still denied them equal intelligence or equal rights on the basis of their evolutionary history, see Moruno 2010, 156, 158–61; Sigusch 2008.

23. Reade to Darwin, May 14, 1873; DAR 176:67.

24. Reade's position on the correspondence of black and white notions of beauty was independently endorsed by Richard Burton and the explorer Gerhard Rohlfs; *CCD*, 19:13, 19, 25, 28, 29, 53, 92, 431, 566–67, 577, 580–82, 588–89, 599; Burton was interested in meeting Darwin to discuss their differences on beauty (*Calendar* 8160), but it does not seem that the meeting ever took place.

25. Reade 1872; P. Morton 1984, ch. 3; Driver 1999, 90–116; Hesketh 2015.

26. Reade 1873, 2:253, 306–8, 316, 521–23.

27. Ibid., 2:314; DAR 89:159. Darwin's copy of *The African Sketch Book* is annotated, and he cited it in the 2nd ed. of the *Descent*, though without reference to Reade's alternative explanation of skin color; *Descent* 1879, 649n68; *Marginalia*, 1:697–98. By the time *The African Sketch Book* appeared, Reade was back in Africa as *Times* correspon-

dent on the second Anglo-Asante War and their correspondence had lapsed. He died soon after, in 1875.

28. *CCD*, 19:190–91, 375, 542, 597–98; *Descent* 1879, 656, 658, 663; McLennan 1896, 50–51; Dawson 2007, 34–35, 75.

29. *CCD*, 19:440–44, 512–13, 517–18, 521; *Descent* 1879, 360.

30. *Calendar* 8143, DAR 171; *CCD*, 19:529; *Variorum*, 242; Wallace 1905, 2:43–45, 300–301; Desmond and Moore 1991, 592.

31. Prodger 2009; J. Smith 2006, 125–36; Browne 2002, 359–368; Marchant 1916 1:272. Cobbe reviewed the *Descent* in 1872 and had no theological difficulties with tracing "Man to the Ape," but forcefully opposed Darwin's "Simious Theory of Morals," his "most dangerous" utilitarian interpretation of the evolution of the human mind and morality from animal instincts. Her doctrine of "Theistic Ethics" neatly removed the mental and moral differences between men and women from the biological to the spiritual domain and so, in Cobbe's view, guaranteed the moral autonomy and authority of women, giving them a special responsibility to "those who have no free-will—the lower animals." Her abundant energies and political skills increasingly were channeled into her anti-vivisection crusade, which brought her into direct conflict with Darwin and Huxley; see Cobbe 1872, 1–33; E. Richards 1997.

32. Marchant 1916 1:278; Browne 2002, 370–406; DAR 96:141.

33. Marchant 1916,1:281–83; DAR 96:161–62, 164; DAR 106:B118–19.

34. G. Darwin 1872; F. Darwin and Seward 1903, 2:43–44.

35. G. Darwin 1873, 414; Galton 1873; Desmond and Moore 1991, 602–3.

36. *Descent* 1879, 3; Browne 2002, 434–36.

37. See for instance *Descent* 1879, 244n4; see also *CCD*, 19:186, where Darwin had made this same point to Wallace.

38. *Descent* 1879, 556; Kottler 1980, 224.

39. *Descent* 1879, 246, 487–88, 495.

40. *Descent* 1871, 2:122; *Descent* 1879, 473, 487–96; Bartley 1994a, 55–56.

41. Dawson 2007, 50 and passim; Bland 1995, 83–84; Richardson 2003, ch. 2.

42. *CCD*, 18:194, 255–56; Dawson 2007, 36–41; *CCD*, 19:221.

43. *Field*, March 18, 1871, DAR 226.2:79; *Saturday Review*, March 11, 1871, 315; *Blackwood's Edinburgh Magazine* 109:517–19, DAR 226.2:71; *The Gardeners' Chronicle and Agricultural Gazette*, May 20, 1871, 649–50, DAR 226.2: 80.

44. *Times*, April 7 and 8, 1871, 3, 5; DAR 226.2:105–8.

45. *Athenaeum*, March 4, 1871, 275–77, 276, DAR 226.2:69–70; *Literary World*, March 17, 1871, 169–71.

46. Mivart 1873, 608; [W. Dawkins] 1871, 234–35; Dawson 2007, 44–74, 78.

47. [Mivart] 1874, 70; T. Huxley 1875, 16–17; Desmond and Moore 1991, 610–13; Browne 2002, 355; Dawson 2007, 74–81.

48. Cf. Dawson 2007, 76; *Descent* 1871, 1:96.

49. Richardson 2003, 56–57; Phillips 2004, ch. 15; Dawson 2007, 50–51; Hamlin 2014, ch. 4. Curiously, the American feminist Antoinette Brown Blackwell, who offered the earliest extended critique of Darwin's interpretations (1875), made no attempt to exploit the feminist potentialities of Darwinian female choice, being more concerned with disputing whether woman's innate mental differences could properly be called inferior to man's, arguing that social progress was dependent on the full expression of both sets of sexually divergent traits; E. Richards 1983, 96–97; Tedesco 1984; Kohlstedt and Jorgensen 1999; Hamlin qualifies this, arguing that Blackwell advocated an equal sexual division of labor; Hamlin 2014, 101–12.

50. Fichman 2004, 268.
51. Wallace 1876c; Mivart 1876.
52. Marchant 19161:290, 291; Wallace 1876a, 1:340.
53. Wallace 1877, 403–4; Wallace 1876b; Darwin's annotated copy of this paper is in DRC.
54. See Kottler 1974, 178–79; Shermer 2002, 187–201; Fichman 2004, 185–86; James Moore 2008, 363–66; Vetter 2010.
55. Weismann 1882, 1:62; Bartley 1994a, 61. Darwin read Weismann's *Studien zur Descendenz-Theorie* in May 1875, and wrote that he was "fully prepared to admit the justice of your criticism on sexual selection of Lepidoptera; but considering the display of their beauty, I am not yet inclined to think that I am altogether in error"; F. Darwin and Seward 1903, 1:356–57.
56. C. Darwin 1876; *Descent* 1871, 2:293, 313; *Calendar* 10335 (DAR 148:346), 10660.
57. Wallace 1877, 398, 399, 401; Cronin 1992, 124–26.
58. Wallace 1877, 399, 400, 401.
59. Ibid., 403–5.
60. Ibid., 408.
61. Wallace 1905, 2:22.
62. Wallace 1877, 471.
63. The Society for Psychical Research included seven other fellows of the Royal Society besides Wallace (and, it should be added, one Prime Minister, William Gladstone); Fichman 2004, 286–91; Lightman 2001. For a recent reevaluation of Victorian scientific naturalism, see Dawson and Lightman, "Introduction," and other contributory papers in Dawson and Lightman 2014.
64. James Moore 1991; Desmond 1997.
65. Darwin's annotated copy of Wallace 1877 is in DPC, marginalia at 405.
66. Marchant 1916, 1:299.
67. Ibid., 1:300, 301–2.
68. Wallace 1889, 295–96; Bartley 1994a, 45; Lightman 2010, 14; G. Romanes 1892–97, 1:391, 2:12–13. On the reprintings of Wallace 1867a and b, see C. Smith 1991, 497–98n272.
69. Browne 2002, 482–83; Fichman 2004, 205–7; Marchant 1916, 1:313–15.
70. Allen 1877, 239–42; Allen 1879a, 4, 25, 155–61, 165–66, 186–93; see also Allen 1886, ch. 9; *Calendar* 10971, 10973, 10996, 12168, 13536, 13600, 13627, MLCD 2:367–68; E. Romanes 1896, 55. On Allen's scientific standing, see P. Morton 2005, 55–56, 96, 99, 106, ch. 5.
71. *Calendar* 11873, 11894; see also 13594. For annotated copies of Allen 1877 and 1879a in Darwin's library, see *Marginalia*, 1:14–15; for Wallace's response to Allen's defense of female choice, see Wallace 1879.
72. *Calendar* 11967, 12062; Allen 1879b.
73. Darwin was of the view that "Trades-Unions" and "Cooperative Societies" which "exclude competition" represented a "great evil for the future progress of mankind." Darwin also thought "foolish" the attempts by various German socialists such as Ludwig Büchner and Friedrich Lange to appropriate Darwinism to their respective causes by reconciling socialism and natural selection; Weikart 1995, 611; Stack 2000, 683.
74. F. Darwin 1888, 3:253; DAR 180:3; *Calendar* 13710, 13753, 13757.
75. Nordenskiöld quoted in C. Darwin 1882, 367, 368; Bajema 1984, 150, 252–54.
76. C. Darwin 1882, 367.

77. Wallace 1905, 2:266–67; Fichman 2004, 268, 271; Bellamy 1888, 179. On Bellamy's views on women and the future society, see A. Morgan 1944, 220–22; Bowman 1958, 246–88; Strauss 1988.

78. Beer 1983; Yeazell 1989; Yeazell 1991, 219–28; J. Smith 1995; Bender 1996; Kaye 2002, 93–108; Bender 2004; Levine 2006, 169–201; Burdett 2009.

79. On the nexus between feminism, social purism and female choice, see Richardson 2003, 48–49 and passim; Bland 1995; Phillips 2004, ch. 15, "The Gynae-centric Universe."

80. Kaye 2002, 91.

81. See Stack 2000, 2003.

82. See B. Taylor 1983, 275ff. and passim.

83. L. Huxley 2:56. On Aveling, see Feuer 1962; Kapp 1972, 1:258–72; Paylor 2005.

84. Biographical detail on Besant is taken from Nethercot 1961 and A. Taylor 1992. On the Secularists, see Royle 1971, 1980; on the history of birth control and its association with free thought, see Fryer 1965; McLaren 1992, ch. 6.

85. DAR 202, transcribed with the kind assistance of Stephen Pocock; see also Peart and Levy 2008, 347–48; this transcription differs from that offered by James Moore 1988, 306, Browne 2002, 443–44, and Dawson 2007, 141. Darwin's letter was partially published by Bradlaugh's daughter Hypatia, giving the erroneous impression that Darwin, had he been able to attend, would have supported Besant and Bradlaugh; Bonner 1895, 28.

86. "Special Trial Number," *National Reformer*, June 23, 1877, 405, 412–13. The trial transcript was published in full in the *National Reformer*, the official organ of the National Secular Society. See also Chandresekhar 1981, 26–54; Peart and Levy 2008.

87. Besant (1877) 1884, reprinted in Chandresekhar 1981, 149–201 (note the Darwin references, 163–64, 196–97). On the "revolutionary truths" of natural and sexual selection, see Aveling 1883b, 4–6; on the relationship between Besant and Aveling, see Nethercot 1961, 163–67.

88. *Calendar* 12757. Darwin's letter to Aveling was long misinterpreted as addressed to Marx; see Colp 1982. *The Student's Darwin* was a revised version of the series "Darwin and His Works" by Aveling, which appeared in twenty-eight installments in the *National Reformer*, between November 16, 1878, and September 19, 1880; see also Royle 1971, 170–72; James Moore 1988, 309.

89. Aveling 1881a, advertisement. Privately, however, in late 1881, Darwin hosted a visit by Aveling and the German materialist and Darwinian Ludwig Büchner to Down House, where he quizzed their atheism; see Aveling 1881b; James Moore 1988, 310–12.

90. The best account of Bradlaugh's campaign is in Arnstein 1984.

91. Arnstein 1984, 55–56, 250–51, 262–63; *National Reformer*, August 28, 1881, 212; September 4, 1881, 234–35; August 20, 1882, 132–33.

92. Fryer 1965, 111, 123–31, 166; [Drysdale] (1855) 1861; Royle 1980, 254ff; Arnstein 1984, 22. On the members of the Drysdale circle (notably Alice Vickery) and the fight for birth control, see Benn 1992.

93. Royle 1980, Bradlaugh quote at 115; Darwin cited in James Moore 1988, 308–9; but see Lightman 2002 on the failure of Huxley's strategy.

94. Aveling 1881a, 2:322–24.

95. Aveling 1882, 5, 7–8.

96. Besant 1882, 58, 60; serialized in the *National Reformer* from February 17, 1878 (XXXI: 1027–28).

97. Aveling 1884; Aveling n.d.; Paylor 2005.
98. Aveling 1883a; letter from Huxley to Crompton, July 16, 1883, in Tribe 1971, 227; Nethercot 1961, 188–89; A. Taylor 1992, 157–60; Desmond 1997, 146–47. The *National Reformer* gave extensive coverage to the whole affair; *National Reformer* 1883, 41:361–62, 380–82, 409–10.
99. Helfand 1977, 170, 174; Dawson 2007, 153–57.
100. L. Huxley 2:56–57; Aveling 1883a.
101. Cited in Kapp 1972, 1:256–58. On Eleanor Marx as a "New Woman" with a new morality, see Brandon 1990, ch. 1. When Aveling announced their unmarried union in late 1884, he was excluded from the Secularist movement by Bradlaugh who alleged a history of debt, fraud, and deceit; Bradlaugh Papers, 1156; Royle 1971, 105–6.
102. Besant 1886, 2; Marx Aveling and Aveling 1886; Draper and Lipow, 1976. Engels's *Origin of the Family, Private Property and the State* (first published in German in 1884) had argued that only when capitalist production and its property relations were removed could the choice of marriage partner be motivated by mutual inclination and affection; extract in Rossi 1974, 478–495, at 493.
103. Bebel 1885, 64–65, 69, 126; see also Rossi 1974, 496–505. Marx Aveling and Aveling, 1886, 3.
104. Marx Aveling and Aveling 1886, 9, 14–16; Dyehouse 1989, ch. 4, 190–92.
105. On the Men and Women's Club, see Bland 1995, ch. 1 and passim; Brandon 1990, ch. 2; Showalter 1995, ch. 3; Dyehouse 1989, 161–66. On the socialist attitude toward "bourgeois feminism," see Boxer 2007.
106. See Yeazell 1989; Yeazell 1990, 229–37; Hamlin 2010; Hamlin 2014, 157–58. Hamlin argues that Ellis's views on human sexuality were inspired by Darwinian sexual selection and that sexual selection in turn gave scientific legitimacy to sexology. However, while early sexology had strong associations with evolutionary theory, Ellis dissented from the fundamental Darwinian element of visually based aesthetic preference; moreover, the rise of sexology coincided with the general rejection of sexual selection in biological theory.
107. Schreiner 1911, 25–27, 281; Love 1983; Showalter 1995, 48–58; Dyehouse 1989, 155–57; Ledger 2007.
108. It was published under the aegis of Havelock Ellis in his Contemporary Science Series and strongly influenced Ellis's own best-selling *Man and Woman* of 1894.
109. Geddes and Thomson 1889, 27–29.
110. Ibid., 267, 271. For critiques, see Mosedale 1978, 32–41; Alaya 1977, 269–71; Russett 1989, 89–92.
111. It should also be noted that just twelve years after the Bradlaugh/Besant trial, Geddes and Thomson risked condemnation for their discussion of sexuality and their advocacy of (and explicit advice on) contraception within marriage ("prudence *after* marriage"), reiterating the Aveling/Besant line that the health of women and the future of society depended on quality, not quantity, of offspring; Geddes and Thomson 1889, 269, 289–98, 311–12; Meller 1990, 81–84.
112. L. Hall 2004, 37–38, 46; Meller 1990, 84; Alaya 1977, 270–71.
113. Schreiner 1911, 25–27; Gamble (1894) 1916; C. Gilman (1898) 1966; Meller 1990, 84; Alaya 1977, 269–71; Love 1983; Egan 1989; Jann 1997; Hamlin 2014, 112–24, 128–141. Gamble offered the most thoroughgoing expropriation of sexual selection for feminist ends: with the onset of wife capture in prehistory, men had seized the power of selection from their natural superiors, women, and inflicted "wasp waists," "furbelows," ill health, and excessive childbearing on them. Women must regain

the power of mate selection and economic independence, thereby elevating their status, curtailing male lust and ridding the world of prostitution and vice; Gamble (1894) 1916.

114. For instance Allen 1886, ch. 9; Allen 1889b.

115. F. Darwin and Seward 1903, 2:50; P. Morton 2005, ch. 5; Paul 2003.

116. Galton (1883) 2004, 39. On Galton's pocket "pricker," see ibid., 38.

117. Galton did think there were some—in 1890 he proposed offering Cambridge women graduates, judged especially physically and intellectually superior, the sum of £50 if they married before the age of twenty-six and £25 on the birth of each child; Parrinder 1997; Paul 2008, 273. Galton opposed the emancipation of women on the grounds that it would reduce the availability of eugenically preferred women.

118. Allen 1890, 54, 56, 58, 61–62; Allen 1889a; 1889b; 1889c; 1889d.

119. Allen 1896, 345; Richardson 2003, 51–52.

120. Allen 1895; P. Morton 2005, ch. 9; Showalter 1995, 51–53; Ledger 2007.

121. Wallace 1890b, 329–30.

122. On Romanes's relationship with Darwin and his views on reason and morality, see R. Richards 1987, 334–408.

123. Wallace 1886; G. Romanes 1889, 251–53; E. Romanes 1896, 141, 210–11; Bartley 1994a, 65–68; Schwartz 1995, 314–16; Milam 2010, 23–24. On Weismann and the neo-Darwinians, see Bowler 1992a, 40–43; Bowler 1992b, 116–17.

124. Romanes 1890, 831; Wallace 1905, 2:314–26, 318; Fichman 2004, 287–88; Elsdon-Baker 2008. On Romanes's later support for sexual selection and female choice, see Bartley 1994a, 68–70.

125. G. Romanes 1887, 661, 664, 672.

126. Wallace 1890b, 337; around this same time, Wallace attacked Romanes's physiological selection in *Nature*, 1890a.

127. Wallace 1890b, 327, 330.

128. Ibid., 331.

129. Ibid., 333–35.

130. Durant 1979, 49–50; Fichman 2004, 268–71, 277–79.

131. Wallace 1890–92 (Wallace's alterations included the advice that the more "reticent" term "marriage" be substituted for "sexual intercourse" where possible, see Dawson 2007, 12); Westermarck 1891, 240–52, 261, 542–43; see also Bajema 1984, 259.

132. See S. Smith (1810) 1991, 509n427; Wallace 1905, 2:209–10; Wallace 1892; Wallace 1893. On Wallace's transatlantic following, see Fichman 2004; Hamlin 2014, 141–44.

133. Wallace 1913, 149; Bland 1995, 229–35; Phillips 2004, 260–61.

134. Huxley's Romanes Lecture of 1893 was supplemented by a "Prolegomena" which expanded on its arguments. The two, usually referred to as "Evolution and Ethics," were published together by Macmillan in 1893; reprinted in T. Huxley 1989, 57–174.

135. The key text is Helfand 1977; Paradis modifies Helfand's analysis, arguing that "Evolution and Ethics" was preeminently Huxley's final defense of Victorian scientific naturalism and the search for a political middle ground that would allow scientific and technical progress, 1989; Desmond adds a significant professional and institutional dimension, 1997, 212–17; Paul White argues further that Huxley's reassertion of an amoral nature cut the traditional ties between nature, morality and scientific practice, absolving professional scientists of social responsibility and legitimating the standard scientific distinction between "facts" and "values"; P. White 2003, "Conclusion."

136. It is argued that his grief over Mady's madness and death heightened Huxley's perception of an amoral, ruthless nature. An alternative interpretation is that Huxley was reacting against the new freedoms that exposed young women to moral danger and psychological damage; that made them (in Mady's case) unfit for motherhood and threatened the puritanical "moralizing naturalism" preached by "Evolution's High Priest." Two younger Huxley daughters, according to a gossipy Beatrice Webb, had caused Huxley recent distress by gadding about to balls and flirting with "inferior young men." Daughter Nettie had seemed set to follow in Mady's footsteps, taking to art and the loose-living art set. On one occasion she introduced Oscar Wilde into the Huxley home, provoking a furious paterfamilial embargo on any future visits from this notorious personification of late Victorian decadence. See Desmond 1997, 158, 172–79, 187, 189; R. Richards 1987, 314n54. Huxley's antagonism toward a sensuous and hedonistic "aestheticism" is discussed by Dawson in the context of Huxley's concern to dissociate respectable Darwinian science from the taint of sensualism and immorality; Dawson 2007, ch. 6.

137. On the journalistic phenomenon of the New Woman as "over-sexed and mannish, over-educated and asinine" and a "sexual decadent," see Ledger 2007, 153–55.

138. T. Huxley (1865) 1968, 68.

139. For the exchange of letters between Romanes and Huxley, see L. Huxley 2:349–57; Desmond 1997, 214–17, 244, 313.

140. T. Huxley 1989, 139–41, 100, 71–72.

141. P. Morton 1985, ch. 5; T. Huxley 1989, 80–81, 94–95. Paradis largely attributes Huxley's anti-eugenicist stance to his familial history of mental instability, which "would not have entitled him to any favors in a Galtonian eugenicist state"; Paradis 1989, 48.

142. T. Huxley 1989, 81, 99, 100, 144. Huxley's conclusion is taken from Tennyson's *Ulysses*. Dawson draws attention to the recourse to Tennysonian rhetoric by Huxley, Tyndall and Darwin, to signify decency and respectability, along with attributes of manliness and notions of evolutionary and social ascent; Dawson 2007, 21–23, 25, 52–54, 113–14, 203, 219.

143. E. Richards 1995; Desmond 1997, 313.

144. For the critical response to "Evolution and Ethics," see Paradis 1989, 42–52, quotation at 52; for Huxley's self-described "egg-dance" and "apples," see L. Huxley 2:355, 356. See also the various contributions in Dawson and Lightman 2014 for new insights into the competing perspectives that constituted scientific naturalism in the late Victorian period.

145. Wallace 1913, 147–52, 163–65; Fichman 2004, 325–27; Showalter 1995; Oppenheim 1989; A. Taylor 1992, 222–25, 253–58; Ledger 2007.

146. Schreiner 1889, 7; Showalter 1995, 56–57.

EPILOGUE

1. Bartley 1994a, 1 and passim.

2. Bowler 1992a, 40–43 and passim; Bowler 1992b, 115–18 and passim; Bajema 1984, 256–62.

3. Bartley 1994a, 61–65, 80–89, 101–2; Bajema 1984, 256–58.

4. J. Cunningham 1900, 28–29, 33, 38; Bajema 1984, 260; Bartley 1994a, 93–101; Bowler 1992a, 89–91.

5. Gould 1980a, ch. 9; Bowler 1992b, 99–102; Bowler 1992a, ch. 7; Cronin 1991, 89, 426.

6. T. Morgan 1903, 221; Morgan attributed sexual dimorphism to hormonal control mechanisms, see Bartley 1994a, 106–20; 1994b, 179; see also Milam 2010, ch.

2. Milam challenges the conventional thesis of the "eclipse" of sexual selection, passim.

7. Cf. Milam's emphasis on Darwin's attribution of "rational" or "conscious" choice to female animals, Milam 2010, ch. 1 and passim.

8. Kaye 2002, 90, 110, 106, 98, 93.

9. Dawson 2007, ch. 6; Hamlin 2010.

10. For Darwin's attempt to commend his own interpretation to Spencer, see F. Darwin and Seward 1903, 1:351–52; Spencer's influential *Study of Sociology* was released in installments in the *Contemporary Review* and across the Atlantic in the *Popular Science Monthly* between 1872 and 1873; Spencer (1877) 1906, 341–42; Geddes and Thomson 1889, 37. Spencer moved from early liberal ideas about women to staunch antifeminism, see Paxton 1991, 43–68; see also Russett 1989, 93; Mosedale 1978, 10; Richardson 2003, ch. 2.

11. Spencer's American followers were not all male. In a lead article in the *Popular Science Monthly* (1882), Miss M. A. Hardaker embellished the Spencerian thesis with her own emphasis on men's greater consumption of food and consequent superior brain size, a sexual disparity of intellect further exacerbated by women's greater energy loss to reproduction. She bolstered this insistence on female intellectual inferiority with apposite quotations from the *Descent*, prompting Caroline Kennard's direct call on Darwin for substantiation and Darwin's reaffirmation of his claim that attempts by women to gain intellectual parity with men could only compromise the "happiness of our homes"; Hardaker 1882; Clarke 1873, 39; Mosedale 1978, 31–32; Russett 1989, 104–5; DAR 185:28, 29, 31; Joy Harvey 2009. For American responses to Clarke, see Hamlin 2014, 73–93.

12. *Descent* 1879, 630–31; Maudsley 1870, 31–33; *Marginalia*, 1:571–72; DAR 89:157–58; CCD, 19:58–59; Russett 1989, 119–22.

13. Maudsley 1874, 468, 471, 477.

14. Hyatt 1897; Bowler 1989, 339–43.

15. See Mosedale 1978, 24–32; Russett 1989, 55.

16. Gould 1992, 54.

17. Wallace 1889, 283.

18. Gould 1980a, ch. 27, "Racism and Recapitulation"; Gould 1981, 113–22.

19. Bartley 1994a; see also MacKenzie 1981, 183–213; Kevles 1985, 176–92; James Moore 2007.

20. Fisher 1915, 1916, 1930; Bartley 1994b; Milam 2010, 43–51: see also Hoquet and Levandowsky 2015, 35–38.

21. J. Huxley 1914, 559; Bartley 1994a, 125–58; Bartley 1995; Milam 2010, 37–39.

22. J. Huxley 1938, 12; J. Huxley 1942, 484; Cronin 1991, 239; Bartley 1994a, 209–14; Frankel 1994; Milam 2010, 39–43.

23. Bartley 1994a, 148; Durant 1993; Frankel 1994.

24. See Bartley 1994a; Cronin 1991, 243–49; Frankel 1994.

25. Milam 2010, ch. 6 and passim; quotations at 8 and 145. Where I would dissent from Milam's thesis is her insistence that Darwin represented mate choice as an act of deliberation, of conscious rational choice, and that it was this that presented its major obstacle to acceptance; ibid., 1 and passim.

26. Frankel 1994, 182–83; see also Gould 1992, where Gould, in his review of Cronin's *The Ant and the Peacock*, argued that "the delay in acceptability for Darwin's well-formulated concept of female choice lies in the social impediments of sexism and speciesism"; see also Bartley 1994a, 228–29.

27. For early critiques and woman-centered reinterpretations of sexual selection, see Zihlman 1974; Hubbard 1979; Tanner 1981; Hrdy 1981; and E. Morgan 1985. For an analysis of these developments, see Haraway 1989; on more recent developments, see Milam et al., 2011; introduction to Hoquet 2015.

28. Grammer et al. 2003; Andersson 1994, 11–13; but see Arnqvist and Rowe 2005.

29. See, for instance, Grammer et al. 2003; Andersson 1994, 11–13, 19–31; Cronin 1991, 195–204; Arnqvist and Rowe 2005, 22–25; Mead and Arnold 2004; Welsch 2004; Milam 2010; Millstein in Milam et al. 2011, 258. Cronin attributes the dissociation of mate choice from the notion of aesthetic taste to have freed sexual selection from its main obstacle to acceptance, but herself argues for an aesthetic component to sexual selection, as does Miller; Cronin 1991, 247; Miller 2001; see also Prum 2012; Hoquet 2015, sec. 3.

30. But see Diamond 1992, ch. 6.

31. Grammer et al. 2003; Welsch 2004; Etcoff 2000, 143–46; Miller 2001. Dixson offers a careful assessment of the evidence, Dixson 2009, ch. 7; see also Haufe 2008.

32. See for instance Miller 2001, 159–76; Cronin 1991, 191–201; Grammer et al. 2003. To my knowledge, Etcoff is the only contemporary writer to make some use of the fashion metaphor in explicating sexual selection; Etcoff 2000, ch. 7.

33. Another possibility, worth further exploration, is that the close ties between artificial selection and eugenics and the postwar revulsion against eugenics and its applications in Nazi genocide discredited the notion of selective breeding as a viable biological metaphor. This association may be a factor in the prolonged "eclipse" of sexual selection until its 1970s "rebirth."

34. Cronin 1991, 247.

35. Roughgarden 2009, 2015; Milam et al. 2011, 253, 275. For a rather idiosyncratic history of evo/devo, see Amundson 2007.

36. The most recent collection of papers on sexual selection is exemplary in this respect, with their ritual invocations of Darwin; see Hoquet 2015.

37. Roughgarden in Milam et al. 2011, 271–72.

38. Levine 2006; Vandermassen et al. 2005.

39. Desmond and Moore 2009.

40. Kaye 2002; Yeazell 1991; Levine 2006.

41. Cronin 1991.

42. On Darwin's iconic status and the historical problems it engenders, see Secord 2009; Kjaergaard 2010; Shapin 2010.

43. Ideology is not to be understood in its classical Marxist sense as entailing the suppression of contradiction, or as "false consciousness" or "bad" science as opposed to "good" science. It is to be understood rather, in the critical sense associated with the writings of Foucault on power as relational and productive. I have discussed this at greater length and illustrated it with reference to the history of teratogeny and evolutionary theorizing; see E. Richards 1994.

44. For a major study and refutation of female choice among peahens, see Takahashi et al. 2008; for an overview, see Cézilly 2015. For reinterpretations of bowerbird aesthetics and bower building, see Diamond 1986; Madden et al. 2004; Madden 2006; but see Prum 2012 for a reassertion of Darwinian arbitrary aesthetic choice as a viable evolutionary interpretation of the diversity of secondary sexual traits and mating preferences that addresses the limits of the explanatory power of natural selection; see also Prum 2015 on how female preferences for bower architecture provide an aesthetic mechanism for reinforcing female sexual autonomy.

Newspaper articles published before 1900 are cited fully in the notes and are not included in this list of references.

Acton, William. (1857)1865. *The Functions and Disorders of the Reproductive Organs*. 4th. ed. London: John Churchill.

Addison, Joseph. 1970. *Critical Essays from the Spectator*. Oxford: Clarendon Press.

Agassiz, Louis. 1849. *Twelve Lectures on Comparative Embryology*. Boston: H. Flanders.

———. 1850. "On the Differences between Progressive, Embryonic, and Prophetic Types in the Succession of Organized Beings through the Whole Range of Geological Times." *Edinburgh New Philosophical Journal* 49:160–65.

———. 1857. *Contributions to the Natural History of the United States of North America*. Vol. 1, part 1. *Essay on Classification*. London: Trübner.

Aiken, R. B. 1982. "Theories of Sexual Difference: The Sexual Selection Hypothesis and Its Antecedents, 1786–1919." *Quaestiones Entomologicae* 18:1–14.

Aimé-Martin, Louis. (1834) 1842. *The Education of Mothers; or, The Civilization of Mankind by Women*. London: Whittaker.

Alaya, Flavia. 1977. "Victorian Science and the 'Genius' of Woman." *Journal of the History of Ideas* 38:261–80.

Allan, J. McGrigor. 1869. "On the Real Differences in the Minds of Men and Women." *Journal of the Anthropological Society* 7:cxcv–ccxix.

Allan, Janice M. 2011. "Mrs. Robinson's 'Day-Book of Iniquity': Reading Bodies of Evidence in the Context of the 1858 Medical Reform Act." In *The Female Body in Medicine and Literature*, edited by Andrew Maugham and Greta Dipledge, 169–79. Liverpool: Liverpool University Press.

Allen, Grant. 1877. *Physiological Aesthetics*. London: Henry S. King.

———. 1879a. *The Colour-Sense: Its Origin and Development*. London: Trübner.

———. 1879b. "A Problem in Human Evolution." *Fortnightly Review* 31:778–86.

———. 1886. *Charles Darwin*. London: Longmans, Green.

———. 1889a. "A Biologist on the Woman Question." *Pall Mall Gazette* 49:1–2.

———. 1889b. "Falling in Love." In *Falling in Love: With Other Essays on More Exact Branches of Science*, 1–17. London: Smith, Elder.

———. 1889c. "Plain Words on the Woman Question." *Fortnightly Review* 46:448–57.

———. 1889d. "Woman's Place in Nature." *Forum* 7:258–63.

————. 1890. "The Girl of the Future." *Universal Review* 7:49–64.

————. 1895. *The Woman Who Did*. London: John Lane.

————. 1896. "Is It Degradation?" *Humanitarian: A Monthly Review of Sociological Science* 9:340–48.

Alter, Stephen G. 2007. "Separated at Birth: The Interlinked Origins of Darwin's Unconscious Selection Concept and the Application of Sexual Selection to Race." *Journal of the History of Biology* 40:231–58.

Altick, Richard. 1997. *"Punch": The Lively Youth of a British Institution, 1841–51*. Columbus: Ohio State University Press.

Amigoni, David. 2007. *Colonies, Cults and Evolution: Literature, Science and Culture in Nineteenth-Century Writing*. Cambridge: Cambridge University Press.

Amrein, Martin, and Kärin Nickelsen. 2008. "The Gentleman and the Rogue: The Collaboration between Charles Darwin and Carl Vogt." *Journal of the History of Biology* 41:237–66.

Amundson, Ron. 2007. *The Changing Role of the Embryo in Evolutionary Thought: Roots of Evo-Devo*. Cambridge: Cambridge University Press.

Anderson, Nancy Fix. 1987. *Woman against Women in Victorian England: A Life of Eliza Lynn Linton*. Bloomington: Indiana University Press.

Anderson, Patricia. 1991. *The Printed Image and the Transformation of Popular Culture 1790–1860*. Oxford: Clarendon.

————. 1995. *When Passion Reigned: Sex and the Victorians*. New York: Basic Books.

Andersson, Malte. 1994. *Sexual Selection*. Princeton, NJ: Princeton University Press.

Anon. 1839. "Mr. Walker on Intermarriage." *British and Foreign Medical Review* 7:370–85.

Anon. 1847. "The Art of Dress." *Quarterly Review* 79:375–76.

Anon. 1860. "Natural Selection." *All the Year Round* 3:293–99.

Appel, Toby A. 1987. *The Cuvier-Geoffroy Debate: French Biology in the Decades before Darwin*. New York: Oxford University Press.

Argyll, Duke of (George Douglas Campbell). 1867. *The Reign of Law*. London: Strahan.

————. 1869. *Primeval Man: An Examination of Some Recent Speculations*. London: Strahan.

Armstrong, Meg. 1996. "'The Effects of Blackness': Gender, Race, and the Sublime in Aesthetic Theories of Burke and Kant." *Journal of Aesthetics and Art Criticism* 54:213–36.

Arnqvist, Göran, and Locke Rowe. 2005. *Sexual Conflict*. Princeton, NJ: Princeton University Press.

Arnstein, Walter L. 1984. *The Bradlaugh Case: Atheism, Sex, and Politics among the Late Victorians*. Columbia: University of Missouri Press.

Audubon, John James. 1831–39. *Ornithological Biography; or, An Account of the Habits of the Birds of the United States of America*. 5 vols. Edinburgh: Adam Black.

Augstein, Hannah Franziska. 1999. *James Cowles Prichard's Anthropology: Remaking the Science of Man in Early Nineteenth Century Britain*. Amsterdam: Rodopi.

Austen, Jane. (1816) 1966. *Emma*. Harmondsworth: Penguin.

Aveling, Edward Bibbins. 1881a. *The Student's Darwin*. 2 vols. London: Freethought Publishing.

————. 1881b. "A Visit to Charles Darwin." *National Reformer*, October 22, 273–74; October 29, 291–93.

————. 1882. *Darwinism and Small Families*. London: Freethought Publishing.

————. 1883a. "At University College." *National Reformer*, July 29, 67–69.

————. 1883b. *The Religious Views of Charles Darwin*. London: Freethought Publishing.

————. 1884. *The Gospel of Evolution*. London: Freethought Publishing.

————. N.d. *The People's Darwin; or, Darwin Made Easy*. London: R. Forder.

Baer, Karl Ernst von. 1853. "Fragments Relating to Philosophical Zoology: Selections from the Works of K. E. von Baer." Translated by Thomas Henry Huxley. In *Scientific Memoirs, Selected from the Transactions of Foreign Academies of Science, and from Foreign Journals: Natural History*, edited by Arthur Henfrey and Thomas Henry Huxley, 176–238. London: Taylor and Francis.

Bajema, Carl Jay. 1984. *Evolution by Sexual Selection Theory Prior to 1900*. New York: Van Nostrand Reinhold.

Bank, Andrew. 1996. "Of 'Noble Skulls' and 'Noble Caucasians': Phrenology in Colonial South Africa." *Journal of Southern African Studies* 22:387–403.

Barbauld, Anna Laetitia, and John Aikin. 1805. *Evenings at Home; or, The Juvenile Budget Opened*. London: J. Johnson.

Barlow, Nora, ed. 1969. *The Autobiography of Charles Darwin, 1809–1882, with Original Omissions Restored*. New York: W. W. Norton.

Barrell, John. 1986a. *The Political Theory of Painting, from Reynolds to Hazlett: "The Body of the Public."* New Haven, CT: Yale University Press.

———. 1986b. "Sir Joshua Reynolds and the Political Theory of Painting." *Oxford Art Journal* 9:36–41.

Barrett, Paul H., et al., eds. 1987. *Charles Darwin's Notebooks, 1836–1844*. Cambridge: Cambridge University Press.

Barta, Tony. 2005. "Mr. Darwin's Shooters: On Natural Selection and the Naturalizing of Genocide." *Patterns of Prejudice* 39:116–37.

Bartley, Mary M. 1992. "Darwin and Domestication: Studies on Inheritance." *Journal of the History of Biology* 25:307–33.

———. 1994a. "A Century of Debate: The History of Sexual Selection Theory (1871–1971)." PhD diss., Cornell University.

———. 1994b. "Conflicts in Human Progress: Sexual Selection and the Fisherian 'Runaway.'" *British Journal for the History of Science* 27:177–96.

———. 1995. "Courtship and Continued Progress: Julian Huxley's Studies on Bird Behavior." *Journal of the History of Biology* 28:91–108.

Barton, Ruth. 1998. "'Huxley, Lubbock, and Half a Dozen Others': Professionals and Gentlemen in the Formation of the X Club, 1851–1864." *Isis* 89:410–44.

———. 2003. "'Men of Science': Language, Identity and Professionalization in the Mid-Victorian Scientific Community." *History of Science* 41:73–119.

Bates, Alan W. 2010. *The Anatomy of Robert Knox: Murder, Mad Science and Medical Regulation in Nineteenth-Century Edinburgh*. Brighton: Sussex Academic Press.

Bates, Henry Walter. 1863. *The Naturalist on the River Amazon*. London: John Murray.

Bebel, August. 1885. *Woman in the Past, Present, and Future*. London: Modern Press.

Beddoe, John. 1863. "On the Supposed Increasing Prevalence of Dark Hair in England." *Anthropological Review* 1:310–12.

———. 1885. *The Races of Britain: A Contribution to the Anthropology of Western Europe*. Bristol: Arrowsmith.

Beer, Gillian. 1983. *Darwin's Plots: Evolutionary Narrative in Darwin, George Elliot and Nineteenth-Century Fiction*. London: Ark Paperbacks.

———. 1996. *Open Fields: Science in Cultural Encounter*. Oxford: Clarendon Press.

Bell, Charles. (1806) 1844. *The Anatomy and Philosophy of Expression*. London: John Murray.

Bellamy, Edward. 1888. *Looking Backward, 2000–1887*. New York: New American Library.

Bender, Bert. 1996. *The Descent of Love: Darwin and the Theory of Sexual Selection in American Fiction, 1871–1926*. Philadelphia: University of Pennsylvania Press.

———. 2004. *Evolution and the "Sex Problem": American Narratives during the Eclipse of Darwinism*. Kent, OH: Kent State University Press.

Benn, J. Miriam. 1992. *The Predicaments of Love*. London: Pluto Press.

Benton, Ted. 2008. "Wallace's Dilemmas." In *Natural Selection and Beyond: The Intellectual Legacy of Alfred Russel Wallace*, edited by Charles H. Smith and George Beccaloni, 368–90. Oxford: Oxford University Press.

———. 2009. "Race, Sex and the 'Earthly Paradise': Wallace versus Darwin on Human Evolution and Prospects." *Sociological Review* 57, issue supplement: 23–46.

Bernstein, Susan David. 2007. "Designs after Nature: Evolutionary Fashions, Animals, and Gender." In *Victorian Animal Dreams: Representations of Animals in Victorian Literature and Culture*, edited by Deborah Denenholz Morse, 65–79. London: Ashgate Publishing.

Besant, Annie. (1877) 1884. *The Law of Population: Its Consequences, and Its Bearing upon Human Conduct and Morals*. London: Freethought Publishing.

———. 1882. *Marriage, as It Was, as It Is, and as It Should Be*. London: Freethought Publishing.

———. 1886. *Why I Am a Socialist*. London: Printed by Annie Besant and Charles Bradlaugh.

Biddiss, Michael D. 1976. "The Politics of Anatomy: Dr. Robert Knox and Victorian Racism." *Proceedings of the Royal Society of Medicine* 69:245–50.

Bindman, David. 2002. *Ape to Apollo: Aesthetics and the Idea of Race in the 18th Century*. London: Reaktion Books.

Birdsall, Amelia. 2002. "A Woman's Nature: Attitudes and Identities of the Bird Hat Debate at the Turn of the 20th Century." Bachelor's thesis. Haverford College. http://hdl.handle.net/10066/675.

Blackburn, Robin. 1988. *The Overthrow of Colonial Slavery, 1776–1848*. London: Verso.

Bland, Lucy. 1995. *Banishing the Beast: English Feminism and Sexual Morality, 1885–1914*. Harmondsworth: Penguin.

Bloch, Maurice, and Jean H. Bloch. 1980. "Women and the Dialectics of Nature in Eighteenth Century Thought." In *Nature, Culture and Gender*, edited by Carol P. McCormack and Marilyn Strathern, 25–41. Cambridge: Cambridge University Press.

Blumenbach, Johann Freidrich. 1865. *The Anthropological Treatises of Johann Freidrich Blumenbach, with Memoirs of Him by Marx and Flourens*. London: Longman, Green, Longman, Roberts and Green.

Blyth, Edward. 1835. "An Attempt to Classify the 'Varieties' of Animals with Observations on the Marked Seasonal and Other Changes Which Naturally Take Place in Various British Species, and Which Do Not Constitute Varieties." *Magazine of Natural History* 8:40–53.

———. 1836. "Observations on the Various Seasonal and Other External Changes Which Regularly Take Place in Birds, More Particularly in Those Which Occur in Britain." *Magazine of Natural History* 9:393–409.

———. 1837. "On the Psychological Distinctions between Man and All Other Animals; and the Consequent Diversity of Human Influence over the Inferior Ranks of Creation." *Magazine of Natural History*, n.s., 1:1–9, 77–85, 131–41.

Bonner, Hypatia Bradlaugh. 1895. *Charles Bradlaugh: A Record of His Life and Work*. London: T. Fisher Unwin.

Bowler, Peter J. 1989. "Holding Your Head Up High: Degeneration and Orthogenesis in Theories of Human Evolution." In *History, Humanity and Evolution*, edited by James R. Moore, 329–53. Cambridge: Cambridge University Press.

———. 1992a. *The Eclipse of Darwinism.* Baltimore: Johns Hopkins University Press.

———. 1992b. *The Non-Darwinian Revolution: Reinterpreting a Historical Myth.* Baltimore: Johns Hopkins University Press.

Bowman, Sylvia E. 1958. *A Critical Biography of Edward Bellamy.* New York: Bookman Associates.

Boxer, Marilyn J. 2007. "Rethinking the Socialist Construction and International Career of the Concept of 'Bourgeois Feminism.'" *American Historical Review* 112:131–58.

Bradshaw, Penny. 2005. "'The Limits of Barbauld's Feminism: Re-reading 'The Rights of Woman.'" *European Romantic Review* 16:23–37.

Brandon, Ruth. 1990. *The New Women and the Old Men: Love, Sex and the Woman Question.* London: Secker & Warburg.

Brandon-Jones, Christine. 1997. "Edward Blyth, Charles Darwin, and the Animal Trade in Nineteenth-Century Britain." *Journal of the History of Biology* 30:145–87.

Brantlinger, Patrick. 2003. *Dark Vanishings: Discourse on the Extinction of Primitive Races, 1800–1930.* Ithaca, NY: Cornell University Press.

Bravo, Michael T. 1996. "Ethnological Encounters." In *Cultures of Natural History*, edited by N. Jardine, J. A. Secord, and E. C. Spary, 338–57. Cambridge: Cambridge University Press.

Bridges, E. Lucas. 1948. *Uttermost Part of the Earth.* London: Hodder & Stoughton.

Briggs, Asa. 1989. *Victorian Things.* Chicago: University of Chicago Press.

Broomfield, Andrea L. 2001. "Much More Than an Anti-feminist: Eliza Lynn Linton's Contribution to the Rise of Victorian Popular Journalism." *Victorian Literature and Culture* 29:267–83.

Browne, Janet. 1989. "Botany for Gentlemen: Erasmus Darwin and *The Loves of the Plants.*" *Isis* 80:593–621.

———. 1994. "Missionaries and the Human Mind: Charles Darwin and Robert FitzRoy." In *Darwin's Laboratory: Evolutionary Theory and Natural History in the Pacific*, edited by Roy MacLeod and Philip E. Rebock, 263–82. Honolulu: University of Hawaii Press.

———. 1995. *Charles Darwin: Voyaging.* London: Jonathan Cape.

———. 2002. *Charles Darwin: The Power of Place.* London: Jonathan Cape.

Bryant, Julius. 2009. "Darwin at Home: Observation and Taste at Down House." In *Endless Forms: Charles Darwin, Natural Science, and the Visual Arts*, edited by Diana Donald and Jane Munro, 28–46. New Haven, CT: Yale University Press.

Buchanan, Roderick D. 2016. "Darwin's 'Mr. Arthrobalanus': Sexual Differentiation, Evolutionary Destiny and the Expert Eye of the Beholder." *Journal of the History of Biology.* Published online April 20. http://link.springer.com/article/10.1007%2Fs10739-016-9444-9.

Budge, Gavin. 2003. *Aesthetics and Religion in Nineteenth-Century Britain.* Bristol: Thoemmes Press.

Bulmer, Michael. 2005. "The Theory of Natural Selection of Alfred Russel Wallace FRS." *Notes & Records of the Royal Society* 59:125–36.

Burdett, Carolyn. 2009. "Sexual Selection, Automata and Ethics in George Eliot's *The Mill on the Floss* and Olive Schreiner's *Undine* and *From Man to Man.*" *Journal of Victorian Culture* 14:26–52.

Burke, Edmund. 1823. *A Philosophical Inquiry into the Origin of Our Ideas of the Sublime and Beautiful.* London: Thomas M'Lean.

———. 1864. *Reflections on the Revolution in France.* In *The Works of the Right Honourable Edmund Burke*, vol. 2. London: Henry G. Bohn.

Burkhardt, Frederick, et al., eds. 1985–. *The Correspondence of Charles Darwin.* 23 vols. Cambridge: Cambridge University Press.

Burkhardt, Frederick, and Sydney Smith, eds. 1994. *A Calendar of the Correspondence of Charles Darwin, 1821–1882*. Cambridge: Cambridge University Press.

Burkhardt, Richard W. 1979. "Closing the Door on Lord Morton's Mare: The Rise and Fall of Telegony." *Studies in History of Biology* 3:1–21.

———. 1985. "Darwin on Animal Behaviour and Evolution." In *The Darwinian Heritage*, edited by David Kohn, 327–65. Princeton, NJ: Princeton University Press.

Burnett, D. Graham. 2000. *Masters of All They Surveyed: Exploration, Geography, and a British El Dorado*. Chicago: University of Chicago Press.

———. 2009. "Savage Selection: Analogy and Elision in *On the Origin of Species*." *Endeavour* 33:120–25.

Burrow, John W. 1966. *Evolution and Society: A Study in Victorian Social Theory*. Cambridge: Cambridge University Press.

Burton, Richard F. 1864. "Notes on Waitz's Anthropology." *Anthropological Review* 7:233–50.

Butcher, Barry. 1994. "Darwinism, Social Darwinism, and the Australian Aborigines: A Reevaluation." In *Darwin's Laboratory*, edited by Roy MacLeod and Philip F. Rehbock, 371–94. Honolulu: University of Hawaii Press.

Butler, Marilyn. 1972. *Maria Edgeworth: A Literary Biography*. Oxford: Oxford University Press.

———, ed. 1984. *Burke, Paine, Godwin, and the Revolution Controversy*. Cambridge: Cambridge University Press, 1984.

Byrde, Penelope. 1992. *Nineteenth Century Fashion*. London: B. T. Batsford.

Caine, Barbara. 1992. *Victorian Feminists*. Oxford: Oxford University Press.

Callanan, Laura. 2006. *Deciphering Race: White Anxiety, Racial Conflict, and the Turn to Fiction in Mid-Victorian English Prose*. Columbus: Ohio State University Press.

Camfield, Gregg. 1988. "The Moral Aesthetics of Sentimentality: A Missing Key to Uncle Tom's Cabin." *Nineteenth Century Literature* 43:319–45.

Camper, Petrus. 1794. *The Works of the Late Professor Camper on the Connexion Between the Science of Anatomy and the Arts of Drawing, Painting, Statuary, etc.* 2 vols. London: C. Dilly.

Carlile, Richard. 1838. *Every Woman's Book; or, What Is Love? Containing Most Important Instructions for the Prudent Regulation of the Principle of Love*. London: A. Carlile.

Carlyle, Thomas. (1833–34) 1987. *Sartor Resartus*. Oxford: Oxford University Press.

———. 1849. "Occasional Discourse on the Negro Question." *Frazer's Magazine for Town and Country* 40:670–79.

Caro, Tim, Sami Merilaita, and Martin Stevens. 2008. "The Colours of Animals: From Wallace to the Present Day: 1. Cryptic Colouration." In *Natural Selection and Beyond: The Intellectual Legacy of Alfred Russel Wallace*, edited by Charles H. Smith and George Beccaloni, 119–38. Oxford: Oxford University Press.

Carpenter, William B. 1854. *Principles of Comparative Physiology*. 4th ed. London: Churchill.

Carter, Michael. 2003. *Fashion Classics from Carlyle to Barthes*. Oxford: Berg.

Cézilly, Frank. 2015. "Preference, Rationality and Interindividual Variation: The Persisting Debate about Female Choice." In *Current Perspectives on Sexual Selection*, edited by Thierry Hoquet, 191–209. Dordrecht: Springer.

[Chambers, Robert]. (1844) 1969. *Vestiges of the Natural History of Creation*. Reprint, Leicester: Leicester University Press.

[———]. 1853. *Vestiges of the Natural History of Creation*. 10th ed. London: John Churchill.

Chandrasekhar, S. 1981. *"A Dirty, Filthy Book": The Writings of Charles Knowlton and Annie Besant on Reproductive Physiology*. Berkeley: University of California Press.

Chapman, Alison. 1999. "Phantasies of Matriarchy in Victorian Children's Literature." In *Victorian Women Writers and the Woman Question*, edited by Nicola Diane Thompson, 60–79. Cambridge: Cambridge University Press.

Chapman, Anne. 1982. *Drama and Power in a Hunting Society: The Selk'nam of Tierra del Fuego*. Cambridge: Cambridge University Press.

———. 2006. *Darwin in Tierra del Fuego*. Buenos Aires: Imago Mundi.

Claeys, Gregory. 2008. "Wallace and Owenism." In *Natural Selection and Beyond: The Intellectual Legacy of Alfred Russel Wallace*, edited by Charles H. Smith and George Beccaloni, 235–62. Oxford: Oxford University Press.

Clarke, Edward. 1873. *Sex in Education; or, A Fair Chance for the Girls*. Boston: James R. Osgood.

Cobbe, Frances Power. 1869. "The Subjection of Women." *Theological Review* 6:355–75.

———. 1872. *Darwinism in Morals, and other Essays*. London: Williams and Norgate.

———. 1894. *Life of Frances Power Cobbe by Herself*. 2 vols. London: Richard Bentley and Son.

Cohen, Deborah. 2006. *Household Gods: The British and Their Possessions*. New Haven, CT: Yale University Press.

Colp, Ralph. 1977. *To Be an Invalid: The Illness of Charles Darwin*. Chicago: University of Chicago Press.

———. 1981. "Charles Darwin, Dr. Edward Lane, and the 'Singular Trial' of *Robinson v. Robinson and Lane*." *Journal of the History of Medicine and Allied Sciences* XXXVI:205–13.

———. 1982. "The Myth of the Darwin-Marx Letter." *History of Political Economy* 14:461–82.

Combe, George. 1840. *Moral Philosophy; or, The Duties of Man Considered in His Individual, Social, and Domestic Capacities*. Edinburgh: Maclachlan, Stewart.

Cooper, Brian, and Margueritte S. Murphy. 2000. "The Death of the Author at the Birth of Social Science: The Cases of Harriet Martineau and Adolphe Quetelet." *Studies in History and Philosophy of Science* 31:1–36.

Cooper, Robyn. 1992. "Definition and Control: Alexander Walker's Trilogy on Woman." *Journal of the History of Sexuality* 2:341–64.

———. 1993. "Victorian Discourses on Women and Beauty: The Alexander Walker Texts." *Gender & History* 5:34–55.

Cooter, Roger. 1984. *The Cultural Meaning of Popular Science: Phrenology and the Organization of Consent in Nineteenth-Century Britain*. Cambridge: Cambridge University Press.

Corsi, Pietro. 2005. "Before Darwin: Transformist Concepts in European Natural History." *Journal of the History of Biology* 38:67–83.

Courtney, C. P. 1975. "Alexander Walker and Benjamin Constant: A Note on the English Translation of *Adolphe*." *French Studies* 24:137–50.

Cowan, Ruth Schwartz. 1972. "Francis Galton's Contribution to Genetics." *Journal of the History of Biology* 5:389–412.

Cowling, Mary. 1989. *The Artist as Anthropologist: The Representation of Type and Character in Victorian Art*. Cambridge: Cambridge University Press.

Cronin, Helena. 1991. *The Ant and the Peacock: Altruism and Sexual Selection from Darwin to Today*. Cambridge: Cambridge University Press.

Crozier, Ivan. 2000. "William Acton and the History of Sexuality: The Professional and Medical Contexts." *Journal of Victorian Culture* 5:1–27.

Cunningham, Andrew, and Nicholas Jardine, eds. 1990. *Romanticism and the Sciences*. Cambridge: Cambridge University Press.

Cunningham, Joseph Thomas. 1900. *Sexual Dimorphism in the Animal Kingdom: A Theory of the Evolution of Secondary Sexual Characters*. London: A. & C. Black.

Curthoys, Ann, and John Docker. 2008. "Defining Genocide." In *The Historiography of Genocide*, edited by Dan Stone, 9–41. Houndsmill: Pallgrave Macmillan.

Dabydeen, David. 1985. *Hogarth's Blacks: Images of Blacks in Eighteenth Century English Art*. Manchester: Manchester University Press.

Darwin, Charles. 1851–54. *A Monograph of the Sub-class Cirripedia*. 2 vols. London: Ray Society.

———. 1859. *On the Origin of Species by Means of Natural Selection; or, The Preservation of Favored Races in the Struggle for Life*. London: John Murray.

———. 1868. *The Variation of Animals and Plants under Domestication*. 2 vols. London: John Murray.

———. 1871. *The Descent of Man, and Selection in Relation to Sex*. 2 vols. London: John Murray.

———. 1872. *The Expression of the Emotions in Man and Animals*. London: John Murray.

———. 1876. "Sexual Selection in Relation to Monkeys." *Nature* 15:18–19.

———. (1876) 2010. "Recollections of the Development of My Mind and Character." In *Charles Darwin: Evolutionary Writings*, edited by James A. Secord, 355–425. Oxford: Oxford University Press.

———. 1877. "A Biographical Sketch of an Infant." *Mind: Quarterly Review of Psychology and Philosophy* 2:285–94.

———. (1879) 2004. *The Descent of Man, and Selection in Relation to Sex*. Reprint of the second revised edition, with an introduction by James Moore and Adrian Desmond. London: Penguin Classics.

———. 1882. "'Preliminary Notice' to 'On the Modification of a Race of Syrian Street-Dogs by Means of Sexual Selection,' by Dr. Van Dyck." *Proceedings of the Zoological Society of London* 25:367–69.

———. 1887. *The Life of Erasmus Darwin by Charles Darwin: Being an Introduction to an Essay on His Scientific Works, by Ernst Krause*. 2nd ed. London: John Murray.

———. 1909. *The Foundations of the Origin of Species: Two Essays Written in 1842 and 1844 by Charles Darwin*. Edited by Francis Darwin. Cambridge: Cambridge University Press.

———. 1972. *The Voyage of the* Beagle. London: Dent.

Darwin, Erasmus.1789–91. *The Botanic Garden: A Poem, in Two Parts*. 2 vols. London: J. Johnston. Facsimile ed. New York: Garland Publishing, 1978.

———. 1794. *Zoonomia; or, The Laws of Organic Life*. London: J. Johnson.

———. 1796. *Zoonomia; or, The Laws of Organic Life*. 2nd ed. 3 vols. London: J. Johnson.

———. 1797. *A Plan for the Conduct of Female Education in Boarding Schools*. London: J. Johnson.

———. 1800. *Phytologia*. London: J. Johnson.

———. 1801. *Zoonomia; or, The Laws of Organic Life*. 3rd ed. 4 vols. London: J. Johnson.

———. 1803. *The Temple of Nature; or, The Origin of Society: A Poem with Philosophical Notes*. London: J. Johnson.

Darwin, Francis, ed. 1888. *The Life and Letters of Charles Darwin*. 3 vols. London: John Murray.

Darwin, Francis, and A. C. Seward, eds. 1903. *More Letters of Charles Darwin: A Record of His Work in a Series of Hitherto Unpublished Letters*. 2 vols. London: John Murray.

Darwin, George, H. 1872. "Development in Dress." *Macmillan Magazine* 26:410–16.

———. 1873. "On Beneficial Restrictions to Liberty of Marriage." *Contemporary Review* 22:412–26.

Darwin, Leonard. 1929. "Memories of Down House." *Nineteenth Century* 106:118–23.

Daunton, Martin J. 1995. *Progress and Poverty: An Economic and Social History of Britain, 1700–1850*. Oxford: Oxford University Press.

Davidoff, Lenore. 1973. *The Best Circles: Society, Etiquette and the Season*. London: Croom Helm.

Davidoff, Lenore, and Catherine Hall. 1987. *Family Fortunes: Men and Women of the English Middle Class, 1790–1850*. Chicago: University of Chicago Press.

Davis, David Brion. 1975. *The Problem of Slavery in the Age of Revolution, 1770–1823.* Ithaca, NY: Cornell University Press.

Dawkins, Richard. 1976. *The Selfish Gene.* Oxford: Oxford University Press.

[Dawkins, William Boyd]. 1871. "Darwin on the Descent of Man." *Edinburgh Review* 134:195–235.

Dawson, Gowan. 2007. *Darwin, Literature and Victorian Respectability.* Cambridge: Cambridge University Press.

Dawson, Gowan, and Bernard Lightman, eds. 2014. *Victorian Scientific Naturalism: Community, Identity, Continuity.* Chicago: University of Chicago Press.

De Beer, Gavin. 1962. "Darwin and Embryology." In *A Century of Darwin,* edited by S. A. Barnett, 153–72. London: Mercury Books.

Desmond, Adrian. 1982. *Archetypes and Ancestors: Paleontology in Victorian London, 1850–1875.* London: Blond and Briggs.

———. 1984. "Robert E. Grant: The Social Predicament of a Pre-Darwinian Transmutationist." *Journal of the History of Biology* 17.189–223.

———. 1987. "Artisan Resistance and Evolution in Britain, 1819–1848." *Osiris* 3:77–110.

———. 1989. *The Politics of Evolution; Morphology, Medicine and Reform in Radical London.* Chicago: University of Chicago Press.

———. 1994. *Huxley: The Devil's Disciple.* London: Michael Joseph.

———. 1997. *Huxley: Evolution's High Priest.* London: Michael Joseph.

———. 2001. "Redefining the X Axis: 'Professionals,' 'Amateurs' and the Making of Mid-Victorian Biology—A Progress Report." *Journal of the History of Biology* 34:3–50.

Desmond, Adrian, and James Moore. 1991. *Darwin.* London: Michael Joseph.

———. 2009. *Darwin's Sacred Cause: Race, Slavery and the Quest for Human Origins.* London: Allen Lane.

Di Gregorio, Mario, and Nick Gill, eds. 1990. *Charles Darwin's Marginalia,* vol. 1. New York: Garland.

Diamond, Jared. 1986. "Animal Art: Variation in Bower Decorating Styles among Male Bowerbirds *Amblyornis inornatus.*" *Proceedings of the National Academy of Sciences USA* 83:3042–46.

———. 1992. *The Rise and Fall of the Third Chimpanzee.* London: Vintage.

Dickens, Charles. N.d. *The Personal History of David Copperfield (The Fireside Dickens).* London: Chapman & Hall.

———. 1853. "The Noble Savage." *Household Words* 7 (168): 337–39.

Dickie, George. 1996. *The Century of Taste.* Oxford: Oxford University Press.

Dillon, Elizabeth Maddock. 2004. "Sentimental Aesthetics." *American Literature* 76:495–523.

Dixon, Edmund Saul. 1851. *The Dovecote and the Aviary.* London: John Murray.

Dixson, Alan F. 2009. *Sexual Selection and the Origins of Human Mating Systems.* Oxford: Oxford University Press.

Docker, John. 2015. "A Plethora of Intentions: Genocide, Settler Colonialism and Historical Consciousness in Australia and Britain." *International Journal of Human Rights* 19:74–89.

Douglas, Bronwen. 2008. "Climate to Crania: Science and the Rationalization of Human Difference." In *Foreign Bodies: Oceania and the Science of Race, 1750–1940,* edited by Bronwen Douglas and Chris Ballard, 33–96. Canberra: ANU E Press.

Draper, Hal, and Anne G. Lipow. 1976. "Marxist Women versus Bourgeois Feminism." *Socialist Register* 13:179–226.

Driver, Felix. 1999. *Geography Militant: Cultures of Exploration in the Age of Empire.* Oxford: Blackwell.

[Drysdale, George]. (1855) 1861. *Elements of Social Science; or, Physical, Sexual, and Natural Religion*. London: E. Truelove.

Duncan, Ian. 1991. "Darwin and the Savages." *Yale Journal of Criticism* 4:13–45.

Durant, John R. 1979. "Scientific Naturalism and Social Reform in the Thought of Alfred Russel Wallace." *British Journal for the History of Science* 12:31–58.

——. 1985. "The Ascent of Nature in Darwin's Descent." In *The Darwinian Heritage*, edited by David Kohn, 283–306. Princeton, NJ: Princeton University Press.

——. 1993. "The Tension at the Heart of Julian Huxley's Evolutionary Ethology." In *Julian Huxley: Biologist and Statesman of Science*, edited by Albert van Helden, 250–67. Houston: Rice University Press.

Dyehouse, Carol. 1989. *Feminism and the Family in England, 1880–1839*. Oxford: Basil Blackwell.

Eagleton, Terry. 1989. "Aesthetics and Politics in Edmund Burke." *History Workshop Journal* 28:53–62.

——. 1990. *The Ideology of the Aesthetic*. Oxford: Blackwell.

Easley, Alexis. 1999. "Gendered Observations: Harriet Martineau and the Woman Question." In *Victorian Women Writers and the Woman Question*, edited by Nicola Diane Thompson, 80–98. Cambridge: Cambridge University Press.

Eaton, John Matthews. 1851. *A Treatise on the Art of Breeding and Managing the Almond Tumbler*. London: J. M. Eaton.

——. 1852. *A Treatise on the Art of Breeding and Managing Tame, Domesticated and Fancy Pigeons*. London: J. M. Eaton.

Edgeworth, Maria, and Richard Lovell Edgeworth. 1798. *Practical Education*. 2 vols. London: J. Johnson.

Edgeworth, Richard Lovell. 1820. *Memoirs* (concluded by Maria Edgeworth). 2 vols. London: R. Hunter.

Edwards, Bryan. 1793. *The History, Civil and Commercial, of the British Colonies in the West Indies*. 2 vols. Dublin: Luke White.

Egan, Maureen L. 1989. "Evolutionary Theory in the Social Philosophy of Charlotte Perkins Gilman." *Hypatia* 4:102–19.

Ehrlich, Paul R., David S. Dobkin, and Darryl Wheye. 1988. "Plume Trade." *Birds of Standford*. http://www.stanford.edu/group/stanfordbirds/text/essays/Plume_Trade.html

Eiseley, Loren. 1979. *Darwin and the Mysterious Mr. X: New Light on the Evolutionists*. New York: E. P. Dutton.

Elsdon-Baker, Fern. 2008. "Spirited Dispute: The Secret Split between Wallace and Romanes." *Endeavour* 32:75–78.

Elston, Mary Ann. 1987. "Women and Anti-vivisection in Victorian England, 1870–1900." In *Vivisection in Historical Perspective*, edited by Nicholas A. Rupke, 259–94. London: Croom Helm.

Endersby, Jim. 2003. "Darwin on Generation, Pangenesis and Sexual Selection." In *The Cambridge Companion to Darwin*, edited by Jonathan Hodge and Gregory Radick, 69–91. Cambridge: Cambridge University Press.

——. 2014. "Odd Man Out: Was Joseph Hooker an Evolutionary Naturalist?" In *Victorian Scientific Naturalism: Community, Identity, Continuity*, edited by Gowan Dawson and Bernard Lightman, ch. 6. Chicago: University of Chicago Press.

Etcoff, Nancy. 2000. *Survival of the Prettiest: The Science of Beauty*. London: Abacus.

Eze, Emmanuel Chukwudi. 1997. *Race and the Enlightenment: A Reader*. Oxford: Blackwell.

Faas, Ekbert. 2002. *The Genealogy of Aesthetics*. Cambridge: Cambridge University Press.

Fagan, Melinda Bonnie. 2007. "Wallace, Darwin, and the Practice of Natural History." *Journal of the History of Biology* 40:601–35.

———. 2008. "Theory and Practice in the Field: Wallace's Work in Natural History." In *Natural Selection and Beyond: The Intellectual Legacy of Alfred Russel Wallace*, edited by Charles H. Smith and George Beccaloni, 66–90. Oxford: Oxford University Press.

Fara, Patricia. 2003. *Sex, Botany & Empire: The Story of Carl Linnaeus and Joseph Banks.* Cambridge: Icon Books.

Feeley-Harnik, Gillian. 2004. "The Geography of Descent." *Proceedings of the British Academy* 125:311–64.

Feuer, Lewis S. 1962. "Marxist Tragedians. A Death in the Family." *Encounter* 19:23–32.

Fichman, Martin. 2004. *An Elusive Victorian: The Evolution of Alfred Russel Wallace.* Chicago: University of Chicago Press.

Finzsch, Norbert. 2005. "'It Is Scarcely Possible to Conceive That Human Beings Could Be So Hideous and Loathsome': Discourses of Genocide in Eighteenth- and Nineteenth-Century America and Australia." *Patterns of Prejudice* 39:97–115.

Fisher, Ronald A. 1915. "The Evolution of Sexual Preference." *Eugenics Review* 7:184–92.

———. 1916. "Racial Repair." *Eugenics Review* 7:204–7.

———. 1930. *The Genetical Theory of Natural Selection.* Oxford: Clarendon.

FitzRoy, Robert. 1839. *Narrative of the Surveying Voyages of His Majesty's Ships* Adventure *and* Beagle *between the Years 1826 and 1836, Describing Their Examination of the Southern Shores of South America, and the* Beagle's *Circumnavigation of the Globe: Proceedings of the Second Expedition, 1831–36, under the Command of Captain Robert FitzRoy, R. N.* Vol. 2, with an appendix. London: Henry Colburn.

Flint, Kate. 2000. *The Victorians and the Visual Imagination.* Cambridge: Cambridge University Press.

Forbes, Linda C., and John M. Jermier. 2002. "The Institutionalization of Bird Protection." *Organization & Environment* 15:458–65.

Foreman, Amanda. 2011. *A World on Fire: Britain's Crucial Role in the American Civil War.* New York: Random House.

Forster, Johann Reinhold. (1778) 1996. *Observations Made During a Voyage Round the World.* Reprint, edited by Nicholas Thomas, Harriet Guest, and Michael Dettelbach. Honolulu: University of Hawaii Press.

Foucault, Michel. 1980. *The History of Sexuality: An Introduction.* New York: Vintage.

———. 2004. *The Archeology of Knowledge.* London: Routledge.

Frankel, Simon J. 1994. "The Eclipse of Sexual Selection Theory." In *Sexual Knowledge, Sexual Science: The History of Attitudes to Sexuality*, edited by Roy Porter and Mikulas Teich, 158–83. Cambridge: Cambridge University Press.

Fryer, Peter. 1965. *The Birth Controllers.* London: Secker & Warburg.

Galton, Francis. 1853. *The Narrative of an Explorer in Tropical South Africa.* London: John Murray.

———. 1865. "Hereditary Talent and Character." *Macmillan's Magazine* 12:157–66, 318–27.

———. (1869) 1950. *Hereditary Genius: An Enquiry into Its Laws and Consequences.* London: Watts.

———. 1873. "Hereditary Improvement." *Fraser's Magazine* 7:116–30.

———. (1883) 2004. *Inquiries into Human Faculty and Its Development.* Whitefish, MT: Kessinger Publishing.

Gamble, Eliza Burt. (1894) 1916. *The Sexes in Science and History: An Inquiry into the Dogma of Woman's Inferiority to Man.* Rev. ed. New York: G. P. Putnam.

Garfinkle, Norton. 1955. "Science and Religion in England, 1790–1800: The Critical Response to the Work of Erasmus Darwin." *Journal of the History of Ideas* 16:376–88.

Gates, Barbara T. 1998. *Kindred Nature: Victorian and Edwardian Women Embrace the Living World*. Chicago: University of Chicago Press.

Gates, Barbara T., and Ann B. Shteir, eds. 1997. *Natural Eloquence: Women Reinscribe Science*. Madison: University of Wisconsin Press.

Gayon, Jean. 2010. "Sexual Selection: Another Darwinian Process." *Comptes rendus Biologies* 333:134–44.

Geddes, Patrick, and J. Arthur Thompson. 1889. *The Evolution of Sex*. London: Walter Scott.

George, Sam. 2005. "The Cultivation of the Female Mind: Enlightened Growth, Luxuriant Decay and Botanical Analogy in Eighteenth-Century Texts." *History of European Ideas* 31:209–23.

Gerzina, Gretchen. 1995. *Black London: Life before Emancipation*. New Brunswick, NJ: Rutgers University Press.

Ghiselin, Michael T. 1972. *The Triumph of the Darwinian Method*. Berkeley: University of California Press.

———. 1974. *The Economy of Nature and the Evolution of Sex*. Berkeley: University of California Press.

———. 1976. "Two Darwins: History versus Criticism." *Journal of the History of Biology* 9:121–32.

Gibbons, Luke. 2003. *Edmund Burke and Ireland: Aesthetics, Politics, and the Colonial Sublime*. Cambridge: Cambridge University Press.

Gilman, Charlotte Perkins. (1898) 1966. *Women and Economics*. New York: Harper and Row.

Gilman, Sander L. 1982. *On Blackness without Blacks: Essays on the Image of the Black in Germany*. Boston: G. K. Hall.

———. 1985. *Difference and Pathology: Stereotypes of Sexuality, Race, and Madness*. Ithaca, NY: Cornell University Press.

———. 1995. *Health and Illness: Images of Difference*. London: Reaktion Books.

Gleadle, Kathryn. 1995. *The Early Feminists. Radical Unitarians and the Emergence of the Women's Rights Movement, 1831–51*. New York: St. Martin's Press.

Gobineau, Joseph Arthur de. 1856. *The Moral and Intellectual Diversity of Races, with Particular Reference to Their Respective Influence on the Civil and Political History of Mankind*. Translated and with notes by H. Hotz[e]. Philadelphia: Lippincott.

Gould, Stephen Jay. 1977. *Ontogeny and Phylogeny*. Cambridge, MA: Harvard University Press.

———. 1980a. *Ever Since Darwin: Reflections in Natural History*. Harmondsworth: Pelican.

———. 1980b. *The Panda's Thumb*. Harmondsworth: Penguin.

———. 1981. *The Mismeasure of Man*. New York: Norton.

———. 1985. "The Hottentot Venus." In *The Flamingo's Smile: Reflections in Natural History*, 291–305. New York: W. W. Norton.

———. 1992. "The Confusion over Evolution." *New York Review of Books*, November 19, 47–54.

Grammer, Karl, Bernhard Fink, Anders P. Moller, and Randy Thornhill. 2003. "Darwinian Aesthetics: Sexual Selection and the Biology of Beauty." *Biological Reviews* 78:387–407.

Gray, Richard T. 2004. *About Face: German Physiognomic Thought from Lavater to Auschwitz*. Detroit: Wayne State University Press.

Greene, John C. 1961. *The Death of Adam: Evolution and Its Impact on Western Thought*. New York: Mentor Books.

———. 1977. "Darwin as a Social Evolutionist." *Journal of the History of Biology* 10:1–27.

Greg, William Rathbone. 1850. "Prostitution." *Westminster Review* 53:448–506.

———. 1862. "Why Are Women Redundant?" *National Review* 14:434–60.

———. 1868. "On the Failure of 'Natural Selection' in the Case of Man." *Fraser's Magazine* 78:353–62.

Grindle, Nicholas. 1997. "'Our Own Imperfect Knowledge': Petrus Camper and the Search for an 'Ideal Form.'" *RES: Anthropology and Aesthetics* 31:139–48.

Gruber, Howard E. 1974. *Darwin on Man: A Psychological Study of Scientific Creativity.* London: Wildwood House.

Guest, Harriet 1996. "Looking at Women: Forster's Observations in the South Pacific." In Forster, *Observations Made during a Voyage Round the World,* edited by Nicholas Thomas, Harriet Guest, and Michael Dettelbach, xli–liv. Honolulu: University of Hawaii Press.

Gurstein, Rochelle. 2000. "Taste and 'the Conversible World' in the Eighteenth Century." *Journal of the History of Ideas* 61:203–21.

Guyer, Paul. 2002. "Beauty and Utility in Eighteenth-Century Aesthetics." *Eighteenth-Century Studies* 35:439–53.

———. 2005. *Values of Beauty: Historical Essays in Aesthetics.* Cambridge: Cambridge University Press.

Haeckel, Ernst. 1866. *Generelle Morphologie der Organismen.* 2 vols. Berlin: George Reimer.

Hale, Piers J. 2012. "Darwin's Other Bulldog: Charles Kingsley and the Popularization of Evolution in Victorian England." *Science and Education* 21:977–1013.

Hall, Catherine. 1992. *White, Male and Middle-Class: Explorations in Feminism and History.* Cambridge: Polity Press.

———. 2002. *Civilizing Subjects: Metropole and Colony in the English Imagination, 1830–1867.* Cambridge: Polity.

Hall, Donald E., ed. 1994. *Muscular Christianity: Embodying the Victorian Age.* Cambridge: Cambridge University Press.

Hall, Lesley. 2004. "Hauling Down the Double Standard: Feminism, Social Purity and Sexual Science in Late Nineteenth-Century Britain." *Gender & History* 16:36–56.

Haller, John S. 1995. *Outcasts from Evolution: Scientific Attitudes of Racial Inferiority, 1859–1900.* Carbondale: Southern Illinois University Press.

Halttunen, Karen. 1982. *Confidence Men and Painted Women. A Study of Middle-Class Culture in America, 1830–1870.* New Haven, CT: Yale University Press.

Hamlin, Kimberly A. 2010. "The Birds and the Bees: Darwin's Evolutionary Approach to Human Sexuality." In *Darwin in Atlantic Cultures: Evolutionary Visions of Race, Gender and Sexuality,* edited by Jeannette Eileen Jones and Patrick Sharp, 53–72. New York: Routledge.

———. 2014. *From Eve to Evolution: Darwinian Science and Women's Rights in Gilded Age America.* Chicago: University of Chicago Press.

Haraway, Donna. 1989. *Primate Visions: Gender, Race and Nature in the World of Modern Science.* New York: Routledge.

Hardaker, Miss M. A. 1882. "Science and the Woman Question." *Popular Science Monthly* 20:577–84.

Harris, George. 1869. "On the Distinctions, Mental and Moral, Occasioned by the Difference of Sex." *Journal of the Anthropological Society* 7:clxxxix–cxcv.

Harrison, Mark. 1996. "'The Tender Frame of Man': Disease, Climate, and Racial Difference in India and the West Indies, 1760–1860." *Bulletin of the History of Medicine* 70:68–93.

Hartley, Lucy. 2001. "A Science of Beauty? Femininity, Fitness and the Nineteenth Century Physiognomic." *Women: A Cultural Review* 12:19–34.

———. 2005. *Physiognomy and the Meaning of Expression in Nineteenth-Century Culture.* Cambridge: Cambridge University Press.

Harvey, John. 1995. *Men in Black*. London: Reaktion Books.

Harvey, Joy. 1997. *"Almost a Man of Genius": Clémence Royer, Feminism and Nineteenth-Century Science*. New Brunswick, NJ: Rutgers University Press.

———. 2009. "Darwin's 'Angels': The Women Correspondents of Charles Darwin." *Intellectual History Review* 19:197–210.

Haufe, Chris. 2008. "Sexual Selection and Mate Choice in Evolutionary Psychology." *Biology and Philosophy* 23:115–28.

Hawkesworth, John. 1773. *An Account of the Voyages Undertaken by the Order of His Present Majesty, for Making Discoveries in the Southern Hemisphere*. 3 vols. London: W. Strachan and D. Cadell.

Hayek, F. A. 1951. *John Stuart Mill and Harriet Taylor: Their Friendship and Subsequent Marriage*. London: Routledge & Kegan Paul.

Healey, Edna. 2002. *Emma Darwin: The Inspirational Wife of a Genius*. London: Review.

Helfand, Michael, S. 1977. "T. H. Huxley's *Evolution and Ethics*: The Politics of Evolution and the Evolution of Politics." *Victorian Studies* 20:157–77.

Herbert, Christopher. 1991. *Culture and Anomie. Ethnographic Imagination in the Nineteenth Century*. Chicago: University of Chicago Press.

Herbert, Sandra. 1974. "The Place of Man in the Development of Darwin's Theory of Transmutation: Part I." *Journal of the History of Biology* 7:217–58.

———. 1977. "The Place of Man in the Development of Darwin's Theory of Transmutation: Part II." *Journal of the History of Biology* 10:155–227.

Herzog, Don. 1998. *Poisoning the Minds of the Lower Orders*. Princeton, NJ: Princeton University Press.

Hesketh, Ian. 2015. "A Good Darwinian? Winwood Reade and the Making of a Late Victorian Evolutionary Epic." *Studies in History and Philosophy of Biological and Biomedical Sciences* 51:44–52.

Hilts, Victor L. 1982. "Obeying the Laws of Hereditary Descent: Phrenological Views on Inheritance and Eugenics." *Journal of the History of the Behavioral Sciences* 18:62–77.

Hipple, Walter John. 1957. *The Beautiful, the Sublime, & the Picturesque in Eighteenth-Century British Aesthetic Theory*. Carbondale: Southern Illinois University Press.

Hobsbawm, Eric John. 1973. *The Age of Revolution*. London: Cardinal.

———. 1975. *The Age of Capital, 1848–1875*. London: Wiedenfield and Nicolson.

Hodge, Jonathan. 1972. "The Universal Gestation of Nature: Chambers' *Vestiges* and *Explanations*." *Journal of the History of Biology* 5:127–51.

———. 1977. "The Structure and Strategy of Darwin's 'Long Argument.'" *British Journal for the History of Science* 10:237–46.

———. 1982. "Darwin and the Laws of the Animate Part of the Terrestrial System (1835–37): On the Lyellian Origins of his Zoonomical Explanatory Program." *Studies in the History of Biology* 7:1–106.

———. 1985. "Darwin as a Lifelong Generation Theorist." In *The Darwinian Heritage*, edited by David Kohn, 207–43. Princeton, NJ: Princeton University Press.

———. 2003. "The Notebook Programmes and Projects of Darwin's London Years." In *The Cambridge Companion to Darwin*, edited by Jonathan Hodge and Gregory Radick, 40–68. Cambridge: Cambridge University Press.

———. 2005. "Against 'Revolution' and 'Evolution.'" *Journal of the History of Biology* 38:101–21.

———. 2009. "Capitalist Contexts for Darwinian Theory: Land, Finance, Industry and Empire." *Journal of the History of Biology* 42:399–416.

———. 2010. "The Darwin of Pangenesis." *Comptes rendus Biologies* 333:129–33.

Hollander, Anne. 1993. *Seeing through Clothes*. Berkeley: University of California Press.

Holmes, Anne Sumner. 1995. "The Double Standard in the English Divorce Laws, 1857–1923." *Law & Social Inquiry* 20:601–20.

Hoquet, Thierry, ed. 2015. *Current Perspectives on Sexual Selection*. Dordrecht: Springer.

Hoquet, Thierry, and Michael Levandowsky. 2015. "Utility vs. Beauty: Darwin, Wallace and the Subsequent History of the Debate on Sexual Selection." In *Current Perspectives on Sexual Selection*, edited by Thierry Hoquet, 19–44. Dordrecht: Springer.

Hrdy, Sarah Blaffer. 1981. *The Woman That Never Evolved*. Cambridge: Harvard University Press.

Hubbard, Ruth. 1979. "Have Only Men Evolved?" In *Women Look at Biology Looking at Women*, edited by Ruth Hubbard, M. S. Henifen, and B. Fried, 7–35. Cambridge: Schenkmen.

Hudson, Nicholas. 2004. "'Hottentots' and the Evolution of European Racism." *Journal of European Studies* 34:308–32.

———. 2008. "The 'Hottentot Venus,' Sexuality, and the Changing Aesthetics of Race, 1650–1850." *Mosaic* 41:19–41.

Hunt, James. 1863. "On the Negro's Place in Nature." *Memoirs of the Anthropological Society of London* 1:1–64.

———. 1864. "Abstract and Discussion of 'On the Negro's Place in Nature.'" *Journal of the Anthropological Society*, published in *Anthropological Review* 2 (4): xv–lvi.

———. 1866. "On the Application of the Principle of Natural Selection to Anthropology, in Reply to Views Advocated by Some of Mr. Darwin's Disciples," *Anthropological Review* 4:320–40.

———. 1867. "On the Doctrine of Continuity Applied to Anthropology," *Anthropological Review* 5:110–20.

Hunter, John. 1837. *Observations on Certain Parts of the Animal Economy*. Vol. 4 of *The Works of John Hunter*, edited by Richard Owen. London: Longman, Rees, Orme, Brown, Green, and Longman.

Huxley, Julian S. 1914. "The Courtship-Habits of the Great Crested Grebe (*Podiceps cristatus*): With an Addition to the Theory of Sexual Selection." *Proceedings of the Zoological Society of London* 35:491–562.

———. 1938. "The Present Standing of the Theory of Sexual Selection." In *Evolution: Essays on Aspects of Evolutionary Theory*, edited by Gavin de Beer, 11–42. Oxford: Clarendon.

———. 1942. *Evolution: The Modern Synthesis*. London: George Allen and Unwin.

Huxley, Leonard. 1900. *The Life and Letters of Thomas Henry Huxley*. 2 vols. London: Macmillian, 1900.

Huxley, Thomas Henry. 1854a. "Contemporary Literature—Science." *Westminster Review* 62:242–56.

———. 1854b. "Vestiges of the Natural History of Creation." *British and Foreign Medico-Chirurgical Review* 13:425–39.

———. 1855a. "On Certain Zoological Arguments Commonly Adduced in Favour of the Hypothesis of the Progressive Development of Animal Life in Time." *Proceedings of the Royal Institution* 2:82–85.

———. 1855b. "Science." *Westminster Review* 63:239–53.

———. 1863. *Evidence as to Man's Place in Nature*. London: Williams & Norgate.

———. (1865) 1873. "On the Methods and Results of Ethnology." *Critiques and Addresses*, 134–66. London: Macmillan.

———. (1865) 1968. "Emancipation—Black and White." *Collected Essays*, 3:66–75. New York: Greenwood Press.

———. 1871. "Mr. Darwin's Critics." *Contemporary Review* 18:443–76.

———. 1875. "*Anthropogenie* by Ernst Haeckel." *Academy* 7:16–18.

———. 1989. *Evolution and Ethics*. Edited by James Paradis and George C. Williams. Princeton, NJ: Princeton University Press.

Hyatt, Alpheus. 1897. "The Influence of Woman in the Evolution of the Human Race." *Natural Science* 11:89–93.

Jacyna, L. Stephen. 1983. "Immanence or Transcendence: Theories of Life and Organization in Britain, 1790–1835." *Isis* 74:311–29.

Janes, R. M. 1978. "On the Reception of Mary Wollstonecraft's *A Vindication of the Rights of Woman*." *Journal of the History of Ideas* 39:293–302.

Jann, Rosemary. 1994. "Darwin and the Anthropologists: Sexual Selection and its Discontents." *Victorian Studies* 37:287–306.

———. 1997. "Revising the Descent of Woman: Eliza Burt Gamble." In *Natural Eloquence: Women Reinscribe Science*, edited by Barbara T. Gates and Ann B. Shteir, 147–63. Madison, London: University of Wisconsin Press.

Jeffrey, Frances. 1811. "Alison on Taste." *Edinburgh Review* 18:1–46.

Jenkin, Fleeming. 1867. "The *Origin of Species*." *North British Review* 46:277–318.

Jones, Greta. 2002. "Alfred Russel Wallace, Robert Owen and the Theory of Natural Selection." *British Journal for the History of Science* 35:73–96.

Jones, Robert W. 1998. *Gender and the Formation of Taste in Eighteenth-Century Britain*. Cambridge: Cambridge University Press.

Jones, Vivien, ed. 1990. *Women in the Eighteenth Century: Constructions of Femininity*. London: Routledge.

Juengel, Scott. 1996. "Godwin, Lavater, and the Pleasures of Surface." *Studies in Romanticism* 35:73–97.

———. 2001. "Countenancing History: Mary Wollstonecraft, Samuel Stanhope Smith, and Enlightenment Racial Science." *ELH* 68:897–927.

Kaiser, David A. 1996. "Whither Kantian Aesthetics?" *Eighteenth-Century Life* 20:101–8.

Kames, Lord (Henry Home). 1774. *Sketches of the History of Man*. 2 vols. Edinburgh: W. Creech and W. Strahan and T. Cadell.

Kapp, Yvonne. 1972. *Eleanor Marx*. 2 vols. London: Lawrence and Wishart.

Kaye, Richard A. 2002. *The Flirt's Tragedy: Desire without End in Victorian and Edwardian Fiction*. Charlottesville: University of Virginia Press.

Kelly, Gary. 1993. *Women, Writing and Revolution, 1790–1827*. Oxford: Clarendon.

Kenny, Robert. 2007. "From the Curse of Ham to the Curse of Nature: The Influence of Natural Selection on the Debate on Human Unity before the Publication of the *Descent of Man*." *British Journal for the History of Science* 40:367–88.

Kevles, Daniel. 1985. *In the Name of Eugenics: Genetics and the Uses of Human Heredity*. New York: Alfred A. Knopf.

Keynes, Randal. 2001. *Annie's Box: Charles Darwin, His Daughter and Human Evolution*. London: Fourth Estate.

Keynes, Richard Darwin, ed. 1988. *Charles Darwin's* Beagle *Diary*. Cambridge: Cambridge University Press.

Kincaid, James R. 1971. *David Copperfield: Laughter and Point of View*. Oxford: Clarendon Press.

King-Hele, Desmond. 1977. *Doctor of Revolution: The Life and Genius of Erasmus Darwin*. London: Faber & Faber.

———. 1981. *The Letters of Erasmus Darwin*. Cambridge: Cambridge University Press.

———, ed. 2003. *Charles Darwin's "The Life of Erasmus Darwin."* Cambridge: Cambridge University Press.

Kirby, Percival R. 1965. *Sir Andrew Smith, M.D., K.C.B.: His Life, Letters and Works*. Cape Town: A. A. Balkemar.

Kjaergaard, Peter C. 2010. "The Scientific Enterprise: From Scientific Icon to Global Product." *History of Science* xlviii:105–22.

Klonk, Charlotte. 1996. *Science and the Perception of Nature: British Landscape Art in the Late Eighteenth and Early Nineteenth Centuries*. New Haven, CT: Yale University Press.

Knox, Robert. 1850. *The Races of Men: A Fragment*. London: Henry Renshaw.

———. 1852a. *Great Artists and Great Anatomists*. London: John van Voorst.

———. 1852b. *A Manual of Artistic Anatomy, for the Use of Sculptors, Painters, and Amateurs*. London: Henry Renshaw.

———. 1855. "Some Remarks on the Aztecque and Bosjieman Children, Now Being Exhibited in London." *Lancet* 1:357–60.

———. 1857. *The Greatest of our Social Evils: Prostitution, as It Now Exists in London, Liverpool, Manchester, Glasgow, Edinburgh and Dublin: An Enquiry into Its Cause and Means of Reformation, Based on Statistical Documents, by a Physician*. London: H. Balliere.

———. 1862. *The Races of Men: A Philosophical Enquiry into the Influence of Race over the Destinies of Nations*. 2nd ed. London: Henry Renshaw.

———. 1863. "Some Additional Observations on a Collection of Human Crania and Other Human Bones, at Present Preserved in the Crypt of a Church at Hythe, in Kent." *Transactions of the Ethnological Society of London* 2:136–40.

Kohlstedt, Sally Gregory, and Mark R. Jorgensen. 1999. "'The Irrepressible Woman Question': Women's Response to Darwinian Evolutionary Biology." In *Disseminating Darwinism: The Role of Place, Race, Religion and Gender*, edited by Ronald R. Numbers and John Stenhouse, 267–93. Cambridge: Cambridge University Press.

Kohn, David. 1996. "The Aesthetic Construction of Darwin's Theory." In *The Elusive Synthesis: Aesthetics and Science*, edited by Alfred I. Tauber, 13–48. Dordrecht: Kluwer Academic Publishers.

———. 2009. "Darwin's Keystone: The Principle of Divergence." In *The Cambridge Companion to the "Origin of Species,"* edited by Michael Ruse and Robert J. Richards, 87–108. Cambridge: Cambridge University Press.

Kohn, David, and M. J. S. Hodge. 1985. "The Immediate Origins of Natural Selection." In *The Darwinian Heritage*, edited by David Kohn, 185–206. Princeton, NJ: Princeton University Press.

Kottler, Malcolm Jay. 1974. "Alfred Russel Wallace, the Origins of Man, and Spiritualism." *Isis* 65:142–92.

———. 1980. "Darwin, Wallace, and the Origin of Sexual Dimorphism." *Proceedings of the American Philosophical Society* 124:203–26.

———. 1985. "Charles Darwin and Alfred Russel Wallace: Two Decades of Debate over Natural Selection." In *The Darwinian Heritage*, edited by David Kohn, 367–432. Princeton, NJ: Princeton University Press.

Kuchta, David. 1996. "The Making of the Self-Made Man: Class, Clothing and English Masculinity, 1688–1832." In *The Sex of Things*, edited by Victoria Di Grazia and Ellen Farlough, 54–78. Berkeley: University of California Press.

Kunzle, David. 1982. *Fashion and Fetishism: A Social History of the Corset, Tight-Lacing and Other Forms of Body-Sculpture in the West*. Totowa, NJ: Rowman and Littlefield.

Kuper, Adam. 2005. *The Reinvention of Primitive Society: Transformations of a Myth*. London: Routledge and Kegan Paul.

Landes, Joan B. 1988. *Women and the Public Sphere in the Age of the French Revolution*. Ithaca, NY: Cornell University Press.

Langland, Elizabeth. 1992. "Nobody's Angels: Domestic Ideology and Middle-Class Women in the Victorian Novel." *PMLA* 107:290–304.

Laqueur, Thomas. 1990. *Making Sex: Body and Gender from the Greeks to Freud.* Cambridge, MA: Harvard University Press.

Larson, Barbara. 2009. "Introduction." In *The Art of Evolution: Darwin, Darwinisms, and Visual Culture,* edited by Barbara Larson and Fay Brauer, 1–17. Hanover, NH: Dartmouth College Press.

———. 2013. "Darwin, Burke, and the Biological Sublime." In *Darwin and Theories of Aesthetics and Cultural History,* edited by Barbara Larson and Sabine Flach, 17–36. Farnham and Burlington: Ashgate Publishing.

Latham, Robert Gordon. 1851. *Man and His Migrations.* London: Van Voorst.

Lavater, Johann Caspar. 1789. *Essays on Physiognomy: For the Promotion of the Knowledge and the Love of Mankind.* 3 vols. London: G. G. J. and J. Robinson.

———. Ca. 1800. *The Whole Works of Lavater on Physiognomy.* 4 vols. London: W. Butters.

———. 1820. *L'Art de connaître les hommes par la physionomie.* 10 vols. Paris: Dépalafol.

Lawrence, William. 1822. *Lectures on Physiology, Zoology, and the Natural History of Man.* London: Benbow.

Leask, Nigel. 2003. "Darwin's 'Second Sun': Alexander von Humboldt and the Genesis of *The Voyage of the Beagle.*" In *Literature, Science, Psychoanalysis, 1830–1970: Essays in Honour of Gillian Beer,* edited by Helen Small and Trudi Tate, 13–36. Oxford: Oxford University Press.

Ledger, Sally. 2007. "The New Woman and Feminist Fictions." In *The Cambridge Companion to the Fin de Siècle,* edited by Gail Marshall, 153–68. Cambridge: Cambridge University Press.

Lessing, Gotthold Ephraim. Ca. 1874. *Laocoon.* London: George Routledge & Sons.

Levine, George. 2006. *Darwin Loves You: Natural Selection and the Re-enchantment of the World.* Princeton, NJ: Princeton University Press.

Levy, Michelle. 2006. "The Radical Education of *Evenings at Home.*" *Eighteenth Century Fiction* 19:123–50.

Lewes, George Henry. 1856. "Hereditary Influence, Animal and Human." *Westminster Review* 10:135–62.

Leypoldt, Gunter. 1999. "A Neoclassical Dilemma in Sir Joshua Reynolds's Reflections on Art." *British Journal of Aesthetics* 39:330–49.

Lightman, Bernard. 2001. "Victorian Sciences and Religions: Discordant Harmonies." *Osiris* 1 6:343–66.

———. 2002. "Huxley and Scientific Agnosticism: The Strange History of a Failed Historical Strategy." *British Journal for the History of Science* 35:271–89.

———. 2007. *Victorian Popularizers of Science: Designing Nature for New Audiences.* Chicago: University of Chicago Press.

———. 2010. "Darwin and the Popularization of Evolution." *Notes and Records of the Royal Society* 64:5–24.

Lindfors, Bernth, ed. 1999. *Africans on Stage: Studies in Ethnological Show Business.* Bloomington: Indiana University Press.

Lindquist, Jason H. 2004. "'Under the Influence of an Exotic Nature . . . National Remembrances Are Insensibly Effaced': Threats to the European Subject in Humboldt's Personal Narrative of Travels to the Equinoctial Regions of the New Continent." *HiN, International Review for Humboldtian Studies* 9:2–17.

[Linton, Eliza Lynn]. 1868. "The Girl of the Period." *Saturday Review,* March 14, 339–40.

Litchfield, Henrietta, ed. (1904) 1915. *Emma Darwin: A Century of Family Letters, 1792–1896.* 2 vols. London: John Murray.

Livingstone, David N. 1999. "Geographical Inquiry, Rational Religion, and Moral Philosophy: Enlightenment Discourses on the Human Condition." In *Geography and Enlightenment*, edited by David N. Livingstone and Charles W. J. Withers, 93–119. Chicago: University of Chicago Press.

———. 2005. "Risen into Empire." In *Geography and Revolution*, edited by David N. Livingstone and Charles W. J Withers, 304–35. Chicago and London: University of Chicago Press.

———. 2011. *Adam's Ancestors: Race, Religion, and the Politics of Human Origins*. Baltimore: Johns Hopkins University Press.

Logan, James Venable. 1972. *The Poetry and Aesthetics of Erasmus Darwin*. New York: Octagon Books.

Logan, Thad. 2001. *The Victorian Parlour*. Cambridge: Cambridge University Press.

Lonsdale, Henry. 1870. *A Sketch of the Life and Writings of Robert Knox the Anatomist*. London: Macmillan.

Lorimer, Douglas A. 1978. *Colour, Class and the Victorians: English Attitudes to the Negro in the Mid-nineteenth Century*. Leicester: Leicester University Press.

———. 1997. "Science and the Secularization of Victorian Images of Race." In *Victorian Science in Context*, edited by Bernard Lightman, 212–35. Chicago: University of Chicago Press.

Love, Rosaleen. 1983. "Darwinism and Feminism: The 'Woman Question' in the Life and Work of Olive Schreiner and Charlotte Perkins Gilman." In *The Wider Domain of Evolutionary Thought*, edited by David Oldroyd and Ian Langham, 113–31. Dordrecht: Reidel.

Lubbock, John. 1865. *Pre-historic Times, as Illustrated by Ancient Remains, and the Manners and Customs of Modern Savages*. Edinburgh: Williams & Norgate.

———. 1870. *The Origin of Civilization and the Primitive Condition of Man*. London: Longmans, Green.

Lucas, Prosper. 1847–50. *Traité philosophique et physiologique de l'hérédité naturelle*. 2 vols. Paris: J. B. Baillière.

Lurie, Edward. 1954. "Louis Agassiz and the Races of Man." *Isis* 45:227–42.

———. 1960. *Louis Agassiz: A Life in Science*. Chicago: University of Chicago Press.

Lyell, Charles. 1830–33. *Principles of Geology, Being an Attempt to Explain the Former Changes of the Earth's Surface, by Reference to Causes Now in Operation*. 3 vols. London: John Murray.

———. 1838. *The Elements of Geology*. First ed. London: John Murray.

———. 1863. *The Geological Evidences of the Antiquity of Man, with Remarks on the Theories of the Origin of Species by Variation*. 2nd ed. London: John Murray.

Lyell, K. M., ed. 1881. *Life, Letters, and Journals of Sir Charles Lyell, Bart*. 2 vols. London: John Murray.

Lyons, Sherrie L. 1999. *Thomas Henry Huxley: The Evolution of a Scientist*. Amherst, NY: Prometheus Books.

MacKenzie, Donald A. 1981. *Statistics in Britain, 1865–1930: The Social Construction of Scientific Knowledge*. Edinburgh: Edinburgh University Press.

Madden, Joah R. 2006. "Interpopulation Differences Exhibited by Spotted Bowerbirds *Chlamydera maculata* across a Suite of Male Traits and Female Preferences." *Ibis* 148:425–35.

Madden, Joah R., Tasmin J. Lowe, Hannah V. Fuller, Kanchon K. Dasmahapatra, and Rebecca L. Coe.. 2004. "Local Traditions of Bower Decoration by Spotted Bowerbirds in a Single Population." *Animal Behaviour* 68:759–65.

Magubane, Zine. 2001. "Which Bodies Matter? Feminism, Poststructuralism, Race, and the Curious Theoretical Odyssey of the 'Hottentot Venus.'" *Gender & Society* 15:816–34.

Malthus, Thomas Robert. 1976. *An Essay on the Principle of Population, Text, Sources and Background Criticism*. Edited by Philip Appleman. New York: W. W. Norton.

Mantegazza, Paolo. 1867. *Rio de la Plata e Tenerife*. Milano: Gaetono Brigola.

———. 1871. "E'Elezione sessuale e la neogenesi (lettera a C. Darwin)." *Archivio per l'Antropologia e la Etnologia* 1:306–23.

Marchant, James, ed. 1916. *Alfred Russel Wallace: Letters and Reminiscences*. 2 vols. London: Cassell.

Martineau, Harriet. (1838) 1989. *How to Observe: Morals and Manners*. New Brunswick, NJ: Transaction.

———. (1877) 1983. *Autobiography*. 2 vols. Reprint, London: Virago.

Marx Aveling, Eleanor, and Edward Aveling. 1886. *The Woman Question*. London: Swan Sonnenschein.

Mason, Michael. 1994. *The Making of Victorian Sexual Attitudes*. Oxford: Oxford University Press.

———. 1995. *The Making of Victorian Sexuality*. Oxford: Oxford University Press.

Masters, William H., and Virginia E. Johnson. (1966) 1981. *Human Sexual Response*. New York: Bantam Books.

Maudsley, Henry. 1870. *Body and Mind*. London: Macmillan.

———. 1874. "Sex in Mind and Education." *Fortnightly Review* 21:468–77.

McCarthy, William. 2008. *Anna Letitia Barbauld: Voice of the Enlightenment*. Baltimore: Johns Hopkins University Press.

McClintock, Anne. 1995. *Imperial Leather: Race, Gender, and Sexuality in the Colonial Contest*. New York: Routledge.

McKnight, Natalie. 2006. "Dickens and Darwin: A Rhetoric of Pets." *Dickensian* 102:131–43.

McLaren, Angus. 1992. *A History of Contraception from Antiquity to the Present Day*. Oxford: Blackwell.

McLennan, John Ferguson. 1865. *Primitive Marriage*. Edinburgh: Adam & Charles Black.

———. 1896. *Studies in Ancient History*. London: Macmillan.

McNeil, Maureen. 1987. *Under the Banner of Science: Erasmus Darwin and His Age*. Manchester: Manchester University Press.

Mead, L. S., and S. J. Arnold. 2004. "Quantitative Genetic Models of Sexual Selection." *Trends in Ecology & Evolution* 19:264–71.

Meijer, Miriam Claude. 1999. *Race and Aesthetics in the Anthropology of Petrus Camper (1722–1798)*. Amsterdam: Rodopi.

Meller, Helen. 1990. *Patrick Geddes: Social Evolutionist and City Pioneer*. London: Routledge.

Mellersh, Howard Edward Leslie. 1968. *Fitzroy of the Beagle*. London: Rupert Hart-Davis.

Menand, Louis. 2001–02. "Morton, Agassiz, and the Origins of Scientific Racism in the United States." *Journal of Blacks in Higher Education* 34:110–13.

Midgley, Clare. 1992. *Women against Slavery: The British Campaigns, 1780–1870*. London: Routledge.

Milam, Erika Lorraine. 2010. *Looking for a Few Good Males: Female Choice in Evolutionary Biology*. Baltimore: Johns Hopkins University Press.

Milam, Erika L., Roberta L. Millstein, Angela Potochnik, and Joan E. Roughgarden. 2011. "Sex and Sensibility: The Role of Sexual Selection." *Metascience* 20:253–77.

Mill, John Stuart. 1850. "The Negro Question." *Fraser's Magazine* 41:25–31.

———. 1869. *The Subjection of Women*. London: Longmans, Green, Reader & Dyer.

———. 1970. *Essays on Sexual Equality*. Edited by Alice S. Rossi. Chicago: University of Chicago Press.

Miller, Geoffrey. 2001. *The Mating Mind: How Sexual Choice Shaped the Evolution of Human Nature*. London: Vintage.

Millstein, Roberta L. 2012. "Darwin's Explanation of Races by Means of Sexual Selection." *Studies in History and Philosophy of Biological and Biomedical Sciences* 43:627-33.

Milne Edwards, Henri. 1844. "Considérations sur quelques principes relatifs à la classification naturelle des animaux." *Annales des sciences naturelles* 1:65-99.

Mitchell, Sally. 2004. *Frances Power Cobbe: Victorian Feminist, Journalist, Reformer*. Charlottesville: University of Virginia Press.

Mitchell, Thomas Livingstone. 1838. *Three Expeditions into the Interior of Eastern Australia*. 2 vols. London: D. Carlisle.

[Mivart, St. George Jackson]. 1871a. "Darwin's *Descent of Man*." *Quarterly Review* 131:47-90.

[———]. 1871b. *On the Genesis of Species*. London: Macmillan.

———. 1873. "Contemporary Evolution." *Contemporary Review* 22:595-614.

[———]. 1874. "Primitive Man: Tylor and Lubbock." *Quarterly Review* 137:40-77.

———. 1876. *Lessons from Nature, as Manifested in Mind and Matter*. London: John Murray.

Montague, Ken. 1994. "The Aesthetics of Hygiene: Aesthetic Dress, Modernity, and the Body as Sign." *Journal of Design History* 7:91-112.

Moore, James R. 1985. "Darwin's Genesis and Revelations." *Isis* 76:570-80.

———. 1988. "Freethought, Secularism, Agnosticism: The Case of Charles Darwin." In *Religion in Victorian Britain*, edited by Gerald Parsons, 274-319. Manchester: Manchester University Press.

———. 1989. "Of Love and Death: Why Darwin 'Gave Up Christianity.'" In *History, Humanity and Evolution*, edited by James R. Moore, 195-229. Cambridge: Cambridge University Press.

———. 1991. "Deconstructing Darwinism: The Politics of Evolution in the 1860s." *Journal of the History of Biology* 24:353-408.

———. 1997. "Wallace's Malthusian Moment: The Common Context Revisited." In *Victorian Science in Context*, edited by Bernard Lightman, 290-311. Chicago: University of Chicago Press.

———. 2001. *Good Breeding: Science and Society in a Darwinian Age: Study Guide*. A426 Study Guide. Milton Keynes: Open University Press.

———. 2007. "R. A. Fisher: A Faith Fit for Eugenics." *Studies in History and Philosophy of Biological and Biomedical Sciences* 38:110-35.

———. 2008. "Wallace in Wonderland." In *Natural Selection and Beyond: The Intellectual Legacy of Alfred Russel Wallace*, edited by Charles H. Smith and George Beccaloni, 353-67. Oxford: Oxford University Press

Moore, Jane. 1999. *Mary Wollstonecraft*. Plymouth: Northcote House.

Morgan, Arthur E. 1944. *Edward Bellamy*. New York: Columbia University Press.

Morgan, Elaine. 1985. *The Descent of Woman*. London: Souvenir Press.

Morgan, Lewis H. 1868. "A Conjectural Solution of the Origin of the Classificatory System of Relationship." *Proceedings of the American Academy of Arts & Sciences* 7:436-77.

Morgan, Thomas Hunt. 1903. *Evolution and Adaptation*. London: Macmillan.

[Morley, John]. 1871. Review of *Descent*. *Pall Mall Gazette*. March 20, 11-12; March 21, 11-12.

Morton, Peter. 1984. *The Vital Science: Biology and the Literary Imagination, 1860-1900*. London: Allen & Unwin.

———. 2005. *"The Busiest Man in England": Grant Allen and the Writing Trade, 1875-1900*. New York: Palgrave Macmillan.

Morton, Samuel George. 1847. "Hybridity in Animals, Considered in Reference to the Question of the Unity of the Human Species." *American Journal of Science and Arts* 3:39–50, 203–12.

Moruno, Delores Martín. 2010. "Love in the Time of Darwinism: Paolo Montegazza and the Emergence of Sexuality." *Medicina & Storia* X:147–64.

Moruzi, Kristine. 2009. "Fast and Fashionable: The Girls in *The Girl of the Period Miscellany*." *Australasian Journal of Victorian Studies* 14:9–28.

Mosedale, Susan Sleeth. 1978. "Science Corrupted: Victorian Biologists Consider 'The Woman Question.'" *Journal of the History of Biology* 11:1–55.

Müller, Fritz. (1863) 1869. *Facts and Arguments for Darwin*. London: John Murray.

Munro, Jane. 2009. "'More Like a Work of Art Than of Nature': Darwin, Beauty and Sexual Selection." In *Endless Forms: Charles Darwin, Natural Science, and the Visual Arts*, edited by Diana Donald and Jane Munro, 253–91. New Haven, CT: Yale University Press.

Murray, Lindley. 1831. *Introduction to the English Reader*. Philadelphia: Robert Christy.

Nead, Lynda. 1990. *Myths of Sexuality: Representations of Women in Victorian Britain*. Oxford: Basil Blackwell.

Neher, Allister. 2011. "Robert Knox and the Anatomy of Beauty." *Medical Humanities* 37:46–50.

Nethercot, Arthur H. 1961. *The First Five Lives of Annie Besant*. London: Hart-Davis.

Nichols, Peter. 2003. *Evolution's Captain: The Dark Fate of the Man who Sailed Charles Darwin around the World*. New York: Harper Collins.

Nicholson, Mervyn. 1990. "The Eleventh Commandment: Sex and Spirit in Wollstonecraft and Malthus." *Journal of the History of Ideas* 51:401–21.

Noakes, Richard. 2004. "*Punch* and Comic Journalism in Mid-Victorian Britain." In *Science in the Nineteenth-Century Periodical*, edited by Geoffroy Cantor, Gowan Dawson, Graeme Gooday, Richard Noakes, Sally Shuttleworth, and Jonathan R. Topham, 91–122. Cambridge: Cambridge University Press.

Norton, Robert E. 1995. *The Beautiful Soul*. Ithaca, NY: Cornell University Press.

Nott, Josiah C., and George R. Gliddon. 1854. *Types of Mankind; or, Ethnological Researches, Based upon the Ancient Monuments, Paintings, Sculptures, and Crania of Races*. Philadelphia: Lippincott & Gambo.

Nyhart, Lynn K. 2009. "Embryology and Morphology." In *The Cambridge Companion to the Origin of Species*, edited by Michael Ruse and Robert J. Richards, 194–215. Cambridge: Cambridge University Press.

Oppenheim, Janet. 1989. "Prophets without Honour? The Odyssey of Annie Besant." *History Today* 39:12–18.

Oppenheimer, Jane M. 1967. *Essays in the History of Embryology and Biology*. Cambridge: MIT Press.

Orton, Reginald. 1855. *On the Physiology of Breeding: Two Lectures, Delivered to the Newcastle Farmer's Club*. 2nd ed. Sunderland.

Ospovat, Dov. 1976. "The Influence of K. E. von Baer's Embryology, 1828–1859: A Reappraisal in Light of Richard Owen's and William B. Carpenter's Palaeontological Application of 'von Baer's Law.'" *Journal of the History of Biology* 9:1–28.

———. 1981. *The Development of Darwin's Theory*. Cambridge, MA: Harvard University Press.

Owen, Alex. 1990. *The Darkened Room: Women, Power and Spiritualism in Late Victorian England*. Philadelphia: University of Pennsylvania Press.

Owen, Richard. 1849. *On the Nature of Limbs*. London: John Van Voorst.

———. 1855. *Lectures on the Comparative Anatomy and Physiology of the Invertebrate Animals*. 2nd ed. London: Longman, Brown, Green & Longman.

Paradis, James. 1989. "*Evolution and Ethics* in Its Victorian Context." In *Evolution and Ethics*, edited by James Paradis and George C. Williams, 3–55. Princeton, NJ: Princeton University Press.

Parrinder, Patrick. 1997. "Eugenics and Utopia: sexual selection from Galton to Morris." *Utopian Studies* 8:1–12.

Paul, Diane B. 2003. "Darwin, Social Darwinism and Eugenics." In *The Cambridge Companion to Darwin*, edited by Jonathan Hodge and Gregory Radick, 214–39. Cambridge: Cambridge University Press.

———. 2008. "Wallace, Women and Eugenics" in *Natural Selection and Beyond: The Intellectual Legacy of Alfred Russel Wallace*. Edited by Charles H. Smith and George Beccaloni, 263–78. Oxford: Oxford University Press.

Paul, Diane B., and Benjamin Day. 2008. "John Stuart Mill, Innate Differences, and the Regulation of Reproduction." *Studies in History, Philosophy, Biology & Biomedical Science* 39:222–31.

Paxton, Nancy L. 1991. *George Eliot and Herbert Spencer: Feminism, Evolutionism and the Reconstruction of Gender*. Princeton, NJ: Princeton University Press.

Paylor, Suzanne. 2006. "Edward B. Aveling: The People's Darwin." *Endeavour* 29:66–71.

Peart, Sandra J., and David M. Levy. 2008. "Darwin's Unpublished Letter at the Bradlaugh-Besant Trial: A Question of Divided Expert Judgment." *European Journal of Political Economy* 24:343–53.

Peckham, Morse, ed. 1959. The Origin of Species *by Charles Darwin: A Variorum Text*. Philadelphia: University of Pennsylvania Press.

Peterson, Linda. 1990. "Harriet Martineau: Masculine Discourse, Female Sage." In *Victorian Sages and Cultural Discourse: Renegotiating Gender and Power*, edited by Thais Morgan, 171–86. New Brunswick, NJ: Rutgers University Press.

Phillips, Melanie. 2004. *The Ascent of Woman: A History of the Suffragette Movement and the Ideas behind It*. London: Abacus.

Pieterse, Jan Nederveen. 1995. *White on Black: Images of Africa and Blacks in Western Popular Culture*. New Haven, CT: Yale University Press.

Pike, Luke O. 1869. "On the Claims of Women to Political Power." *Journal of the Anthropological Society* 7:xlvii–lxi.

Polwhele, Richard. 1798. *The Unsex'd Females: A Poem*. London: Cadell and Davies.

Poovey, Mary. 1984. *The Proper Lady and the Woman Writer: Ideology as Style in the Works of Mary Wollstonecraft, Mary Shelley, and Jane Austen*. Chicago: University of Chicago.

———. 1989. *Uneven Developments: The Ideological work of Gender in Mid-Victorian England*. London: Virago Press.

Porter, Duncan M. 1993. "On the Road to the *Origin* with Darwin, Hooker, and Gray." *Journal of the History of Biology* 26:1–38.

Porter, Roy. 1985. "Making Faces: Physiognomy and Fashion in Eighteenth-Century England." *Études anglaises* 38:385–96.

———. 1989. "Erasmus Darwin: Doctor of Evolution?" In *History, Humanity and Evolution*, edited by James R. Moore, 39–69. Cambridge: Cambridge University Press.

Powell, Baden. 1855. *Essays on the Spirit of the Inductive Philosophy, the Unity of Worlds, and the Philosophy of Creation*. London: Longman.

Preston, Samuel Tolver. 1880. "Evolution and Female Education." *Nature* 22:485–86.

Prichard, James Cowles. (1813) 1973. *Researches into the Physical History of Man*. Edited by George W. Stocking. Reprint, Chicago: University of Chicago Press.

———. 1826. *Researches into the Physical History of Mankind*. 2nd ed. 2 vols. London: John and Arthur Arch, 1826.

————. 1835. *A Treatise on Insanity and Other Disorders Affecting the Mind*. London: Sherwood et al.

————. 1836–47. *Researches into the Physical History of Mankind*. 3rd ed. 5 vols. London: Sherwood, Gilbert & Piper.

————. 1843. *The Natural History of Man*. London: Hippolyte Baillière.

————. 1851. *Researches into the Physical History of Mankind*. 4th ed. 2 vols. London: Houlston & Stoneman.

Priestman, Martin. 1999. *Romantic Atheism: Poetry and Freethought, 1780–1830*. Oxford: Oxford University Press.

Prodger, Phillip. 2009. "Ugly Disagreements: Darwin and Ruskin Discuss Sex and Beauty." In *The Art of Evolution: Darwin, Darwinisms, and Visual Culture*, edited by Barbara Larson and Fay Brauer, 40–58. Hanover, NH: Dartmouth College Press.

Prum, Richard O. 2012. "Aesthetic Evolution by Mate Choice: Darwin's *Really* Dangerous Idea." *Philosophical Transactions of the Royal Society B* 367:2253–65.

————. 2015. "The Role of Sexual Autonomy in Evolution by Mate Choice." In *Current Perspectives on Sexual Selection*, edited by Thierry Hoquet, 237–262. Dordrecht: Springer.

Qureshi, Sadiah. 2004. "Displaying Sara Baartman, the 'Hottentot Venus.'" *History of Science* 42:233–57.

————. 2011a. *Peoples on Parade: Exhibitions, Empire, and Anthropology in Nineteenth-Century Britain*. Chicago: University of Chicago Press.

————. 2011b. "Robert Gordon Latham, Displayed Peoples, and the Natural History of Race, 1854–66." *Historical Journal* 54:143–66.

Radick, Gregory. 2000. "Language, Brain Function, and Human Origins in the Victorian Debates on Evolution." *Studies in History and Philosophy of Biological and Biomedical Sciences* 31:55–75.

————. 2002. "Darwin on Language and Selection." *Selection* 3:7–16.

————. 2003. "Is the Theory of Natural Selection Independent of Its History?" In *The Cambridge Companion to Darwin*, edited by Jonathan Hodge and Gregory Radick, 143–67. Cambridge: Cambridge University Press.

————. 2007. *The Simian Tongue: The Long Debate about Animal Language*. Chicago: University of Chicago Press.

————. 2008. "Race and Language in the Darwinian Tradition (and What Darwin's Language-Species Parallels Have to Do with It)." *Studies in History and Philosophy of Biological and Biomedical Sciences* 39:359–70.

————. 2010a. "Darwin's Puzzling *Expression*." *Comptes rendus Biologies* 333:181–87.

————. 2010b. "Did Darwin Change His Mind about the Fuegians?" *Endeavour* 34:50–54.

Rae, Isobel. 1964. *Knox the Anatomist*. Edinburgh: Oliver and Boyd.

Raverat, Gwen. 1960. *Period Piece: A Cambridge Childhood*. London: Faber and Faber.

Razek, Rula. 1999. *Dress Codes: Reading Nineteenth Century Fashion*. Stanford, CA: Stanford University Press.

Reade, William Winwood. 1864. *Savage Africa: Being the Narrative of a Tour in Equatorial, South-western, and North-western Africa*. New York: Harper & Brothers.

————. 1872. *The Martyrdom of Man*. London: Trübner.

————. 1873. *The African Sketch-Book*. 2 vols. London: Smith, Elder.

Rehbock, Philip F. 1983. *The Philosophical Naturalists: Themes in Early Nineteenth-Century British Biology*. Madison: University of Wisconsin Press.

Rendall, Jane. 2004. "Bluestockings and Reviewers: Gender, Power, and Culture in Britain, c. 1800–1830." *Nineteenth-Century Contexts* 26:355–74.

Reynolds, Sir Joshua. 1798. *The Works of Sir Joshua Reynolds, Knight; . . . Containing His Discourses, Idlers, etc. (with His Last Corrections and Additions)*. 2nd ed. 3 vols. London: T. Cadell, Jun. and W. Davies.

Reynolds, Kimberley, and Nicola Humble. 1993. *Victorian Heroines: Representations of Femininity in Nineteenth-Century Literature and Art*. New York: Harvester Wheatsheaf.

Richards, Evelleen. 1976. *The German Romantic Concept of Embryonic Repetition and Its Role in Evolutionary Theory in England up to 1859*. PhD diss., University of New South Wales.

———. 1983. "Darwin and the Descent of Woman." In *The Wider Domain of Evolutionary Thought*, edited by David Oldroyd and Ian Langham, 57–111. Dordrecht: Reidel.

———. 1987. "A Question of Property Rights: Richard Owen's Evolutionism Reassessed." *British Journal for the History of Science* 20:129–71.

———. 1989a. "Huxley and Woman's Place in Science: The 'Woman Question' and the Control of Victorian Anthropology." In *History, Humanity and Evolution*, edited by James R. Moore, 253–84. Cambridge: Cambridge University Press.

———. 1989b. "The 'Moral Anatomy' of Robert Knox: The Interplay between Biological and Social Thought in Victorian Scientific Naturalism." *Journal of the History of Biology* 22:373–436.

———. 1990. "'Metaphorical Mystifications': The Romantic Gestation of Nature in British Biology." In *Romanticism and the Sciences*, edited by Andrew Cunningham and Nicholas Jardine, 130–43. Cambridge: Cambridge University Press.

———. 1994. "A Political Anatomy of Monsters, Hopeful and Otherwise: Teratogeny, Transcendentalism, and Evolutionary Theorizing." *Isis* 85:377–411.

———. 1995. "Gendering the Romanes Lecture: The Sexual Politics of T. H. Huxley's *Evolution and Ethics*." Paper delivered at Huxley Conference, Imperial College, London.

———. 1997. "Redrawing the Boundaries: Darwinian Science and Victorian Women Intellectuals." In *Victorian Science in Context*, edited by Bernard Lightman, 119–42. Chicago and London: University of Chicago Press.

Richards, Robert J. 1987. *Darwin and the Emergence of Evolutionary Theories of Mind and Behavior*. Chicago: University of Chicago Press.

———. 1992. *The Meaning of Evolution: The Morphological Construction and Ideological Reconstruction of Darwin's Theory*. Chicago: University of Chicago Press.

———. 2002. *The Romantic Conception of Life: Science and Philosophy in the Age of Goethe*. Chicago: University of Chicago Press.

———. 2003. "Darwin on Mind, Morals and Emotions." In *The Cambridge Companion to Darwin*, edited by Jonathan Hodge and Gregory Radick, 92–115. Cambridge: Cambridge University Press.

———. 2008. *The Tragic Sense of Life: Ernst Haeckel and the Struggle over Evolutionary Thought*. Chicago: University of Chicago Press.

———. 2009. "The Descent of Man." *American Scientist* 97 (5): 415–17.

———. 2012. "Darwin's Principles of Divergence and Natural Selection: Why Fodor Was Almost Right." *Studies in History and Philosophy of Biological and Biomedical Sciences* 43:256–68.

Richards, Thomas. 1991. *The Commodity Culture of Victorian England: Advertising and Spectacle, 1851–1914*. Stanford CA: Stanford University Press.

Richardson, Angelique. 2003. *Love and Eugenics in the Late Nineteenth Century: Rational Reproduction and the New Woman*. Oxford: Oxford University Press.

Ritvo, Harriet. 1987. *The Animal Estate: The English and Other Creatures in the Victorian Age*. Cambridge, MA: Harvard University Press.

———. 1988. "Sex and the Single Animal." *Grand Street* 1:124–39.

Robbins, Sarah. 1993. *"Lessons for Children* and Teaching Mothers: Mrs. Barbauld's Primer for the Textual Construction of Middle-Class Domestic Pedagogy." *Lion and the Unicorn* 17:135–51.

Romanes, Ethel, ed. 1896. *The Life and Letters of George John Romanes.* London: Longmans, Green.

Romanes, George J. 1887. "Mental Differences between Men and Women." *Nineteenth Century* 21:654–72.

———. 1888. "Recent Critics of Darwinism." *Contemporary Review* 53:836–54.

———. 1889. "Mr. Wallace on Darwinism." *Contemporary Review* 56:244–58.

———. 1892–97. *Darwin and after Darwin.* 3 vols. London: Longmans, Green.

———. 1896. *Animal Intelligence.* New York: D. Appleton.

Rosen, David. 1994. "The Volcano and the Cathedral: Muscular Christianity and the Origins of Primal Manliness." In *Muscular Christianity: Embodying the Victorian Age,* edited by Donald E. Hall, 17–44. Cambridge: Cambridge University Press.

Rossi, Alice S., ed. 1974. *The Feminist Papers: From Adams to de Beauvoir.* Toronto: Bantam Books.

Roughgarden, Joan. 2009. *The Genial Gene: Deconstructing Darwinian Selfishness.* Berkeley: University of California Press.

———. 2015. "Sexual Selection: Is Anything Left?" In *Current Perspectives on Sexual Selection,* edited by Thierry Hoquet, 85–102. Dordrecht: Springer.

Rousseau, Jean-Jacques. (1761) 1971. *A Discourse on the Origin and Foundation of the Inequality among Mankind.* London: R. and J. Dodsley. Facsimile edition. New York: Franklin.

———. (1762) 1981. *Émile; or, On Education.* Translated by Allan Bloom. New York: Basic Books.

Rowland, Peter. 1996. *The Life and Times of Thomas Day, 1748–1789.* Lewiston: Edward Mellon.

Royle, Edward. 1971. *Radical Politics, 1790–1900: Religion and Unbelief.* London: Longman.

———. 1980. *Radicals, Secularists and Republicans: Popular Freethought in Britain, 1866–1915.* Manchester: Manchester University Press.

Royle, Edward, and J. Walvin. 1982. *English Radicals and Reformers 1760–1848.* Brighton: Harvester Press.

Rudolphi, Carl Asmund. 1812. *Beyträge zur Anthropologie und allgemeinen Naturgeschichte.* Berlin: Haude & Speur.

Rupke, Nicolaas. 1999. "A Geography of Enlightenment: The Critical Reception of Alexander von Humboldt's Mexican Work." In *Geography and Enlightenment,* edited by David N. Livingstone and Charles W. J. Withers, 319–39. Chicago: University of Chicago Press.

———. 2005. "Alexander von Humboldt and Revolution." In *Geography and Revolution.* Edited by David N. Livingstone and Charles W. J. Withers, 336–50. Chicago: University of Chicago Press.

Ruse, Michael. 1979. *The Darwinian Revolution.* Chicago: University of Chicago Press.

———. 2004. "The Romantic Conception of Robert J. Richards." *Journal of the History of Biology* 37:2–23.

———. 2015. "Sexual Selection: Why Does It Play Such a Large Role in *The Descent of Man*?" In *Current Perspectives on Sexual Selection,* edited by Thierry Hoquet, 3–17. Dordrecht: Springer.

Russell, E. S. 1916. *Form and Function: A Contribution to the History of Animal Morphology.* London: John Murray.

Russell, Nicholas. 1986. *Like Engend'ring Like: Heredity and Animal Breeding in Early Modern England.* Cambridge: Cambridge University Press.

Russett, Cynthia Eagle. 1989. *Sexual Science: The Victorian Construction of Womanhood*. Cambridge, MA: Harvard University Press.

Sanders, Valerie. 2013. "'Mady's Tightrope Walk': The Career of Marian Huxley Collier." In *Crafting the Woman Professional in the Long Nineteenth Century: Artistry and Industry in Britain*, edited by Kyriaki Hadjiafxendi and Patricia Zakreski, ch. 11. London: Ashgate.

Sandow, Alexander. 1938. "Social Factors in the Origin of Darwinism." *Quarterly Review of Biology* 13:315–26.

Schiebinger, Londa. 1993. *Nature's Body: Gender in the Making of Modern Science*. Boston: Beacon Press.

Schmitt, Cannon. 2009. *Darwin and the Memory of the Human: Evolution, Savages, and South America*. Cambridge: Cambridge University Press.

Schofield, Robert E. 1963. *The Lunar Society of Birmingham*. Oxford: Clarendon Press.

Schreiner, Olive. 1889. "Life's Gifts." *Women's Penny Paper* 1 (47): 7.

———. 1911. *Woman and Labour*. London: T. Fisher Unwin.

Schwartz, Joel. 1984. *The Sexual Politics of Jean-Jaques Rousseau*. Chicago: University of Chicago Press.

———. 1995. "George John Romanes's Defense of Darwinism: The Correspondence of Charles Darwin and His Chief Disciple." *Journal of the History of Biology* 28:281–316.

Sebright, John Saunders. 1809. *The Art of Improving the Breeds of Domestic Animals*. London: John Harding.

———. 1831. *County Reform: Opinion of Sir J. S. Sebright in Reply to a Letter from a Freeholder in Nutford . . . on the Subject of Votes Being Given to, or Withheld from, Farmers Being Tenants at Sell, etc.* London: Roake and Varty.

———. 1836. *Observations upon the Instincts of Animals*. London: Gossling & Egley.

Secord, James A. 1981. "Nature's Fancy: Charles Darwin and the Breeding of Pigeons." *Isis* 72:163–86.

———. 1985. "Darwin and the Breeders: A Social History." In *The Darwinian Heritage*, edited by David Kohn, 519–42. Princeton, NJ: Princeton University Press.

———. 1989. "Behind the Veil: Robert Chambers and *Vestiges*." In *History, Humanity and Evolution*, edited by James R. Moore, 165–94. Cambridge: Cambridge University Press.

———. 2000. *Victorian Sensation: The Extraordinary Publication, Reception, and Secret Authorship of* Vestiges of the Natural History of Creation. Chicago: University of Chicago Press.

———. 2009. "Focus: Darwin as a Cultural Icon. Introduction." *Isis* 100:537–41.

———. 2014. *Visions of Science: Books and Readers at the Dawn of the Victorian Age*. Oxford: Oxford University Press.

Sedgwick, Adam. 1845. "Vestiges of the Natural History of Creation." *Edinburgh Review* 82:1–85.

Semmel, Bernard. 1963. *The Governor Eyre Controversy*. Boston: Houghton Mifflin.

Seward, Anna. 1804. *Memoirs of the Life of Dr. Darwin, Chiefly during His Residence at Lichfield*. London: J. Johnson.

———. 1811. *Letters of Anna Seward, 1784–1807*. 6 vols. Edinburgh: Archibald Constable.

Shanafelt, Robert. 2003. "How Charles Darwin Got Emotional Expression Out of South Africa (and the People Who Helped Him)." *Comparative Studies in Society and History* 45:815–42.

Shapin, Steven. 2010. "The Darwin Show." *London Review of Books* 32:3–9.

Sheets-Pyenson, Susan. 1981. "Darwin's Data: His Reading of Natural History Journals, 1837–1842." *Journal of the History of Biology* 14:231–48.

Shermer, Michael. 2002. *In Darwin's Shadow: The Life and Science of Alfred Russel Wallace*. Oxford: Oxford University Press.

Showalter, Elaine. 1995. *Sexual Anarchy: Gender and Culture at the Fin de Siècle*. London: Virago Press.

Shteir, Ann B. 1996. *Cultivating Women, Cultivating Science: Flora's Daughters and Botany in England. 1760–1860*. Baltimore: Johns Hopkins University.

Sigusch, Volkmar. 2008. "The Birth of Sexual Medicine: Paolo Montegazza as Pioneer of Sexual Medicine." *Journal of Sexual Medicine* 5:217–22.

Sinha, Mrinalini. 1995. *Colonial Masculinity: The "Manly Englishman" and the "Effeminate Bengal" in the Late Nineteenth Century*. Manchester: Manchester University Press.

Skotnes, Pippa. 2001. "'Civilised Off the Face of the Earth': Museum Display and the Silencing of the /Xam." *Poetics Today* 22:299–321.

Sloan, Phillip R. 1985. "Darwin's Invertebrate Program, 1826–1836." In *The Darwinian Heritage*, edited by David Kohn, 71–120. Princeton, NJ: Princeton University Press.

———. 2003. "The Making of a Philosophical Naturalist." In *The Cambridge Companion to Darwin*, edited by Jonathan Hodge and Gregory Radick, 17–39. Cambridge: Cambridge University Press.

Smith, Alison. 1996. *The Victorian Nude: Sexuality, Morality, and Art*. Manchester, New York: Manchester University Press.

———. 2005. "The 'Snake Body' in Victorian Art." *Early Popular Visual Culture* 3:151–64.

Smith, Andrew. 1831. "Observations Relative to the Origin and History of the Bushmen." *Philosophical Magazine* 9:119–27, 197–200, 339–42, 419–23.

Smith, Charles Hamilton. 1852. *The Natural History of the Human Species, Its Typical Forms, Primaeval Distribution, Filiations, and Migrations*. London: Bohn.

Smith, Charles H., ed. 1991. *Alfred Russel Wallace: An Anthology of His Shorter Writings*. Oxford: Oxford University Press.

———. 2008. "Wallace, Spiritualism, and Beyond: 'Change' or 'No Change'?" In *Natural Selection and Beyond: The Intellectual Legacy of Alfred Russel Wallace*, edited by Charles H. Smith and George Beccaloni, 391–423. Oxford: Oxford University Press.

Smith, Jonathan. 1995. "'The Cock of Lordly Plume': Sexual Selection and *The Egoist*." *Nineteenth-Century Literature* 50:51–77.

———. 2006. *Charles Darwin and Victorian Visual Culture*. Cambridge: Cambridge University Press.

Smith, Samuel Stanhope. (1788) 1995. *An Essay on the Causes of the Variety of Complexion and Figure in the Human Species*. Reprint of the Edinburgh edition, with an introduction by Paul B. Wood. Bristol: Thoemmes Press.

———. (1810) 1965. *An Essay on the Causes of the Variety of Complexion and Figure in the Human Species*. Reprint of the second edition, edited by Winthrop P. Jordan. Cambridge, MA: Harvard University Press/Belknap Press.

Spencer, Herbert. (1854) 1966. "Personal Beauty." In *Essays: Scientific, Political & Speculative*, vol. 2, 385–99. Osnabrück: Otto Zeller.

———. 1855. *The Principles of Psychology*. London: Longman, Brown, Green, and Longmans.

———. 1859. "Physical Training." *British Quarterly Review* 29:362–97.

———. (1877) 1906. *The Study of Sociology*. New York: D. Appleton.

Stack, David A. 2000. "The First Darwinian Left: Radical and Socialist Responses to Darwin, 1859–1914." *History of Political Thought* 21:682–710.

———. 2003. *The First Darwinian Left: Socialism and Darwinism, 1859–1914*. Cheltenham: New Clarion.

Stanton, William. 1960. *The Leopard's Spots: Scientific Attitudes toward Race in America, 1815–59*. Chicago: University of Chicago Press.

Stauffer, Robert C., ed. 1975. *Charles Darwin's* Natural Selection: *Being the Second Part of His Big Species Book Written from 1856 to 1858*. Cambridge: Cambridge University Press.

Steele, Valerie. 1985. *Fashion and Eroticism: Ideals of Feminine Beauty from the Victorian Era to the Jazz Age*. New York: Oxford University Press.

Stepan, Nancy. 1982. *The Idea of Race in Science: Great Britain 1800–1960*. London: MacMillan.

———. 2001. *Picturing Tropical Nature*. Ithaca, NY: Cornell University Press.

Stephens, Lester. 2000. *Science, Race, and Religion in the American South: John Bachman and the Charleston Circle of Naturalists, 1815–1895*. Chapel Hill: University of North Carolina Press.

Sterrett, Susan G. 2002. "Darwin's Analogy between Artificial and Natural Selection: How Does It Go?" *Studies in History and Philosophy of Science Part C: Studies in History and Philosophy of Biological and Biomedical Sciences* 33:151–68.

Stock, John Edmonds. 1811. *Memoirs of the Life of Thomas Beddoes, M.D.* London: John Murray.

Stocking, George W. 1971. "What's in a Name? The Origins of the Royal Anthropological Institute (1837–71)." *Man* 6:369–90.

———. 1987. *Victorian Anthropology*. New York: Free Press.

Story, Graham, and Kathleen Tillotson, eds. 1995. *The Pilgrim Edition of the Letters of Charles Dickens, 1856–58*. Vol. 8. Oxford: Oxford University Press.

Stott, Rebecca. 2003. *Darwin and the Barnacle*. London: Faber and Faber.

Strauss, Sylvia. 1988. "Gender, Class and Race in Utopia." In *Looking Backward, 1988–1888: Essays on Edward Bellamy*, edited by Daphne Patai, 68–90. Amherst: University of Massachusetts Press.

Strzelecki, Paul Edward de. 1845. *Physical Description of New South Wales and Van Diemen's Land*. London: Longman, Brown, Green & Longman.

Sulloway, Frank J. 1982a. "Darwin and His Finches: The Evolution of a Legend." *Journal of the History of Biology* 15:1–53.

———. 1982b. "Darwin's Conversion: The *Beagle* Voyage and Its Aftermath." *Journal of the History of Biology* 15:325–96.

Summerscale, Kate. 2012. *Mrs Robinson's Disgrace: The Private Diary of a Victorian Lady*. London: Bloomsbury.

Takahashi, Mariko, Hiroyuki Arita, Mariko Hiraiwa-Hasegawa, and Toshikazu Hasegawa. 2008. "Peahens Do Not Prefer Peacocks with More Elaborate Trains." *Animal Behaviour* 75:1209–19.

Tammone, William. 1995. "Competition, the Division of Labor, and Darwin's Principle of Divergence." *Journal of the History of Biology* 28:109–31.

Tanner, Nancy Makepeace. 1981. *On Becoming Human*. Cambridge: Cambridge University Press.

Taylor, Anne. 1992. *Annie Besant: A Biography*. Oxford: Oxford University Press.

Taylor, Barbara. 1983. *Eve and the New Jerusalem: Socialism and Feminism in the Nineteenth Century*. New York: Pantheon Books.

———. 1992. "Mary Wollstonecraft and the Wild Wish of Early Feminism." *History Workshop Journal* 33:197–219.

Taylor, Miles. 1995. *The Decline of British Radicalism, 1847–1860*. Oxford: Clarendon Press.

Tedesco, Marie. 1984. "A Feminist Challenge to Darwinism: Antoinette L. B. Blackwell on the Relation of the Sexes to Society and Nature." In *Feminist Visions: Toward a Transformation of the Liberal Arts Curriculum*, edited by Diane J. Fowlkes and Charlotte S. McClure, 53–65. University, Ala.: Alabama University Press.

Tegetmeier, William Bernhard. 1856–57. *The Poultry Book.* 11 parts. London: Orr.

———. 1866–67. *The Poultry Book.* 15 parts. London: Routledge.

Theunissen, Bert. 2012. "Darwin and His Pigeons: The Analogy between Artificial and Natural Selection Revisited." *Journal of the History of Biology* 45:179–212.

Thomas, Julia. 2004. *Pictorial Victorians. The Inscription of Values in Word and Image.* Athens: Ohio University Press.

Thompson, Edward P. 1975. *The Making of the English Working Class.* Harmondsworth: Pelican.

Thompson, William [and Anna Wheeler]. (1825) 1983. *Appeal of One Half of the Human Race, WOMEN, against the Pretensions of the Other Half, MEN, to Retain Them in Political, and Thence in Civil and Domestic Slavery.* London: Virago.

Tipton, Jason A. 1999. "Darwin's Beautiful Notion: Sexual Selection and the Plurality of Moral Codes." *History and Philosophy of the Life Sciences* 21:119–35.

Tomalin, Claire. 1974. *The Life and Death of Mary Wollstonecraft.* London: Weidenfeld and Nicolson.

Tosh, John. 1994. "What Should Historians Do with Masculinity? Reflections on Nineteenth-Century Britain." *History Workshop Journal* 38:179–202.

———. 1999. *A Man's Place: Masculinity and the Middle-Class Home in Victorian England.* New Haven, CT: Yale University Press.

Tribe, David. 1971. *President Charles Bradlaugh, M.P.* London: Elek.

Trivers, Robert. 1972. "Parental Investment and Sexual Selection." In *Sexual Selection and the Descent of Man, 1871–1971,* edited by Bernard Campbell, 136–79. Chicago: Aldine.

Turner, Frank M. 1978. "The Victorian Conflict between Science and Religion: A Professional Dimension." *Isis* 69:356–76.

Tylor, Edward B. 1865. *Researches into the Early History of Mankind.* London: John Murray.

Tyrrell, Alex. 2000. "Samuel Smiles and the Woman Question in Early Victorian Britain." *Journal of British Studies* 39:185–216.

Uglow, Jenny. 2002. *The Lunar Men: The Friends Who Made the Future, 1730–1810.* London: Faber and Faber.

Uphaus, Robert W. 1978. "The Ideology of Reynolds' Discourses on Art." *Eighteenth-Century Studies* 12:59–73.

Van Amringe, William Frederick. 1848. *An Investigation of the Theories of the Natural History of Man, by Lawrence, Prichard, and Others, Founded upon Animal Analogies: And an Outline of a New Natural History of Man, Founded upon History, Anatomy, Physiology, and Human Analogies.* New York: Baker & Scribner.

Vandermassen, Griet, Marysa Demoor, and Johan Braeckman. 2005. "Close Encounters with a New Species: Darwin's Clash with the Feminists at the End of the Nineteenth Century." In *Unmapped Countries: Biological Visions in Nineteenth Century Literature and Culture,* edited by Anne-Julia Zwierlein, 71–81. London: Anthem Press.

Vetter, Jeremy. 2010. "The Unmaking of an Anthropologist: Wallace Returns from the Field, 1862–70." *Notes & Records of the Royal Society* 64:25–42.

Vickery, Amanda. 2001. *Women, Privilege and Power.* Stanford, CA: Stanford University Press.

Vogt, Carl. 1864. *Lectures on Man: His Place in Creation and in the History of the Earth.* London: Anthropological Society.

Walker, Alexander, ed. 1809. *Archives of Universal Science.* 3 vols. Edinburgh.

———. 1834. *Physiognomy Founded on Physiology and Applied to Various Countries, Professions and Individuals.* London: Smith, Elder.

———. 1836. *Beauty: Illustrated Chiefly by an Analysis and Classification of Beauty in Woman.* London: Effingham Wilson.

———. 1838. *Intermarriage; or, The Mode in Which, and the Causes Why, Beauty, Health and Intellect, Result from Certain Unions, and Deformity, Disease and Insanity from Others.* London: John Churchill.

———. 1839. *Woman Physiologically Considered as to Mind, Morals, Marriage, Matrimonial Slavery, Infidelity, and Divorce.* London: A. H. Bailey.

———. (1839) 1973. *Documents and Dates of Modern Discoveries in the Nervous System.* Facsimile of original edition, London, edited by Paul F. Cranefield. Metuchen, NJ: Scarecrow Reprint Corporation.

Walker, Mrs. A. 1837. *Female Beauty, as Preserved and Improved by Regimen, Cleanliness and Dress.* London: Thomas Hurst.

Walkowitz, Judith, R. 1980. *Prostitution and Victorian Society.* Cambridge: Cambridge University Press.

Wallace, Alfred Russel. 1853. *A Narrative of Travels on the Amazon and Rio Negro, with an Account of the Native Tribes, and Observations on the Climate, Geology, and Natural History of the Amazon Valley.* London: Reeve.

———. 1855. "On the Law Which Has Regulated the Introduction of New Species." *Annals & Magazine of Natural History* 16:184–96.

———. 1858. "On the Tendency of Varieties to Depart Indefinitely from the Original Type." *Journal of the Proceedings of the Linnaean Society London* 3:53–62.

———. 1864. "The Origin of the Human Races and the Antiquity of Man Deduced from the Theory of Natural Selection." *Journal of the Anthropological Society of London* 2:clvii–clxxxvii.

———. 1865a. "How to Civilize Savages." *Reader* 5:671a–672a.

———. 1865b. "On the Phenomena of Variation and Geographical Distribution, as illustrated by the Papilionidae of the Malayan Region." *Transactions of the Linnean Society* 25:1–71

———. 1866a. "Presidential Address." In "Anthropology at the British Association," *Anthropological Review* 4:386–408.

———. 1866b. *The Scientific Aspect of the Supernatural: Indicating the Desirableness of an Experimental Enquiry by Men of Science into the Alleged Powers of Clairvoyants and Mediums.* London: F. Farrah.

———. 1867a. "Creation by Law." *Quarterly Journal of Science* 4:471–88.

———. 1867b. "Mimicry, and Other Protective Resemblances among Animals." *Westminster Review* 32 (n.s. 1): 1–43.

———. 1867c. "The Philosophy of Birds' Nests." *The Intellectual Observer* LXVI:413–20.

———. 1868. "A Theory of Birds' Nests: Shewing the Relation of Certain Sexual Differences of Colour in Birds to Their Mode of Nidification." *Journal of Travel & Natural History* 1:73–89.

———. 1869a. *The Malay Archipelago: The Land of the Orang-Utan, and the Bird of Paradise.* 2 vols. London: Macmillan.

———. 1869b. "Sir Charles Lyell on Geological Climates and the Origin of Species." *Quarterly Review.* 126:359–94.

———. 1870. *Contributions to the Theory of Natural Selection: A Series of Essays.* London and New York: Macmillan.

———. 1871. Review of *Descent. Academy* 2:177–83.

———. 1876a. *The Geographical Distribution of Animals; with a Study of the Relations of Living and Extinct Faunas as Elucidating the Past Changes of the Earth's Surface.* 2 vols. London: Macmillan.

———. 1876b. "Opening Address by the President, Alfred Russel Wallace, Glasgow Meeting of the British Association." *Nature* 14:403–12.

————. 1876c. "Review of *Lessons from Nature* by St. George Mivart." *Academy* 9:578–88.

————. 1877. "The Colours of Animals and Plants: I and II." *Macmillan's Magazine* 36:384–408, 464–71.

————. 1878. *Tropical Nature and Other Essays.* London: Macmillan.

————. 1879. Review of *The Colour-Sense,* by Grant Allen. *Nature* 19:501–5.

————. 1886. "Romanes *versus* Darwin: An Episode in the History of the Evolution Theory." *Fortnightly Review* 40:300–316.

————. 1889. *Darwinism: An Exposition of the Theory of Natural Selection with Some of Its Applications.* London: Macmillan.

————. 1890a. "Dr. Romanes on Physiological Selection." *Nature* 43:79, 150.

————. 1890b. "Human Selection." *Fortnightly Review* 48:325–37.

————. 1890–92. "Letters to Edward Westermarck (S712: 1940)." The Alfred Russel Wallace Page. Accessed February 27, 2010. http://people.wku.edu/charles.smith/wallace/S712.htm.

————. 1892. "Human Progress: Past and Future." *Arena* 5:145–59.

————. 1893. "Woman and Natural Selection: Interview with Dr. Alfred Russel Wallace." *Daily Chronicle,* December 4, 1893. The Alfred Russell Wallace Page. Accessed February 27, 2010. http://people.wku.edu/charles.smith/wallace/S736.htm.

————. 1905. *My Life: A Record of Events and Opinions.* 2 vols. London: George Bell & Sons.

————. 1913. *Social Environment and Moral Progress.* New York: Cassell.

Waller, John C. 2001. "Ideas of Heredity, Reproduction and Eugenics in Britain, 1800–1875." *Studies in the History of Biological & Biomedical Sciences* 32:457–89.

————. 2003. "Parents and Children: Ideas of Heredity in the 19th Century." *Endeavour* 27:51–56.

Wark, Robert R. 1975. Introduction to *Sir Joshua Reynolds: Discourses on Art.* Edited by Robert R. Wark. New Haven, CT: Yale University Press.

Watts, Ruth. 1998. "Some Radical Educational Networks of the Late Eighteenth Century and Their Influence." *History of Education* 27:1–14.

Wedgwood, Barbara, and Hensleigh Wedgwood. 1980. *The Wedgwood Circle, 1730–1897.* London: Studio Vida.

Weeks, Jeffrey. 1981. *Sex, Politics and Society: The Regulation of Sexuality since 1800.* London: Longman.

Weikart, Richard. 1995. "A Recently Discovered Darwin Letter on Social Darwinism." *Isis* 86:609–11.

Weismann, August. 1882. *Studies in the Theory of Descent.* 2 vols. London: Sampson Low, Marston, Searle, & Rivington.

Wells, Kentwood D. 1971. "Sir William Lawrence (1783–1867): A Study of Pre-Darwinian Ideas on Heredity and Variation." *Journal of the History of Biology* 4:319–61.

Welsch, Wolfgang. 2004. "Animal Aesthetics." *Contemporary Aesthetics* 2. Accessed January 21, 2009. http://www.contempaesthetics.org/newvolume/pages/article.php?articleID=243.

Westermarck, Edward. 1891. *The History of Human Marriage.* London: Macmillan.

Westwood, John Obadiah. 1839–40. *An Introduction to the Modern Classification of Insects.* 2 vols. London: Longman, Orme, Brown, Green & Longmans.

Whale, John. 2000. *Imagination under Pressure, 1798–1832: Aesthetics, Politics and Utility.* Cambridge: Cambridge University Press.

White, Charles. 1799. *An Account of the Regular Gradation in Man, and in Different Animals and Vegetables; and from the Former to the Latter.* 2 vols. London: C. Dilly.

White, Paul. 2003. *Thomas Huxley: Making the "Man of Science."* Cambridge: Cambridge University Press.

———. 2007. "Acquired Character: The Hereditary Material of the 'Self-Made Man.'" In *Heredity Produced: At the Crossroads of Biology, Politics, and Culture, 1500–1870*, edited by Staffan Muller-Wille and Hans-Jorg Rheinberger, 375–97. Cambridge, MA: MIT Press.

Wilson, Edward O. 1975. *Sociobiology: The New Synthesis*. Cambridge, MA: Harvard University Press/Belknap Press.

Wilson, Leonard G., ed. 1970. *Sir Charles Lyell's Scientific Journals on the Species Question*. New Haven, CT: Yale University Press.

Winch, Donald. 1987. *Malthus*. Oxford: Oxford University Press.

Wood, Paul B. 1996. "The Science of Man." In *Cultures of Natural History*, edited by N. Jardine, J. A. Secord, and E. C. Spary, 197–210. Cambridge: Cambridge University Press.

Wyhe, John van. 2013. *Dispelling the Darkness: Voyage in the Malay Archipelago and the Discovery of Evolution by Wallace and Darwin*. Singapore: World Scientific Publishing.

Yarrell, William. 1843. *A History of British Birds*. 3 vols. London: John van Voorst.

Yeazell, Ruth Bernard. 1989. "Nature's Courtship Plot in Darwin and Ellis." *Yale Journal of Criticism* 2:33–53.

———. 1991. *Fictions of Modesty: Women and Courtship in the English Novel*. Chicago: University of Chicago Press.

Young, Robert J. C. 2006. *Colonial Desire: Hybridity in Theory, Culture and Race*. London: Routledge.

———. 2008. *The Idea of English Ethnicity*. Oxford: Blackwell.

Young, Robert M. 1985. *Darwin's Metaphor*. Cambridge: Cambridge University Press.

Zihlman, Adrienne. 1974. "Review of *Sexual Selection and the Descent of Man*, ed. Bernard Campbell." *American Anthropologist* 76:475–78.

INDEX

Note: all references to figures are italicized; the name Darwin used alone refers solely to Charles; the name of his grandfather Erasmus appears in full or as "E. Darwin." Familial vocabulary in parentheses after proper name entries indicates relationships to Charles Darwin unless otherwise noted.